THE

# PUBLICATIONS

OF THE

# 𝕷incoln 𝕽ecord 𝕾ociety

FOUNDED IN THE YEAR

## 1910

VOLUME 84

ISSN 0267–2634

# THE PRINTED
# MAPS OF LINCOLNSHIRE
## 1576–1900

## A CARTO-BIBLIOGRAPHY

### WITH AN APPENDIX
### ON ROAD-BOOKS
### 1675–1900

R. A. CARROLL

The Lincoln Record Society

The Boydell Press

First published 1996


A Lincoln Record Society Publication
Published by The Boydell Press
an imprint of Boydell & Brewer Ltd
PO Box 9, Woodbridge, Suffolk IP12 3DF, UK
and of Boydell & Brewer Inc.
PO Box 41026, Rochester, NY 14604–4126, USA


ISBN  0 901503 57 6


British Library Cataloguing in Publication Data
Printed Maps of Lincolnshire, 1576–1900:
With an Appendix on Road-books,
1675–1900. – (Publications of the Lincoln
Record Society, ISSN 0267–2634; Vol. 84)
I.  Carroll, R. A.   II.  Series
016.9124253
ISBN  0–901503–57–5


This publication is printed on acid-free paper

Printed in Great Britain by
St Edmundsbury Press Ltd, Bury St Edmunds, Suffolk

To my father-in-law
Ernest Sharman
who looked after the dogs

# CONTENTS

## THE PRINTED MAPS OF LINCOLNSHIRE, 1576–1900

# APPENDIX. ROAD BOOKS, 1675–1900

# PREFACE

As a boy at school I was always keen on maps and later, at college as a student librarian, historical bibliography aroused the greatest interest; the development of printing from moveable type and its geographic spread throughout Europe with a further focus on type-faces and illustration methods touched a lasting chord. Only when I was nearing retirement did the fusion of the two develop into the collection of maps and deepening enquiry into their origins, history and differences. Following a suggestion by Dr. Mary Finch of the Lincoln Archives Office I undertook some work on the manuscript maps under her control in 1983; through that work my attention was caught by the variety of the printed maps of the county scattered among the deposited collections; the quality of their engraving especially so. A further suggestion by Mary led to my starting serious research.

When retirement came I began to put my notes into order. I had bought Skelton's *County Atlases Of the British Isles* (1970), which gives a detailed picture of the printed county maps from Saxton to 1703, when it first came out; I knew of Chubb's lists, of course, and had one or two other volumes on the subject. Luckily, while at the Archives Office, two new bibliographies were published almost simultaneously; Mrs. Betty Chambers' on Bedfordshire and Mr. David Kingsley's on Sussex maps. They both wrote encouraging letters at that time and I was able later to have the privilege of seeing the latter's fine collection of atlases. Close study revealed the need to invest in Mr. Donald Hodson's work on Hertfordshire maps (1978). At one time in my working life I had used quite often Chubb's lists of the maps of Gloucestershire (1913) but even a cursory look at the Hertfordshire volume made it clear that here the study of county maps had been brought to new levels of refinement. In 1984 the first book of Hodson's continuation of Skelton's work appeared with its fresh approach to the study of eighteenth century maps, putting all future researchers further in his debt. Donald and his wife, Yolande (a notable student of maps herself and whose writings figure in the footnotes here) provided much hospitality at their home during visits to study his wonderful collection of atlases. The frequency of reference in this book to the works of the above writers is evidence of my debt to their researches.

I had just reached what turned out to be the half-way mark in my own work when Mr. Roger Fairclough (Cambridge University Library) drew my attention to the new work on Berkshire maps privately published by Mr. Eugene Burden. Mr. Burden gladly agreed to let me have a copy; continuing correspondence and frequent meetings have been fruitful in fresh sources of material, including access to his own large collection of maps, particularly strong on nineteenth century maps sold in covers. His friend, Mr. Tony Burgess is researching Kent maps and has welcomed me in his home to inspect his wide range of atlases and maps. Mr. Clive Burden and his son, Philip, who now have the most extensive private collection of British atlases, full of the greatest treasures, have also made me welcome on numerous visits. Mr. David Webb has put at my disposal much useful

information, based on his very fine collection of road books. To all of them I offer
my sincere thanks for their help and kindness.

Librarians and curators have welcomed my wish to inspect the materials in their
care and seemed only too pleased to dig into the far corners of their stacks. They
have often been pleased that valuable resources are being made use of (and, in
some cases, bolstered resistance to suggestions that limited funds are being
wasted on their preservation and conservation). The feeling of camaraderie among
all in the map/book world has been a part of the pleasure in compiling the present
work. The sharing of new findings has been part of the fun; I can only hope that
now I add my share to the pool of knowledge in this field and have maintained
the required standards.

I have visited so many libraries (and so many times) that I must content myself
with a similar expression of thanks to them all. The librarians and their staff in
the three English copyright collections have borne the brunt of my enquiries; if I
single out the map librarians and their colleagues the efforts of the staff in the
book sections are not the less cause for gratitude. The many unheralded fetchers
of books from the stacks also deserve their due recognition. In the order, in which
their holdings are shown in these lists, they are: Mr. Tony Campbell of the British
Library, with Mr. Geoff Armitage a noted friend of all researchers; Mr. Roger
Fairclough and his staff, particularly Mr. Tony Rawlings, of Cambridge University
Library; Mr. Fairclough has always taken a great interest in my work and has put
me further in his debt by undertaking to read the manuscript and make further
suggestions for its improvement; Miss Betty Fathers, now retired, of Bodleian
Library, Oxford University. The largest collections of British county atlases
outside the big three are held in Leeds University Library (the Harold Whitaker
collection) and Mr. P.S. Morrish, its curator has always been helpful; and, in the
National Library of Scotland, whose Map Librarian, Miss Margaret Wilkes, has
also put its full resources at my disposal.

The staff of three local collections have given me much in making their maps
available; in Lincolnshire County Archives Office, where the staff were ever
helpful; the same applies to Miss Eleanor Nannestad (and her predecessor, Mrs.
Liz Melrose) in the Local History Department of Lincoln Central Library; Mr.
Derek Whettam of Grimsby Reference Library was always helpful.

The following public libraries and their staff have given much assistance;
Birmingham (Miss Margaret Green has shown me the greatest kindness and
assistance); Bournemouth; Brighton; Cambridge; Cardiff (Mr. Richard Phillips);
Chester; Chesterfield; Derby (Mrs. Linda Owen); Edinburgh; Glasgow Mitchell
Library; Hull; Leeds; Leicester; Liverpool (Mr. Martin Walker); Maidstone;
Manchester; Middlesbrough (Mrs. Joan Waites); Norwich (Mr. Wilkins-Jones);
Nottingham; Peterborough; Scunthorpe (Ms J. Briody); Sheffield (Mrs. R.
Humble); Stafford, William Salt Library; York; and the staff of many Lincolnshire
County Library branches.

To the librarians and their staff at the following university libraries I am also
much indebted: Birmingham; Edinburgh; Leicester; Liverpool; London (notably
Miss Moira Courtman); Loughborough; Manchester, including John Rylands
Library (Mr. Chris Perkins and Mr. David Riley); Nottingham (Mr. John Briggs);
Sheffield (Miss Joan Chibnall). The staff at a variety of specialist organisations
have also made valuable contributions: the Admiralty Library; British Geological
Survey; Canterbury Cathedral; Edinburgh Signet Library; Guildhall Library (Mr.

Ralph Hyde); Holkham Hall; Lincoln Cathedral (its former librarian, Miss Joan Williams, especially); Liverpool Athenaeum (Mr. J.D. Rogers); Manchester Chetham's Library (Dr. Michael Powell); National Maritime Museum (especially Mrs. Rita Bryan); Northamptonshire Archives Office; Nottinghamshire Archives Office; Peterborough Cathedral (Canon Jack Higham); Public Record Office; Royal Geographical Society (Mr. Francis Herbert has been a great support during many visits); Scunthorpe Museum (Mr. D. Taylor); Spalding Gentlemen's Society (in particular its President, Mr. Norman Leveritt); West Suffolk Record Office; West Sussex Records Office; Wisbech Fenland Museum. Finally, a number of Dutch librarians answered my questions and kindly provided photocopies of maps from their collections in Amsterdam, Rotterdam, Utrecht, Leeuwarden and Groningen. To all who have provided copies for the illustrations in this volume I extend my thanks.

I owe a very large debt for the friendship and hospitality during my frequent visits to London to Ralph and Tina Gee; Helen Gristwood also provided bed and board, while her husband, Philip, put a good deal of his computing expertise to great effect in setting up my word processor and solving the (to me) insoluble problems that have come up from time to time. My family have, of course, provided the conditions for my work over the last eleven years and my debt to my wife, my mother and father-in-law is incalculable.

I am greatly indebted to the British Academy and the Marc Fitch Fund, both of which provided financial contributions towards the many expenses incurred during my travels; to Dr. Kathleen Major, who supported my application to the British Academy, and Dr. Dorothy Owen, who has been a constructive and generous editor; she and her husband, Arthur, have added to my sense of gratitude by their friendly hospitality.

In a work of this kind it is impossible to regard it as finished and complete. Of maps before 1800 it is unlikely that any completely new types will come to light; perhaps a few volumes hitherto unknown will turn up that contain a plate already described. Of nineteenth century county maps one can not be so sanguine. These pages contain many references to maps of other counties of which the Lincolnshire counterpart has not yet been discovered, that is, if they were ever produced. From such notes future collectors can at least provide themselves with their own lists of *desiderata*. Many of the maps of the railway and cycling ages, regarded as ephemeral, have not survived in great numbers and new finds are likely. As for atlases, 'new' examples will occasionally appear; I would expect several editions of the Walker atlas series of 1835 and its derivative, the Hobson Fox-Hunting atlases, to come to light and not all the Philips' Handy Atlas editions have yet been found and analysed. There may be others as not all the atlases in Whitaker's lists have yet been found. A line has to be drawn eventually in spite of the failure of my original hope to explore the collections of all British universities and major town and city libraries! The errors and omissions are my responsibility. I would like to hear from users of this text of any additions and corrections.

# INTRODUCTION

The availability of maps is now so widespread and their use so commonplace, particularly by motorists, walkers and cyclists, that it is difficult to believe that the origins of modern map-making are only a little more than four hundred years old. Professor Harvey has written of the 'maplessness' of the Middle Ages.[1] In the catalogue of surviving maps produced in England before 1500 only thirty examples were recorded,[2] later increased to thirty-five.[3] A similar picture emerges in other western European countries; but the pattern varies, particularly in Italy and the Netherlands. In Italy the maps are often the works of artists and are often signed; this practice occurred to a lesser degree in the Netherlands; nevertheless, since there are no scales or signs of scientific measurement their value as maps is somewhat limited. After the earliest dated map in Italy (1291 from Asti in Piedmont) the fourteenth and fifteenth centuries saw a greater number of maps in Italy than elsewhere. The Italians were also the leading producers of the Portolano; the word means written sailing instructions, which are separate from the sea-charts which are nowadays called portolans. The latter consist of basic coastal outlines with very few inland features, distinctively marked with rhumb lines as a practical aid to navigation. About thirty survive from the fourteenth century but the total from the following century is five times as great.[4] The majority show areas of the Mediterranean.

The earliest maps were often produced to help settle disputes over boundaries. Previous to the introduction of such sketches all surveys were written; the earliest, in England, date back to c.900. These were not concerned so much with the description of boundaries and notable features, but with assessments for rents and services. Where more measured returns were made they were an expedient aid when settling disputes. In any case, the knowledge of and the instruments for accurate surveying did not exist until, at least, the thirteenth century, when a number of treatises on surveying circulated in manuscript. Even then any measurements made were still incorporated in a written description and were not the basis for mapping. To understand the measurements described in surveys it is necessary to know the local standards that were being used. Some local versions survived into the last century. Local variants in Lincolnshire are discussed in the consideration of the maps of Saxton (**1**); Morden (**19**) gives three scales for the mile in his 1695 map, over a hundred years after the mile had been standardised by Elizabeth I.

Of the earliest surviving maps, the two oldest are sketches of buried water systems in London and Canterbury. The third concerns the bounds of land in

[1]  Skelton, R.A. and Harvey, P.D.A. Local Maps and Plans from Medieval England. (Oxford, 1986), p. 6.

[2]  Ibid.

[3]  Harvey, P.D.A. Medieval Maps. (London, 1991), p. 91.

[4]  Ibid., pp. 39–45.

Lincolnshire, the subject of dispute between the abbey of Kirkstead and the lords of the baronies of Scrivelsby, Horncastle and Bolingbroke. The plan was produced between 1224 and 1249 to define lands, whose dimensions appear in a survey written between 1141 and 1148.[5] From the late fourteenth century a map has survived[6] of lands under dispute between the Bishop of Ely and the people of Long Sutton.[7] Two further examples, from the fifteenth century, survive of maps related to lands in Lincolnshire. One is concerned with territory in Barholm, the accompanying text being a supposed history of the parish, possibly written by a cleric from Bourne abbey.[8] The other probably shows the northern part of Deeping Fen, but the map was badly damaged by fire in 1731.[9] Apart from six maps or plans of London or the Thames Valley four of the pre-1500 survivors are concerned with property in Lincolnshire, while three more involve fen lands just outside the county. This leads one to think that there might have been a local surveying tradition in the Wash area and perhaps, if there were, it owed something to a Dutch influence, since local mapping there was developing in a roughly parallel fashion at that time.

By 1500 only two surveys can be associated with properly measured plans; at Shouldham in Norfolk (c.1440–1441) and Tanworth in Arden (Warwickshire, c.1500). During the period from 1500 to 1570 there grew up a body of men qualified to make surveys, largely following the stimulus provided by the earliest printed manuals. The issue of Ptolemy's *Geographia* (Bologna, 1477), illustrated with twenty-six copper-plate maps created enormous interest, sufficient for its re-issue in more than twenty further editions in Italy, France and Germany. It is very doubtful if the maps used in the editions of Ptolemy are earlier than the twelfth century, the dates of the manuscripts from which the text was set. The world map is stated to have been drawn by Agathodaimon of Athens, possibly a contemporary of Ptolemy.[10] The earliest manuscript work on surveying dates from the fourteenth century.[11] This work describes the use of the quadrant and the plane-table for measuring heights and distances. In 1523 John Fitzherbert issued a manual for the stewards of manors and this included the elements of surveying. That was followed in 1537 by *The Boke sheweth the maner of measurynge of all maner of lande* by Richard Benese. Various other advances in the production of more precise instruments, especially in Germany, developments in mathematics

---

[5] Kirkstead Psalter, Fo.4ᵛ. See also the article on the map by Hallam, H.E. in Skelton and Harvey, 1986, op. cit., pp. 78–81, which gives the Psalter's text and a reproduction of the map; Webb, B. An early map and description of the inquest on Wildmore Fen in the twelfth century in Lincolnshire. *Architectural and Archaeological Reports and Papers*. New Series. ii (1938), pp. 141–156.
[6] PRO MPC 45.
[7] Owen, A.E.B. (Skelton and Harvey, 1986, op. cit. pp. 87–98) suggests the map is based on Thorney Abbey records; Owen, A.E.B. A Fenland Frontier: the establishment of the boundary between Cambridgeshire and Lincolnshire. *Landscape History*, iv (1982), pp. 41–46 (copy in LAO – FL Box L346.0432 OWE).
[8] LAO Lindsey Deposit 32/2/5/1 Fo. 17ᵛ (BRA 437); Cripps, J.A. (in Skelton and Harvey, 1986, op. cit., pp. 272–288) gives the surviving text.
[9] BL Cotton Ms. Otho B. xiii, fo. 1ᵛ. Owen, A.E.B. in Skelton and Harvey, 1986, op. cit., pp. 289–291; Searle, W.G. Ingulf and the Historia Croylandensis: an investigation attempted. *Cambridge Antiquarian Society*. Octave Publications. Vol. XXVII (1894), pp. 550–52.
[10] Skelton, R.A. Explorers' Maps. (London, 1958), pp. 3 and 23.
[11] BL Sloane Ms. 213, ff. 123 b–124.

and the publication of (Rainer) Gemma Frisius' text from *Cosmographicus Liber Petri Apiani Mathematici* (Antwerp, 1533) setting out the principles of triangulation, provided the materials for accurately measured surveys. The spread of the texts and use of the instruments was slow and the profession of surveying was still in its infancy when Saxton undertook to survey all the English and Welsh counties (**1**).

1570 marks a watershed in the world of cartography since in that year Ortelius issued his *Theatrum Orbis Terrarum* and Saxton began his survey of the fifty-two English and Welsh counties. Ortelius (1528–1598) produced a world map in 1564 on eight sheets; there followed several maps of individual areas and the gathering of maps from his travels as a book dealer through many parts of Europe. By 1570 he had his collection of maps all re-engraved to a uniform plate size and issued the first 'atlas' (although the word was not introduced until 1595 by Mercator). At first only a general map of Great Britain was included; in 1572 individual maps of the four home countries were introduced and up to the final edition (there were forty-two editions with texts in seven different languages) no other maps of British subjects were included.[12] Gerard Mercator (1512–1594) had already been active in map-making for more than thirty years when Ortelius produced his map volume. Responsible for surveying, production of drawings and engraving he had produced large-scale maps of Europe (1564), the British Isles (1564) and, in 1569, a world map on eighteen sheets, which used the projection named after him for his use of it on navigational charts. The last years of his life were spent in preparing his three-volume atlas, covering Europe in great detail and providing large maps of the other continents. The third volume, seen through the press in 1595 by his son Rumold, contained sixteen maps relating to the British Isles. Lincolnshire was included on the map of *Eboracum, Lincolnia, Derbia, Staffordia, Notinghamia, Lecestria, Rutlandia, et Norfolcia.* More plates were added in the various expansions of this work and when the plates finally passed to Jansson they became the basis for his *Novus Atlas* (**13**).[13] The supremacy of Dutch map engraving remained unchallenged for over a century. Most of Saxton's plates were engraved by Dutch craftsmen and several other series of British maps were the work of engravers from the Netherlands; Van den Keere (**4**) and their derivatives issued by George Humble (**7**); Jodocus Hondius, who prepared Speed's plates, including Lincolnshire (**6**); Van Langeren (**10**) and the enlarged versions by Jenner (**11**) and culminating in the highly prized plates of Jan Blaeu (**12**) and Jan Jansson (**13**).

It is likely that Sir William Cecil, Queen Elizabeth's Secretary of State, was aware of this Dutch mapmaking activity; he would certainly have known of Mercator's wall-map of the British Isles of 1564, which was the source of the general map and the maps of Scotland and Ireland in Ortelius' *Theatrum*.[14] It is against this background of cartographic work in Britain's seafaring rival and, in this country, an immature surveying profession that Saxton's work must be set. In 1570 he was set to make his survey of all fifty-two English and Welsh counties and Lord Burghley (as William Cecil had become) began to receive proof copies

---

[12] Skelton, pp. 219–220; Moreland and Bannister, pp. 99–100.
[13] Skelton, pp. 220–226 gives details of the changes and additions to the British plates.
[14] Skelton, p. 219.

of the maps from 1575 (now in BL as Royal MS.18.D.iii), indicating his interest and awareness of the maps as a national asset.

Considering the difficulties of transport and the state of surveying knowledge the quality of Saxton's work is remarkable. No single county had been surveyed previously and to complete all fifty-two counties and have the results engraved within ten years to a very high standard remains a wonderful accomplishment. The plates were drawn and engraved over the five years, 1574–1578; the surveying took not more than nine years. One indication of their value is the fact that, for the next hundred years all published county maps are based very closely on Saxton, there being no innovation in British mapping until Ogilby's *Britannia* of 1675 (Appendix **1** (**1**)) provided the most notable detail lacking on Saxton's plates – the roads. No other survey of the counties as a whole took place until Cary undertook the work for the Postmaster-General two hundred years after Saxton.

Other attempts to make county surveys followed Saxton; Symonson produced a map of Kent in 1596; William Smith drew maps of twelve English counties in 1602–1603,[15] including plates based on the maps by Norden of Surrey (1594) and Hertfordshire (1598); Norden also produced a map of Hampshire (1595), which was not published at that time. Smith is credited with the introduction of tables of symbols and the use of a grid or graticule; Skelton has suggested that he learnt these practices from German maps while he lived for ten years in Nuremburg.[16]

Curiously, the next maps of the English and Welsh counties after Saxton were included on a set of playing cards, since there were (until 1974) forty English counties and twelve Welsh, thus making the maps a numerical match. This first set of maps on playing cards appeared c.1590 (**2**) and their assumed producer, William Bowes, produced a new different set in c.1605 (**3**). The use of maps in this way occurs on only a few further occasions. In 1676 appeared a most significant set from Robert Morden (**15**) in that he incorporated details of roads culled from Ogilby's *Britannia*, which had appeared in the year before the cards. John Ogilby's great work, in which the major roads from London and the most important cross-roads are drawn at the scale of one mile to an inch, was the result of the most notable surveying work since Saxton, a century earlier. What makes it the more remarkable is the then state of the roads he travelled. Only one road had been turnpiked at the time; the first act in 1663 covered the Great North Road between Wadesmill and Royston. That this important artery was not fully turnpiked until 1776 is an indication of the slowness of the process of road improvement. These cards are, therefore, the first county maps to show roads, albeit in a miniature format. A rival set of cards appeared in the same year from William Redmayne (**16**). His cards were issued later by John Lenthall, who, in 1717, produced his own set (**22**). Morden's survived until c.1773 when they were used in atlas form but without the suit marks. No other use of the idea of matching normal playing cards to maps of the counties occurs after Lenthall.

On some of Saxton's plates two or more counties share some plates, Lincolnshire with Nottinghamshire being one example. With the maps accompanying the

---

[15] Skelton, pp. 19–22 discusses the sources of the maps drawn by Smith, including manuscript maps made by Norden.
[16] Skelton, p. 21.

1607 edition of Camden's *Britannia* (**5**) each county has a folio plate to itself for the first time. The maps are not otherwise distinguished in content or quality of engraving. The maps produced in the first decade of the seventeenth century by John Speed to accompany his *Theatre of Great Britain* (**6**) are of much greater significance. The innovatory feature is the inclusion of town plans of the most important place(s) in each county and largely surveyed personally by Speed himself. Lincoln appears on the county plate and Stanford (sic) on the map of Rutland. These are the earliest maps of towns in the county and the next (Stukeley's plans of Lincoln and Horncastle in his *Itinerarium Curiosum*) did not appear until 1724. Another novelty on most county plates is the inclusion of the names and boundaries of the wapentakes, hundreds, etc. These had been included on some of the earlier individual county surveys noted above and, in five cases, on Saxton's plates. For Lincolnshire this was the first time such detail had been given. Speed also introduced his own decorative features, notably the coats of arms of prominent noblemen with county associations, an ornate compass and notes on historical matters used to fill otherwise blank spaces. His work shared with Saxton a long life; the plates of both men were still being used as late as 1770, though it is doubtful if Saxton would have recognised his plates immediately so many alterations having taken place in the intervening two hundred years.

Meanwhile, a number of surveys of especial local significance had been produced. One is the map of the Fens, attributed to John Hexham in 1589 and known as the Hatfield House map, since that is its present location.[17] Another fen map is that of John Hayward of 1604. For many years this map was only presumed to exist, but it seems to have been the basis of the maps prepared by Sir Jonas Moore (1684 and after); Badeslade recorded the use of Hayward's map on his *PLAN and DESCRIPTION of the FENNS*, dated 1724, included in *The History Of . . . the Port of King's-Lyn* by John Armstrong and published in 1725. Henry Valentine surveyed the *Manour of Toynton* in 1614 in thirteen separate maps. Made for Lord Willoughby de Eresby they are probably the first attempt in Lincolnshire at an estate map and by a trained surveyor.[18] The best known maps of the Fens of this period are, of course, those by Hondius and Blaeu. Hondius prepared *A generall Plott and description of the Fennes* (dated 1632) for the issue in 1633 of the Mercator-Hondius atlas. Skelton says that it is copied from a manuscript map, which was 'perhaps a copy of a lost map by William Hayward'.[19] The plate appeared subsequently in later editions of the above atlas and all the atlases of Jansson, whose property it became (**13**). Only part of the southern end of Lincolnshire appears on maps of the Fens and the same is true of the map *Regiones Inundatae* which Blaeu issued (**12**) in his first atlas (1645). Although an almost exact copy of Jansson's Blaeu does include one or two names not in the other version and has shown a degree of originality. The two plates appeared in all the later editions of their respective atlases.

While the maps produced in 1636 by van Langeren (**10**) have no original points (they are too small to include more than a small selection of place-names or any physical detail) the plates are the first to include a mileage chart. Even the idea

[17] Photographic copy in Cambridgeshire AO. (R 69/22).
[18] LAO 5 Anc 4/A/14.
[19] Skelton, p. 222. The manuscript (BL Cotton Ms. Aug. I.i.78) has a suggested date of 1604.

was not his invention, since his data come from John Norden's *England: an intended Guyde* of 1625. Now the inclusion of such charts is a *sine qua non* of the modern road atlas, but this is the only occasion when such charts were incorporated on the same plates as the county maps. Following the acquisition of the plates by Thomas Jenner (**11**) the plates were redrawn and had a life of nearly forty more years.

One other small volume of maps deserves mention. In 1626 John Bill issued a single edition of small oblong maps, similar to those of Van den Keere (**4** and **7**). They were the first maps to show longitude, which was then measured from a meridian set in the Azores, because it was there that the magnetic north seemed most nearly to coincide with the true north. Morden's maps on playing cards record (in words) Lincoln at 53° North. Bill's map shows Lincoln at 53° 15′ North; the present Ordnance Survey maps show that 53° 15′ is a tenth of a mile north of Bishop Grosseteste College. The margin is, therefore, negligible when the lack of modern surveying equipment over three hundred years ago is considered. It may be mentioned that the Azores meridian had been replaced by a London meridian aligned with St. Paul's and then by the adoption of the Greenwich meridian by the end of the seventeenth century. John Seller, on some of his general maps but not on any county plate in *Anglia Contracta* (1694) (**18**), marked the sides of the maps with the degrees of latitude and longitude. On the map of Great Britain and Ireland in that work Seller used a meridian adopted by French cartographers placed some 20° west of Paris. In the following year, Morden's plates for the Gibson version of Camden's *Britannia* used the meridian calculated then from St. Paul's Church. Kingsley has shown, however, that the Sussex plate, amongst others, used a meridian set at Greenwich.[20] Quite what meridian was used by Morden for the map of Lincolnshire is unclear. The Greenwich meridian passes through Holbeach and Louth but Morden shows his meridian passing some 5 minutes further east at the southern end of the county and more than that at the northern end. In his 1701 map (**20**) Morden's error worsens; in the south the meridian passes slightly to the west of Holbeach but, at the northern end, it passes through Manby, approximately 6 minutes east of Louth. Fordham placed the first general acceptance of the Greenwich meridian at 1794[21] and its first use in an English county atlas seems to have been Smith's *New English Atlas*, published in 1801 (**56**). Sutton Nichols' map of 1712 (**21**) uses the meridian of the Azores, but Moll in 1724 uses a London meridian, as indicated in the upper frame (**24**). Armstrong may have been the first cartographer to specify Greenwich as the meridian on his map of Lincolnshire (**44**), i.e. some fifteen years before Fordham's suggested date.

The superiority of Dutch engravers reached, perhaps, its summit in terms of quality of workmanship and harmony of design in the maps of the counties prepared by Blaeu (**12**) and his great rival Jansson (**13**). The content of their maps is taken from Saxton through Speed and has no claim to originality. Their force and collectability stem from their engraving, the harmonious balance of the different elements and the use of various decorative features; in the map of Lincolnshire these include images of fishermen and wildfowlers to signify two

[20] Kingsley, pp. 93–94.
[21] Fordham, F.G. Studies in carto-bibliography. (Oxford, 1914), p. 15.

of the county's important occupations (in Jansson) or Italianate *putti* (Blaeu) and the inclusion of banks of coats of arms in both (taken directly from Speed). No other plates achieved such distinction of execution, enhanced in many cases, by contemporary hand colouring. The supremacy of the Dutch as engravers and sellers of English county maps faded thereafter; undoubtedly the imposition of a tax in 1701 on all imported maps speeded the process. In the eighteenth century the lead in mapping using more scientific methods, based on astronomical observations and exact triangulation, passed to the French. However, their efforts were concentrated on the single project of mapping the whole of France on a uniform basis and to an agreed national standard; the only French maps of the English counties, those of Perrot issued in 1823 (**83**), are not a serious indication of the best qualities of French mapping.

Reference has already been made to the addition of roads to the map cards of Robert Morden. Morden also engraved the first folio size maps of the counties to include such material copied from Ogilby. The two sets of plates he produced for Camden's *Britannia* show roads; the smaller set, intended for Gibson's translation of Camden and finally issued in 1701, marked the roads by single fine lines but the folio plates actually used in the *Britannia* show roads as double lines in three styles. Morden was copying Ogilby directly. The adoption by Morden of this system of differentiation of types of roads helped establish the method followed by later mapmakers though the delineation and interpretation changed in later years. On his larger county map Morden deserves credit for his use of other sources since he shows some roads not taken from Ogilby, the most prominent being that from Grimsby to Barton.

Ogilby's work was issued between 1675 and 1698 but, because of its folio size, was of little practical use to travellers. The impracticality of Ogilby's volume was probably recognised but only some 20 years after its last edition was a useful alternative attempted. Then four rivals vied to produce a pocketable version of the original material. Herman Moll was the first in the field in 1718, but only issued loose sheets, including one of the Great North Road (Appendix **1 (2)**). In the following year Thomas Gardner issued his *Pocket-Guide To The English Traveller* (Appendix **1 (3)**) only a few days before John Senex put out *An Actual Survey Of all the Principal Roads . . .* (Appendix **1 (4)**). Gardner's work was issued in two forms – a single volume bound as an oblong quarto and, in two volumes, where the plates are folded down the centre to create small pocket books. This work only came out in a single edition while the Senex plates, also in two smaller oblong volumes, had a life of nearly 50 years through the efforts of his wife, who issued three editions, and Thomas Kitchin, John Bowles and Robert Sayer and partner, through all of whose hands the plates passed. The value and popularity of the Senex volume is reflected in the issue of French editions from 1759 prepared in Paris by Le Rouge (Appendix **1 (8)**). They were also re-issued by Desnos, the later owner of Le Rouge's plates. An equally long life was the lot of the plates that came out the following year (1720) from the fourth rival, Emanuel Bowen, under the title *Britannia Depicta* (**23** and Appendix **1 (5)**). None of these road-books added anything of material significance to the data already shown in Ogilby, to whom they all signified their indebtedness.

Herman Moll, who failed to complete a wide-ranging road-book as above, did complete an atlas of the counties that appeared in 1724 (**24**). The maps are not particularly significant but their decorative finish has points of interest. In

Lincolnshire's case there are two particular elements; the first is the inclusion of roman coins (other county maps by Moll also showed coins or other ancient artefacts) and the other is a tide table for the crossings of the rivers at Fosdyke and the modern Sutton Bridge, which feed into the Wash. The table was prepared by William Stukeley (1687–1765), the noted polymath, born at Holbeach. His tide table also appeared on the map engraved by Sutton Nichols (**21 (ii)**) for Henry Overton; this was probably its first use prior to its adoption by Stukeley's friend, Moll. The use of the other representations of coins and other ancient curiosities may have owed their use to suggestions and/or drawings provided by Stukeley, whose tours of the country yielded so much of our knowledge of the appearance of our ancient monuments. Stukeley also prepared the first quality map at the scale of one inch to a mile of a portion of the county. Dated 1723 it depicts the Holland division of the county in much detail and clearly owed a great deal to his own youthful wanderings in that area with his friends, as an early version drawn in 1705 makes clear.[22] Certainly, in his map there is a wealth of names of roads, dykes and other sites[23] that do not occur on the next map of the area at that scale – Armstrong's county map in eight sheets of 1779 (**44**).

From 1725 until Armstrong's work appeared no maps were issued that added any further original features. Series of maps were introduced into works issued in parts, which became a popular publishing practice in mid-eighteenth century England. Examples are those engraved by Read (**27**), Robert Walker (**28**) and George Bickham (**30**), although the latter's efforts hardly deserve to be described as maps. Others came out in periodicals, of which there were increasing numbers in the period; an interesting example of these is the series engraved by Thomas Kitchin for *The London Magazine* (**32**), which appeared in a Dublin edition simultaneously but with the plates re-engraved by Sarah and John Exshaw (**33**). Other maps appeared in books of a didactic nature, such as those of John Cowley (**26**) for his *Geography of England* (1743), Thomas Hutchinson, which appeared in *Geographia Magnae Britanniae* (1748) (**29**) and John Gibson, whose little book of *New and Accurate Maps* . . . (1759) was published by the first serious publisher of books especially designed for children, John Newbery (**36**).

Of the maps of the period only those of Emanuel Bowen have any claim to special merit. The engraver of the very crude county maps that were included in *Britannia Depicta* in 1720 produced, from 1751, three series of high quality county maps, declining in size as each group appeared. In spite of the claims in the titles to each map that they were drawn 'from the best Authorities . . . with . . . improvements' (**34**) on the map for the *Large English Atlas*, and repeated on the map for *The Royal English Atlas* (**39**) and extended to 'Improvements not inserted in any other . . . County Map' in the title of the map in *Atlas Anglicanus* (**42**) they include no new geographical features. These three series are distinguished, but not for the novelty of geographical detail; errors found on Saxton's plate are still in evidence, such as the inclusion of Sutton St. Leonards (instead of Sutton St. James) and placing Scampton north of *Aistrope* (i.e. Aisthorpe). Their distinction lies in the quality of the engraving and the inclusion of numerous notes on the history of the various towns in the county to create a harmonious

---

[22] Bod Ms. Top. Gen. d.14(3) is the original drawing for an early version and is dated 1705.
[23] Copies in BL Maps 1308 (4) and K.19.20; Bod Gough Maps. Lincs 5, etc.; LAO Cragg 1/17.

whole. In the map for the *Large English Atlas* Stukeley's tide table reappears in its original form exactly. Bowen also engraved other smaller maps but of no particular value or showing his best qualities (**34, 35, 37**).

The popularity of the various plates by Bowen and the works they adorned suggests a satisfaction, somewhat ill-founded, with the quality of the geographical information they provided. There was, however, a gathering movement towards new and accurate mapping of the counties. Early in the eighteenth century a number of counties were surveyed at one inch to a mile or similar large scales. The pattern had been set by Gascoyne with his nine-sheet map of Cornwall in 1700; later proposals led to the appearance of similar surveys of English counties in the 1720s;[24] Sussex by Richard Budgen came out in 1723, followed by Middlesex by Warburton, Bland and Smyth (1724), Warwickshire by Henry Beighton (1728) and Surrey by John Senex (1729), the same man who had produced the successful book of road maps. In the 1730s large scale maps of three more counties made their appearance, viz., Huntingdon (1731), engraved by Emanuel Bowen, and Bedfordshire (1736), both surveyed by W. Gordon and Suffolk by John Kirby (1736). After this, almost hectic, series of larger scale maps no further serious activity occurred for almost twenty years. Then, in 1752, John Rocque produced two preliminary sheets of Berkshire at the scale of two inches to one mile. He followed this with a similarly scaled set of four sheets covering Middlesex in 1754 and nine sheets mapping Surrey; the latter were not issued in his lifetime and their publication was delayed until c.1768 when his widow saw the sheets through the press. These maps by Rocque and two further examples, covering Hertfordshire and Oxfordshire at the same larger scale, were so detailed in their indications of land use (woods, parklands, etc.) and houses that a clear picture emerges of the extent and nature of the economy of the areas mapped.

In 1759 the Royal Society of Arts (RSA) offered a gratuity of not more than £100 to any person, who provided an accurate survey of any county at the scale of one mile to an inch. An extra payment would be possible for surveyors who showed on their maps the depths of rivers and their suitability for navigation. Several of the above maps were shown to the RSA for their approbation (or otherwise, in the case of Budgen's Sussex map, for instance).[25] For a further fifty years the Society continued to repeat its offers of financial reward; however, in all that period, only £460 was paid out in respect of thirteen different county surveys. The first map submitted – Isaac Taylor's Dorset (1765) – failed to obtain an award, but Donn's map of Devonshire (1765) and Burdett's of Derbyshire (1767) were both rewarded with 100 guineas each. Andrew Armstrong, who undertook surveys of four English and three Scottish counties, including Lincolnshire, only received £50 for his Northumberland map and only then after a number of submissions of further details to satisfy the scrutineers. Armstrong's methods were closely examined by Sir Joseph Banks when Armstrong approached the great man for his support. There are other suggestions that his work, and that of his son, Mostyn John, might not have been of the very highest quality (**44**).

Andrew Armstrong, according to the title of his map of Lincolnshire, spent

[24] William Williams produced a map of Flintshire and Denbighshire in c.1720; it was engraved by John Senex.
[25] Gough, R. British Topography. (London, 1780) records the Royal Society of Art's views on several maps. Their view of Budgen's map is in Vol. II, p. 297; Kingsley, p. 57.

three years, 1776–1778, on his survey of the county. The only advertisements in the county paper, *Lincoln, Rutland and Stamford Mercury*, appeared in 1777, when subscriptions were invited for the eight sheet map at two guineas, half to be paid in advance. This method of raising money towards the costs of the work followed a pattern of which Richard Blome (**14** and **17**) had been an early proponent a century earlier; the system had also been pursued by the early pioneers of mapping counties at the one inch scale. Beighton (Warwickshire) and Senex (Surrey), in particular, had obtained sponsorship by means of adorning their maps with the coats of arms of those landed gentry who were willing to pay a premium for the privilege; on the map of Warwickshire Beighton left space for 250 such coats of arms but only just over half have been completed. Senex had even less success as only 43 spaces out of 152 were taken up. If Armstrong knew of this way of raising extra funds and he probably did, the lack of success by early practitioners would have been a deterrent to its adoption. Armstrong may also have been aware of the earlier plan by Edward Weaver in the 1740s to prepare a large-scale map of the county, which failed because the necessary subscription income was not forthcoming. He raised only £48 after over ten years of advertising his plans and, in his case, the prepayment fee was only five shillings, later raised to ten shillings, with similar sums to be paid on delivery. Whether Armstrong made anything from his map of the county is doubtful. The final list of subscribers shows orders for nearly five hundred copies and a few more sets may have been sold to people who had not booked their copies in advance. The total income was probably no more than £1200–1300, therefore, while estimates for the real costs of producing such a series of maps could be anything over £2000. Thomas Yeakell and William Gardner (1778) estimated that their map of Sussex at one inch to a mile would entail expenditure of more than £2400 when surveying, drawing, engraving costs, paper and printing were all taken into account.[26]

Armstrong's map of the county was issued in 1779 on eight large sheets with a reduction on a single sheet appearing in the following year. The map made little real impact on mapping of the county in later years although it, no doubt, filled a gap for the county's landed gentry at the time. The spread of subscribers covers exactly that intended audience. Errors from earlier maps were now corrected, although Armstrong introduced some of his own, most notably the 'creation' of *Hughington* near Heighington, south-east of Lincoln. This mistake was adopted by many mapmakers well into the following century; they were not necessarily following Armstrong, however, since the error was shown by Cary on the large maps he engraved for the edition of Camden's *Britannia*, prepared by Richard Gough in 1789. It is clear that the majority of later engravers incorporating this particular error on their maps were copying Cary and, almost certainly, had never seen a set of Armstrong's maps. There is nothing very original in the contents of Armstrong's map, apart, of course, from the fact of its complete coverage of the county at the larger scale. One claim for originality resides in Armstrong being the first to issue a map of the county, which acknowledged Greenwich as the meridian. However, the line on the map passes between Whaplode and Moulton, several miles west of its true position; in fact, it is just where one would expect a meridian based on St. Paul's. Its lack of detail in giving, for instance, the names

[26] Kingsley, p. 49.

of individual farms or the larger areas of woodland indicates the lesser standards that Armstrong worked to compared with those who had gone before in surveying counties at one inch to a mile. In the only comparison that can be made with an earlier map of part of the county at a comparable scale – Stukeley in 1723 – Armstrong is notably second best.

The deficiencies for one species of landowner – the fox-hunting fraternity – are revealed when one compares Armstrong with the map prepared for the Duke of Rutland in 1806 by W. King.[27] *A MAP of a Tract of Country Surrounding BELVOIR CASTLE* . . . gives the type of information of use to the fox-hunter; there are many notes on the nature of enclosed areas, such as gorse, thorns, pasture, while many more of the minor roads are clearly delineated, all features generally lacking in the earlier sheets. The generality of fox-hunters of Lincoln-shire still felt that their needs were not met by any available maps, when, in 1817, the county gentlemen asked the Ordnance Survey to survey Lincolnshire out of turn so that they might the earlier be provided with mapping that would meet their requirements. Further reference to that event is made later in this survey.

It is to John Cary that we should look for greater accuracy. He was commis-sioned by the Postmaster-General in 1794 to survey the post-roads of England and Wales. It will be recalled that postage was paid, until the introduction of the penny postage in 1840 by Rowland Hill, by the recipient at rates based on the mileage the missive had travelled. It was felt necessary, to ensure accuracy in the calculations of postage, that the roads should be re-surveyed. It should also be remembered that the roads had not been surveyed since Ogilby prepared the issue of *Britannia* in 1675. The results of Cary's work appeared in his *New Itinerary* of 1798. This volume does not have county maps; there are several general maps but the main thrust of the book lies in its detailed descriptions of all the roads with exact mileages between each post town. Many more roads were necessarily surveyed than ever before since correspondence is not confined to people living on the main and cross roads of earlier roadbooks, always assuming that the mileages were accurate anyway. In fact, the Postmaster-General clearly believed otherwise in commissioning Cary. As noted before, Cary set the standards that all later map-makers followed for the next thirty years or more. Only the Ordnance Survey and the rival private surveyors Andrew Bryant and the Greenwood brothers measured themselves against Cary; the maps of these three appeared within the five year period 1825–1829.

Cary was not only to be praised for the new accuracy but also for the quality of his engraving. He set new standards early in his career with the plates for *New and Correct English Atlas* (**47**), which finally appeared in 1788, the year before the Gough translation of Camden's *Britannia*, which had folio plates engraved by Cary (**49**) and two years before *Cary's Traveller's Companion* (**50**). The latter was a small pocket-sized book with comparatively simple maps; its usefulness is evidenced by the need to replace the plates twice as a result of the wear and tear on the original copper-plates (**61** and **80**). There were numerous editions spanning forty-five years; later, after Cary's death in 1835, the plates passed to George Cruchley and had a further life through the medium of lithographic transfers.

---

[27] BL Maps 4810 (1); CUL Maps aa.69.80.5-; Bod Maps (E) C17: 5 (50); Nottingham UL Em B8.E 06; Leicester UL Local Studies Dept. Library.

There are even indications of preparations for a new issue by the Edinburgh firm of Gall and Inglis (**80 (iv)**) in the 1880s. Cary's work was much copied; the plates for *Britannia* inspired two series of county maps of similar folio plate size. Charles Smith (**56**) issued a fine series of county plates in 1801 in *Smith's New English Atlas* . . . His maps show a high standard of engraving and the incorporation of new detail, much of it taken from the many one inch county surveys then available to the enterprising mapmaker. Also in 1801 John Cary began to issue a new series of county sheets; these came together as a complete work entitled *Cary's New English Atlas* . . . in 1809 (**57**). Cary's plates exhibit the same engraving qualities as before, but, disappointingly, they are almost exact copies of the 1789 set and show no large use of information gathered during his recently completed survey work undertaken for the Postmaster-General. Smith's plates and those of Cary both had long lives, extended by their use for lithographic transfers.

At the same time as Cary was active a number of interesting maps appeared of purely local interest. Several related to the originally abortive schemes for a canal from Grantham to Nottingham (1791 and 1792);[28] others refer to more successful applications to provide waterways, such as the Bain Navigation and the extension to Horncastle, surveyed (1792) by Robert Stickney and Samuel Dickinson.[29] Much work, spread over the previous fifty years, concerned with the improvement of fen drainage and improved navigation on the Witham had been undertaken by the Grundys, father and son, who had their base in Spalding.[30] John Grundy, junior, dominated the scene for forty years from c.1739 in surveying potential engineering works in the county and the East Riding particularly. John Rennie was especially prominent in drawing the preliminary maps when the draining of the East, West and Wildmore Fens was being promoted (and successfully accomplished at last). After working in Grimsby in 1798 on improvements to the port he was approached to prepare a report on the above drainage scheme; the map appeared in 1800 engraved by Aaron Arrowsmith (q.v. **72**).[31] Other maps associated with his work in the county include those for the improvement of the Witham's navigation (1803[32] and 1812[33]), the drainage of East and Wildmore Fens (1811)[34] and work on the Wisbech Outfall with drainage in South Holland (1814).[35]

---

[28] Examples of the 1791 map are BL Maps C.10.c.24 (25) and C.25.d.4 (51); Bod Maps (EC) C 17 (451) [37]; RGS 1 C 88 (52); LAO Brace 119/8; of the 1792 map BL Maps C.25.d.4 (52) and C.10.c.24 (54).

[29] BL Maps C.25.d.4 (87) and 1265 (14); RGS 1 C 88 (58); LAO WRO 2/1; Grimsby RL (RM 98).

[30] Wright, N.R. Lincolnshire Towns and Industry, 1700–1914. (Lincoln, 1982), pp. 39–42; Skempton, A.W. The Engineering Works of John Grundy (1719–1783). *Lincolnshire History and Archaeology*, Vol. 19 (1984), pp. 65–82; Wright, N.R. John Grundy of Spalding . . . (Lincoln, [1983]).

[31] BL Maps 3365 (4); BL B 267 (3) includes the report also; Bod Gough Lincolnshire 3; RGS 1 C 88 (53); LAO 2 BNL 12/12. etc.; Cambridgeshire AO R 59/31.

[32] BL Maps C.25.d.4 (81–82); LRL Map 650. Copies of a second version (1804) are LAO 3 Cragg 1/37 and 2 BNL 11/4).

[33] LRL Map 672.

[34] LAO Cragg 1/7; Grimsby RL (RM 96); a revised version appeared (also in 1811) – LRL L. Fens 627.5 (with the report) and Cambridgeshire AO R 59/31/40/157.

[35] BL Maps 3365 (2) and, with the report, LRL L. Fens 627.5.

Another sequence of local maps relating to the county's towns span the period from 1785 to 1839. Stamford and Lincoln, which figure on Speed's plates, Lincoln and Horncastle, which Stukeley drew (1723),[36] Spalding, which John Grundy, the elder, surveyed in 1732 and Hall's map of Boston (1741)[37] represent the total number of plans of towns in the county in the two centuries from Saxton's first county map. A map of Stamford, which purports to be that of Speed (dated erroneously on the plate 1600, instead of 1610) made a freshly engraved appearance in 1785 in Harrod's *The Antiquities Of Stamford . . .*[38] By 1817 there had been seventy-three different Lincolnshire county maps; in that year the first newly surveyed map of a Lincolnshire town, Lincoln, came from William Marrat at the large scale of 10 inches to a mile.[39] Very soon after, in 1819, Padley (q.v. **91**) issued his map of Lincoln. In the next few years a number of town histories were issued, two of which included maps, viz., George Weir's *Historical And Descriptive Sketches Of . . . Horncastle* (1820)[40] and *Notitiæ Ludæ* [by R.S. Bayley] (1834).[41] Also in 1834 James A. Knipe issued a newly prepared map of Stamford[42] and, during the 1830s two series of maps of towns appeared related to the Reform Bill of 1832 and the consequential town boundary changes (1835) from R.K. Dawson (q.v. **96**) and Robert Creighton (q.v. **104**) respectively. A rare group of maps at two inches to a mile by Dewhirst and Nichols covered the same five towns (Lincoln, Grimsby, Stamford, Boston and Grantham) with a sixth dealing with Louth.[43]

In the years immediately after the completion of *Cary's New English Atlas* (1809) a large number of small atlas volumes appeared, all, more or less, attempting to emulate the success of *Cary's Traveller's Companion*, which had appeared with its second set of engraved plates in 1806. None is especially distinguished. There were also imitators of Cary and Smith's larger format plates. Several of these had their own distinction in terms of engraving quality, although they add very little in the way of novelty. Robert Rowe's maps of 1813 were the first; the 1832 re-issue shows the new crossing at Fosdyke and of the Nene at Sutton Bridge, although the delineation fails to suggest the two new bridges accurately (**71 (iv)**). In 1820 Thomas Dix put out a similarly well executed map of the county in the atlas published by William Darton; almost simultaneously with Rowe the two crossings above were added to his plates (**75 (ii)**), together with the new roads across the former East, West and Wildmore Fens. William Ebden's map of 1828 (**85**) made use of the large Cary as well as Rowe and, in its turn, Ebden's was the model for T.L. Murray's sheet of the county issued in 1830 (**89**).

---

[36] BL K.19.29; Bod Gough Gen Top 55; Nottingham UL DA 620; Sheffield RL 914.2 STQ. (LAO Ex 30/4).
[37] BL K 19.22; Bod Gough Maps Lincs 8 and Gough Maps 41D, LAO FL Maps 24; Nottingham UL Dept. of Geography – drawer 121 (F 133).
[38] BL 980 e 13; CUL Ll.8.37–38; Bod Gough Lincoln 2; LRL STAM 9; Grimsby S810:942 HAR; Stamford Town Hall. Phillips Collection T.65.
[39] BL 2367 BB 15; CUL Maps c.70.81.1; LAO Brace 19/1; LRL LIN 9 and Map 882.
[40] BL 190 e 4 and G 1292; CUL A.8.35³; Bod G.A. Fol. A 72; Nottingham UL Li 1.D14.WEI; LRL L. GRAN; Grimsby RL H630:942 WEI.
[41] Bod G.A. Lincs 8° 9; LRL LOUT 9; Grimsby RL L606:942 BAY; Louth RL LOUT 9.
[42] BL Maps 3425 (1).
[43] LAO Brace 19/10–15.

In the 1820s three new surveys of the county were published and the first of these came as part of a process that was to alter the whole approach to mapping in this country and, eventually, by the turn of the nineteenth century, had ensured that all worthwhile county sheets were based on their products. The Ordnance Survey (O.S.) developed from the need to furnish the government of the day with quality mapping of the coastal areas under threat from Napoleonic invasion; from that start a plan to survey the whole country in a uniform style and to a high standard of surveying was devised. The intention was to work northwards from the southern base until the whole land was covered. By 1817, as noted above, the Lincolnshire landowners were impatient to have a more detailed map of the territory over which they hunted. Sir Joseph Banks spoke directly to Col. Mudge, the Master General of the Ordnance, who accepted the suggestion that the gentlemen of the county would raise a sufficient subscription to defray costs. Surveying of the county – out of turn – thus began, in 1817, at a time when only one county north of the Thames had been surveyed and issued (Essex); the counties between Lincolnshire and the Thames were not completed until 1838. In fact, the subscriptions were not forthcoming at the expected rate; changes were made in the methods of surveying at the O.S. when Major Colby succeeded Mudge, who died in 1820. The net result was that the county's eight sheets (**84**) did not appear until 1825 (in spite of the date 1824 in their imprints).[44]

Just as the O.S. were publishing their Lincolnshire maps Andrew Bryant (**86**) was beginning his own survey, which took the three years to 1827 and resulted in the issue of a six sheet map in 1828. As Bryant finished the Greenwood brothers entered the county to spend two years in survey work, which yielded a six sheet version of the county, published eventually in 1830 (**87**). It is a remarkable period for original map making at county level and proved to be the last decade in which such activity could be undertaken by private individuals. The unequal struggle is shown by Bryant's case. He began, perhaps optimistically in view of the progress the Greenwoods had already made, in 1820 to survey Hertfordshire with a view to covering all the English counties. Such was the emphasis on meticulous work combined with careful engraving that, when the project finally fizzled out in 1835, only twelve complete counties plus East Riding had been finished and published and only two in the final six years of this sustained activity. In terms of content, the details of names of even small farms as well as the larger properties, marking the course of rivers, dikes and minor roads Bryant produced the best of the three versions of the county's topography issued between 1825 and 1830. In Lincolnshire's case it is probable that the landowners would have been best served by Bryant but, having paid for their O.S. maps only a few years earlier, sales were probably low. Judging by the number of copies that have survived the Greenwoods, whose map came out in 1830, fared better. Their map contains more detail than the Ordnance Survey and is only deficient in respect of the delineation of higher ground. The Greenwoods quickly re-issued a revised version and nearly all newly added place-names are to be found on Bryant's maps. Even the Greenwoods, who had begun with a map of Yorkshire,

---

[44] The circumstances of the whole survey are fully described in: The Old Series Ordnance Survey Maps Of England And Wales . . . Introductory Essay by J.B. Harley . . . Volume V. Lincolnshire, Rutland and East Anglia . . . (Lympne, [1987]), pp. vii–xxxvi.

issued on nine sheets in 1818, did not quite fulfil their intention to cover all the English counties through lack of financial resources. They had at first probably shown more business acumen in starting with the larger northern counties, which were in the very distant future of activity for the Ordnance Survey and most of which had not been surveyed at a large scale since the activity of the mid-eighteenth century. It should be noted, perhaps, that neither Bryant nor the Greenwoods consistently followed the Ordnance Survey in mapping counties at an inch to a mile. Only Lincolnshire and East Riding of Bryant's maps are at an inch to a mile; all the others are at a larger scale (mostly 1½ inches to a mile). All counties were surveyed by the Greenwoods at an inch to a mile, with the exceptions of Middlesex (2 inches) and Yorkshire (¾ inch to a mile).

Thereafter, there were no further private surveys of all the English counties. Very little large-scale mapping of individual counties took place after the Green-woods had gone out of business and the O.S. had spread over the United Kingdom.[45] However, there were plenty of atlases, which included full sets of maps of the counties. Reference has already been made to the use of the new process of lithography in the re-use of the plates of Cary by G.F. Cruchley. Lithography came along at exactly the right technical moment. The first two decades of the nineteenth century saw the development of higher speed mechani-cal processes in the whole printing industry; paper had been made by the same handmade system since before Gutenberg had developed (1455) the art of printing from moveable type. Type itself was still set by hand and the printing press was identical to those of the earliest period of printing. The introduction of steam and the production of paper using materials other than cotton rags (wood processed mechanically) contributed to a speeding up of the production of print, especially when larger web machines allowed paper to be produced in rolls instead of single sheets. This chimed very well with the development of new methods of reproduc-ing illustrations, e.g. mezzotint, aquatint, to feed an increasing demand from the art world. In the map world there was great reliance on the use of copper plate engraving. Woodcuts and wood engraving were widely used media for illustrative purposes and had, of course, reached great heights in the hands of masters from Dürer to Bewick; copper-plate engraving, however, had been the sole means of reproduction of maps since Mercator. Copper-plate had two great advantage; it allowed great amounts of fine detail to be incorporated and, as the lists below make clear, the engraver or publisher was afforded the facility of introducing new matter to the plate or the correction of erroneous or out of date material. It was merely necessary to either turn the plate over and carefully hammer out the offending details or touch up the surface before entering the replacement material. Copper engraving is a fine art calling for skilled craftsmanship; it has to be remembered that the design is incised on the plate in reverse, so that when the plate passes through the press the printed result is the right way round. As can be seen in the history of the plates, particularly those of Saxton and Speed in the early years of national cartography and those of J. and C. Walker in the nineteenth century (105) maps could be revised and upgraded as long as the plates withstood

---

[45] W.C. Hobson's 4-sheet map of Yorkshire appeared in 1843, but at half inch to a mile. C. Sanderson's 4-sheet maps each of Derbyshire and Nottinghamshire at one inch to a mile both appeared in 1836, shortly after the Greenwoods' main business collapsed.

the strain of passing through the press; Saxton's plates lasted nearly two hundred years and the Walkers' plates went through perhaps two dozen editions spread over nearly sixty years.

Into this world of change in the publishing trade lithography had made slow progress initially, following Senefelder's invention in Germany in the 1790s. The problem lay in the use of quite large and heavy stones on which the work had to be drawn or laid down and the processes required to obtain impressions from the stone; the use of the usual presses was, comparatively speaking, short-lived in the adoption of lithographic processes. By the middle of the nineteenth century faster steam-driven presses handled the printing of thousands of copies to enable publishers to cope with the demands for serialised novels, the many newly established newspapers and all the other printing requirements of a rapidly industrialised society. The first use of lithography for the printing of a map appears to be James Wyld, senior's, plan of Bantry Bay as early as 1808.[46] In 1832[47] the first set of maps using the process – by R.K. Dawson (**96**) – appeared, the map being drawn first and then transferred from the original to the stone. Creighton's maps, based on Dawson's, appeared in 1835 also using the same system of lithographic transfer (**104**). The process was used increasingly from then on, particularly for transferring images taken from plates originally engraved, thus saving wear and tear of the plate. More than one transfer could be made as demand required and life for the bibliographer becomes difficult in deciding if changes in new issues were made on the original plate before the transfer was taken or on the stone after the transfer had been made. Through the transfer facility it was possible to move round elements of the design; G.W. Bacon particularly seemed fond of moving the title and other blocks of data from one position to another in this way (**120**). About 1820 William Say introduced the use of steel plates instead of copper plates for mezzotint work. The first use of steel in preparing county maps came in the 1840s, when Slater recorded their use after taking over the materials of Pigot (**88 (viii)**). The cumbersome nature of the stones used in lithography led to the search for alternatives, the most successful being zinc. Senefelder also envisaged the use of rotary presses for litho printing, although they were not fully developed until many years after his death. The invention in 1852 by Sigl of the lithographic machine press and then in 1875 by Robert Barclay of the offset process permitted the faster production of all printed materials and that included multi-coloured maps, a feature many map producers, such as George Philip (**123** and **130**) and Bacon, made substantial use of. The introduction of photography added further dimensions. Not only could photographs of maps be transferred directly to the plate but the various elements could be enlarged or reduced. The issue of Hall's plates in 1860 (**94 E**) is a notable early example of the dramatic difference an enlargement could make, although photography played no part in that particular enlargement. The O.S. was a body particularly enthusiastic about the advantages of combining photography with zincography (zinc plates with lithography). In all this record of change in the methods of making maps one oddity of 1833 should be mentioned. Joshua Archer (**98**) issued a map

---

[46] Twyman, M. Early Lithographed Books. (London, 1990).
[47] Hodson, 1978, p. 122 records a map of Hertfordshire drawn directly on the stone and dated 1826.

carved on a wood block in reverse, so that when printed white lines appear on black instead of the usual result of black lines on the white substrate. This was not a new technique – it had been practised for a very long time; it was the use of the technique in map-making at such a late stage that made it unique. Curiously, when the blocks passed in 1847 to Thomas Johnson he used them to make an intervening transfer so that the result became 'normal', i.e. black lines on white.

After the trio of new surveys of the counties in the 1820s all the remaining maps of the century are based on earlier models and their only originality is in their presentation. There are few examples that show a distinctiveness that deserve mention. In 1831 there began to appear a sequence of maps, with county texts, later gathered up to form Moule's *The English Counties Delineated* (**95**). All of these maps are popular among collectors; hand-coloured they are attractive pieces with their frames made up of gothic pillars and carvings, inset illustrations of scenes in the county and coats of arms with county associations. One particular curiosity occurs on the county map in the picture of the Cathedral; the engraver must have copied from an old engraving, since he shows the spires still *in situ*, when they had been removed over twenty years earlier. The plates of J. and C. Walker have already been referred to; they made their first appearance in 1835 and were still being used in their sixtieth year. Part of their interest lies in their issue as a parallel group of transfers, from 1849, overprinted with details of the county's hunts. A final pair of transfers made in the 1880s by Letts yield maps so different again from the Walker plates and their derivatives, the Hobson hunting maps, that the novice might be forgiven for not recognising their use of the same basic source. During the period from 1830 to 1843 a number of maps appeared, which owed their origin and production to local cartographers and craftsmen. The well-known county surveyor of Lindsey, James Sandby Padley, had already prepared a map of Lincoln and in 1830 he produced a map of Lincolnshire for use in a report produced by Lincolnshire Coast Shipwreck Association (**91**). In the same year the first parts of Allen's *History Of The County Of Lincolnshire* appeared and they were accompanied by a county map, possibly engraved by Joseph Noble of Hull for John Saunders (then of Leeds), who later took over the completion of the enterprise (**90**). Saunders was also responsible for a second map issued in 1833 (**101**) and was involved in the preparation of a county map, engraved by James Stevenson in 1838 and issued by J. and J. Jackson of Louth in a new version the following year (**107**). In 1843 the small firm of Victor and Baker, which had a printing house in Lincoln, issued a small plate to accompany a series of local guide books (**113**). Only the 1838 plate of Stevenson enjoyed a long life and had any particular distinction.

Only one series of new road maps of note appeared in the nineteenth century, those of Laurie and Whittle, first issued in 1806 and passing through many editions over the following forty years (Appendix **I** (**16**)). They produced map plates that look quite different from the earlier road strip maps of their predecessors. Their appearance can be seen as the precursor of the modern road map books; the main roads are clearly shown and the indications of the minor roads and their destinations, with mileages all add up to a series that could have easily been used by motorists in the days before motorways and by-passes.

The advance of railways led to the development of county maps overprinted with railway lines and the names of stations. Many series came out from the 1850s. G.F. Cruchley and H.G. Collins were noted providers in the formative years of the railway network, using plates of Cary and others for their transfers. The plates of Sidney Hall (**94**) were used to provide a popular series of travelling atlases of pocket size, another set of plates that had a life of nearly sixty years. One problem, shown particularly in the Hall series, but he was by no means alone, was that as railway acts proliferated so astute mapmakers/businessmen added lines to their maps. Often the result was a map showing railways already in operation, railways approved but not built in the position initially agreed and lines approved but never built. Using railway data, therefore, as a way of dating maps may not be entirely reliable without other supporting evidence. From the 1880s a new life was given to some series of maps by the likes of G.W. Bacon who were market conscious men and sold the county maps in covers (as before) but now aimed at the new breed of travellers – cyclists – with suitable illustrations on the covers. Before the century was out Bacon replaced these covers to show the first motor cars and widen their appeal to another new audience. It is, perhaps, unnecessary to say the maps were not especially different.

In the later nineteenth century a wide variety of genre or thematic maps was produced. Maps for railway travellers, cyclists and early motorists were only the final manifestations of the ways that maps could be adapted for various new markets. Earlier in the century other uses for maps came to the fore, turning the circle round completely. From a time when the written word seemed to make a map unnecessary the map could be used to replace or, more especially, make the written word more immediately intelligible. Reference has been made to the maps of Dawson and Creighton, which showed suggested boundary changes, with electoral information in the form of the siting of polling stations and the number of MPs to be elected. At the same time other engravers added stars to maps to indicate the number of MPs each town or division sent to Parliament. Particularly after the 1885 Act that led to the redrawing of electoral divisions colour printing permitted the clearer delineation of the boundaries of the new electoral areas.

In the first two decades of the century the first maps appeared showing geological information. Cary produced a version of his large plates with coloured overprinting to delineate the various strata; Lincolnshire was one of those not so treated but the notes on the soils, etc. were engraved on the county plate (**57 (ix)**) in 1820, based on the findings of William Smith 'the father of English Geology'. Earlier still, the pioneer, William Farey, produced the first geological cross-section in 1807; the line he chose ran from the coast at Trusthorpe, passing near Revesby (Sir Joseph Banks was the dedicatee) and terminating at Overton in Derbyshire, where Sir Joseph had his seat.[48] An interesting map, usually found with contemporary hand colouring is to be found in Weir's volume on Horncastle (1820 – see note 38); the map only covers the geology of the area around that town. Only one atlas of all the counties showing geological data was issued in the nineteenth century and that used plates not prepared particularly for the purpose; James Reynolds issued his *Travelling Atlas* in 1848 (**114**), but it was only in 1860 the plates were colour printed with geological strata. Taken over by Stanford's this pocket atlas remained on sale for another seventy years.

In 1821 the maps in *Valor Ecclesiasticus* (**79**) showed only the boundaries of the diocese, the archdeaconries and parishes. Arrowsmith issued a similar set in

---

[48] British Geological Survey AM 1112 FAR is the original hand-coloured section.

1836 when the Church Commissioners considered the creation of new dioceses
(**106**). Another series by W.J. Sackett (**124** and **125**) was issued in 1864 to show
the diocesan boundaries for students at Lichfield Theological College. With a
slightly different slant is the curious map by W. Wright, junior, which showed the
disposition of churches before the Conquest and was used to illustrate a book on
church history (**109**); perhaps even more curiously Lincolnshire is the only county
in the book in question singled out for such treatment. In the 1880s and 1890s
the Ordnance Survey issued large maps of the counties over-printed with information devoted to a wide range of local government services. Maps dealt with
the extent of the civil parishes, etc. (**143**), the Sanitary Districts (**143 B**) and the
physical make-up of the county's parishes (**141**). At the end of the century, maps
were used to illustrate the results of clerical research, whether R.E.G. Cole's
(1886) *Glossary Of Words Used In South-West Lincolnshire . . .* (**142 A**), Peacock's (1889) *Glossary Of Words Used In The Wapentakes Of Manley And
Corringham* (**142 B**), E. Venables and G.G. Perry's *Lincoln* in the Diocesan
Histories series, 1897 (**150**) or the map used for *The Natural History Of Lincolnshire* (1898) by E.A. Woodruffe-Peacock (**151**). Finally, reference might be made
to the humorous use of maps in the puzzle pictures for children (**128** and **134**)
and in an educational card game (**147**).

# ARRANGEMENT AND SCOPE

The catalogue of Lincolnshire maps is confined to those maps issued between 1576 and 1900 that show the whole county and to other significant maps that show the county and one other whole county. Such maps may be complete on one sheet or comprise several sheets that are intended to be put together to show the whole county. Nearly all the maps that qualify for inclusion will show parts of most of the counties which have a common boundary with Lincolnshire. The only significant maps that are excluded are the Ordnance Survey sheets from 1825, because there is a recently published bibliography devoted to that series, and multi-sheet maps that cover the entire country, two or more sheets of which could be put together to form a map of Lincolnshire. Several interesting series are, therefore, excluded; however, there are substantial references to others whether for their own intrinsic merits or because later maps, that are included, are based on the excluded model. Additionally there are numerous references to maps of areas of the county when discussing the products of various surveyors or engravers of county maps.

The entries are arranged in chronological order, based on the first appearance of the map sheet. Under each main entry subsequent issues of the map are then shown, also in chronological array, before the next main entry. In this manner, the genealogy of each original is shown as it developed with full indications of the changed features that show how succeeding states differ from their predecessors.

All material taken from the face of the map and from the title-page of the book or atlas, in which the map appeared, is given in italics.

## The Form of the Entries

**HEADING**   At the head of each main entry are (a) the catalogue **serial number** at the top, followed by (b) the name of the engraver or, if not known, the name of the person responsible for the drawing or the first publisher; (c) the first **date of issue**. The date of issue may be derived from the map itself or from the title-page of the book in which it first appeared. In cases where there is any doubt the date is expressed in the form c.1840.

The **size** of the map is then given from the outer edges of the frames. As a librarian, originally trained as a cataloguer, I have followed the system prescribed by the Anglo-American Joint Code. The height from top to bottom precedes the width. British Standard (B.S. 5195; Part 1, 1975) decrees the opposite. All measurements are given in centimetres. Minor variations in size between one issue and another are not noted. Paper sizes vary over a period of time depending on how the sheets have been treated and the type of paper; there are bound to be slight insignificant differences, depending on the paper stock used (many atlases are printed on papers of different makers or produced at different times).

**NOTES**   The first main section contains notes on the surveyor, the engraver and the publisher. These are intended to provide useful background information on the people involved in the map's appearance and, at the same time, give a picture of some aspects of the state, at the time, of mapmaking, engraving, printing and publishing. Some effort has been made to refer to other maps produced by the men described, particularly when they have been involved in mapping parts of Lincolnshire. At the risk of overloading the paraphernalia of footnotes, as a consequence, locations of examples of such local mapping are recorded. The later history of the plates is given in the notes with references to other owners or publishers of the map plates. An attempt has been made to relate the map under discussion to its predecessors and the source(s) of its contents.

**EDITIONS**   Each time a map appeared with changes to the plate a new 'state' was created. Under the heading **Edition(s)** each new state is given a roman number, as an identifying symbol. If the new state is a lithographic transfer the state is separated from other states by upper case letters (in bold); maps that appeared in both forms, i.e. from an engraved plate (or wood block) and lithographic transfer, are listed with the states from the intaglio original first, followed by all litho transfers. Usually, of course, once a plate has been used for a lithographic transfer the plate is never used for direct printing thereafter; its sole use, generally but not always, is to generate fresh transfers. Each state is described in as full detail as necessary to separate the map being described from those that precede or follow it.

When a map makes a fresh appearance but remains unchanged this represents a new **issue**. The significant detail necessary to describe the circumstances of its new appearance, usually the reprint of the book or atlas that contains it, is given in a separate block of information immediately below all the details of the state to which the issue relates.

The detailed description of each map sheet begins, usually, with the **title** on the map; the **imprint** and **signatures** (i.e. the names and addresses, if given, of the people who drew, engraved or published the item). The title and all other data taken from the map are given in the form as they appear on the map; all capital letters are given and the ends of lines are marked by a slash (/). No attempt has been made to reproduce the variety of lettering styles adopted by many engravers. The long s (shown in many cases as f) is transcribed as s. The presence of **explanations** (the usual heading on maps for a list of the symbols employed) is described. The **scale** is given in the form as shown on the map, with the measurement in centimetres of (usually, the whole) scale-bar, in square brackets [ ]; if the scale is given in the form 1+10 that means that the scale shows 1 mile to the left of the zero and 10 to the right. No attempt has been made to give the map's scale in the modern style, e.g. 1:50000 as used in O.S. Landranger maps. Since in so many early maps there is little consistency in the application of the scale on the map, and there is a good deal of doubt often enough in the sort of mile being used by the surveyor, it is felt the provision of such a scale would be of dubious value. Where the map is very small, as so many of the derivatives of Cary's maps for his *Traveller's Companion* were, it seemed pointless to give figures such as 1:576000 (the figure quoted for Cary's map).

The presence of a **compass**, and its type, only if distinctive, is noted. The term may be taken to mean in general anything from a simple cross to an elaborate

32-point indicator; it is assumed that north is to the top of the plate and a suitable note is given when this is not the case. There may be references to any **other features**, such as illustrations, what they are and how they are described on the map; whether there is a graticule and if there are railways (using the numbers that identify the different lines as given in the separate listing – see below). Other features include: lists of **wapentakes** (including hundreds, sokes, liberties) – these are frequently provided and in three groups, which correspond to the parts of Lincolnshire, i.e. Lindsey, Kesteven and Holland (and usually in that order); the borders of maps are frequently engraved to produce an alternating pattern of shaded and unshaded sections to indicate **latitude** or **longitude** (usually both) in various numbers of minutes. The symbol (′) is used to indicate minutes. The presence of notes on the meridian, which passes through the county and whether measured from London, Greenwich, St. Pauls or, on earlier maps, from Paris or the Azores is indicated, whether as a note or in the form provided on the map. Not all maps, of course, have these or other features, but, where they exist, they are recorded. **Place-names** which form a feature in the description of the particular state are given, in italics, in the form as shown on the map. If a name has been altered the former method of spelling is usually given in brackets. Again, the long s, signified by f, is transcribed as s, except in one case where the change from one form to the other is a significant part of the difference between one state and another. The siting of these elements is described with two letters, which correspond to an imaginary grid placed over the map face. The grid idea, devised by Mr. David Kingsley, has 25 'rectangles' lettered A to E along the top and a to e down the sides:

```
Aa  Ba  Ca  Da  Ea
Ab  Bb  Cb  Db  Eb
Ac  Bc  Cc  Dc  Ec
Ad  Bd  Cd  Dd  Ed
Ae  Be  Ce  De  Ee
```

If material is outside the frame the position is marked with the nearest part of the frame, followed by 'outside', abbreviated to (OS). Thus a plate number outside the top right hand corner is shown as Ea, (OS). Where a feature spreads into other 'squares' the outer limits are given; for instance, a simple heading may appear at the centre of the top and is described Ca; if the heading were extended but still centred at the top it would be described as Ba-Da or Aa-Ea.

**SOURCE**   After the description of the map itself its **origin** is then made clear. If the map comes from a **printed book** the title-page is transcribed, usually in full and always in italics; if the title is shortened, enough of its detail is given to separate that particular edition from others similarly titled or from others with the same title but with new imprint particulars. No attempt has been made to show line endings or the variety of types that can appear within a single title-page. Words beginning with or containing all capital letters have an initial upper case letter only in the transcription. The letter f has been transcribed as s in every case. In many cases, both on the maps and on title-pages words are abbreviated in the form S$^t$. with a dot under the superscript final letter(s). It has not proved possible to transcribe this formula here; where superscript letters occur followed by a full point (.) in the succeeding space this should, in all cases, be taken as appearing under the final superscript character(s).

If the map was issued as a **loose sheet** (only) a note to that effect is provided; if it was published as a sheet **folded in a cover** enough of the cover's title is given to separate the edition from other appearances in other types of cover. The **date** is given if it appears on the title-page. If a date has to be supplied from another part of the book or outside source it is shown in square brackets ([ ]). For instance, from c.1790 paper had the date of manufacture as part of the watermark; such dates are readily seen on the larger map sheets (and often somewhere in small books that employ paper folded several times). These watermarks give, at least, a date before which the map could not have been printed. In the general way hand-made paper ceases to be used after c.1835 for the maps described in this work and watermark evidence accordingly disappears. The exception to this is the paper used by Ordnance Survey, some examples of which continue to show watermarks until the early 1870s. If the date is supplied from another, external, source it may be shown with a question mark or in the form [c.1899]. A serious attempt has been made to give dates to every map; the reasoning or authority for all suggested dates is given in a footnote. In the second half of the nineteenth century many issues appeared for which dating can only be suggested on the basis of the railways marked on the map. Such a method has its particular pit-falls since many publishers wanted to appear as up to the minute as possible; to that end railways were drawn on some maps as soon as the bill went before parliament; but many lines were not approved or were approved but not built or, if built, adopted a different alignment. All dates assigned on the basis of railway data only are shown in the style [1889**] to reduce the number of foot-notes.

**LOCATIONS** The **location of copies** of the work described is then given. The convention adopted in recent cartobibliographies has been to provide three locations of copies. Most of such studies have been devoted to counties within easy reach of one of the three copyright libraries, which are the usual source of copies – British Library, Cambridge University Library and Bodleian, Oxford. For a county like Lincolnshire it has seemed a useful addition for the local student of the county's maps to widen the scope. A maximum of six locations is, therefore, given. The locations of copies are noted in two forms; firstly, the availability of the book form with the name(s) of the library or society having copies with the shelf number and, secondly, loose copies. If there are copies in the three copyright libraries these are given first and in the above order, followed by any other copies examined with the emphasis on copies in or near the county. Copies are, of course, only noted if they *include the map*. This is the usual explanation also for the omission of local copies, particularly in the case of county directories, many of which should have maps but are often enough missing. Loose copies of the map are then recorded, again with the appropriate shelf or drawer number used by the holding library, the whole being enclosed in round brackets ( ). The majority of the loose sheets are held in libraries within the county; by providing this information a key is thereby provided for the many fine collections in local libraries. If a library holds two copies both shelf numbers are given; if three or more are held the shelf number of one example has been chosen and that is followed by 'etc.' A separate list of the abbreviations used is given below. It is to be remembered that location numbers may be changed; many of those given in the British Museum Printed Catalogues have been changed (some, more than once) and the Lincolnshire County Archives Office is also in the process of giving new computer catalogue numbers to some items, without, apparently, retaining a record of the previous location symbols. Whether the cited works located in Norwich RL have survived the disastrous fire of 1 August, 1994 remains unclear.

**LATER STATES or ISSUES**   After the listing of locations the next separate re-issue of the volume is recorded when no change to the map has occurred. In such cases the title(s) of the works in which the map appears is given and the locations of copies recorded in separate paragraphs. Where the map has been altered the new state is given a new roman number or, in the case of lithographic transfers, a (bold) upper case letter and the variations are described in exactly the same way as outlined above, and so on until all variant states have been described in the order of their issue.

**NOTES**   At the end of the whole entry there may be any of three further sections. The first consists of further notes on the series as a whole, perhaps, as in the case of the Van den Keere/George Humble plates (7) guidance on the separation of loose sheets using the text on the verso where the map itself has remained unchanged. A second possible section, headed 'Reproduction(s)' (in bold), indicates some of the reissues of the atlas or map volume or uses of the county map in illustrating articles in books or journals. Finally a section headed 'References' (in bold) gives the sources for statements or adds further information to matters referred to in the catalogue text. In the final line(s) appear references to the catalogue numbers used by Skelton (atlases issued by 1703), Hodson (atlases from 1703–1763) and Chubb (after 1763 and before 1870 – his cut-off date). In a few cases references are given from Skelton and Hodson where the latter revises detail previously given in the former. Chubb numbers are applied to pre-1870 works in the road book section, since such works are not dealt with by Skelton or Hodson.

**APPENDIX**   There is only one appendix. In it, the plates in **road-books** or **road maps** in loose sheet form from 1675 to 1900 with a Lincolnshire significance are described in the same manner as the county maps above. Two important differences will be apparent in this section; the chief one lies in that the great majority of road-books had more than one plate that dealt with roads in or passing through the county; all the plates are listed after the heading and notes before the title of the work in which they appear. Secondly, the descriptions are considerably longer, compared to the county map section, since many plates in later editions were altered to reflect the names of new owners of the plates and to incorporate corrections and fresh information. It was felt important to provide enough of the details of plate alterations to allow the collector/student to match a loose sheet to its parent volume/edition. All the road-books of English roads, from John Ogilby onwards, include the Great North Road, usually on three or more plates. Only those plates, which describe the section that passes through the county are described in detail. A distinction is made in the case of the important series *Laurie and Whittle's New Traveller's Companion* (Appendix **16**); three plates contain the whole or parts of the county and these are described, irrespective of the fact that a large part of some plates includes a number of other counties.

As stated above there are many notes on other types of maps showing parts of the county in the discussion of the works of surveyors, engravers or publishers. No attempt has been made here to deal with town plans, canal, drainage, railway maps and all the other types of printed maps relating to the county in a systematic way. Full descriptions of such items would require another substantial volume. **Coastal charts** are not covered in the descriptions either. The bulk of charts that show any part of the Lincolnshire coast fall into two main groups. By far the larger group consists of charts of the whole or large stretches of the East coast of England, some on such a scale that parts of the French, Belgian, German or Dutch coasts or all of these are delineated. A second type, still very substantial, includes charts for the mariner entering the Wash and, more precisely, the Boston Deeps, and Grimsby and the entrance to the Humber. They have been excluded because of the space their description would have occupied; if the same approach to the larger charts were made as with the road-books, i.e. only describing that part of the chart of Lincolnshire significance, a very unbalanced view of the chart would result. The title would not have, in most cases, any reference to the county; the differences between one edition and a successor would rarely relate to anything changed in the county portion; and, in general, the greatest amount of significant detail occurs on the most used waterways, e.g. the entrance to the Thames or in the Straits of Dover. There are very few such maps and charts in the county's library collections or those of neighbouring libraries; additionally, they are not the stuff of private collecting and few map sellers are willing to develop the special expertise necessary to deal with this very complex form of mapping. The few map sellers that do venture in tend to confine themselves to the plates from the earlier books of charts from the Greenvile Collins period (1693) onwards and the many foreign charts issued in the following century. The Admiralty charts alone would make a sizeable supplement.

# ABBREVIATIONS

AO       Archives Office (preceded by a town/county)
BL       British Library
Bod      Bodleian Library, Oxford
CRO      County Record Office
CUL      Cambridge University Library
DNB      Dictionary of National Biography
L        Library (preceded by a name or place)
LAO      Lincolnshire Archives Office
LRL      Lincoln Reference Library
NLS      National Library of Scotland
NLW      National Library of Wales
NMM      National Maritime Museum
O.S.     Ordnance Survey
PRO      Public Record Office
RGS      Royal Geographical Society
RL       Reference Library (preceded by a place-name)
UL       University Library (preceded by a name/town)
W        Whitaker collection (Leeds University Library)

The abbreviations for railway companies are noted in the lists of railway lines and their numbers, as used in this catalogue – see next section.

Many works have been regularly cited throughout the text. The details of these works are readily found in the bibliography under the surname of the writer. Where writers have written more than one work the date of publication in the citation makes clear the intended reference.

There are many references to articles in books and journals. Their details are given in full in the footnotes; their authors' names can be picked up in the general index.

# RAILWAYS SHOWN ON LINCOLNSHIRE MAPS

The railways noted below are arranged chronologically and include three types of line. Railways built and operated before 1900 are indicated by an arabic number; those proposed, but never finally built, are shown by roman numbers; and, railways operating in adjoining counties are given capital letters. Only railways indicated in one form or another on the county maps recorded in these lists are noted here; for clarity some lines are described by their proposed route where the company's name is not sufficiently descriptive. Some railways are referred to by a series of numbers, separated by slashes (/), where the line was opened in sections spread over a period of time. Many other lines were proposed but not built, both in Lincolnshire and in other counties. The initials by which many lines are referred to in the text are given after the railways' names. A few lines not listed here appear on some maps with a wider coverage than usual; they are described in words in the appropriate section of the text. I have made use of many books, the details of which are given in the bibliography.

| A | Hull & Selby Railway (HSR) | 1 July 1840 |
| i | Cambridge & York Railway | 1841 |
| ii | London & York Railway | 1844 |
| iii | Ely, Spalding, Boston, Lincoln | 1844 |
| iv | Cambridge–Lincoln, via M. Deeping, with a branch to Spalding | 1844 |
| v | Sheffield, Bawtry & Gainsborough | 1844 |
| vi | Caistor branch from the GGSJR (later MSLR) | 1845 |
| vii | Nottingham–Kings Lynn, with a branch from | |
| viii | Folkingham to Boston and a branch at | |
| ix | Long Sutton to Wisbech | 1845 |
| x | London–Grantham–Gainsborough | 1845 |
| xi | Grimsby, Louth, Horncastle, Grantham | 1845 |
| xii | Tattershall–Horncastle | 1846 |
| 1 | Nottingham–Lincoln (Midland Railway; MR) | 3 August, 1846 |
| 2 | Peterborough–Stamford (Syston & Peterborough R) | 2 July, 1846 |
| B | Wisbech–Thorney–Helpston, joining 2 above | 1846 |
| C | Hull–Beverley–Scarborough | 6 June, 1846 |
| D | Kings Lynn–Swaffham–Dereham | 27 October, 1846 |
| E | Kings Lynn–Kings Lynn harbour | 29 October, 1846 |
| F | Wisbech–March | 1847 |
| xiii | Stainton–Wragby | 1847 |

| G | Beverley–Market Weighton–York | 4 October, 1847 |
|---|---|---|
| H | Kings Lynn–Downham Market–Ely | 26 October, 1847 |
| I | Wisbech–Magdalen Road (on railway F) | 1 February, 1848 |
| 3 | Grimsby–Louth (East Lincolnshire R; ELR) | 1 March, 1848 |
| 4 | Grimsby–New Holland (Manchester, Sheffield and Lincolnshire; MSLR) | 1 March, 1848 |
| J | Wakefield–Pontefract–Goole | 1 April, 1848 |
| 5 | *Stamford–Melton Mowbray | 1 May, 1848 |
| 6 | Louth–Firsby (ELR) | 1 September, 1848 |
| 7 | Firsby–Boston (ELR) | 1 October, 1848 |
| 8 | Great Northern R loop (GNR; Peterborough–Boston Lincoln) | 17 October, 1848 |
| 9 | Ulceby–Brigg (MSLR) | 1 November, 1848 |
| 10 | Barnetby–Market Rasen (MSLR) | 1 November, 1848 |
| 11 | Lincoln–Market Rasen (MSLR) | 18 December, 1848 |
| 12 | New Holland–Barton on Humber (MSLR) | 1 March, 1849 |
| 13 | Brigg–Gainsborough (MSLR) | 2 April, 1849 |
| 14 | Lincoln–Gainsborough (GNR) | 9 April, 1849 |
| K | Gainsborough–Sheffield | 17 July, 1849 |
| 15 | Nottingham–Grantham; part of Ambergate, Nottingham, Boston & Eastern Junction Railway (ANBEJR) | 15 July, 1850 |
| 16 | Sykes Junction–Clarborough Jn. | 7 August, 1850 |
| 17 | Peterborough–Retford (GNR) | 1 August, 1852 |
| 18 | Grimsby Town–Royal Dock (MSLR) | 1 August, 1853 |
| L | Hull–Withernsea | 26 June, 1854 |
| M | Goole–Doncaster | 4 Sept., 1854 |
| 19 | Horncastle–Kirkstead (HKR) | 11 August, 1855 |
| 20 | Stamford–Essendine (SER) | 1 November, 1856 |
| 21 | Barkston–Sleaford (GNR) | 16 June, 1857 |
| 22 | Edenham–Little Bytham (This line was abandoned in 1873) | 8 December, 1857 |
| 23 | Spalding–Holbeach (Norwich & Spalding R; NSR) | 15 November, 1858 |
| 24 | Sleaford–Boston (GNR) | 12 April, 1859 |
| 25 | Thorne–Keadby (South Yorkshire R; SYR) | September, 1859 |
| 26 | Bourne–Essendine (GNR) | 16 May, 1860 |
| N | Rolleston Jc.–Southwell–Mansfield (Opened 1 July, 1847 but no regular service until 1860). | 1860 |
| 27 | Holbeach–Sutton Bridge (NSR) | 1 July, 1862 |

| | | |
|---|---|---|
| O | Kings Lynn–Hunstanton | 3 October, 1862 |
| 28 | Grimsby–Cleethorpes | 6 April, 1863 |
| P | Hull–Hornsea | 28 March, 1864 |
| 29 | Sutton Bridge–Kings Lynn (NSR) | 1 November, 1864 |
| xiv | Lincoln–Keadby or Althorp – act passed | 5 July, 1865 |
| 30 | Keadby–Scunthorpe (SYR) | 14 April, 1866 |
| 31 | Sutton Bridge–Wisbech–Peterborough | 1 August, 1866 |
| 32 | Spalding–Bourne (SBR) | 1 August, 1866 |
| Q | Heacham–Wells next the Sea | 17 August, 1866 |
| 33 | Scunthorpe–Barnetby | 1 October, 1866 |
| 34 | Spalding–March (GNGEJR) | 1 April, 1867 |
| 35 | Lincoln–Honington (GNR) | 15 July, 1867 |
| 36 | Gainborough–Doncaster (GNGEJR) | 15 July, 1867 |
| R | Stamford–Wansford | 8 August, 1867 |
| xv | Peterborough–Crowland | 1867? |
| 37 | Spilsby–Firsby (SFR) | 1 May, 1868 |
| S | Goole–Staddlethorpe (on the HSR) | 2 August, 1869 |
| 38 | Firsby–Wainfleet (FWR) | 24 October, 1871 |
| 39 | Bourne–Sleaford (GNR) | 3 January, 1872 |
| xvi | Louth–Mablethorpe with branches to Saltfleet Haven & North Somercotes (Louth & East Coast R; LECR) | July, 1872 |
| 40 | Wainfleet–Skegness | 28 July, 1873 |
| xvii | Alford–Mablethorpe | 1873 |
| 41 | Bardney–Donington on Bain | 27 September, 1875 |
| 42 | Barkston–Sedgebrook | 29 October, 1875 |
| 43 | Donington on Bain–Louth | 1 December, 1876 |
| 44 | Louth–Mablethorpe (LECR) | 17 October, 1877 |
| T | Newark–Bottesford | 1 April, 1878 |
| U | Bottesford–Melton Mowbray | 30 June, 1879 |
| V | Kings Lynn–Massingham | 16 August, 1879 |
| 45 | Spalding–Ruskington (GNGEJR) | 6 March, 1882 |
| 46 | Ruskington–Lincoln (GNGEJR) | 1 August, 1882 |
| W | Wisbech–Upwell | 20 August, 1883 |
| xviii | Lincoln–Skegness | 1884 |
| 47 | Alford–Sutton on Sea Tramway (AST) (Closed 1889) | 2 April, 1884 |
| X | Hull–Barnsley | 27 July, 1885 |
| 48 | Sutton on Sea–Willoughby (SWR) | 4 October, 1886 |

| 49 | Sutton on Sea–Mablethorpe (SWR) | 14 July, 1888 |
| 50 | Bourne–Little Bytham (MGNJR) | 15 May, 1893 |
| 51 | Little Bythem–Saxby (MGNJR) | 1 May, 1894 |
| 52 | Chesterfield–Lincoln (LDECR) | 8 March, 1897 |
| 53 | Goole–Reedness | 8 January, 1900 |

5 is only noted when the line appears on the map in Rutland, usually with Oakham and/or Melton Mowbray; railways 1 and 5 shared the same station in Stamford.

# ILLUSTRATIONS

The following maps are reproduced by kind permission of the British Museum (fig. 1); Leeds University Library (3); Mr. E. Burden (10).

# THE PRINTED MAPS OF LINCOLNSHIRE
## 1576–1900

# CHRISTOPHER SAXTON

Size: 40.2cm x 53.25cm.

Saxton described himself as from Dunningley on a number of occasions. Dunningley is a hamlet in the parish of West Ardsley in the West Riding of Yorkshire. It seems likely, however, that he was born in the neighbouring parish of Dewsbury in 1543 and moved with his family in his childhood or early youth.[1] The evidence for the family tree of Saxton worked out by Evans and Lawrence is largely based on the Hearth Tax returns for Dunningley.[2] Thomas Saxton, whose name occurs in the returns between 1567 and 1599, was probably the father of the cartographer; at the next return in 1606 Christopher Saxton is the only person of that surname recorded; in that following – 1620–1621 – Robert Saxton is named, and Christopher's eldest son was named Robert.

Saxton was himself quite unsure of his own date of birth; in 1596 he described himself as 'of Dunningley . . . fiftye twoo yeares or thereabouts' but ten years later he says he is '64 yeres or thereabouts'. The date of his death is equally uncertain since no will can be found. His last recorded work was in 1608. He is presumed to be still alive when his brother Thomas' will was proved in June, 1610. His 'epitaph' was discovered in a notebook written by Dr. John Favour, Fellow of New College, Oxford,[3] which Fordham has suggested was written between 1603 and 1611 and thereby gives us a date of 1610 or 1611 for Saxton's death.[4]

Saxton is likely to have come under the influence of John Rudd, Vicar of Dewsbury from 1554 to 1570. Rudd had a life-long interest in maps and cartography and, since Saxton described Rudd as his master in a receipt signed in 1570 on the occasion of Saxton going to Durham on Rudd's behalf, we may assume Rudd kindled a similar interest in Saxton and gave him his early training.[5] No other source for the practical knowledge he had of surveying has come to light. No example of any survey work undertaken by Saxton in this early period has survived either. Yet, within four years of that receipt he had made such progress in his survey of all the English and Welsh counties that the first two plates had been engraved.

In July 1573 Saxton was appointed 'by speciall direccion & commandment from the Queenes Majesty' to survey the English and Welsh counties. The appointment, on the Queen's behalf, was made by Thomas Seckford of Woodbridge in Suffolk, a Master of Requests to the Queen. Seckford not only financed the enterprise but he also obtained for Saxton a letter of introduction to the Privy Council. Seckford was surveyor to the Court of Ward and Liveries; it has been suggested[6] that this may have been the inspiration for the project. The national interest and fear of invasion in the 1570s are at least as likely reasons, especially as the Secretary of State and Lord Treasurer, Sir William Cecil, later Lord Burghley, had a special interest in maps and believed very strongly in the value of accurate maps in strategic terms. Laurence Nowell in 1563 wrote in a letter to Cecil '. . . you take especial pleasure in geographical maps, and . . . know how to make good use of them . . . to render unceasing service to the state.'[7] Lord Burghley received each of Saxton's plates as they were engraved and added his own notes on the versos.[8] Cecil's support may also have been inspired by the appearance of Mercator's atlas in 1570 and the impression that this may have aroused that the Dutch sea-faring rivals had secured some advantage.

Of Saxton's survey methods we can infer a certain amount from the wording of the Privy Council's orders. On 10th July, 1576 An Order Of Assistance was issued in conjunction with the survey work in Wales. The Justices were required 'to see him [Saxton] conducted unto any towre Castle highe place or hill to view that countrey and that he may be accompanied w[th] ij or iij honest men such as do best know the cuntrey

... do set forth a horseman that can speke both Welshe and englishe to safe conduct him to the next market Towne, etc.'[9] An earlier act of the Privy Council in March, 1575, had been far less specific in the forms of assistance to be rendered in order that the work be furthered. It is probable that some form of triangulation formed the basis for the survey work and, since the plates incorporate the phrase *Christophorus Saxton descripsit*, it is clear that Saxton produced the original drawings on which the engravings were based. It is equally evident that an atlas was envisaged from the start. The plates are of a uniformity of size that could be printed on a single royal sheet of paper, i.e. about 20 x 25 inches in size. In some cases several counties appear on the same plate and only Yorkshire was engraved on two plates to be printed as a folded map.

Thirty-four plates were engraved between 1574 and 1578 and all but one are dated. Two were engraved in 1574 (Norfolk and Oxford); six in 1575; eight in 1576; twelve in 1577, and five in 1578.[10] The undated one was probably produced in 1576. Of the several cases where two or more counties were engraved on the same plate Lincolnshire is one such, sharing its plate with Nottinghamshire.

It has been suggested that a possible itinerary for Saxton can be discerned from the dates – that he worked roughly across England from East Anglia to Cornwall and finished the south of the country by some time in 1576, having filled in the gap in the section by moving to Essex at the end of the process; he then, it is suggested, dealt with the Midland counties before surveying the northern and Welsh areas.[11] There is enough other circumstantial evidence to support some of this hypothesis; the Welsh counties are all in the 1578 batch; Seckford had a new motto on his coat of arms from 1576 and its inclusion on certain plates helps to separate the plates produced during that period.

In terms of accuracy of detail Saxton's maps reveal an extraordinary surety in their overall depiction of large geographical areas, particularly when one considers the difficulties of means of transport, the comparatively primitive nature of the instruments available and the state of surveying knowledge. Except in a few cases Saxton started from scratch and, taking into account the speed with which the work was carried out, the resultant maps are a remarkable achievement. Suffice it to record that there was no other comparable survey carried out until Ogilby's survey of roads that yielded *Britannia* in 1675 and then, over two hundred years after Saxton, Cary's work for the Postmaster-General in the 1790s and that all intervening county plates take something from Saxton in their basic mapping. Taking the measurements on Saxton's map of Lincolnshire on two axes and comparing them with those obtained from a scaled-down modern single-sheet Ordnance Survey map one obtains some idea of the degree of variation and thereby a crude idea of Saxton's accuracy. Taking a measurement from [Skitter Ness] in a direct line, north to south and a direct east to west reading from Newton [on-Trent] to the coast (which passes through Lincoln) yields a proportion of 1:1.53; a similar measurement using the Ordnance Survey map gives a 1:1.57 relationship and that seems a small difference overall.[12] However, when one examines more closely other aspects the differences with a modern map may seem larger when taking the positions of two places in relation to each other. Newton-on-Trent and Lincoln are not on the same east-west line as they appear to be in Saxton; at the southern end of the axis on Saxton's map the line passes through Uffington while in reality the line should pass through West Deeping, some six miles to the east of Uffington. The curve of the coast is shown as an almost smooth arc and there is no definition of the promontories at Gibraltar Point or Donna Nook.

The prominence of rivers on the plate supports the belief that Saxton used boats to survey the rivers and thus provide the basis for the placing of towns and villages on or near them.[13] It seems less likely that Saxton took to the sea to survey the coast since the treacherous sandbanks would have made the work very difficult. It is reasonable to assume that Saxton relied on what he could discern from dry land and accepted the general outline of a regular arc as shown in Mercator in 1570.

For his maps Saxton used the old English mile, equal to about ten furlongs; although the mile differed a good deal from area to area the evidence is not strong that he used the mile pertaining to the area he was surveying. Using the scale Saxton provides, Lincoln to Sleaford is 13 miles (or about 16.5 statute miles) and from Lincoln to Skegness is 28 miles (in a direct line in each case), i.e. about 35 statute miles; the true distances are 16.5 and 38.5 miles so there is an error rate of about 10% in one measurement only. From that one may deduce that the county as depicted is 'stretched' by about 10% on its east-west axis in relation to its north-south axis.

It is very difficult to state firmly what other sources of information Saxton might have used. Speed (6) says he obtained information from 'the Parliament Rowles' or, where they proved unhelpful, 'the Nomina Villarum in their Sheriffes bookes'. Such a source might have helped Saxton in determining parish boundaries but the greatest informational source must have been the landowners with whom he must have had contact. The prominent physical features shown on the map are the rivers, with double lines crossing them to suggest a bridge or ford, and the hills; the latter are indicated by 'humps' with shading on their eastern slopes. This method does not, of course, provide any indication of relative heights but the method as applied here does give a fair impression of a range or line of hills, such as the Cliff from Wellingore northwards to Kirton in Lindsey. On the other hand, parts of Kesteven would appear to be as hilly as the Wolds.

The sites of former religious institutions are a regular feature of Saxton's map. Although many have disappeared or are in a very ruinous condition now they must still have been prominent landmarks in the county landscape less than forty years after the dissolution. *Catley* (east of Digby) and *Molwood* (= Melwood, and south-east of Epworth) are two such sites of religious foundations no longer visible at ground level. Other sites of an earlier religious settlement shown on the map include *Irford ab* (shown as west of Binbrook when it is more accurately placed to the north-west), *Folethorp* (north-west of Hannay and on the site of Hagnaby Abbey) and *Gokewell* (north-west of present-day Brigg).

*Foresee* (= Eresby) is an example of another kind of site that has lost its former significance. There are quite a few other names which are marked as villages, which are now the sites of a large house only; examples are *Madenhowse* (near Cranwell), *Inglebye* (north of Saxilby), *Hawdonby* (= Holdenby, west of Burton-on-Stather) and *Hawarabye* (= Hawerby, and west of North Thoresby). At least two sites are shown which are now marked on the Ordnance Survey as deserted villages; these are *Somerbye* (north-east of Gainsborough) and *Wikam* (north-west of Burgh on Bain and actually two lost villages – East and West Wykeham). Of the list of lost villages produced by Canon Foster[14] thirteen are shown on this plate; this is more than a tenth of those in that list when one discounts those names which never were villages and those 'monastic granges and vills which have not been traced other than in Domesday Book'.[15] *Skinnand* (west of Navenby) and *Besebye* are but two examples of other lost villages which are noted by Beresford but not included in Foster's listing.[16]

There are also a number of sites marked that are either erroneous or indicate something whose meaning is hard to decipher now. Some are simple blunders; the presence of *Sct:Leonard* (south of Gedney) is surely an error for Sutton St. Edmund, which is not marked. *N: Feribye* is clearly a mistake in transcription for South Ferriby. *Bolystow* (north of Holbeach), *Haerbothe* (on the south bank of the Witham and east of Southrey), and *Cawthorp* (south of Fulstow)[17] are examples of places not now to be found on the modern Ordnance Survey map. The prominence given to *Lymberghe Magna* was copied by many later cartographers.

The market towns are usually given prominence by the use of enlarged lettering. Holbeach is not so marked but *Quaplode* (i.e. Whaplode) is awarded the honour. There are no roads on the plate, a feature that did not appear on county maps for a century. There is no compass but the four main directions are named, in Latin, in the centre of the respective borders. One feature that does add to one's pleasure is the inclusion of ships and sea monsters in the Wash area and off Spurn Head. The ship off Spurn Head has a remarkable resemblance to that on the Lhuyd and Ortelius map of England produced in Antwerp in 1573 and in almost the identical spot too; similarly there is a sea monster on the earlier map which is the twin of that threatening to swim up to Boston on the Saxton map. The chief decorative features of the map are all down the left hand side. Reading from the top downwards these are the royal arms of Queen Elizabeth; the map's title in a plain rectangle but surrounded by leaves and 'classical' features from architecture; the arms of Thomas Seckford with the new motto *Industria Naturam Ornat* which he took in 1576, the year of the Nottinghamshire and Lincolnshire plate; and, finally, a large pair of dividers opened at the bottom to permit a scroll bearing *Christophorus Saxton descripsit* and the scale to be inserted. In very small letters below the scale bar appears the name of the engraver: *Remigius hogenbergius sculpsit*. Seven engravers signed their plates and of these Hogenberg was the most prolific with nine to his credit. He was one of four engravers from the Low Countries who engraved plates for Saxton, while the other three who signed plates were English, namely Francis Scatter (who contributed two county plates), Nicholas Reynolds (one) and Augustine Ryther (four, plus the general map).

On 22 July, 1577 Saxton was granted royal letters patent, which gave him a ten-year privilege for printing and selling his maps. The full collection of all the plates was completed by the production of a general map of England and Wales, which is dated 1579, and a number of preliminary pages though there is not a title-page in the modern sense. The volume was entered at Stationers' Hall in 1579 and it is assumed from that the atlas was placed on sale simultaneously. A further royal letter patent issued in July, 1579 provided Saxton with a grant of arms and it is tempting to believe that this was official recognition of his achievement on the occasion of the publication of the complete work.

More than fifty copies have survived but it is clear that they were not all assembled at the same time. The probability is that Saxton compiled atlases as and when they were called for during the period of his privilege. The sale of later copies seems likely to have been the responsibility of Augustine Ryther, who had previously been an engraver (see above). Some of these copies can be safely dated to the 1590s since a plate, which contains eighty-four coats of arms (one is left blank), records the arms of Sir Thomas Heneage as Vice-Chancellor, a post he did not take up until 1589; a further reference to Sir Christopher Hatton as Chancellor can only have been valid up to 1591, the year of his death.

Before discussion of the later history of the Saxton plates reference must be made to his later career, since several surviving records of his surveying activity show that he worked in Lincolnshire. In 1596 Saxton was a witness in a boundary dispute between crown tenants in Haxey and Owston in the Isle of Axholme.[18] Two plans were apparently submitted to a special commission arranged by the Court of the Exchequer, but only that of Saxton survives.[19] Saxton appeared before the commission personally on September 18th, 1596 held at Stockwith. It was at this appearance that Saxton made the deposition that he was aged 'fiftye twoo yeares or thereabouts'.[20] The map is now incomplete but probably measured 55.5cm. x 47.5cm. originally. It shows an area bounded by the Rivers Idle, Eae and Trent on the northern, southern and eastern sides respectively and includes the villages of Newbie, Haxey, Owston, Mysterton and Kelfeilde. The details of enclosures and natural features are quite considerable.[21]

### Later history
After the expiry of Saxton's privilege it seems likely that the plates passed to Augustine Ryther (*fl.*1576–95), since two copies of the atlas have survived in their 'final' state (c.1590) which include his Armada charts of 1590. Nothing more is heard of Ryther after

1595 when he was in the Fleet Prison for debt[22] nor of the plates until the edition by William Web put out possibly to meet a demand for maps during the Civil War. Web (*fl.*1629–52) was an Oxford bookseller although the address in the imprint is in London; Skelton suggests that the London address was that of the printer and that as three plates have now been re-engraved with the arms of Charles I a loyal gesture of support for the royalist cause is implied.[23]

Once more the plates go underground and only make a firm reappearance in the hands of Philip Lea in c.1689. It has been argued with some force that an edition was planned in c.1665, the date on one plate in Lea's edition, while several other plates have borne that date only for it to be altered, somewhat inefficiently, later. The clinching argument is based on the addition to eight of the plates of the royal cypher C R which do not appear in Web's edition but must have been added before 1685, that is, at least four years before the presumed date of Lea's edition.[24]

Philip Lea (*fl.*1666–1700) specialised in cartographic works and making globes and instruments. He was at one time Pepys' map colourist and acted as his consultant on cartographical matters. A lot of his stock was produced from plates made in conjunction with many of the notable map-makers of the period but he acquired the Saxton plates in about 1685 and systematically began to revise them. He died in 1700 and a number of different editions can be identified where the Saxton plates were used or, where there were gaps, an alternative plate of the county was substituted. After his death, his wife continued the business for three more years in association with Robert Morden and then with other map-sellers until c.1725.[25] After Anne Lea's death in 1730 her plates including the Saxtons were auctioned.[26]

George Willdey is first recorded in 1707 and he and his partner built up a business in making spectacles and optical instruments.[27] Soon, however, he was acquiring maps and dealing in this type of stock and at the above auction came into possession of the Saxton plates. After Willdey's death in 1737 the plates probably passed into the hands of his brother Thomas, and on his death in 1748, Thomas Jefferys obtained them.[28] Jefferys (*fl.*1732–71) seems not to have made very great use of the plates but he still retained them in 1765 since he lists them in a catalogue of his stock published in that year.[29] Jefferys went bankrupt in the following year but with the help of friends was able to resume trading with an enforced partner.[30] At his death when his stocks were auctioned[31] the plates were acquired by the Dicey family business. William Dicey was a Northampton publisher of prints and chap-books but by 1736 the business had grown so well that his son Cluer took charge of the London 'branch'. The only issue that can be ascribed to the Diceys is now tentatively dated c.1772[32] and that was their final appearance, nearly 200 years after they had been originally engraved.

## EDITIONS

**(i)** *LINCOLNIÆ NOTINGHAMIÆQ*/*Comitatuū noua vera et*/*accurata descriptio. Anno*/*Domini 1576*/ (Ab–Ac in an ornamental rectangular frame). *Scala Miliarium*/*10* [= 6.55cm] (Ae). *Remigius hogenbergius sculpsit* (Ae, below scale). The decorative borders contain in panels placed centrally the compass points *OCCIDENS, SEPTENTRIO, ORIENS,* and *MERIDIES.* The royal coat of arms of Queen Elizabeth (Aa). The arms of Thomas Seckford (Ad).

**Copy** Proof copies – the 'Burghley atlas'. BL Royal Ms. 18 D iii.

**(ii)** *Christophorus Saxton descripsit.* has been added on a ribbon (Ae, above the scale). *Wyne flu* has been corrected to *Nyne flu,* [i.e. Nene]. *Cast=* after Tattershall has been deleted although some traces are still visible. About 180 places have had their circle symbols replaced by a church symbol. The 'cross' symbols below *Tumbye* and *Reasbye* *ab:* have been deleted. *Saltfletbye* has been deleted but its church symbol remains; the village name has been re-entered below *Est Saltfletbye. Mabneton* has been re-engraved as *Malmeton* [= Manton, north of Kirton in Lindsey].

Issued in a volume without title-page, firstly in 1579 but in its definitive form in 1590.

**Copies**   BL Maps C.7.c.1, etc.; CUL Atlas 4.57.6; Bod Fol. B5.45, etc.; Leeds UL W 1; Wisbech Museum Press I.47; RGS 263 G 7, etc. (LRL Maps 574 and 67; Gainsborough RL Map 12; Grimsby RL – SAXTON 1576 (RM 102)).

**(iii)**   The title now reads: *LINCOLNE & NOTINGHAMSHIRE/with theire severall Hundreds/and Wapontakes most Exactly/drawne and described./Anno Domini.1642./* (Ab–Ac). The decoration around the rectangular tablet that contains the title has been considerably retouched and is most noticeable in the face and headdress of the woman drawn immediately below the title. The royal coat of arms (Aa) has been changed to that of King Charles I. The names of the hundreds and wapentakes have been added to the map in capital letters with short dashes used to mark their boundaries.

*The Maps Of All The Shires In England, and Wales. Exactly Taken And Truly Described by Christopher Saxton. And graven at the Charges of a private Gentleman for the publicke good. Now newly Revised, Amended, and Reprinted . . . for William Web at the Globe in Cornehill, 1645.*

**Copies**   BL Maps C.7.c.3; CUL Atlas 4.64.3; Bod Gough 96*.

**(iv)**   Most of the left hand side of the plate has been re-engraved. At the top is a plan of *LINCOLNE*, which is based on that of Speed (see no. **6**) but is orientated now with north at the top. The plan is enclosed within a larger rectangular frame, the remainder of which is taken up with eleven shields (one is blank) that show the coats of arms of county noblemen (Aa–Ac). *LINCOLNE SHIRE/AND/ NOTTINGHAM SHIRE/ By. C SAXTON/* (Ad), enclosed within an oval wreath. The scale with dividers survives (Ae). The galleons, boats and mythical beast off the coast have been deleted; the boat with oars south of Spurn Head remains but has been redrawn on a larger scale; as a result of the redrawing much of the stipple in the North Sea has been removed and the names of the adjoining counties have been re-engraved, especially *PARTE OF/YORKE/SHIRE; PARTE/of* above *DARBY/SHIRE* has been removed. Crosses have been added to the church symbols to indicate market town status; a mitre has been added to the Lincoln symbol; crowns have been added at Grantham, Lincoln, Grimsby and Stamford to denote parliamentary boroughs (but not at Boston and Newark).

*All The Shires of England And Wales Described by Christopher Saxton Being the Best and Original Mapps With many Additions and Corrections by Philip Lea* [A four-column list of the county names with an explanation of the new symbols]. *Sold by Phillip Lea at the Atlas and Hercules in Cheapside near Friday Street and at his Shop in Westminster Hall near the Court of Common Pleas where you may have all sorts of Globes Mapps &^{ct}* [1689].[33]

**Copies**   BL Maps C.21.e.10; Bod Douce Prints b. 28; RGS 264 H 17.

**(v)**   Roads have been added in four grades; double continuous lines denote the Great North Road (only); main (post) roads are double lines, one continuous and one of short dashes; cross roads are double lines of short dashes; minor roads are single continuous lines. *LIBERTIE/ WAPONTAK* has been changed to *LANGO/WAPON/TAK; P^T OF/ WEVE/BRID* has been added (west of Grantham). In Nottinghamshire there are two additions: *SOUTH/CLEY* and *WAPONTAKE* below *BASSET/LAW. Strugshil* has been added (above Wigtoft). A crossing of the Humber has been added with the words *from Flambour Head.* At Boston there is now a crown.

*The Shires Of England And Wales . . . Corrections viz: y$^e$ Hund$^{ds}$, Roads. &c. by Philip Lea Also the New Surveis of Ogilby. Seller. &c.* [List of counties, etc.] *Sold by Phillip Lea* . . . [1693].[34]

**Copies**  BL Maps C.7.c.4; CUL Atlas 4.69.2; Leeds UL W 2.[35] (LAO Exley 32/2/161; Louth RL – Map A 13).

*Atlas Anglois Contenant Les Cartes Nouvelles & tres exactes Des provinces, duchés, Comtés, & Baronies du Royaume D'Angleterre. le tout Enrichi des Plans des Villes & des Armes de la Noblesse. London, Par P. Lea.* [1694].

**Copy**  The private collection of Mr. C.A. Burden.[36]

*The Shires of England and Wales . . .* [1695].[37]

**Copy**  NMM 912.44(42) ″16″:094 [E6837].

(vi)  A new imprint has been added (Be): *Sold by Geo: Willdey at the Great Toy, Spectacle,/China ware, and Print Shop, the Corner of/ Ludgate Street near S$^t$. Pauls London/* The plate is now quite worn and previously 'deleted' names are making a shadowy re-appearance, viz. *Cast* (after Tattershall) and *Bratoft* (this name had been engraved twice at the outset and one of the versions had been deleted at the proof stage).

*The Shires of England and Wales . . . by Philip Lea Also the New Surveis of Ogilby. Seller, &c.* [list of county names] *Sold by Geo: Willdey at the Great Toy, Spectacle, China-ware, and Print Shop, the Corner of Ludgate Street near S$^t$. Pauls London.* [1732].[38]

**Copies**  BL Maps C.29.c.3; Bod Gough Maps 90. (LRL Maps 65 and 68; Lincoln Cathedral L.).

(vii)  An attempt to remove the Willdey imprint has been made; parts of it are still visible, particularly the word *Sold*.

*The Shires of England and Wales . . . Sold by Thomas Jefferys, Geographer to his Royal Highness the Prince of Wales; in Red Lyon Street near S$^t$ John's Gate.* [1749].[39]

**Copies**  BL Maps C.29.d.31 and C.29.e.12; Leeds UL W 3.[40]

An atlas without title-page, but issued probably by Cluer Dicey and Co. in about 1772 may be noted here. The only recorded copy (Leeds UL W 4) lacks three maps; viz. Lincolnshire-Nottinghamshire, Anglesey-Caernarvonshire and Yorkshire. It is only through the inclusion of a substitute for the missing Yorkshire map, which bears the Dicey imprint, that their ownership of the Saxton plates is confirmed.

## LITHOGRAPHIC TRANSFER

(A)  A Lithographic transfer of state (ii) with the imprint added (Ce OS): *JOHN HEYWOOD LITHOG. MANCHESTER.* In this version Nottinghamshire has been removed (but not the small portion of Leicestershire as far as *Lughborugh* [= Loughborough] that was on the original plate. Rutland has been separately transferred and no attempt has been made to 'fit' it to Lincolnshire. It is 'floating' in the space west of Newark.

*A Translation Of That Portion Of Domesday Book Which Relates To Lincolnshire And Rutlandshire by Charles Gowan Smith. London: Simpkin, Marshall, & Co., Manchester: John Heywood. Lincoln: J. Williamson And George Gale. Caistor: Geo. Parker. Boston: J. M. Newcomb, And J. Morton.* [1870].[41]

**Copies**  BL 2367 b 1; CUL Ll.53.27; Bod G.A. Lincs. 8°; Nottingham UL LH.EM.D22 DOM; LAO Sc.942.53; Grimsby RL X000:333.332 SMI; Reference Libraries in Lincoln, Louth, Boston, Grantham, Stamford, Skegness and Gainsborough (shelf mark 333.32).

**Reproductions**
(i) The maps were reproduced using collotypes made by Emery Walker on behalf of the Trustees of the British Museum in: An atlas of England and Wales The maps of Christopher Saxton engraved 1574–1579. (London, 1936). The BL copy used for this is Maps C.7.c.1. Lithographic copies of the individual counties from the same source have been on sale for many years at the BL. (ii) Christopher Saxton's Sixteenth Century Maps . . . Introduction by William Ravenhill . . . (Shrewsbury, 1992) depicts all the maps in full colour from the Chatsworth House copy. (iii) A reduced version of the Lincolnshire (only) map appeared in *Lincolnshire Life* Vol. 13, no. 8 (October, 1973), pp. 55–56 together with a descriptive note by John E. Holehouse. (The editor's heading for the third section of the text incorporates a notable error).

**References**

1  Evans, I.M. and Lawrence, H. Christopher Saxton, Elizabethan map-maker. (Wakefield and London), 1979. p. 1.
2  ibid., p. xvi and Appendix 1.
3  Bod. MS. Wood. D.13.
4  Fordham, Sir George. Christopher Saxton of Dunningley . . . *Thoresby Society, Miscellanea* vol. 28 (1928), pp. 356–384 and additional note on p. 491.
5  Marcombe, D. Saxton's apprenticeship: John Rudd, a Yorkshire cartographer. *Yorkshire Archaeological Journal*, vol. 50. (1978), pp. 171–175; summarised in Marcombe, D. John Rudd: a forgotten Tudor mapmaker? *Map Collector*, no. 61 (Spring, 1993), pp. 34–37.
6  Briscoe, A.D. A Tudor worthy. (Ipswich, 1979), p. 42.
7  BL Lansdowne MS 6/54 and transcribed in Harley, J.B. The map collection of William Cecil, First Baron Burghley,1520–1598. *Map Collector*, no. 3 (June, 1978), pp. 12–21.
8  BL Royal MS. 18D iii. Some comment on the manuscript additions made to the proof set by Cecil is given in: A Description of Maps . . . in the Collection made by William Cecil . . . Hatfield House (Roxburghe Club, 1971), pp. 26–27; Cecil added a list of the justices to the verso of every map and, to the important counties along the English Channel, he also added lists of ordnance stores, the divisions of the counties for military purposes, etc.
9  Skelton, pp. 7 & 14–16.
10  PRO Privy Council Register 2:11.
11  Evans & Lawrence, 1979 op. cit., pp. 12–15.
12  Kingsley, p. 5.
13  ibid., p. 5 draws attention to Lythe's use of boats in his survey of Ireland in 1572.
14  Foster, C.W. and Longley, T. The Lincolnshire Domesday And The Lindsey Survey . . . *Lincoln Record Society*, vol. 19. (Lincoln, 1924), pp. xlvii–lxxii; Quinn, J. West Wykeham. (In: Land, People And Landscapes . . .; ed. by D. Tyszka *et al.* Lincoln, 1991).
15  Beresford, M.W. Lost villages of England. (London, 1954), p. 360.
16  ibid., pp. 308–310 and 360–364.
17  ibid., p. 363 refers to Cawthorp in Covenham [O.S. reference c. 350960]; there has never been an indication of such a place in that area from the first edition of the Ordnance Survey. Beresford's source is the tax list of 1334 and he is here referring to villages which were junior partners in a pair taxed together. Mr. Maurice Tennant of Holbeach has very plausibly suggested that Bolystow is intended to mark the site of Elloe Stone; while Bolystow is marked on the map north of Holbeach its true site is west of that town on the Spalding road.
18  Evans & Lawrence, 1979, op. cit., pp. 99–100.
19  PRO MPB 16.
20  PRO E.134(39 Eliz).E.14.
21  Illustrated in full in Dunston, G. The rivers of Axholme. (London, 1912); a complete account of the case is given on pp. 114–119.
22  Skelton, p. 240; Hind, A. Engravings in England in the sixteenth and seventeenth centuries, Vol. 1. (Cambridge, 1952), pp. 138–149.
23  Skelton, p. 244; Plomer, 1907, p. 191 (under Webb – sic).
24  Evans & Lawrence, 1979 op. cit., pp. 47–48 summarises the argument first put forward in detail in: Whitaker, H. The later editions of Saxton's county maps. *Imago Mundi*, vol. 3, 1939,

pp. 72–86, after the suggestion made by E. Lynam in the introduction to the BM facsimile edition of Saxton's first atlas (London, 1936).

25  Skelton, p. 246; Tyacke, pp. 120–122; Tanner, J.R. (editor) Pepys' naval minutes. (London, 1926), p. 120.
26  Advertised in *Daily Journal*, 5 Aug. 1730.
27  Tyacke, pp. 146–148.
28  Hodson, 1984, pp. 147–148.
29  Bod. Gen. Top. 363. Vol.1, f 485$^v$. Undated but it lists Donn's map of Devon published in 1765. (Cited in Hodson, 1984, p. 148).
30  Hodson, 1984, pp. 147–148; Harley, J.B. The bankruptcy of Thomas Jefferys', *Imago Mundi*, vol. 20 (1966), pp. 30–31.
31  A ten day sale beginning 29 Jan. 1772 was announced in *Public Advertiser* of that date. (Cited by Hodson, 1984, p. 148n).
32  Hodson, 1984, pp. 148–150.
33  The Somerset and Wiltshire plates are both dated 1689.
34  Whitaker, H. 1939. op. cit., pp. 72–86 infers the date of publication as earlier than 1694 since the Kent map is dedicated to John Tillotson, Archbishop of Canterbury, 1691–1694.
35  The Whitaker copy in Leeds UL is endorsed by its first owner Narcissus Luttrell who has, as Whitaker says 'in accordance with his custom, his autograph and the date of the acquisition, 1693, written on the title-page'. Tyacke, no. 255 quotes the advertisement for this atlas placed in *London Gazette* for 8–12 Feb, 1693/4, which may suggest a later date of publication or a fresh re-issue.
36  A copy formerly in Blackburn RL but later stolen is collated in *Map Collector*, no. 5, Dec., 1978, p. 49. Skelton, 1970, p. 180 notes the Blackburn copy; there are minor differences in the title in his transcription.
37  Skelton, 1970, p. 177 notes that the contents of the recorded copies all vary; the NMM copy includes Lea's map of the Channel dedicated to Lord Berkeley and dated 1695. This map may have been bound in later by a subsequent owner, since the collection is now housed in a modern binding.
38  Hodson, 1984, pp. 141–144 notes that Anne Lea's effects were to be auctioned on August 14, 1730 (*Daily Journal*, 5 Aug. 1730) and it is argued, therefore, that Willdey acquired the Saxton plates at that time and issued his atlas shortly afterwards.
39  Willdey died in 1737 (Hodson, 1984, pp. 144–147) and his business was carried on by his son Thomas until he died in 1748. It seems likely that Jefferys had the plates soon after and began using them almost immediately.
40  The Whitaker copy may have been issued by Willdey; the title-page was bound in after Whitaker had acquired the atlas and Hodson, 1984, p. 146 places this atlas with the Willdey issue.
41  Date in preface – November, 1870.

Skelton numbers 1, 27, 80, 110, 112, 113 refer to the editions between 1579 and 1693.
Hodson numbers 183–185 refer to the editions from 1732 to 1772.

## 2

### W.B. [WILLIAM BOWES]                                1590

Size: 5.45cm x 5.5cm.

This is one of a set of playing cards and is the first map of Lincolnshire by itself. Since there were forty English counties and twelve Welsh – before the local government re-organisation of 1974 – the possibility of using a pack of fifty-two playing cards to depict maps of each county was soon recognised. The title card of this pack is engraved *W B inuent 1590*. A second but different pack of cards (3) came to light in 1972, which bore the inscription *W.Bowes Inventor* and it has been assumed, therefore, that the two sets come from the same source.[1]

1. 1590   MAP 2 – PLAYING CARD BY W.B.
[WILLIAM BOWES].
By permission of the British Museum.

William Bowes has not been identified. One possibility is that he was related to Ralph Bowes, who held the monopoly for the importation of playing cards from 1578 until his death in 1598. Another candidate is a descendant of Sir Marten Bowes, a goldsmith and Lord Mayor of London in 1546. He had a son born in 1543 called William and another son (or, possibly, a grandson) also called William who was born in 1556.[2] There is a cartographic connection since, after Sir William's death, his (third) wife married the same Thomas Seckford who had been Christopher Saxton's patron.[3]

The whole card measures 9.5cm. by 5.75cm. The cards were printed from four plates, each plate bearing the cards of the four 'suits'; in this case, Eastern and Midland counties, Southern and Western counties, Northern counties, and Welsh counties correspond to plates one, two, three and four (although none of the plates is actually numbered). That is to say that Lincolnshire would have been printed on the 'third' plate. There are no suit marks but Lincolnshire was the 12 of North. There were fifteen cards to each plate with text and other cards, arranged in three rows of five cards each. The cards are arranged on the plates without any numerical or geographical plan.[4] The map shows rivers and hills; only market towns are marked by their initial letters; B marks the site of Belvoir.

See also Addenda, pp. 411–12.

**EDITION**

(i)   *LINCOLNSHIRE y^e 12 of y^e North hath Miles/In Quantitie superficiall 1575 In Circuit 200/In length from Yorkshire to Cambridgeshire 62/In Bredth from Nottingham to y^e Germaī sea 48/* (this inscription occupies the space at the top of the card while the map is in a box in the centre of the card). In the lower panel appears: *LINCOLN: plenty of corne, fruite, & cattell/Numbers of townes, rivers, with store of fishe/ Havinge the German sea East Nottingham=West/Yorkshire North Cambridge= & Northamptō South/.* On the map: *XII* (Aa and Ee). Compass (Ea). Dividers and mileage *10* [= 0.6cm.] (Ae). The cardinal points *N, S, E, W* are in the borders; eight lines are drawn from the borders of the map to the county boundaries to indicate the eight main compass bearings.

**Copy**   B.M. Print Room. Willshire E.178.B. This set has been cut up and mounted ready for play.

### Reproductions
The Lincolnshire card is illustrated in Mann, S. and Kingsley, D. Playing cards depicting maps of the British Isles and of English and Welsh counties. *Map Collectors' Series. Map Collectors' Circle*, no. 87. 1972. Plate 1 (b). The title card is shown in Whitaker, 1949, p. 2.

### References
1   Mann, S. & Kingsley, D. 1972. op. cit., pp. 5–15, with discussion on the identity of Bowes on pp. 15–16.
2   Transcript of the Registers of the united parishes of S. Mary Woolnoth and S. Mary Woodchurch Haw, 1538 to 1760. London, 1866.
3   Calendar of Wills. Court of Hustings, London. Part III, 1358–1688. London, 1890.
4   A set in RGS (264 H 16) consists of three complete sheets in their uncut state; the missing sheet is that for the Northern counties, which would have included Lincolnshire.

Skelton no. 2.

**3**

**WILLIAM BOWES**                                             **1605**

Size: 2.55cm x 2.4cm.

This is a second set of playing cards[1] and the inscriptions: *W. BOWES Inventor* and *W. Hole Sculp* appear on two of the title cards. The cards are marked with the traditional suit marks and Lincolnshire is the Queen of Clubs. The card is divided into ten panels, three each along the top and bottom and two pairs in the central area. The map occupies the bottom right-hand panel of the four in the middle of the card. The whole card measures 9.03 x 5.25cm. The suit mark appears in the central panel of the top row, the Queen being depicted in the left hand portion and the club sign in the top right hand corner. On other cards the central panel in the bottom row repeats the suit mark, but, in Lincolnshire's case, there is no suit mark and only the lower drapery of the queen's dress is drawn. The corner panels contain various texts but only the lower right-hand corner panel contains geographical information: *LINCOLN S.:/plenty of Corne,/Fruict, and Catle,/ numbers of townes/ruiers* (sic) *with store/of Fish./* William Hole was a well-known engraver of portraits in the early seventeenth century, who was born in Leeds. His map work was also considerable for he not only engraved this set of cards, but also many of the maps in Camden's *Britannia* (**5**) – though not Lincolnshire – and the plates, which illustrated Drayton's *Polyolbion* (**8**).

### EDITION
(**i**)   The panel containing the Lincolnshire map has eight tiny rectangles placed around the edge of the panel to correspond to the eight main compass points; those in the centre of each side contain the letters *N, W, E, S*. All eight squares have two horizontal lines across their centres, between which are engraved, in abbreviated form, information about the county and the adjoining county names. In the boxes containing the letters for the cardinal points this information is superimposed on the initials N,S,E and W. The arrangement produces this effect:

| *575.200/SC* | *Yorke=/N* | *62.48/LB* |
|---|---|---|
| Nottin:/W | [County map] | *Sea/E* |
| North 20 10/dividers | *Cam=/S* | *Shire/XII* |

The symbols have these meanings: 575 refers to Superficial area; 200 to the Circuit; 62 is the length; 48 the breadth; XII is a further reference to the card value (that is, 12 = Queen); 20 and 10 refer to mileages on a putative scale; Cam. is short for Cambridgeshire, North is an abbreviation of Northamptonshire and Nottin equals Nottinghamshire.

The map has the initial letters of the market towns, with B for Belvoir Castle included, while Stamford is omitted. Some hills are indicated.

**Copy**   British Museum Print Room. Willsher E.179/4. 1878–6–8–46.

#### Reproduction
(i) The Lincolnshire card is illustrated in Mann, S. and Kingsley, D. 1972. op. cit., Plate IX (b).

#### Reference
1   Mann, S. and Kingsley, D. 1972. op. cit. pp. 5–15.

Not in Skelton.

4

## PIETER VAN DEN KEERE                             **c.1605**

Size: 8.5cm x 12cm.

Born in Ghent Pieter van den Keere (1571–c.1646) came to London in 1584 or 1585 as a protestant refugee. By 1593 he had returned to Amsterdam where he died. The plate size in this collection of maps is closely related to that of two miniature atlases published in Holland between 1598 and 1600 and for which van den Keere engraved many of the plates. It has been suggested, therefore, that a companion atlas of the British Isles was projected shortly afterwards.[1] What has come down to us is a collection of 44 maps of the English and Welsh counties (following the example of Saxton several counties share the same plate) with regional maps of Scotland and Ireland. It is not known when the engraving was completed; from the condition of the earliest surviving examples the maps appear to be proofs, since the map books have text in manuscript; it has been deduced that one is an incomplete copy of the other.[2]

Three plates are dated 1599 and 21 are signed *Petrus Kaerius caelavit* or variant forms, i.e., Pieter van den Keere. Crone[3] has suggested that the use of paper water-marked in the same way as the re-issue of 1617 would provide a date somewhat later than 1600. References in the text suggest compilation between 1605 and 1610 and Skelton suggests the earlier date.[4] The plates came through Janssons' father into the possession of George Humble, who issued them, as the so-called 'miniature Speeds' from about 1619. In their refurbished form they were issued for more than 60 years; the Latin titles for the counties were altered to their English form and new plates were prepared to replace the Van den Keere plates showing two counties together (7).

## EDITIONS

(i)   *Lincolnia et/Notingham.*(Ab). *Scala miliarium 20* [= 2.8cm].(Ae).

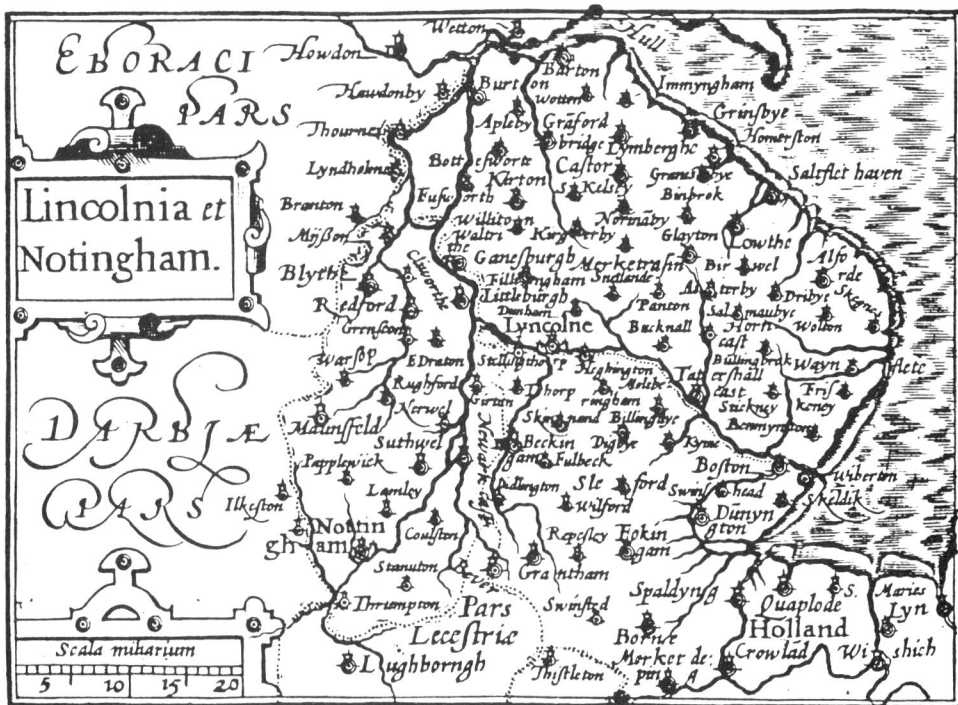

2.  1605   MAP 4 (i) – PIETER VAN DEN KEERE.

In volumes without title-pages but with manuscript texts; the text in the BL copy seems to be a fair and less complete copy of the version in the RGS example.

**Copies**   BL Manuscripts Dept. Harley Ms. 3813; RGS 264 a 28.[5]

(ii)   Page number 539 added in type with the Latin text of page 340 printed on the verso. Signature Y 2 appears below the map.

*Gulielmi Camdeni, Viri Clarissimi Britannia, Sive Florentissimorum Regnorum Angliae, Scotiae, Hiberniae, & Insularum adjacentium ex intima antiquitate descriptio. In Epitomen Contracta a Regnero Vitellio Zirizæo, & Tabulis Chorographicis illustrata. Amstelredami, Ex Officina Guilielmi Ianssonij, M.DC.XVII.*

**Copy**   Leeds UL W 14.

(iii)   Page number corrected to *339*.

*Gulielmi Camdeni . . . M.DC.XVII.*

**Copies**   BL 577.a.2 and 796 a 2; CUL Atlases 7.61.2 and 7.61.3; Bod. Allen 1 and Vet B2 f.22 (1617); Birmingham UL Wigan 16.C256; RGS 264 C 20; Liverpool UL Y 61.1.6.

**Reproductions**
(i) The copy belonging to Mr. David Kingsley with the maps in their first state was issued as: Atlas of the British Isles, c. 1605. (Lympne, 1972). (ii) The map from the 1972 re-issue illustrates an article in *Lincolnshire Life*, Vol. 14, no. 7 (Sept., 1974), written by John E. Holehouse.

**References**

1   Skelton, p. 22; Skelton, R.A. Pieter van den Keere. *The Library*, series V, vol. V. (1950), pp. 130–132.; Keuning, J. Pieter van den Keere. *Imago Mundi*, vol. XV (1960), pp. 66–72.
2   Skelton, p. 23.
3   Crone, G.R. Early atlases of the British Isles. *Book Handbook*, no. 6, 1948, pp. 341–4.
4   Skelton, p. 23.
5   The page containing the Lincolnshire and Nottinghamshire map is decorated with hand-drawn coats of arms and the names of the families to which they belong. The eight along the top are, from left to right, Perpayn, Burin, Stähop, Maners, Dimok, Pail, Fitzwilliam, and Tyrwitt. Similarly along the bottom: Markhã, Lassells, Whaley, Chauwort, Thimelby, Wraye, Aylrough, and Salxil. Along the sides there are four each; down the left are Candish, Clifton, Paultney and Stapletõ; and down the right are Tharold, Baclles, Disney and Henach. On the verso are 18 manuscript lines on Lincolnshire and 12 on Nottinghamshire.

Skelton nos. 4 and 12.


# 5

## WILLIAM KIP                                                          1607

Size: 29.95cm x 35.05cm.

The maps in this series are variously known by the engravers; Kip who was responsible for Lincolnshire and 33 other plates (although several of these are also referred to as Norden's as Kip's work is based on the plates prepared for an aborted county atlas by John Norden)[1] or William Hole, who engraved 21 plates or by the author of the text, William Camden. Some dealers also refer to them as Saxton/Hondius. Two plates are unsigned. This is the first time that separate plates for each English and Welsh county were included in volume form.

Camden's Britannia was first published in 1586; the 1607 edition is the sixth but the first to have maps although Camden did contemplate a series of county maps as early as 1589.[2]

William Camden (1551–1623) was Clarenceux King of Arms, who, in 1593, was appointed headmaster of Westminster School. The text of the first six editions is in Latin and the sources of his historical and antiquarian material are discussed in Copley.[3] Camden founded a chair of history at Oxford and the Camden Society was founded in his honour in 1838.

The history was very influential, being the basis for the abbreviated version referred to in **4(ii)** above; it was also produced in fresh translations in 1695 (with maps by Robert Morden – **19**) and again in 1789 (with maps by John Cary – **49**) each of which had lives extending over many decades. The popularity of the work is evident from the large number of copies of the three editions with Kip's maps which have survived.

The map is closely based on Saxton and is the first large sheet map of the county by itself. There are no roads and the county's rivers are a prominent feature. William Kip was active from 1598 to 1610. He was responsible for engraving the circular map used by van Langeren (**10**).


### Later history
*Britannia* appeared in three editions with the plates by Kip and Hole, the second and third being with English text translated by Philemon Holland. Holland (1552–1637) was well known as a classical scholar who produced translations of Livy, Plutarch, Pliny *inter alia*. After the edition of 1637 the plates were not used again.

# EDITIONS

(i)  *LINCOLNIÆ/Comitatus vbi olim/insederunt/CORITANI/* (Ea). *Scala/Miliarium Anglicorum/10* [=4.75cm.]/*Christophorus Saxton descrip:/William Kip Sculp:/* (Ec). On the verso one half has the signature *Bbb2* (only) and the other half the text of page 398.

*Britannia, Sive Florentissimorum Regnorum Angliæ, Scotiæ, Hiberniæ, Et Insularum adiacentium ex intima antiquitate Chorographica descriptio: Nunc postremò recognita, plurimis locis magna accessione adaucta, & Chartis Chorographicis illustrata. Gulielmo Camdeno Authore. Londini, Impensis Georgii Bishop & Ioannis Norton. M.DC.VII.*

**Copies**  BL Maps C.7.b.1 and 576 m 7; CUL R.8.26 and L* 8.30; Bod Gough Gen Top. 51 etc; Leeds UL W 5; Lincoln Cathedral L. S.2.8; Birmingham RL Q094/1607/8. (LRL Map 201; Grimsby RL).

There is no text on the verso of the map leaf, which is bound between pages 527 and [528].

*Britain, Or A Chorographical Description Of The Most flourishing Kingdoms, England, Scotland, and Ireland, and the Ilands adioyning, out of the depth of Antiquitie: Beautified With Mappes Of The severall Shires of England: Written first in Latine by William Camden Clarenceux King of Arms. Translated newly into English by Philmon Holland . . . Finally, revised, amended, and enlarged with sundry Additions by the said Author. Londini, Impensis Georgii Bishop & Ioannis Norton. M.D.C.X.*

**Copies**  CUL Syn.3.61.25 and Peterborough B.5.15; Bod Fol. Θ. 676; Birmingham RL F094/1610/8; Leicester UL H 942F CAM; Liverpool UL H17 10A; Maidstone (Centre for Kentish Studies) 942. (LAO – LD 87/7 and ATH 2/274; Grimsby RL – 1610; Boston RL Map 27; Gainsborough RL Map 8; Grantham RL – 2 copies).[4]

(ii)  Compass added (Db). There is no text on the verso.

*Britain . . . M.D.C.X.*

**Copies**  BL Maps C.7.b.2; Bod Gough Gen Top. 50; Manchester RL BR q942 C50; Nottingham UL Li 1.B8 C10 (Sir Francis Hill's copy). (LRL L9 – the county text and the plate, contained in a volume of maps and county texts owned and annotated by Sir Joseph Banks).

(iii)  Plate number 26 added (Ae).[5]

*Britain . . . M.D.C.X.*

**Copies**  Liverpool UL H17 10; West Suffolk CRO (Cullum Library 942).

*Britain . . . Printed by F.K. R.Y. and I.L. for Andrew Crooke. 1637.*[6]

**Copies**  CUL Adams 3.63.7; Bod Allen 3.

*Britain . . . for Ioyce Norton and Richard Whitaker. 1637.*

**Copies**  Birmingham RL F/094/1637/10/(942); Chethams L. (Manchester) DD.3.62; NLS R.290 b; London UL D-L L. Dºc. Camden Fol.; Manchester RL BR f942 C52.

*Britain . . . for Andrew Heb. 1637.*

**Copies**  BL 1321 k 1; CUL Syn 2.63; Bod G.A. fol.A3; RGS 264 H 19; Lincoln Cathedral L. Dd 1.7; Liverpool Athenaeum 914.2f.

*Britain . . . for W. Aspley. 1637.*

**Copies**  Bod Gough Gen. Top. 52; Manchester RL f914.2 C 50; Birmingham UL q16 C.2.

*Britain . . . for George Latham. 1637.*

**Copy**  Glasgow Mitchell L. B 862707. (LAO Scorer Map 1 and Box L910; LRL Maps 124, 153 and 802). The loose copies come from any one of the post-1610 editions).

Chubb records a later edition issued by Christopher Browne but no copy has ever come to light.[7]

### References

1  Six maps are based on Norden – Hampshire, Surrey, Kent, Hertfordshire, Middlesex and Sussex.
2  Skelton, p. 26.
3  Copley, G.J. (editor). Camden's Britannia. Surrey and Sussex. (London, 1972). For the origins of the work and an account of Camden's life see: S. Piggott, William Camden and the Britannia, *Proceedings of the British Academy*, vol. 37, 1951, pp. 199–217. Later editions are also discussed – Gibson's on pp. 209–213 and Gough's on pp. 215–216.
4  Grantham RL Newcome Collection aa.5 is an example of the 1610 edition but there are no maps, which had almost certainly gone when Newcome bought (and bound?) the volume in 1706.
5  The addition of compasses to the plates was probably soon followed by the addition of plate-numbers; the double process could have been completed by 1622, since copies of William Burton's *Description of Leicestershire* (London, 1622) contain the Leicestershire map in the final completed state.
6  The initials in the imprint have been identified as Felix Kingston, Richard Young and John Legatt.
7  Chubb, 1927, No. XXI.

Skelton nos. 5, 6 and 23.

# 6

## SPEED, JOHN                                                              1611

Size: 38.15cm x 50.5cm.

John Speed (1552–1629) was born in Cheshire. Originally he was a tailor and a member of the Worshipful Company of Merchant Taylors by the age of 28. His interest in antiquities was such that he became a member of the Society of Antiquaries and befriended Camden. As a result he came to the notice of Sir Fulke Greville, who eventually made him an allowance. That permitted him to continue the researches which took him all over the country and gave him time to make the surveys of towns that are incorporated within the framework of his county map plates.

He published *History of Great Britaine* in 1611 and the issue of *Theatre of the Empire of Great Britaine* was completed in the following year. With a total of 67 maps it constitutes the first atlas of the whole of the British Isles. The *Theatre* was designed as the topographical section of the *History,* being the first four books of the larger work. The task of compilation took Speed fifteen years from 1596. The idea of including maps probably came about half way through the project. Five sets of proofs have survived (a sixth was recorded before 1937 but broken up subsequently). In one of the sets there is a map of Cheshire that is radically different from that included in the work's final form.[1] This map was engraved by William Rogers, who died soon after 1604 and, it is believed, was the intended engraver for the whole series. The Oxfordshire map is dated 1605 and various other dates appear on other plates although the majority, in their original published state, are dated 1610.

All the maps were engraved by Jodocus Hondius (1563–1612) or under his direction in his workshop; thirty-three maps are signed by Hondius, including Lincolnshire. Hondius was Dutch and the plates were made in his Amsterdam shop. The name Hondius is to be found on a number of other important maps and he was responsible for the reissue in 1607 of the third part of Mercator's atlas. His son Henricus carried on his father's business and produced several later editions of Mercator's atlas. Of special local interest is Hondius' map: *A generall Plott and description of the Fennes and surrounded grounds . . . Amstelodami, sumptibus Henrici Hondii, 1632;* – see **13** below.

None of the maps is wholly original. Speed borrowed from as many available sources as possible, although Saxton's plates provided the bulk of his material. He did admit to his borrowings in a much-quoted phrase from his address to the reader: 'I have put my sickle into other mens corne'. What seem to be original are the town plans which are incorporated in the majority of the plates. Of the 73 inset town plans 50 have the '*Scale of paces*' and he claims that phrase as the distinguishing sign of his own surveying activity. Their general similarity of style and the preference for the ground plan as opposed to the bird's eye view testify to the same hand at work. The plan of Lincoln fills the lower left hand corner, but drawn with north to the right. A plan of *Stanford* is included on the plate for Rutland. These are, of course, the first plans of these two Lincolnshire towns that we have. Skelton notes that among the more important sources for the remaining town plans are manuscript maps in the library of Sir Robert Cotton, which was '. . . set open to my free accesse' and the plans in *Civitates orbis terrarum* by G. Braun and F. Hogenberg (Volumes 1–5, 1572–c.1600). Ironically the sixth volume (1618) contained nine plans or views taken from Speed.[2]

The creation of the wood-blocks used in the preparation of the town plans is ascribed by Skelton to Christopher Schwytzer.[3] In the 'Conclusion' to the *History* Speed pays tribute to his work and, incidentally, to that of earlier cartographers and 'for the matters of Herauldry' to William Smith.

On the maps there is one important innovation for which Speed should be credited. He marked the boundaries of the various counties' internal divisions and indicated their names. In Lincolnshire's case he also introduced a compass, thus anticipating the incorporation of such a feature in the Kip plates by a few years. Speed also filled in odd spaces within the frame with historical notes – in this case, an account of the wars which have involved the city of Lincoln. This type of 'space-filling' took many forms with later cartographers; Emanuel Bowen, in the second half of the eighteenth century, notably surrounded his maps with topographical and historical notes (**34** is one example of several).

None of the broad mass of information on the map is original and the extent of Speed's borrowings from Saxton is most clearly to be seen when checking the mistakes found on the earlier map. *S. Leonard* is still indicated in the fens; *Skampton* is drawn north of *Asthorpe* (instead of south); *bever cast[le]* is shown in Lincolnshire and mistaken names continue to abound and are clearly copied from Saxton; *Billingore* instead of Billingborough is just one example. The indication of woods by the liberal use of tree symbols in Saxton is not to be seen in Speed; but parklands are more clearly shown by a sort of palisade instead of Saxton's faint dotted line. A few clumps of trees are shown here – the two largest are north of *North Kyme* and south of *Reasbye Abbey* and are hardly an accurate indication of the wooded nature of some parts of the county in the early seventeenth century.

Some of the decorative features of the plate have already been touched on – the plan of Lincoln, the text on that city's history and the compass. The right hand side of the map has, reading from top to bottom, the arms of the monarch, the map's title in an elegant alphabet and enclosed in a quite plain box from which is 'suspended' the compass and from that a pair of dividers supported by *putti* who are 'sitting' on the scale tablet. A sportive whale spouts in the area south of Spurn Head and a small galleon is beating its way towards Boston.

On the backs of the maps are short descriptive texts, taken from Camden, and lists of the parishes with their respective wapentakes.

On 29 April, 1608 George Humble obtained a royal privilege which gave him the right to print and sell the atlas for 21 years. It is assumed, therefore, that publication would have taken place soon after. In the event, probably because of Hondius' illness the first two books were printed in 1611 and the third and fourth in 1612. Clearly a large number of copies of the text were printed off before the type was distributed; the large number of surviving copies bears witness to this. The text was reset for later issues and the variations help to identify loose sheets with their appropriate edition. In 1616 an edition was put out with the text translated into Latin; this seems to have been partly, at least, aimed at the overseas market since there is evidence of its being sold in Amsterdam, the centre of the map trade in Europe at that time (see below). Humble (*fl.* before 1603 until his death in 1640) was the son of a London print-seller and was in partnership with his uncle John Sudbury (*fl.*1599–1618; d.1621) in the publication of the *History*. Humble later acquired the plates of Pieter van den Keere from which he prepared the atlas often referred to as 'little Speed' (7).

### Later history
The plates remained in the ownership of Humble till his death, when they passed to his eldest son, William (*fl.*1640–1659). He prepared the last edition of *The History* . . . (1650) and the 1646 edition of *The Theatre* . . . . During the Civil War he was a supporter of the King and was rewarded for his services with a baronetcy in 1660 and, possibly at this time, he retired and the plates passed to Peter Stent and Roger Rea. Rea and Stent both died during the plague in 1665 and the edition that Rea the younger prepared was largely lost in the Great Fire of London in 1666. The Speed plates were sold on to Thomas Bassett (*fl.*1659–1693) and Richard Chiswell (1639–1711) at some time after 1668 and before 1674. Their publication of *The Theatre* . . . in 1676 was the last issue in this form. The plates passed to Christopher Browne after the expiry of the ten-year privilege of the previous owners had lapsed in 1686. Browne probably intended to produce an atlas and he certainly added his name to six of the plates (but not Lincolnshire), but no complete atlas bears his name. The next stage in the life of the plates after Browne had them is still not clear. The earlier assumption[4] that John Overton had the plates from about 1700 and that he passed them on to his son, Henry, after 1707[5] is now discounted.[6] Hodson has proposed the sequence of ownership thus; Browne had retired in or by 1713 when Henry Overton bought the Speed plates. Overton put out atlases in small quantities for over thirty years and perhaps up to his death in 1751 when his son, also Henry, carried on for a further short period but had probably sold the Speed plates by 1754 to W. and C. Dicey, who already had the Saxton plates. The last date of use is probably 1782,[7] by which time the atlas had been superseded by those of Bowen, Kitchen and their eighteenth century contemporaries.

### EDITIONS

(i)  *THE/COUNTIE AND/CITIE OF LYN/COLNE DESCRI=/BED WITH THE/ARMES OF THEM/THAT HAVE BENE/EARLES THEREOF/since the Conquest./* (Eb–Ec). The title is surmounted by the arms of James I and has a compass below (Ed). *The Scale of English Miles/10* [= 6.6cm] (Ee). *The armes of such No=/ble Familyes as have/borne the dignitye and/title of Earles of Lyncolne,/since the tyme of the Nor=/ mane conquest./* (Aa). 10 shields with the names of the bearers underneath (Ab–Ad). *LINCOLNE* [a plan of the city] (Ae), with *A scale of Pases 200* [= 2.7cm].The following villages are marked by symbols but they are not named: *Doddington* (north of Grantham); *Waddington; Aslackby; Thedlethorp* (west of Mablethorpe); *Boultham; Stokirith* (i.e. Stockwith); *Swafeild* (i.e. Swayfield); *Cramwell* (sic); in some cases the map shows signs of disturbance where these names should be. There are no compass point names in the borders.

A set of proofs probably put together in 1608 or shortly afterwards and lacking a title-page.[8]

**Copy**    BL Maps C.7.c.5.

(ii)    The place-names in (i) above have been added. The wapentakes, etc. have been added and their boundaries shown by dotted lines. *Haven* has been extended to *Haven-holme*. *Strobby, Firbye, Habro/ugh, Abin/gham* [= Alvingham] have all been added with symbols. *Scarle* has been added after *North*. *Water* has been added in front of *Willoughby*. *Witham flu* has been added south of Southrey. The four cardinal points are named in the centres of the respective borders.

Proofs without title-page put together after (i) above but before 1610.

**Copies**    CUL Atlas 2.61.1; RGS 265 G 25.

(iii)    *Jodocus Hondius cælavit/Anno Domini 1610* added (Ed). Trees with shadows have been inserted in the three parks in Kesteven. The verso of the map has two text pages. The page numbered 63 consists of two columns of text on the county's history and geography, which is derived from Camden. The signature is P 2 and the page is headed *Book.1. Chap. 33*. There are marginal references to the text, including the sources of some of the information and dates of events. The page numbered 64 has eight columns filled with the names of the parishes with the abbreviated names of their wapentakes; the first column is headed by a list of the wapentakes sub-divided into the three parts of the county, *viz.* Lindsey 1–17; Kesteven 18–27; and, Holland 28–30. Lincoln City and Liberty is not included.

*The Theatre Of The Empire Of Great Britaine: Presenting An Exact Geography of the Kingdomes of England, Scotland, Ireland, and the Iles adioyning: With The Shires, Hundreds, Cities and Shire-Townes, within y[e] Kingdome of England, divided and described By Iohn Speed. Imprinted at London Anno Cum Privilegio 1611 And are to be solde by Iohn Sudbury & Georg Humble, in Popes-head alley at y[e] signe of y[e] white Horse.* [1612].[9]

**Copies**    BL G 7884; CUL L* 7.8; Bod Gough Maps 92, etc; RGS 264 H 10; Sheffield RL 914.2 STF; Wisbech Museum – Presses 1.50 and I.49. (LRL Map 741; Grimsby Speed 1610 (RM 118); Nottingham UL Dept. of Geography F 133).

(iv)    The bottom left hand corner of the plate has broken off; a triangular piece with a vertical height of 1.1cm. and a horizontal measurement of 1.5cm. is missing. In the edition of 1616 with Latin text the area broken off is slightly larger and there is a new crack 2.6 cm. further up from the earlier break and extending c.1.8 cm. into the plan of Lincoln. Another crack has also appeared in the top frame 2.8 cm. from the upper right corner. None of these cracks shows any notable change during the remaining life of the plate.

An untitled map volume. Each page consists of the county plate set on a larger sheet and surrounded by text; in the case of Lincolnshire the material originally on the map's verso (i.e. pages 63 and 64) appears without the running titles and headings in galley form to the left and right and below the map. The text is in the setting of the edition ascribed to 1623. The imprint in the lower corner of the final galley reads: *Imprinted at London by Thomas Snodham for John/Sudberie and George Humble, and are to be sold/in Popes head Palace. 1615./* .

**Copy**    Bod Map C.17.b.1.

*The Theatre Of The Empire . . . 1614 . . .* [1616].[10]

**Copy**   BL Maps C.7.c19.

(v)   With text in Latin on the verso. Signature Nn on page 63.

*Theatrum Imperii Magnæ Britanniæ; Exactam Regnorum Angliæ, Scotiæ, Hiberniæ et Insularum adiacentium Geographiā ob oculos ponens: una cum Comitatibus, Centurijs, Urbibus et primarijs Comitatuᵘm* (sic) *oppidis, intra Regnum Angliæ, divisis et descriptis. Opus, Nuper quidem à Iohanne Spedo cive Londinense, Anglicé conscriptum: Nunc verò, a Philemone Hollando ... Latinitate donatum. Imprinted at London Anno Cum Privilegio 1616 ... by Iohn Sudbury & Georg Humble in Popes-head alley at yᵉ signe of yᵉ white Horse.*

**Copies**   BL Maps C.7.c.20; CUL Atlas 4.61.1; Bod Douce S. subst. 50; Leeds UL W 9; RGS 263 G 16; NMM 912.44 (42) ″16″: 094 (E6830). The Bodleian copy has a label pasted on the title-page, which carries the same title as above but with the imprint *Amsterdami, Apud Iohannem Blaeu. MDCXLVI.* There are, in fact, four title-pages, the other three bearing the dates 1627, 1611 and 1614. Another copy, previously in the collection of the late Dr. Eric Gardner, had a label stuck over the original imprint which read *Amstelodami, Anno ... 1621 Ex officina Iudoci Hondij*. A third copy has an overslip pasted on the title-page with the imprint of Jodocus Hondius and date 1616.[11]

(vi)   *Coritani* has been added (Cc, above *The Fenn/ASWARDHURNE WAPONTAK*). In the entrance to the Humber *ABUS ÆSTUARIUM* has been added and *METARUS ÆSTUARIUM* in the Wash. Other additions are: *Trent flu* (south-west of Lincoln) and *Lindum* above *LINCOLNE. The text on the verso is in English.*

*The Theatre Of The Empire ... 1614 ... [1623].*[12]

**Copy**   Leeds UL W 9.

*The Theatre Of The Empire ... 1627. Are to be sold by George Humble at the Whit horse in Popes-head Alley.*[13]

**Copies**   BL Maps C.7.c.13; CUL L* 7.10; Canterbury Cathedral L. W2/x–13. (LRL Map 816).

*The Theatre Of The Empire ... 1627 ... [1631].*[14]

**Copies**   Bod Douce Prints b.24(2); Glasgow Mitchell L. B 763203.

*The Theatre Of The Empire ... 1627. Are to be sold ... [1632].*[15]

**Copies**   BL Maps C.7.c.6; CUL Atlas 4.62.2 and Bury 18.2; Bod Douce Prints b.24; RGS 263 F 2.

*The Theatre Of The Empire... 1627... [1646].*[16]

**Copies**   BL Maps C.27.c.2; CUL Atlases 7.64.4 and 7.62.2; Liverpool Athenaeum 910 11F.

*The Theatre Of The Empire ... 1650 ...*[17]

*The Theatre Of The Empire ... 1650 ... [1651]*

**Copies**   Leeds UL W 10. (LAO Exley 32/2/157; Grimsby RL).

*The Theatre Of The Empire ... 1650 ... [1652].*[18]

*The Theatre Of The Empire ... 1650 ... [1653].*[19]

*The Theatre Of the Empire ... 1650 ... [1654].*[20]

(vii)   Under the imprint the following has been added: *Are to sold be by Roger Rea the/Elder and younger at yᵉ Golden/Crosse in Cornhill against yᵉ Exchange/*

*The Theatre Of The Empire . . .1650 Are to be sold by Roger Rea the Elder and Younger at the Golden Crosse in Cornhill aga^t: y^e Exchange.* [1665].[21]

**Copies**  BL 118 e 8(2); Leeds UL W 11. (LAO FL Maps 1).

A set of maps without title-page.[22]

**Copy**  Manchester RL BR f912.42 S1.

A set of maps without title-page, put together by John Overton. [1670].

**Copy**  Admiralty Library Ve 117.[23]

**(viii)**  The imprint in (vii) above has been erased and in its place is: *Are to be sold by Thomas Bassett/in Fleetstreet and Richard Chiswell/in S^t Pauls Church yard./*

*The Theatre of the Empire Of Great-Britain . . . As also A Prospect of the most famous Parts of the World By Iohn Speed With many Additions never before Extant London Printed for Thomas Bassett and Richard Chiswell. 1676. R. White sculp.*[24]

**Copies**  BL Map C.7.e.5, etc; CUL Atlas 4.67.7; Bod B.1.14 Jur; Leeds UL W 12; Sheffield UL *F914.2(S); RGS 264 H 11. (LAO FL Maps 1; LRL Map 192 is mounted and it is impossible to detect if it has any text on the verso. Grimsby RL has a copy of the Rutland plate with the plan of *Stanford* and a copy of Lincolnshire with blank verso).

*The Theatre of the Empire . . . 1676 . . .*

**Copy**  CUL Ely Gg.1(5).[25]

A collection of maps without title-page.[26]

**Copy**  BL 1 TAB 18.

A collection of 55 maps without title-page or text, published by Christopher Browne. [c.1695].[27]

**Copies**  NMM 912.44(42) ″16″:094 (E6841); PRO M21.1 Press 23.

*England Fully Described in a Compleat Sett of Mapps of y^e County's of England and Wales, with their Islands, Containing in all 58 mapps Printed & Sold by Henry Overton at y^e white Horse without Newgate London R: White sculp* [c.1713].[28]

**Copy**  NLW Fo. 77.

*England Fully Described . . .Newgate, London* [c.1716].[29]

**Copies**  NMM 912.44(42) ″16″:094 [E6840]; BL Maps C.7.e.12 is a similar volume, which lacks a title-page.[30]

**(ix)**  The imprint now reads: *Jodocus Hondius cælavit/Anno Domini 1610/Sold by Henry Overton at the/White Horse without Newgate/ London./* Roads have been added in the form of two parallel lines. The Great North Road is shown as far north as Newark, where it makes a junction with the Fosse Way from Grimsby through Lincoln. The road from Deeping to Barton on Humber is drawn in an almost straight line between those places, passing east of *Borne* and, north of *LINCOLNE*, through *Skampton* and *Ingham.*

*England Fully Described . . . Newgate London* [c.1720].[31]

**Copies**  Bod 2027 a.73 (13–70) and C.17.b.2.[32]

**(x)**  *1610* has been erased from the imprint.

An atlas without title-page compiled from the stock of Henry Overton. [c.1738].[33]

**Copy**  BL Maps 145 c.19.

*England Fully Described . . .58 Mapps by John Speed. Reprinted Anno, 1743. Printed & Sold by Henry Overton . . .*

**Copies**   CUL Atlas 3.74.3; Sheffield RL 912.42 STF. A copy in BL (Maps C.7.e.7) lacks the title-page but is otherwise identical.

An atlas without a printed title-page comprising maps in the hands of Cluer Dicey and Co. [post-1756].[34]

**Copy**   John Rylands UL 15205. Three other 'Dicey atlases' all differ in their make-up but, in each, Lincolnshire is represented by map (**21**).

(**A**)   A lithographic transfer of the map in state (iii), with the addition of a new imprint (Ce OS): *RE-PRINTED AND PUBLISHED BY KELLY & CO., POST OFFICE DIREC-TORY OFFICES, 51, GREAT QUEEN STREET, LINCOLN'S INN FIELDS, LONDON, W.C.* [1877].[35]

Only issued as loose sheets.

**Copies**   BL Maps 3355 (36); CUL Maps bb.30.01.70; Bod (E)C17: 39(15); NLS Map 1.23.40; LAO FL Maps 2 and Exley 32/2/160; Grantham RL.[36]

The text on the verso of the map was re-set a number of times and it is possible to assign loose sheets to their appropriate edition in most cases. In the 1611 edition the signature on page 63 was P2 and the chapter number was 33; all later editions had the signature Nn and the chapter was re-numbered 32. In the 1614/1623 edition at the top left of page 63 appears Booke 1; in the 1627 this is now Book.1. (i.e. two dots). The editions of 1631/2 are distinguished by the addition of *Commodities* in the marginal notes opposite para-graph 6. In the 1646 version the heading at the top of page 63 has now become Book 1. which, in 1650, has become Book 1 (i.e. no punctuation). There are, of course, many more varieties especially in the settings of the names of villages on page 64.

**Reproductions**
(i) The plates from the 1616 edition (the BL Maps C.7.c.10 example) are used in: The Counties of Britain: A Tudor Atlas by John Speed. Introduction by Nigel Nicolson; County Commentaries by Alasdair Hawkyard. (London, 1988). The original texts are not included. (ii) John Speed's England A Coloured Facsimile . . . Edited . . . By John Arlott. (London, 1953). Lincolnshire is in the (final) Part IV; the original text in English is also reproduced. (iii) A reduced version of the plate appears in *Lincolnshire Life*, vol. 13, no. 6 (Aug. 1973), pp. 48–49, with a note by John E. Holehouse. (iv) The lithographic transfer (A) was reissued coloured c. 1924; the imprint reads: *REPRINTED AND PUBLISHED BY KELLY'S DIRECTORIES LTD., 186, STRAND, W.C.2.* (Copy: CUL Maps bb.30.02.70).

**References**
1  Now CUL Atlas 2.61.1 (ex-Gardner collection); Skelton, pp. 33–36 and 210–211.
2  Skelton, p. 32; Skelton, R.A. Tudor town plans in John Speed's Theatre. *Archaeological J.* CVIII (1952), pp.109–120.
3  Skelton, pp. 32–33.
4  ibid., pp. 36, 245 and 249.
5  ibid., pp. 137–138.
6  Hodson, 1984, pp. 41–42.
7  ibid., pp. 66–70.
8  The date of the plate for Anglesey.
9  Books III and IV are dated 1612; the title-pages of Books II and IV bear the name of the printer, William Hall. Hall was in partnership at the time with John Beale (Skelton, p. 33).
10  Dated 1616 in Books II, III and IV.

11  Skelton, p. 49 for details of the first two copies; the copy dated 1616 was offered as lot 219 at Christie's sale on 27 April, 1994.

12  ibid., p. 53 refers to a report in *Times Literary Supplement* (6 Dec. 1947) by Dr. Eric Gardner on the discovery of such an atlas. The plates are in a post-1616 but pre-1627 state (when the map of Surrey, for instance, had lost its top left corner). The presumption was that it had been prepared as a companion volume to the re-issue in 1623 of Speed's *History*. Further support was lent by the discovery of a copy in Wadham College, Oxford with the title-page dated 1623 (Skelton, p. 212).

13  This edition was issued with Speed's *A Prospect of the . . . World . . . 1627*. The CUL copy has no main title-page but Books II, III and IV are dated 1627. This is the last edition issued by Speed himself.

14  Issued with *A Prospect . . .* dated 1631.

15  The map of England is a new plate with the imprint *Abraham Goos Amstelodalmensis Sculpsit Anno 1632./*

16  Bound with *A Prospect . . .* with the imprint *London, Printed by John Legatt for William Humble . . . 1646*. The copy in Liverpool Athenaeum has a label stuck to the inside cover on which is written 'Empt: ex Gulielmo Humble 9 Junij 1647'.

17  Skelton, p. 94. Although Skelton records Leeds UL W 10 as an example of the 1650 version it has, in fact, the characteristics of the [1651] edition (i.e. the maps of the World and of Ireland are both dated 1651). The atlas is listed under 1651 in these notes. Skelton recorded a second copy of the 1650 edition in the collection of the late Dr. Eric Gardner, but its present location is not now known.

18  The map of Scotland is dated 1652 and has been considerably reworked.

19  The maps of France and of Great Britain in *A Prospect . . .* are now dated 1653.

20  The map of XVII Provinces (in *A Prospect . . .*) is now dated 1654. Skelton records a copy in this state owned by Dr. Eric Gardner but its present location is now unknown.

21  This edition is bound with *A Prospect . . .* dated 1662 on its title-page. Skelton, pp. 123–125 argues that the whole work appeared in 1665 on the basis of an entry in *Term Catalogue* (Hilary Term, 1675) which refers to 'the greatest part of an Impression, then newly printed, being destroyed by the late dreadful Fire, 1666'. There is evidence to suggest that Rea the younger planned an edition after the fire as the dates on the plates for Sussex, Buckingham and Derbyshire were changed to 1666. Tyacke, p. 12 records an advertisement in *London Gazette* for 1–4 Feb, 1674/5, which refers to the availability of the maps 'without description, if so wished'. See also note 26.

22  This volume contains maps by Saxton, Norden, Speed (mostly 1610, but also later states; none is later than the issues by the Rea family), and Jansson; it could have been compiled c.1666.

23  Skelton, pp. 129–138 collates the so-called Overton atlases, of which this was his first example. Overton acquired 22 plates from various sources which he supplemented with copies of other maps to construct a complete set of all the counties. The map of Ireland engraved by Wenceslas Hollar is dated 1669 and this suggests that the compilation was made shortly afterwards.

24  There are two title-pages, of which the first (transcribed here) is engraved and the second, much larger, is type-set.

25  Although the Lincolnshire plate is not affected Mr. R. Fairclough (Map Librarian, Cambridge University) has shown in a note attached to the Ely copy that it was produced after the other 1676 editions. Changes to several plates were noted, the reworking of the upper left frame of the Staffordshire map being particularly outstanding.

26  Bassett and Chiswell offered copies of their maps without text. *Term Catalogue* (10 Feb 1676) had an announcement that 'the Particular Maps of . . . any Shire . . . will be sold by themselves, without the Description . . . each map, 6d.' The BL copy seems to be a set made up from such maps as were available, although not all the maps are Speed originals. Sotheby's sold (13 April, 1989, lot 157) an example with all the maps by Speed; in this case the maps were bound with a full set of Ogilby's *Britannia* which had been issued in 1675.

27  Bassett and Chiswell had the privilege for printing *The Theatre* for ten years and this expired in 1686. It is likely that the plates passed to the prolific seller of maps and prints, Robert Walton, but he had died by July, 1688, when Browne was advertising the 'maps, prints and copy books as sold by the lately deceased Robert Walton'. A collection of plates put together

after 1691 (the date on the map of Ireland) and owned by Francis Edwards Ltd in 1967 provides evidence that Browne had not yet added his name to the six county plates which appear in the cited [1695] copy. It is assumed, therefore, that the addition of his name was undertaken after 1692; Skelton, p. 191 suggests 1695 for the appearance of this edition.

28  Hodson, 1984, pp. 43–44 argues that H. Overton produced this edition soon after he obtained the plates c. 1713 from his father and from Christopher Browne, who retired in c.1713. Browne last advertised in the *Daily Courant* of 14 Aug.1712.

29  ibid., pp. 44–45. Several plates now have Henry Overton's imprint and the Northumberland plate incorporates elements taken from Warburton's map which was first advertised in *Evening Post* 23–25 Aug.1716. Hodson reviews Warburton's life and work on pp. 169–179.

30  Hodson, 1984, p. 45. It is pointed out that there are changes to the Dorset plate and this edition might have come out later.

31  Hodson, 1984, pp. 46–47 suggests that it probably took Overton several years to change all the imprints on his plates, the state achieved in this example.

32  Hodson, 1984, p. 46, records that a third copy, Dr. E. Gardner's, was sold by Sotheby's on 12 Feb. 1968 (lot 232); it is believed that it was broken up. It was stated that the copy included a general map dated 1722, thus suggesting a slightly later re-issue. However, this general map has not survived in any other recorded copy and there must be some doubt about the ascribed dating.

33  The map of Huntingdon is dated 1738.

34  Hodson, 1984, pp. 66–70. It is possible that one of the three other copies (that in Pembroke College, Cambridge – e.3.1) was put together as late as 1782.

35  Similar transfers for other counties exist with the rubric added that they were made available to subscribers to Kelly's Directories. Burden pp. 9–10 records the Berkshire example in two different county directories both published in 1877. The map in Kelly's Directory of Lincolnshire for 1876 has the normal Becker plate (**116**) only. A collection in Leeds UL (W 13a) has 31 maps of a similar type (but not Lincolnshire), which Whitaker ascribed to c.1883 but the evidence is not clear.

36  There is a copy in Stamford RL; it has been bound into a copy of Howlett's *A Selection of Views of the County of Lincoln . . .* (1805).

Skelton nos. 7, 10, 11, 14, 16, 18, 36, 48, 50, 51, 55, 57, 81, 92, 116.
Hodson nos. 135–138.

7

## GEORGE HUMBLE                                                      c.1619

Size: 8.2cm x 12.3cm.

George Humble has already been noted for his connection with the publication of Speed's works. The plates originally engraved by van den Keere were probably acquired shortly after their use in the 1617 edition of the reduced *Britannia* (**4**).[1] He had new plates prepared, which were similar in size and character, for those counties which had previously shared a plate with an adjoining county. The new plates are much cruder in their general effect than their predecessors; all the county titles are now in English with the scales drawn in a chequer board style. It is not known which engraver Humble employed in this.

The text is, with minor alterations, taken from that which accompanied the plates in Speed's *Theatre*. Although the title of the earlier states of the book does not refer to Scotland, nevertheless there are maps of that country. Simultaneously with the publication of the new maps, which are often referred to as 'miniature Speeds', Humble issued a miniaturised version of Speed's *Prospect* in 1627. The two works are frequently found bound together and the dates on the latter are used to separate the three 1627 editions of the former. The history of these plates mirrors that of the larger Speed plates. After thirty years or more of use by Humble the plates passed to the Reas, father and son, who produced several editions before a final edition produced by Bassett and Chiswell.

## EDITIONS

(i) *LINCOLNE/SHIRE./* in a plain box (Ad). *10/The Scale of/Miles* [10 = 1.75cm] (Ae). 29 (plate number, Be). The verso is blank. No compass.

Proof copies, in volumes without text or title-page.

**Copies** BL Maps C.7.b.29; Derby RL (Local Studies Dept. 5755).[2]

*England, Wales, And Ireland: Their Severall Counties, Abridged from A farr larger voulume: by John Speed And are To be sold by George Humble in popehead, alley* [1619].[3]

**Copy** BL Maps C.21.a.4.

An atlas without title-page, but with text that, but for a few minor changes, is an advanced proof for the 1627 edition.

**Copy** The private collection of Mr. T. Burgess.

(ii) The plate number 29 has been deleted and a new one 31 added in a new position (Ee). The county map is on the verso of the last leaf of Leicestershire text, while the last page of the Lincolnshire text has the map for Nottinghamshire on its verso. All later editions follow this pattern.

*England Wales Scotland and Ireland Described and Abridged with y^e Historie Relation of things worthy memory from a farr Larger Voulume Done by John Speed And are to bee sould by Georg Humble at y^e Whithorse in popeshead Alley Anno Cum Privilegio 1627*

**Copies** BL Maps C.7.a.27; CUL Atlas 7.62.1; Bod Allen 2; Liverpool UL Y 62.3.22. (Scunthorpe Museum has an item which comprises the map and text of Lincolnshire).

*England Wales Scotland and Ireland . . . 1627. . .* [c.1632].[4]

**Copies** BL Maps C.7.c.6; CUL Syn 7.62.91; Leeds UL W 18.[5]

*England Wales Scotland and Ireland . . . 1627. . .* [1646].[6]

**Copies** BL Maps C.7.a.6 and C.7.a.8; CUL Atlas 7.64.4; Bod Tanner 2(1) and Gough Maps 147 (2); Leeds UL W 19. (Grimsby RL has a copy of the Lincolnshire text and map).

*England Wales Scotland and Ireland Described . . . 1662 And are to be sould by Roger Rea the Elder & younger at y^e Golden Crosse in Cornhill against y^e Exchange*

**Copies** Leeds UL W 20; Birmingham RL 094/1662/19.

Skelton[7] refers to a separate issue of these volumes by Rea with the original 1627 title-page and date but with a label bearing a slightly varied version of the new imprint pasted over the Humble imprint.

**Copy** Liverpool RL G 3063.[8]

Because of its worn state the title-page plate was not used in a printing probably produced in 1665. With re-set text the work is preceded with a title-page for *A Prospect of the most Famous Parts of the World . . . Printed by M.S.* [i.e. Mary Simmons] *for Roger Rea . . . 1665.*

**Copy** The only recorded copy was in the collection of the late Dr. Eric Gardner.[9]

*England Wales Scotland and Ireland . . . Done by John Speed Anno Cum privilegie* (sic) *1666*

**Copy**   The private collection of Mr. C.A. Burden.

*England Wales Scotland and Ireland . . . Cum privilegio 1666*

**Copies**   Bod Allen 8; Hull RL 912.42.

*England Wales Scotland and Ireland . . . 1666* [1668].[10]

**Copies**   BL Maps C.7.a.4; CUL Atlas 7.66.3 and 7.66.4; RGS 264 a 16; Derby RL (Local Studies Dept. 3601). (LRL Map 845a).

*England Wales Scotland and Ireland . . . 1676*

**Copies**   BL Maps C.7.a.5; Leeds UL W 21. (LAO Scorer Map 11 consists of the map and text of Lincolnshire; LRL Map 804).

Loose copies of the map can be dated by reference to the settings of the text – the map is on the verso of the last page of the Leicestershire text. In the 1627 setting the eleventh line has *Kerkby-Bellers*; in 1632 this has been changed to *Kerby-Bellers* and, additionally, the paragraph numbers now have round brackets; in 1646 there are no round brackets; the 1662 edition is identical to the 1646; in 1666 signature M2 has been added; the [1668] edition also has the signature M2 but the text has been re-set again. The clearest signs of their differences occur in line 2 where the final letter (s) is raised and, on line 4, 'neer Bosworth' has been altered to 'near Bosworth'. Finally in 1676 the catchword *LINCOLN-* has been added to the features of the 1668 edition.

### Reproduction
(i)   The plates are reproduced, but without their plate numbers in: An Atlas Of Tudor England And Wales . . . Introduced and Described By E.G.R. Taylor. (London, 1951).

### References
1   Skelton, p. 57. Skelton notes that the van den Keere plates were not in the engraver's stock in 1623 and cites Kleerkooper, M.M. and van Stockum, W.P. *De Boekhandel te Amsterdam.* (Amsterdam, 1914–1916), pp. 1205–8.
2   The copy in Derby is the earlier version since there are two plates notably changed in the BL set.
3   The new work was prepared after 1617 and before 1623 and it is assumed that Humble acquired the plates soon after van den Keere's last use of them in 1617; allowing time for the engraving of the 16 plates that were freshly prepared 1619 seems a likely date for the proof version.
4   The text has been re-set and Skelton p. 62 suggests simultaneous issue with *The Theatre . . .* dated 1632.
5   Because the title-page on the BL copy is mutilated Chubb thought the work was published in 1620. Chubb no. XI.
6   Copies are bound with the miniature *Prospect . . .* dated 1646.
7   Skelton, p. 215.
8   Mr. C.A. Burden owns now the only copy known to Skelton.
9   Skelton, p. 125 assumes that the scarcity of copies is because the bulk of the printing was lost in the Great Fire of London. The issue of a new edition in 1666 would support this, although the 1666 edition is almost equally rare.
10  All the recorded copies are bound with *A Prospect . . .* dated 1668.

Skelton nos. 17, 19, 37, 69, 82, 83, 86 and 93.

# 8

**WILLIAM HOLE**[1]                                                      **1622**

Size: 23.8cm x 30.9cm.

The first edition [1612] and the second, dated 1613, of Michael Drayton's *Polyolbion* contained 18 maps. Lincolnshire only appeared in the third edition of 1622 as one of a further ten new plates. The plates are purely decorative and have no claim to consideration as maps with serious topographical intent. That has not reduced their favour with collectors.

Drayton (1563–1631) wrote a kind of miniature *Polyolbion* as early as 1594 and he had made known his intention to produce a larger work by 1598. The first 18 songs, as they are called, were not ready until 1612. It has been suggested by Skelton[2] that, although registered at the Stationers' Hall on 7 Feb 1612, the first edition was issued prematurely. Certainly Drayton complained that this was the case in the preface to the enlarged 1622 edition. There are 12 more songs added to the book in its final form, making almost 15,000 lines of alexandrines.

Hole's work in producing 21 maps for the 1607 edition of Camden's *Britannia* has already been referred to (**5**). He engraved the title-page of the first issue of *Polyolbion* and his portrait of Prince Henry is found in three different versions in the three issues.

## EDITION

(**i**)  *LINCOLNE/SHYRE* is engraved across the centre. (Cc). There is neither scale nor compass. Only nine towns are marked, together with the parts of Lincolnshire, *The Isle of Axholme and Ancaster heath.* All places are indicated by mythological figures. Several rivers are marked and named, forming a particularly prominent feature. *105* (Ea) – represents the page which follows the map.

*A Chorographicall Description Of All The Tracts, Rivers, Mountains, Forests, and other Parts of this Renowned Isle of Great Britain With intermixture of the most Remarkable Stories, Antiquities, Wonders, Divided into two Bookes, the latter containing twelve Songs; never before Imprinted. Digested into a Poem by Michael Drayton . . . London, Printed for Iohn Mariott, Iohn Grismand, and Thomas Dewe. 1622.*[3]

**Copies**   Bod Malone 13, etc; Leeds UL W 16. (LRL Map 795).

*The Second Part, Or A Continuance Of Poly-Olbion From The Eighteenth Song. Containing all the Tracts, Rivers, Mountains, and Forrests: Intermixed with the most remarkable Stories, Antiquities, Wonders, Rarities, Pleasures, and Commodities of the East, and Northerne parts of this Isle, lying betwixt the famous Rivers of Thames, and Tweed. London, Printed by Augustine Mathews for Iohn Marriott, Iohn Grismand, and Thomas Dewe. 1622.*[4]

**Copies**   BL 838 m 1; CUL SSS.21.10; Birmingham RL 094/1613/3; John Rylands UL 821.34 (R 4570); London UL [D–L L] Bc.

### Reproductions
(i) The Poly-Olbion: A Chorographicall Description Of Great Britain, By Michael Drayton. Part 3. Printed For The Spenser Society. 1890. (ii) The map is reproduced in *Lincolnshire Life*, Vol. 6, no. 10 (Dec. 1966), p. 30 and Vol. 16, no. 12 (Feb. 1977), p. 40. The captions and text contain inaccuracies.

### References
1   Hole engraved the royal portrait in all the editions of *Polyolbion* and Skelton (1970, p. 45)

says that Hole probably engraved all the plates; Cope, G. The puzzling aspects of Drayton's Polyolbion. *Map Collector* 17, (Nov. 1981), pp. 16–20 discusses the relationship between Hole and Drayton; Hind, A.M. Engraving in England in the sixteenth and seventeenth centuries. Part 2 (Cambridge,1955), pp. 316–350 discusses Hole's output.

2   Skelton, p. 46.
3   The title quoted is that usually taken to be the new title-page for the 1622 edition; it is letterpress and illustrated by Cope (see note 1).
4   The engraved title noted here is that inserted between the end of the first 18 songs and the start of the second part. The first engraved title- page reads: *Poly-Olbion Great Britaine By Michael Drayton Esq$^r$ London printed for M Lownes. I Browne. I Helme. I Busbie. Ingrave by W Hole*. All the copies noted here have these two title-pages.

Skelton no. 13.


# 9

## JOHN BILL                                                     1626

Size: 8.5cm x 2.7cm.

The map is a close copy of van den Keere's plates (**4**) and George Humble's (**7**) in particular, with Lincolnshire separated from Nottinghamshire. Its one distinguishing feature is the inclusion, for the first time, of longitude measured at five minute intervals and latitude. Longitude is measured from the Azores.

John Bill was born in Shropshire but flourished as a bookseller in London from c. 1592. He was commissioned by Sir Thomas Bodley to travel abroad to buy books for his library. He was the King's Printer from 1604 until he died in 1630.[1] This collection of maps was never re-issued and it may safely be assumed that the success of Humble's little volume would have militated against a re-issue.

The text is based on Camden's *Britannia* and is arranged on a single leaf facing the map which occupies the right hand half of an opening. The result of this method is that the verso of the *Norfolke* map has the Lincolnshire text while the Lincolnshire map has the text for *Rutland Shire* on its verso. While there is only one edition of the maps the text was being corrected as printing proceeded.[2]

### EDITION

(i)   *THE COUN/TIE OF/LINCOLNE/* (Ab, in a rectangular box). Dividers occupy the bulk of the left side and incorporate a cross bearing the names of the four cardinal points (Ac–Ad) . [Scale] 20 [= 2.7cm] (Ae). The left hand frame is marked by shading (and numbered) at 5' intervals from [52°] 45' to 53° 45' and the bottom frame is similarly shaded and numbered from [20°] 25' to 24° 55'. G2 [signature outside and below the map].

*The abridgment of Camden's Britañia With The Maps of the seuerall Shires of England and Wales Printed by Iohn Bill Printer to the Kings most excellent Maiestie, 1626.*

**Copies**   BL Maps C7.a.9; CUL Syn 7.62.57; Bod Antiq. f.E 11(1); Leeds UL W 17. A copy in Nottingham UL (DA 130.C2) lacks the Lincolnshire plate. (LRL A9 (UP 1658) is a very incomplete copy, but it has *inter alia* the title-page, the Lincolnshire map and county text).

### References
1   Skelton, p. 243; Kingsley, p. 26; Plomer, 1907, p. 24.
2   Mr. C.A. Burden has three copies and there are differences in all three; one notable change

appears at the end of the dedicatory leaves, in one of which Bill's name lacks the final letter; another copy is a large paper edition.

Skelton no. 15.


## 10

### JACOB van LANGEREN                                      1635

Size: 3.9cm x 2.2cm (both measurements are approximate). The plate size is 10.3cm x 10.3cm.

This is the earliest road book with maps. There is evidence to suggest that the work was conceived some years before the earliest extant copy. The title-pages of the earliest copies show clear signs of having been reworked; the frontispiece was already over thirty years old and the basis of the maps' designs suggests a long gestation period.

However, the preparation of the plates would probably not have begun before 1625. In that year John Norden's *England: an Intended Guyde* was published and it is from the mileage tables in that work that the present plates were copied.[1] The plates are divided into two portions with the thumb-nail map occupying a position in the lower right hand segment; along the top and down the left-hand side are the names of the principal towns of the county with a mileage grid taking up the upper left portion (roughly two-thirds of the whole) of the plate.

The maps are drawn to a uniform scale and have radiating lines drawn from the county boundaries to the eight principal compass points. The map shows the county boundary, the chief towns (indicated by initial letters) and rivers. Although *Bourne* is one of the places on the mileage chart it is not marked on the map. B (for Belvoir) is marked on the map but is not included in the chart (and is not, of course, in Lincolnshire). All these features show a notable likeness to the playing card maps of 1590 (**2**).

The circular map, which serves as a frontispiece and is signed by W. Kip and H.W. (Hans Woutneel), was probably engraved in 1602–3 and Skelton suggests that it may be a small version of the larger map that Kip engraved for Woutneel in 1603.[2] However, this is the first book entered to Simmons in the Stationers' Register and it had not apparently been assigned to him from another bookseller. For that reason the reworking of the original title-page is not clear evidence that there was an earlier edition that has been lost and it may simply mean that errors had been made while the plate was being prepared for the first edition.

Matthew Simmons began his career as a bookseller with this work and remained active until his death in 1654. He undertook printing work after 1641 and printed books and broadsides for Jenner (**11**) and a number of Milton's works. After his death his widow Mary (**11 (iii)**) and his son Samuel carried on the business until 1678.[3]

Jacob van Langeren was probably the second son of a family of cartographers. Little is known of him, however, and it may be that he never visited this country since the only record of his home is in Brussels. Skelton suggests that the plates for this work were prepared in the South Netherlands for printing in this country.[4] He seems to have worked for Simmons and for Jenner and a work printed for the latter and dated 1656 is signed 'Ja.v.L.'[5]

### Later history

After three editions by Simmons within two years the plates are next found in the possession of Thomas Jenner and he revised them so extensively that they are treated as a separate series (**11**). However, the changes to the plates were not all undertaken

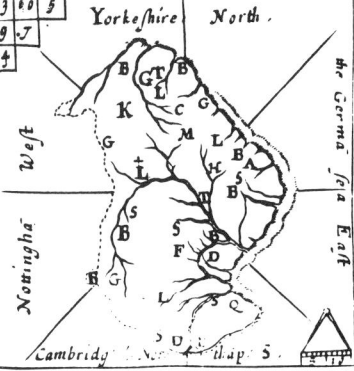

3. 1636   MAP 10 (i) – JACOB VAN LANGEREN.
By permission of Leeds University.

simultaneously. The first editions under Jenner's auspices reveal a piecemeal approach to the reworking of the plates. The order in which the maps were revised seems to correspond to the course of the Civil War with the areas involved in the first battles being the first to be revised.

## EDITIONS

(i)   *Lincoln/shire/*(Aa). 24 (plate number) (Be). [Scale] (and dividers) 10 [= 0.6cm] (Ee). The map is 3.9cm from north to south and 2.2cm at its widest and occupies about one twelfth of the total plate area which is largely occupied by the mileage chart.

*A Direction For The English Traviller By which he Shal be inabled to Coast about all England and Wales. And also to know how farre any Market or noteable Towne in any Shire lyeth one from an other, and whether the same be East, West, North, or South from yᵉ Shire Towne As also the distance betweene London and any other Shire or great towne: with the scituation thereof East, West, North, or South from London. By the help also of this worke one may know (in what Parish, Village, or Mansion house soeuer he be in) What Shires, he is to passe thorough & which way he is to trauell, till he come to his*

*Journies End.* [Motto] *Are to be Sold by Mathew Simons at the golden Lion in Ducke laine, A° 1635. Jacob van Langeren sculp.*

**Copies**   BL 291 a 46; CUL Syn 7.63.303.

*A Direction For The English Traviller . . . 1636 . . .*

**Copies**   BL Maps C.7.a.29 and C.7.a.12; Bod G.A.Gen.Top. 8° 628 and G.A. Fol. A3; Leeds UL W 26.

(ii) an extra line of figures has been added to, but outside, the mileage chart. They represent the distances from London of the places named at the left.

*A Direction For The English Traviller . . . 1636 . . .*

**Copy**   Only one copy is recorded – BL Maps C.7.a.30 – but it lacks the Lincolnshire plate. Two plates in that copy – Essex and Middlesex – do not have the additional line of figures, The additional line was not added to the Lincolnshire plate until later in 1643 (11(ii) ).

### References

1   The Lincolnshire plate from Norden's book  is shown in: Elias, W. Road Maps for Europe's Early Post Routes 1630–1780. *Map Collector*, No. 16 (September, 1981), p. 31. Interestingly, Norden's original is set from type, with printers' leads separating the individual figures. The result is a greater legibility than van Langeren's engraved plate.
2   Skelton, pp. 63–64.
3   ibid., p. 243; Plomer, 1907, p. 164.
4   Skelton, p. 64.
5   Keuning, J. The Van Langeren family. *Imago Mundi*, XIII (1956), pp. 101–109.

Skelton nos. 20, 21, and 22.

## 11

**THOMAS JENNER**                                          **1643**

Size: 6.3cm x 5.5cm (maximum space occupied by the map).

Thomas Jenner (fl. 1618–1673) was a printseller and the publisher of a large number of engraved plates. As a puritan and parliamentarian some of his publications were produced opportunely for use in the Civil War. The so-called 'Quartermaster's map' was etched for him by Wenceslas Hollar in 1644. He is likely to have come into ownership of the van Langeren plates at the start of the war and the mixed states of the plates in the first editions suggest a rushed process in the reworking of the plates to catch the new demand for a book of maps in a handy pocketable form. When he signed his will in 1666 he was 'sickly in body' but carried on working until he died seven years later.

One of the most frequent visitors to his shop was Samuel Pepys. About 1666 Pepys bought a collection of sixteen maps, including an example of the Quartermaster's map, which he had bound up as a pocket atlas.[1] Pepys in his work in the Navy Office realised the value of maps as a source of useful knowledge, particularly during periods when the country faced the belligerent activities of other nations.[2]

Jenner left a bequest to John Garrett of twenty shillings to look after his (Jenner's) wife.[3] Garrett (fl. 1667–1718) who assumed business at Jenner's address was probably the son of the bookseller William Garrett and was the brother-in-law of John Overton. He carried on a thriving business re-issuing Jenner's works; determining his exact output is difficult because there were a number of John Garretts active in this period.[4]

**Later history**

The plates were in a mixed state when they came into his ownership. Some were already in a second state while under van Langeren's control, while others, including Lincoln-shire, were still in their original state. The county plate was revised during 1643 when an extra line of engraved figures was added to the mileage chart; these figures related to the distances from London of the towns listed down the left side of the chart (i.e. the newly added figures did not form a part of the chart from a strictly purist point of view).

As stated above the plates remained with Jenner until his death in 1673. After his death there were a number of editions brought out by Garrett, the last probably being that entered in the *Term Catalogue* for Easter, 1680.

## EDITIONS

(i)   The map is in the state described in **10** (**ii**) above, i.e. there is an extra line of figures added to but outside the grid of the original mileage chart.

*A Direction for the English Traviller . . . Are to be sold by Thomas Jenner at the South entrance of the Exchange 1643.*

**Copy**   Leeds UL W 27.

(ii)   The map has been completely re-engraved. It now fills the entire space not used by the mileage chart, but because of the confines of the available space the map has been turned with north-west at the top. The initial letters have been replaced by the full spelling of place-names. Skelton says that in spite of the tilting of maps to fit the space 'in every case the names are engraved so that the map is to be read with north uppermost.'[5] This is certainly not the case on the Lincolnshire map, particularly with places near the coast. There are a number of curiosities of spelling also, *Scalding* and *Below* (=Belleau) being the most notable. The dividers have been erased and a new pair form two sides of a triangle over the plate number *24* (Be).

*A Direction for the English Traviller. . . 1643.*

**Copies**   BL Maps C.7.a.31 etc.; CUL Atlas 7.64.5; Bod Gough Maps 132 etc.;[6] Leeds UL W 28;[7] RGS 262 a 34; Birmingham UL r G 1854.

*A Booke of the Names of all the Hundreds contained in the Shires of the Kingdom of England; Together with the Number of the Towns, Parishes, Villages, and Places of every Hundred . . . Usefull for Quarter-masters, Brief-Gatherers, and all such as have to doe the Shires of England. Printed for Thomas Jenner, and are to be sold at the Royal Exchange, 164[ ].*

**Copy**   Previously owned by the late Dr. Eric Gardner it now belongs to Mr. C.A. Burden.[8] Lincolnshire is one of the 17 missing plates.

(iii)   The map is set on p. 99 with part of a letter-press list of towns and villages below the map and on its verso. The signature on p. 99 is *N2*. The pagination is faulty in a number of places in this setting. The Lincolnshire map has page 101 on its verso, while the following leaf has 102 on both sides. The following leaf is numbered 103 with 105 on its verso. What should be page 89 is also erroneously numbered 99.

*A Book Of The Names Of All Parishes, Market Towns, Villages, Hamlets, and smallest Places, In England and Wales. Alphabetically set down, as they be in every Shire. With the Names of the Hundreds in which they are, and how many Towns there are in every Hundred . . . A work very necessary for Traveilers, Quartermasters, Gatherers of Breefs, Strangers, Carriers, and Messengers with Letters . . . London: Printed by M.S. for Tho: Jenner, at the South-entrance of the Royall Exchange. 1657.* [9]

**Copies**   BL Maps C.7.b.6; CUL Syn. 7.65.19; Bod Wood 468, etc.; RGS Fordham 130; Lincoln Cathedral L. Ff.7.18; Leicester UL H942.2 SIM.

(**iv**)   The map remains unchanged but the table of distances has been retouched; this can be most readily seen in many of the figures 7, which have what appears to be a serif at the front. The text has been re-set. The errors of pagination have not been altered but the list shows a number of notable changes on p. 99: viz. *BOURNE* (previously *BOVRNE*), the abbreviated names of the wapentakes after several village names have lost their initial capital letter. Examples are man (after *Botsford*) and *booth* (after *Boothbye* and *Boultham*). Similar changes appear in the lists on the verso. The signature now appears as N 2 (i.e. a space has been inserted).

*A Book Of The Names Of All Parishes . . . 1662.*

**Copies**   BL Maps C.7.b.7; Bod Tanner 273(3); RGS Fordham 131.

The map remains unchanged but the text has again been re-set. In this case the errors of pagination which affect the Lincolnshire pages have been corrected; new errors have been introduced, but the only one that concerns the Lincolnshire text is in the running title on p. 98 – (*Linclon-shire*). In the text on p. 99 some of the alterations from the 1662 setting now read: *BOVRNE; Boothby, Booth;* the initial letter of Branston is in italic and from a different, smaller font.

*A Book Of The Names Of All Parishes . . . 1668.*

**Copy**   Leeds UL W 34.

The chart has now been re-worked. Several of the dividing lines between the names along the top of the chart have been redrawn, resulting in doubling between *Spalding* and *Crowland,* which continues down into the first two lines of the chart, and the curve at the top of the line between *Lincolne* and *Grantham* has been straightened although remnants of the original curve can be seen. 70 has been added at the end of the Crowland line of the chart. The right hand frame line has been extended downwards and almost touches *Welton* above the map.

*A Book Of The Names Of All Parishes . . . 1668.*

**Copies**   BL Maps C.7.b.8; Guildhall L. Bay H2.4.no.21.

*A Book Of The Names Of All Parishes, Market Towns, Villages, Hamblets* (sic), *and Smallest Places. in England and Wales. Alphabetically set down . . . Together with a Catalogue of the Market Towns, Castles, Parish Churches, Rivers, Forrests and Parks; and also the Length, Breadth, and Circumference of every particular Shire. A work very necessary for Travellers . . . London, Printed by S.S. for John Garret, at his Shop as you go up the Stairs of the Royal Exchange in Cornhill: Where is also sold . . . 1677.* [10]

**Copies**   BL Maps C.7.b.9; CUL Atlas 7.67.2 and R.II.30; Bod Redcliffe e 55; RGS Fordham 132; Guildhall L. Bay H3.4.no.24.

Without text. The plate is in the revised state of 1668.

*A Direction for the English Traviller . . . Printed and are to be sold By John Garrett, at the south Entrance of y^e Royall Exchange in Corn-hill, where you have a most exact Mapp of England . . .* [1677].[11]

**Copies**   Bod Allen 5; Leeds UL W 26a; RGS 265 A 31.

*A Direction for the English Traviller . . . Where is also sold, a Book of y^e names of all Parishes, Market Townes, Villages, Hamletts, & smallest Places in England Alphabetically sett down as they bee in every Shire.* [1680][12]

**Copies**   BL Maps C.7.a.4; CUL Atlas 7.64.6; Bod Gough Maps 142.

**References**
1   Pepys Library. Magdalen College, Cambridge. No. 1601.
2   Tyacke, S. Samuel Pepys as Map Collector. In: Myers, R. and Harris, M. (eds.). Aspects of the English book trade. (Oxford, 1984), pp. 1–29.
3   Skelton, p. 69 and p. 244; Tyacke, p. 118.
4   Tyacke, p. 118; plate 9 (p.115) of this work is an illustration of the Royal Exchange building, where Jenner leased the shop that Garrett later took over.
5   Skelton, p. 69.
6   This copy has lost its title-page but since eight maps (Cumberland, Durham, Northumberland, Westmoreland, Yorkshire, Suffolk, Nottinghamshire and Herefordshire) are in the earlier thumb-nail size there can be no doubt of its dating. This list of counties would lend some support to the idea that the plates were revised quickly and in accordance with the progress of the war across the country.
7   This is a large paper edition. The leaves now measure 18.1cm x 14.1cm compared with 12.45cm x 12.45cm. in the usual format.
8   Only one copy has been recorded of this title. The title-page has been trimmed during binding and has thus lost the last digit of the date as well as the final letters of some words. It is assumed that it was issued soon after the 1643 edition and, since there is a reference to the work's usefulness to Quartermasters in the title, it is thought to be contemporary with Hollar's Quartermaster map of 1644.
9   M.S. is assumed to be Mary Simmons, widow of Matthew Simmons (**10**). Most copies have two title-pages, i.e. the 1657 transcribed above and that originally engraved for 1643.
10  S.S. is Samuel Simmons, son of Mary Simmons.
11  An announcement in the *Term Catalogue* for Easter, 1677 is assumed to refer to this issue. The title-page is that of 1643 with the lower portion reworked as quoted. Chubb, No. XLVII conjecturally dated this to 1645.
12  Announced in *Term Catalogue* for Easter, 1680. Chubb, No. XLVIII had assigned this to 1650.

Skelton nos. 25, 26, 62, 70, 87, 88, 98, 99 and 101.


# 12

**JAN BLAEU**                                                              **1645**

Size: 41.9cm x 49.8cm.

Jan Blaeu (1596–1673) was the son of Willem Janszoon Blaeu (1571–1638), who published the 1617 volume, which contained the maps of Van den Keere (**4**) and his first world atlas in 1630. This single volume work was increased to two volumes in 1635 and, when his son took over the business, a third volume appeared in 1640. At this stage only four general maps covering the British Isles, Scotland, Ireland and England appeared. In contrast there were sixteen area maps of the various parts of the British Isles available in the German language version of the rival Mercator/Hondius atlas of 1636 issued by Jansson (**13**) and its range was extended further in the English language edition (also 1636). More seriously from Blaeu's viewpoint Jansson had begun to produce maps of the individual English counties.[1] In the race, however, to be the first to produce a volume of separate plates of all the English and Welsh counties Blaeu won by 14 months, although there are tell-tale signs of haste in the preparation (see note 7). Blaeu was also the first to produce an atlas which included plates for each of the Scottish counties – in 1654.[2]

The English and Welsh maps usually occur as volume IV in the Latin, French and Spanish editions or V in the other language sets (German and Dutch) which had expanded to twelve volumes by the time of their final appearances. Usually the verso has a text derived from Camden in one or other of these five languages, but maps without text on the verso were sold separately and included in atlases.[3] Blaeu also follows Camden in his ordering: the counties with their maps start with Cornwall and Devon and work firstly eastwards and then northwards.

The maps are based on Speed plates, mostly in their post-1623 manifestation. The names of the roman stations are included as are the names of the British tribes. The coats of arms are taken from Speed but the town plans and historical notes are omitted. The other general geographical features are also taken from the same source (rivers, trees, compass points in the borders, etc.). There are two little galleons in the North Sea as decorative elements.

The engraved title-page was of a standard design used in all editions; individual titles were created by pasting a printed label over the blank central panel. Dates in the imprint were altered as necessary but were treated in a rather cavalier manner and are not always the actual date of issue; a later date often appears at the end of the notice to the reader.

As the lists make clear Blaeu was also 'careless' when assembling his atlases. He took whatever was to hand to compile his sets. Reference is made in the notes to elements of particular interest in what are often (bibliographically speaking) unstable works.

The atlas also contained the Fen map: *REGIONES/INUNDATÆ/In finibus Comitatus/ NORFOLCIÆ, SUFFOLCIÆ,/ CANTABRIGIÆ, HUNTINGTONIÆ,/ NORTHAM-TONIÆ, et/LINCOLNE./* This plate is most frequently found without text on its verso and remained unaltered during its lifetime.

### Later history
There was a disastrous fire at the Blaeu printing-house in 1672, when most of the plant and plates were destroyed. Blaeu died in the following year. Enough copies of sheets survived in other premises belonging to Blaeu that, as late as 1689, atlases comprising full sets of Blaeu sheets were being offered by the bookbinder Albertus Magnus and a choice of texts in Latin, French, German or Dutch. Blaeu's sons carried on business after 1672 and loose sheets were supplied to dealers in London and Amsterdam who could thereby make up composite atlases. Some plates survived[4] but their whereabouts until 1714 have not been recorded. At least one county appears in more than one revised state in works produced by David Mortier in London from 1714.[5] Hodson's analysis of an *Atlas Anglois* prepared in London in 1714 suggests that 19 plates survived the fire in 1672. This title occurs again with a mixture of Blaeu and Jansson plates; the last being in 1728 or just after.[6]

### EDITIONS

(i)  *LINCOLNIA/COMITATUS. Anglis/LINCOLN-SHIRE/* (Ee, in a decorated panel upon which there are two *putti*; one holds a pair of dividers and the other a further panel which has the scale and the words (reading upwards) *8* [miles = 5.2cm]/*Octo Milliaria Anglice, sive IIII Germanica*). The royal shield with *Anglice* below (Aa). Ten shields with eight coats of arms and two left blank (Ac–Ae), with the wording *LINCOLNIA Comitum Insignia* above. The royal coat of arms (Ea). The cardinal points are named in Latin (Ac, Ca, Ce and Ec).

*Guil. Et Ioannis Blaeu Theatrum Orbis Terrarum, sive Atlas Novus. Pars Quarta. Amsterdami, Apud Iohannem Blaeu. MDCXLV.* (Latin text, with pages 289–290 on the verso; there is no signature).

Copy   BL Maps C.5.d.3.

[*Le Theatre du Monde, ou Nouvel Atlas, Mis en lumiere Par Guillaume et Iean Blaeu Quatriesme Partie. Amsterdami, Apud Iohannem Blaeu. M DCX LV.*] (French text, with pages 245–246 on the verso, signature Hhhh on p. 245; p. 245 is the last page of the Rutland text).

**Copies**   Bod 2027 a.46;[7] Amsterdam UL 1800 A 4. (BL K 19.16).

*Novus Atlas Das ist Welt=beschreibung mit schönen newen aussführlichen Land=Taffeln in Kupffer gestochen Und an den Tag gegeben Durch Wilhelm und Joan Blaeu. Vierter Theil. Amsterdami, Apud Iohannem Blaeu. M DC XLV.* [1646].[8] (German text, with pages 282–283 on verso).

**Copy**   The only recorded copy is in Tübingen University Library.[9]

*Theatrum Orbis Terrarum, Sive Atlas Novus: in quo Tabulae et Descriptiones omnium Regionum, Editæ a Guiljel: et Ioanne Blaeu. Amsterdami, Apud Iohannem Blaeu. M DC XLVI.* (Latin text, with pages 289–290 on the verso, signature Yyyy on p. 289).

**Copy**   Rotterdam Municipal Archive Office Bibl/GAR 58 A 11.[10]

*Le Theatre du Monde, ou Novvel Atlas, mis en lumiere . . . M DC XLVI.* (French text numbered as before and signature Hhhh on p. 245).

**Copy**   London UL D–L L.[11]

*Novus Atlas . . . M DC XLVI.*

**Copy**   Amsterdam Scheepvaartsmuseum – A 111 46 IV.[12]

*Toonneel des Aerdrycks, oft Nieuwe Atlas, uitgegeven door Wilhelm en Jan Blaeu. Vierde Deel. Amsterdami. Apud Iohannem Blaeu. M DC XLVI.* [1647][13] (Dutch text, with pages 271–272 on the verso and signature Ssss on p. 271).

**Copy**   The private collection of by Mr. C.A. Burden. (Grimsby RL Blaeu 1645 – RM 113).

*Novus Atlas . . . M DC XLVII.*

**Copy**   No copies recorded in the UK.[14]

*Guil. Et Ioannis Blaeu Theatrum Orbis Terrarum . . . M DCX LVIII.* (The Latin text now has pagination 249–250, signature Nnnn on p. 249).

**Copies**   BL 9 TAB 6 and Maps C.5.d.2; Bod 2027 a.41; Leeds UL W 32; NMM 912.44 (100) "16":094 [E5256]. (LRL Map 73b).

*Le Theatre du Monde . . . Quatrieme Partie . . . M DC XLVIII.*

**Copy**   Leeds RL SRF 912 B569.

*Vierde Stuck der Aerdrycks-Beschryving, welck vervat Engelandt . . . MDCXLVIII.* (Dutch text on verso, pages 231–232, signature Ffff on p. 231).

**Copies**   BL Maps C.4.b.5. (LRL Map 73).

*Toonneel Des Aerdrycks, oft Nieuwe Atlas . . . M DC XLVIII.* (Dutch text, pages 231–232 on verso, with signature Ffff on p. 231).

**Copy**   Sheffield UL LF 5.

*Novus Atlas, Das ist Welt-Beschreibung . . . MDCXLVIII.* (The verso is blank in this German language edition).

**Copy**   Leeds UL W 31.

*Theatrum Orbis Terrarum, Sive Atlas Novus; in quo Tabulae et Descriptiones omnium Regionum, Editæ a Guiljel: et Ioanne Blaeu. Amsterdami, Apud Iohannem Guiljelmi F. Blaeu. Anno M DC XLVIIII.* [1650]. The verso is blank.[15]

**Copy**   BL 9 TAB 6.

*Toonneel Des Aerdrycx, oft Nieuwe Atlas . . . M DC XLVIII.* [1655].[16]

**Copy**   RGS 1 C 215–220.

Various editions with Dutch, Latin, French and German texts issued between 1653 and 1655 are noted.[17] The texts have not been revised, the maps remain unchanged and no copies are recorded in UK.

*Nuevo Atlas Del Reyno De Ingalaterra. En Amsterdam, A costa y en Casa de Juan Blaeu. M DC LVIII.* [1659].[18] (Spanish text, with pp. 265–266 on verso, signature Ssss on p. 265).

**Copies**   Leeds UL W 32a; Glasgow Mitchell L. SR. 314 (B106596).

*Geographiæ Blavianæ Volumen Quintum, Quo Anglia, Quæ Est Europæ Liber Undeci-umus, Continetur. Amstelædami, Labore & Sumptibus Ioannis Blaeu, M DC LXII.* (Latin text, pp. 249–250 on the verso, but re-set with several differences from before; a simple distinction is the use of W now instead of VV).

**Copies**   BL Maps C.4.c.1; CUL Atlas 2.66.5; NMM 912.44 (100) ″16″:094 [D5255]; John Rylands UL R912 (8069); NLS WD 3 B; Edinburgh UL JZ 34.[19]

*Angleterre, Qui Est Le XI Livre De L'Europe. Amsterdami, Apud Iohannem Blaeu. M DC LXII.* (French text with the pages now numbered 209–210 on the verso and signature Cccc on p. 209).

**Copies**   Bod Allen 7; Liverpool RL GF 153.[20]

*Nuevo Atlas Del Reyno De Ingalterra. En Amsterdam, En la Officina Blaviana. Amster-dami, Apud Iohannem Blaeu. M DC LXII.*[21]

**Copies**   BL Maps C.5.c.2. (LRL Map 789).

*Cinquiéme Volume De La Geographie Blaviane, Contenant L'Angleterre, Qui Fait L'Onziéme Livre De L'Europe. A Amsterdam, Chez Jean Blaeu. M DC LXII.* [1663]. (French text).

**Copies**   (LAO FL Maps 4; LRL Map 73a).[22]

*Cinquiéme Volume De La Geographie Blaviane . . . M DC LXIII.*

**Copies**   BL Maps C.5.b.1; NMM 912.44 (100) ″16″:094 [E5253].

*Vierde Stuck Der Aerdrycks-Beschryving, Welck Vervat Engelandt. Amsterdami, Apud Iohannem Blaeu. M DC LVIII.* [1664]. (Dutch text).

**Copies**   BL Maps C.4.d.1; BL Maps C.4.d.2.[23]

*Angleterre, Qui Est Le XI Livre De L'Europe . . . M DC LXII.* [1665].[24]

**Copy**   Holkham Hall.

*Cinquième Volume De La Geographie Blaviane . . . M DC LXVII.*

**Copies**   BL 114 h 5; Birmingham RL AE 094/1667/18E.

Although the Latin texts in the second and third settings have the same page numbers they are easily separated; from 1648 the upper case W is printed as double V while in the 1662 it is replaced by W. Although the Dutch text of 1664 is substantially the same as in 1648 extra marginal notes have been added, the catchword at the foot of column two on page 231 is *schoone* and on page 232 is *het*.

## Reproduction
(i) A facsimile in twelve volumes was published in 1963 by Theatrum Orbis Terrarum of Amsterdam. The text is that of the 1663 French edition; (ii) Atlas Of England, Scotland, Wales And Ireland, with an introduction by R.V. Tooley. (London, [1970]). There are no text pages from the atlases; plates are taken from various issues to provide maps of the four countries.

## References

1 Skelton, pp. 73–75; 81–84; 220–225; Stevenson, E.L. *Willem Janszoon Blaeu*. (New York, 1914); Keuning, J. Blaeu's Atlas. *Imago Mundi*, XIV, 1959 pp. 74–89.

2 Skelton, pp. 97–109 gives a full description and analysis of the contents of the Scottish plates as does Keuning; Koeman, J. *Atlantes Neerlandici*. Vol. 1 (Amsterdam, 1967). For the most detailed account of all atlases produced by Dutch cartographers the five volumes of Koeman are indispensable. Volume 1 contains full details of all Blaeu atlases with locations on a worldwide basis.

3 Louth RL Map A7 is an example of the county map without text on the verso; Grimsby RL Blaeu 1645 (RM 113) and LRL Map 811 are examples of the Fen map with plain backs.

4 Kingsley, p. 30; Hodson, 1984, p. 32.

5 Burden, p. 17 records three variant states of the Berkshire plate between 1714 and 1715.

6 Hodson, 1984, pp. 34–40. The Atlas Anglois is discussed below under Jansson (**13** (**ii**)).

7 This copy lacks the title-page but in all other respects matches the description in Koeman, 1967, pp. 173–175; the pagination of the text for Lincolnshire and succeeding texts is haywire; that for the county goes from 246–250, followed by 245–246; Nottinghamshire follows with pp. 153–154 then Derbyshire pp. 155–156, followed by 257, 259–262, 167, 264–265, etc.

8 The preface is dated 1 March 1646.

9 Koeman, 1967, p. 183 (BL 47A).

10 Koeman, 1967, p. 178 (BL 43B). The Rotterdam copy has two title-pages; the first is from the 1640 atlas and the second is blank except for the imprint and date MDCXLVI. In these notes the 1640 title is quoted with the 1646 imprint.

11 Koeman, 1967, p. 175 (BL 42B) cites no copies in the UK.

12 Koeman, 1967, pp. 183–185 (BL 47B) records four copies in German universities and one in the Royal Library, Stockholm.

13 The preface is dated 12 November, 1647.

14 Koeman, 1967, p. 185 (BL 47C) – copies in Innsbruck and Berne ULs.

15 This is the fourth volume of a 6-volume set. Volume I has the imprint date MDCXLVIII. Its 'Vorrede' is dated 21 July 1650.

16 At the back of this example is the 'original' Latin title-page dated MDCXLVIII. There is no text in this volume although the index sheets refer to a complete text. A copy of the Arnold Colom map of the British Isles 1654 has been bound in. The six volumes are variously dated; the first four were issued in 1650, the fifth is dated 1654, while the sixth is undated but includes 3 privileges dated 1655 in each case.

17 Koeman, 1967, (BL 48, 50–51, 53–55) refers to these atlases.

18 The privilege for the States of Holland and West Friesland is dated 24 March 1659 and the address to the reader is also dated 1659. The copy in Glasgow and one other copy (privately owned) also have title-pages in Latin dated 1648.

19 The NMM, Manchester and Edinburgh copies have two title-pages; the first is as here transcribed; the second reads: *Anglia, Quae Est Europæ Liber XI. Amsterdami, Apud Iohannem Blaeu. M DC LXII* (NMM), the other two being dated *M DC XLVIII*.

20 The title-page of the Liverpool copy has nothing in the central title panel but the imprint panel is complete.

21 This is an extraordinary edition. The Spanish text leaves were printed single-sided and then stuck on the versos of map leaves, which had already been provided with the French text. In such cases it is possible to view the French text beneath the superimposed Spanish text.

22 These loose sheets could come from the French text editions of 1662, 1663 or 1667 since the text was identical in each issue.

23 The title-page belongs to the 1648 edition, but this volume is part four of a 9-volume set. The first volume has two title-pages, the first with the title: *Grooten Atlas, Oft Werelt-Beschryving . . .*; the second bears the title *Geographia Blaviana*. Neither is dated but the address to the reader is dated 1 Sept., 1664. The volume in the second BL set (Maps C.4.d.2) is a composite work, which consists of French maps and text, a map of Portugal and then the English county maps with the same Dutch text on the versos but without the intervening text pages. The title is *Seste Stuck Der Aerdrycks-Beschryving, Welck verwat Frankryck* and there is no place or date of issue. The sixth (Seste) volume contained maps of China on its first appearance in 1655.

24  This title-page is from a French edition; a second is in the Latin form of 1662. The title-page
    of volume one is dated 1665. The text is in Latin throughout.

Skelton nos. 28–33, 38, 42–45, 64, 71–73, 75, 77 and 84.

# 13

## JAN JANSSON                                                        1646

Size: 39.75cm x 50.3cm.

Jan Jansson (or Johannes Janssonius, the Latin form often found in his work) was born
in 1588 and worked as engraver and publisher until his death in 1664. He seems to have
had the rights of publication of the Mercator-Hondius *Atlas minor* from its first publication in 1607. He married the daughter of Jodocus Hondius the elder and was thus the
brother-in-law of Jodocus, the younger (**6**) and Henricus. With the latter he published
later editions (from 1633 onwards) of Mercator's Atlas.[1] As we have seen, in discussing
the work of Blaeu, Jansson had begun to expand the number of plates relating to the
British Isles while simultaneously enlarging the work from two volumes with each extra
volume being devoted to particular geographical areas.

In 1638 the Mercator Atlas was expanded into a third volume and from about 1640
Jansson began work on a fourth volume to be devoted solely to the British Isles. He
already had sixteen maps depicting broad areas from the first Mercator Atlas of 1595; as
the number of plates gradually increased he also initiated the creation of fresh plates of
individual counties as replacements of the earlier area maps.

With the knowledge that Blaeu was creating a volume of individual county maps
Jansson pressed ahead with his own efforts and a year after the appearance of Blaeu's
volume of plates devoted to the English and Welsh counties Jansson's came out. In terms
of presswork his production was inferior to Blaeu's; errors in numbering occur, particularly in the French language version that appeared simultaneously with the Latin edition
in November, 1646. The engraving of the maps was, however, of the same excellent
quality as Blaeu's and the two series are regarded by many as the most attractive county
plates ever produced.

Many of the features are taken directly from Speed or from Blaeu's Speed-based
plates. There are eleven coats of arms (one blank) in one corner and in the other corners
are the title, a scale and the royal coat of arms; the bottom corners have *putti* and hunters
and fishermen to typify two of the county's chief resources. Like Blaeu he has two
galleons. His own distinctive touch on the plates of maritime counties is the inclusion of
a compass in the sea with radiating lines reaching to the coastline. Like Blaeu (and Speed)
there are no roads and Speed's town plans are not included.

The publication of his atlas followed the same pattern as Blaeu's and finally expanded
into eleven volumes and a similar spread of languages for the text, although the Spanish
text only appeared in two editions of an *Atlas contractus* that did not include the full
range of plates. For that reason that work is not referred to in the following descriptions.
The texts are rarely reset, unlike those of Blaeu.

One area map that appeared originally in the Mercator Atlases of 1633 and continued
to appear throughout the life of the Jansson atlases was *A general Plott and description
of the Fennes* . . . which appears variously dated (1632 or 1636) and with the imprint of
Henricus Hondius; Jansson removed the dates and references to Hondius for his 1646
atlases. Blaeu clearly based his *Regiones Inundatae* on this Hondius/Jansson map.

## Later history
After Jansson's death in 1664 the sheets were used by Dutch and English booksellers in

composite collections for a few years. Of interest to English cartographers are the atlases put together by John Overton, using the county plates which were available to him at the times of compilation. Overton (1640–1713) had as the nucleus of a complete set of county plates the twelve William Smith plates which formed part of an abortive atlas prepared in 1602–1603. He acquired other plates and for those counties for which he had no plates he used sheets from disparate sources to make up full sets from c.1670 onwards. By definition no two sets are the same, since the sources of his maps varied so much. Only atlases which contained maps of Lincolnshire in the Jansson version are noted here.[2]

On behalf of Jansson's heirs his son-in-law, Joannes van Waesberghe, carried on the business. Apart from an *Atlas Contractus* (1666) which only included 12 British maps he produced no other atlas. He died in 1681 and it was after two auctions in 1694 that the plates were acquired by Pieter Schenk (1660–1719) and his partner Gerard Valck (1651/2–1726). They added their own imprint to the plates, which they seem to have sold as separate sheets but never as a complete atlas under their own name. The plates do appear in the works of others.[3] Joseph Smith had use of a complete set of plates by Jansson in 1724 and Hodson suggests that it is reasonable to assume that David Mortier had had access to them in his *Atlas Anglois*.[4]

## EDITIONS

(i)  *LINCOLNIA/COMITATVS/Anglis/LYNCO LNE SHIRE./* (Ae). 11 shields, with one blank (Aa–Ac). Royal coat of arms (Ea). [Scale] *8/Milliaria Anglica/* [= 5.2cm] (Ee). The cardinal points are named in Latin (Ac, Ca, Ec and Ce). Compass with radiating lines to the inner frame line or to the Lincolnshire coast. (Db).

*Ioannis Ianssonii Novus Atlas, Sive Theatrum Orbis Terrarum: In quo Magna Britannia, seu Angliæ & Scotiæ nec non Hiberniæ, Regna exhibentur. Tomus Quartus. Amstelodami, Apud Ioannem Ianssonium. Anno M DC XLVI.* (Latin text pp. 223 (signature Dddd) & 224 on verso).

**Copies**   BL Maps C.6.b.1; CUL Atlas 3.65.4. (LRL Map 790 and Grimsby RL Jansson 1646 (RM 1); Grimsby RL Jansson 1646 (RM 37) is a copy of the Fen map).

*Le Nouvel Atlas ou Theatre du Monde, Auquel est representée la Grande Bretagne Contenant les Royaumes D'Angleterre, d'Ecosse & d'Irlande. Par Jean Jansson. Tome Quatriesme. Amstelodami . . . Anno M DC XLVI.*

**Copies**   No copy is recorded in the UK.[5] (CUL Maps bb.36.64.3 is a copy of the Fen map).

*Le Nouvel Atlas . . . M DC XLVII.* (French text pp. 243–244 on the verso; signature Llll on p. 243).

**Copies** BL Maps C.6.b.2;[6] Leeds UL W 446. (LAO Dixon 21/4/1/1).

*Nieuwen Atlas, ofte Werelt-Beschrijvinghe, Vertonende Groot Britannien, Verwattende De Koninghrijcken van Engelandt, Schotlandt ende Yrlandt. Het Vierde Deel. Amstelodami . . . Anno M DC XLVII.* (Dutch language text on the verso, the pages are unnumbered; signature Bbb 2 on the first leaf of the county text; Fraktur type is used with roman for county place-names).

**Copy**   Provincial Library of Friesland, Leeuwarden – 125 A.

*Novus Atlas, Oder Welt-Beschreibung, in welcher ausführlichen abgebildet die Königreiche Engelland, Schotland, Und Irland. Das Vierdte Theil. Durch Johan Janssen. Amstelodami . . . M DC XLVII.* (German text pp. 285–286 on verso; signature Bbbb on p. 285; the text is set in Fraktur with roman for place-names).

**Copy**   Marburg UL.

*Novus Atlas, Das ist: Welt-Beschreibung mit schönen neuen Geographischen Figuren. Inhaltende Gross-Britannien und Irland. Vierter Theil. Amstelodami . . . 1649.* (German text on verso).

**Copies** BL 9 TAB 4. (Boston RL Map 28).

*Nieuwen Atlas . . . M DC LII.*

**Copy** Royal UL Groningen.[7]

*Nieuwen Atlas . . . M DC LII.*

**Copy** Admiralty Library. Vc 10.

*Nieuwen Atlas . . . M DC LIII.*

**Copy** No copy is recorded in UK.[8]

*Le Nouvel Atlas . . . M DC LVI.*

**Copy** Birmingham RL AE 094/1656/16D.

*Ioannis Ianssonii Novus Atlas . . . M DC LII.* [1656].[9]

**Copies** CUL Atlas 3.65.10. (LAO Scorer Maps 2).

*Ioannis Ianssonii Novus Atlas . . . M DC LIX.* (Latin text in a close resetting; distinguished from the 1646 edition by the catchword on p. 223; now it is *finum* while it was *cymbu-* in earlier editions).

**Copies** BL Maps C.6.b.3; CUL Huntingdon 8.7; Bod Gough Gen. Top. 224; RGS 1.C.35; Liverpool UL H49 44. (CUL Maps bb.36.64.2 is the Fen map with Latin text from this edition).

*Novus Atlas Absolutissimus. Das ist generale Welt-Beschreibung mit aller=ley schönen und neuen Landskarten geziret. Das Siebende Theil begreiffet Engelands erstes Stuck. Amstelodami, Apud Ioannem Ianssonium. MDCXLVII.* [1659].[10]

**Copies** NMM 912.44:094 [4692]. (Manchester UL Map C 271.2 is the Fen map).

*Novus Atlas, Oder Welt-Beschreibung in welcher ausführlich abgebildet die Königreiche Engeland, Schotland, Und Irland. Das Funffte Theil. Amstelodami, Apud Ioannem Ianssonium. Anno M DC XLVII.* [1659].[11]

**Copies** BL 9 TAB 21; NMM 912.44:094. [E4693].

*Guilielmi Cambdeni Britannia Magna Illustrata. Amstelodami, Apud Ioannem Ianssonium.* [1659].

**Copy** The private collection of Mr. C.A. Burden.[12]

*Nieuwen Atlas . . . M DC LIX.*

**Copy** Amsterdam UL I–5–4–7.[13]

Atlas published by John Overton. [c.1675].

**Copy** Formerly in the collection of Dr. Eric Gardner.[14]

*Atlas Major, Ex Novissimis, Selectissimisque, A Quoris Auctore Editis, Cum Generalibus Omnium Totius Orbis Terrarum Regnorum. Rerum-publicarum Et Insularum, Tum Particularibus in hisce sitarum Regionum Tabulis Geographicis, Juxta perfectissimam Itinerariam Seriem in hoc opere positis, consistens. Ad numerum 521 Tabularum; in Tres Tomos Divisus. Tomus I. Ex collectione Caroli Allard, Amstelo-Batavi. Amstelodami, Prostat apud Caroli Allard, Cum Privilegio Ordinum Hollandiæ & Westfrisiæ.* [1705].[15]

**Copy** BL Maps C.6.c.4.

**(ii)** *Amstelodami apud/P. SCHENK. et G Valk./Cum Privit* (sic)/ has been added (Ea).

The inner border has been shaded at 2′ intervals; figures for latitude/longitude appear at 10′ intervals between the inner and outer frames.

*Atlas Anglois, Ou Description Generale De L'Angleterre, Contenant Les cartes Geographiques De chaque Province, avec les Genealogies des plus Illustres Familles, & les Archevêchés & Evêchés. A Londres, Chez David Mortier, Libraire. M.DCC.XV.*

**Copies**   (LRL Map 89; Grimsby RL Map RM 140).[16]

*Nova Totius Geographica Telluris Projecto. Amsterdam, Gerard Valk.* [1717].[17]

*Atlas Anglois . . . A Londres, Chez Joseph Smith, Marchand Libraire à L'Enseigne d'Inigo Jones proche Exeter=Exchange dans le Strand. MDCCXXIV.* [1728].[18]

**Copy**   BL G 2715.

**(iii)**   Plate number *22* has been added (Ee, between the inner and outer frame-lines.

*Atlas Anglois . . . Joseph Smith . . . MDCCXXIV.* [1728].

**Copy**   Leeds UL Brotherton Collection.

### References

1   Skelton, pp. 81–83.
2   For a full description and analysis of the contents of Overton's atlases see Skelton, pp. 129–138; 145–147; 167–168; 176; 200–202; 216; and p. 245 for a biographical note and further references.
3   Skelton, p. 113–114 and 189.
4   ibid., p.83; Hodson, 1984, pp. 36–40.
5   Koeman, C. Atlantes Neerlandici. Vol. 2. Amsterdam, 1970. (Me 159) records copies in the BL and the Bibliothéque Nationale; the BL copy is actually 1647 (q.v.).
6   The BL copy has the Latin title-page of 1646; the text is, however, French.
7   A copy of this atlas was sold at Sotheby's (lot 181) on 8 May, 1986.
8   Skelton, p. 56 is the source of a record not listed by Koeman.
9   The CUL volume has English county maps without text; bound in after them is a title-page: *Escosse, Qui Est Le XII. Livre De L'Europe. Amstelodami, Apud Ioannem Blaeu. MDCLXII*, followed by maps of Scottish counties and a set of Irish maps both with French text. A final (index) leaf refers to Scottish and Irish plates. Some maps are by Blaeu. The other volumes in this set are dated 1656, hence the placing in these notes; the date on the Blaeu Scotland title-page must be ignored. Skelton's entry (no. 52) quotes the presumed French title only and places it under 1652. The LAO copy has a blank verso also, but it may have been separately issued thus.
10   This atlas is ascribed to 1659 because by then Jansson had expanded his atlas series to its final length of eleven volumes of which England is the seventh and Scotland and Ireland together form the eighth. Volume 1 of the NMM set is dated 1658.
11   The BL copy has the 1647 German title-page but the contents are otherwise identical. The NMM copy is the fifth of a six-volume set; the first volume is dated 1658. Both sets have the 6 new maps prepared for a new edition in 1659 – Skelton, p. 95.
12   This is the 1659 Latin edition with a label stuck on the title-page with the quoted title. See Koeman, 1970, item Me 152.
13   The Amsterdam copy has a Blaeu title-page dated 1659, although the contents are by Jansson.
14   This is the second of Overton's atlases (Skelton, pp. 145–147) and only contains 22 plates, some of which are new plates of counties for which he had no source. Most of the later John Overton atlases had Lincolnshire in the Blaeu version, but they are indistinguishable from 'normally' issued Blaeu's.
15   Hodson, 1984, pp. 32–33 suggests 1705 since that is the date on the plate of *Regnorum Hispaniæ et Portugalliæ . . .* (by F. de Witt) and Allard died in c.1706; his stock was sold in 1708. He cites Koeman, C. 1967 p. 31 for a note on the advertisement of the sale of copies of the atlas on 24 Feb. 1706/7. The atlas is made up, like Overton's, from sheets bought in – a

so-called *atlas factice* – and such works are normally very unstable. In this case the only other copy (Leiden UL) is identical to the BL example.

16   One copy is noted by Hodson, 1984, pp. 38–39; it is in the Library of Congress (G1808.N8 1715 Vault). The Lincolnshire plate is illustrated in Beresiner, Y. British County Maps. (Antique Collectors Club, 1983), p. 121. An example of the work, but dated 1716, was offered by Dulau & Co. in their catalogue of 1813 for 2 guineas (copy of the catalogue in Bod. 2593 e 58).

17   A set in six volumes was sold by Sotheby's (lot 199, 13 April, 1989); of the maps by Schenk and Valck most had dates from 1703 to 1717.

18   The county map is in an intermediate state; the Leicestershire plate had 21 engraved in the lower corner and, similarly, Nottinghamshire had 23; Lincolnshire, bound between these two, remained without a number – see state (iii). All the plates in this volume have marks of their former owners, Schenk and Valck (variously spelt); besides the map of Lincolnshire 17 others have no plate numbers. Sotheby's sold an example of the set, dated 1714–1716 (Lot 175, of 13 April, 1989), in which the plates had all been numbered by hand, following the pattern later carried through by the engraver. The BL set and the set in Leeds (state iii) were issued after 1724. The atlas formed the second (and last) part of the third (and last) book of *Nouveau Theatre*. The first part of Book 3 has a separate title-page dated 1728. Another example of the 6-volume work was sold by Christie's (lot 65, 13 April, 1988), but it is not known whether the county maps all had plate numbers.

Skelton nos. 34, 35, 39–41, 46–47, 52–54, 56, 61, 63, 66–68.
Hodson nos. 130–134.


## 14

### RICHARD BLOME                                            1673

Size: 31.95cm x 26..6cm.

Blome (*fl.*1660–1705; d.1705) was described as 'a publisher and compiler of some celebrity, who, by the aid of subscriptions adroitly levied, issued many splendid works'.[1] He produced much geographical and heraldic work, the latter stemming, no doubt, from his being originally an heraldic painter. He was perhaps a descendant of Jacob Blome who had the copyright of Guillim's *Display of Heraldry*, the fifth edition of which Richard Blome published in 1679.[2] He made no claims to originality for the contents of his works but Gough, in 1780, described his *Britannia* as 'a most notorious piece of plagiarism'.

Blome's *Britannia* was intended to be the third volume of an *English Atlas*; Blome had entered the title at the Stationers' Company in 1668. The first volume was to be a translation of Varenius with 100 maps; the second a world atlas and the third a *Description of England*. The second volume appeared as *A Geographical Description of the World* in 1670. The Varenius did not appear until 1682 and a volume four (only announced in the preface to *A Geographical Description . . .*) was abandoned. The publication history of the third volume, which eventually became the work considered here, is complicated. A prospectus and subscription form issued in 1670[3] carried a promise of readiness in 1671. In February of that year it was supposed to be ready but a second prospectus[4] promised it by Michaelmas Term of that year. The *Term catalogue* of 24 November 1673 advertised the work at 30s. and a further advertisement, which appeared on 9 Feb. 1674, must have appeared after publication of *Britannia*.

The large number of copies that have survived shows that it found a ready market in spite of its drawbacks. The author of the text is unknown but was probably a hack employed to take what was required from Camden. Only four of the map plates are signed (by 3 different engravers) but they are all of the poor quality typical of contemporary journeyman work. Some are dated and the series probably occupied the years 1669–1673. The maps are based on Speed but lacking the distinguishing features. Skelton draws attention to the map of Shropshire being dedicated to Francis, Lord Newport, a title only created in 1675.[5]

Two sets of map plates were being engraved simultaneously for Blome, judging by the dates on some of the plates issued in the new work of 1681 (**17**). No explanation of the duplication has yet been convincingly put forward. It has been suggested that Blome had hoped to produce a miniature book, analogous to the 'miniature Speeds' but for some unknown reason had to abandon the project.[6]

Thomas Roycroft (*fl.*1651–1677; d.1677), the printer, was noted for the high quality of his work by seventeenth standards. He produced a polyglot Bible and fine editions of the classics, edited by John Ogilby. In 1660 Charles II appointed him the King's printer in oriental languages, but he lost his printing shop and stocks in the Great Fire.[7]

### Later history

There were only two issues of the text and maps with the second made up by John Wright from remaindered sheets. It is probable that the first edition was on sale for the four years before the re-issue of 1677 and the copies were being extended in the interim. Skelton has drawn attention to the variations between the number of coats of arms, which are left blank on leaf [12]. The last numbered space on leaf [12] is 812 for 'Benefactors & Promoters of this worke . . .', but the numbering stopped variously at 806, 807, 808, 811 and 812 with any remaining spaces left blank.[8] He assumed that these 'incomplete' copies were earlier examples. It is now, of course, impossible to know how many shields were completed on the day of publication. The finding of a copy containing a leaf [13] with coats of arms filled in up to number 826 must mean it is a later example and that some or all of those above may be post-1673 copies also.[9]

### EDITION

(i)  *A MAPP/OF  Y$^E$  COUNTY  OF/LINCOLNE./W$^{th}$  its  Diuisions/&  Hundreds: or/Wapontacks: by/Ric: Blome./* (Aa). Dedication (with coat of arms) *To y$^e$ Hon$^{bl}$ Rob Earle of Lindsey Baron Willoughby of/Eresby Leiutenant* (sic) *of this County. Lord Great/Chamberlane of England & one of y$^e$ Lords of his/Matys most Hon$^{ble}$ privy Councell etc/This Mapp is Humbly/Dedicated by Ric/Blome* (Ea). *A TABLE OF Y$^E$ DIVISION$^S$/ &* (drawn reversed) *WAPONTACKS OR HUNDREDS/* is followed by a two column table (Ae). *A Scale of 10 Miles* [=5.2cm] (Ee). Compass (Eb).

*Britannia; or, A Geographical Description of the Kingdoms of England, Scotland, and Ireland, with the Isles and Territories thereto belonging. And for the better perfecting the said Work, there is added an Alphabetical Table of the Names, Titles and Seats of the Nobility and Gentry that each County of England and Wales is, or lately was, enobled with. Illustrated with a Map of each County of England, besides several General ones. The like never before Published. London, Printed by Tho. Roycroft for the Undertaker, Richard Blome. MDCLXXIII.*

**Copies**   BL 577 k 2, etc.; CUL Atlas 5.67.1 and Peterborough T.5.1; Bod Allen 9 and Fol.Δ.777; Leeds UL W 35; Leicester UL H942 F (Hatton 17); John Rylands UL Q 942.006 B 621 (R16072). (LRL Map 582a; Grimsby RL BLOME 1673 – RM106 and RM 40).

An atlas without title-page and without text.

**Copy**   BL Maps C.24.g.6.

*Britannia; or, A Geographical Description of the Kingdoms of England, Scotland, and*

*Ireland, with the Isles and Territories thereunto belonging . . . London, Printed for John Wright at the Crown on Ludgate-Hill. MDCLXXVII.*

**Copy**   The only copy recorded was owned (in 1970) by Sir Gyles Isham.[10]

**References**

1  DNB. Vol. V. (London, 1885), pp. 225–226. Very uncomplimentary remarks about Blome's publishing methods are also quoted.
2  Plomer, 1922, p. 39.
3  Example in BL Maps 187.1.1 (16), reproduced in Skelton, p. 141.
4  Bod Wood 658, f.815. Blome's methods are discussed in: Clapp, S.L.C. The subscription enterprises of John Ogilby and Richard Blome. *Modern Philology,* Vol. 30. 1933, pp. 365–379.
5  Skelton, p. 144.
6  ibid., pp. 140 and 158–159.
7  Plomer, 1907, p. 158.
8  Skelton, p. 139.
9  Mr. C.A. Burden purchased this example in November, 1992. I am grateful to him for showing me his copy.
10  Skelton, p.156.

Skelton nos. 90 and 100

## 15

**ROBERT MORDEN**                                                                **1676**

Size: 9.15cm. x 5.6cm overall; the map portion being 5.75cm x 5.6cm.

This is the first map on a card that is recognisably a playing card. The title and suit mark appear in a small panel at the top, with the map forming the larger central part above another panel containing five lines of geographical data.

Robert Morden (*fl.*1650–1703; d.1703) first came to notice as a globe-maker in 1668[1] but by 1671 he was advertising his own printed products.[2] Throughout a long working life he produced a stream of geographical works but he never seems to have made much of a fortune from his various enterprises. His reputation as a mapmaker does not stand high but that has not harmed the collectability of his larger county map series. The sets of maps he prepared for the new edition of Camden's *Britannia* are discussed below (**19** and **20**).

He first advertised these cards in 1676[3] as 'The 52 Countries (sic) of England and Wales, described in a Pack of Cards . . .' and they were to be sold by Morden and three other booksellers – Will Berry, Robert Green and George Minikin. Over the next few years the cards appeared in a number of formats. The name of the engraver is not known. The chief claim to particular notice is that the county card is the first Lincolnshire map to show roads. Ogilby's *Britannia* had appeared in 1675 and Morden was sufficiently enterprising to make first use of the new information that was now widely available; there were, of course, no copyright laws. The roads that are shown are not all taken from Ogilby or, at least, not accurately. A double line connects *Stanford* to Newark while a single fine line joins Deeping to Barton (via *Shelford* (=Sleaford) and *Fakingham)* and another from Deeping misses out Spalding by proceeding by way of *Dunington* and *Boston* to *Wenfleet.* One more road runs from Lincoln to *Marketrasen* with branches to Grimsby and Tattershall. The latter is not in Ogilby.

The latitude for Lincoln is given as 53 and this seems to be taken from Ogilby's general map of Britain. On the modern Ordnance Survey map Lincoln Cathedral is 53° 14′.

## EDITIONS

(i)   8 Lincolne Sh: VIII (in upper panel with the Club mark stencilled over the first letter of the county name). Compass (Ea, on the map portion). *10/Miles* [= 1cm] (Ae, on the map). *Lenght.* (sic) *60./Bredth. 99./ Circumference. 170./Lincolne { D. from Lon:* C [computed miles] *103 128./Lattitude 53°./*(all this text is in the lower panel).

*The 52 Counties of England and Wales, Geographically described in a pack of Cards, Whereunto is added yᵉ Length. Breadth. & Circuit. of each County the Latitude the Scituation and distance from London of yᵉ principle Cities. Towns. and Rivers. with other Remarks as plaine and ready for the playing of our English games as any of yᵉ Common Cards.* [1676].[4]

**Copies**   BM Dept. of Prints and Drawings. Schreiber E 45[5]; Leeds UL W 37.

(ii)   Without suit marks.

Issued as a bound atlas. [c.1676].

**Copies**   BL Maps C.7.a.33; Leeds UL W 38.

(iii)   With suit mark and the names of adjoining counties added in the form *Pᵗ/of . . .* in four cases or *P./Rut/* in a fifth. Instead of Leicestershire the abbreviation *Lin.* (sic) appears.

*The 52 Counties of England and Wales . . .* [1676].[6]

**Copy**   BM Dept. of Prints & Drawings. Schreiber E46 (1896.5.1.926).

(iv)   The road from Deeping to Barton now consists of a double line; two roads have been added from Lincoln: to *Thorncaster* (=Caistor) and also to Saltfleet via Horncastle and Burwell. The circle symbols have been replaced by tower-like symbols for all except five towns. *Spilsby, Holbeach* and *Burgh* have been added with new symbols. *Rasen* has been added on the road to *Thorncaster.*

Issued as a pack of cards. [c.1680].

**Copy**   The only copy was in the collection of Dr. Eric Gardner.[7]

(v)   As (iv) but in book form with two cards mounted to a page and trimmed so the suit marks have been lost.

*A Pocket Book Of All The Counties Of England and Wales: Wherein are describ'd, the Cheif* (sic) *Cities, Market-Towns and others; With the Rivers and Roads From London. To which is added, a Compass, shewing the Bearing and a Scale for the Distance of Places. There is also given the Length, Breadth and Circumference of each County: The Latitude of each City or Town, and its Distance from London. Being a necessary and plain Direction for Travelling to any Place or Town in all England or Wales. Sold by Robert Morden at the Atlas in Cornhil, and Joseph Pask Stationer, at the three Ink-bottles in Castle-ally, under the west-end of the Royal Exchange.* [c.1680].[8]

**Copy**   BL Maps C.24.aa.11.

(vi)   As (iv) but in book form, with one map to the page and text on the county. In Lincolnshire's case the map faces a text beginning on p. 53 and ending on p. 56. There is no suit mark.

*A Brief Description Of England and Wales; Containing A particular Account of each County; With its Antiquities, Curiosities . . . Embellished with Maps of each County. very useful for Travellers and others, and very proper for Schools, to give Youth an idea of Geography, and the Nature of his own Country, and each County. London: Printed for H. Turpin, No. 104, St. John's Street, West Smithfield.* [1773].[9]

**Copies** BL Maps C.27.b.34; Leeds UL W 64.

### Reproductions
(i) Playing Cards Of Various Ages And Countries Selected From The Collection Of Lady Charlotte Schreiber. Vol. 1 England And Scotland . . . (London, 1892). (ii) A set of the first issue of the cards, belonging to Mr. David Kingsley, was re-issued in facsimile in 1972 by Harry Margary, Lympne Castle, Kent. The suit marks have been printed in black and red. (iii) The Lincolnshire card is shown in Mann, S. and Kingsley, D. op. cit., Plate XI (c). (iv) The card for Lincolnshire from the facsimile set has been reproduced in *Lincolnshire Life* Vol. 13, no. 3 May, 1973, p. 26, with a note on p. 24; and, Vol. 14, no. 8 Oct., 1974, p.41; the map portion only is used on the covers of *Lincolnshire History and Archaeology* [the Journal of Society for Lincolnshire History and Archaeology].

### References
1   DNB. Vol. XXXVIII (London, 1894), pp. 410–411; Skelton, p.247 says he flourished from 1650 but Tyacke, p. 123 says that the earliest recorded book with his name was published in 1669.
2   Plomer, 1922, p. 210.
3   *Term catalogue.* 5 May, 1676.
4   The title is taken from the first of two preliminary cards. The date is assumed from the advertisement in *Term catalogue.* Wadsworth, F.A County Maps as Playing Cards. *Transactions of the Thoroton Society of Nottinghamshire.* Vol. XLV (1941) pp. 18–23 discusses the topic generally with the map illustrations naturally of Nottinghamshire cards.
5   The Schreiber collection was formed by Lady Charlotte Bertie of Uffington House, near Stamford. She became Lady Charlotte Guest on marrying John Guest and gained fame for a translation of *Mabinogion,* which is still in print. In 1855, three years after Guest's death she married Charles Schreiber. She formed several collections, some of which she gave to the Victoria and Albert Museum and several, including the playing cards, went to the British Museum.
6   The only set lacks the two preliminary cards and the cards with the maps of Cambridgeshire and Worcestershire. This set was advertised in *Term catalogue,* Michaelmas, 1676 as 'The Second Edition, whereunto is added the adjacent Counties in each card, with other amendments. Sold by Robert Morden . . . W. Berry . . . Robert Green . . . and G. Minikin.'
7   Skelton, p. 157.
8   Skelton, p. 157 suggests 1680 but cites no evidence.
9   Turpin lived at this address from 1764–1787. His catalogues of 1770 and 1772 do not list this item and it is assumed, therefore, that it appeared after 1772.

Skelton nos. 94, 94a, 95, 102 and 103.

# 16

## WILLIAM REDMAYNE                                                          1676

Size: 9cm x 5.3cm overall.

William Redmayne was a printer active from 1674 until 1719 when he was imprisoned for printing a libel on the government. He died in Newgate of fever in April of that year.[1] He advertised this set of cards in the *Term catalogue* (Trinity, 1676) as 'Recreative pastime by Card-play; Geographical, Chronological, and Historiographical, of England and Wales; shewing the commodities and rarities of each County. Very useful for all Travellers . . . By a long Student in the Mathematicks. Sold by W. Redmayne at the Crown on Addle Hill; Henry Mortlock at the Phoenix, Robert Turner at the Star, in St. Paul's Church-yard; H. Cox in Holborn; and B. Billingsley at the Printing Press in Cornhill'.

Of Redmayne's associates only Mortlock achieved any distinction; a prolific publisher of theological works he became Master of the Company of Stationers. His relative, John Mortlock, had bookshops in Nottingham and Newark and, for him, Henry Mortlock published a book by John Twells, Master of the Free School in Newark.[2]

The cards are filled with printed information about the county with a thumb-nail map in the centre. Superimposed on the map are the suit mark and most of what detail there is on the map is consequently covered. Off the coast are two galleons and a sea mammal. There are neither scale nor compass and Lincoln is the only town marked.

### Later history
After the two editions that Redmayne issued the whereabouts of the plates is unknown until they were acquired by John Lenthall in about 1716. However, the cards made an appearance again in 1711 and it is possible that they were issued by Lenthall.[3] Lenthall was first the apprentice, then son-in-law and, finally, successor to the stationer William Warter, having become his partner in 1708. Warter retired in 1709. Between them they developed a considerable business in issuing playing cards of all sorts. Lenthall's business seems to have declined when rival firms started issuing card sets and he appears to have retired in c.1734. His cards were still being sold by other retailers until the mid-1750s.[4]

### EDITIONS

(i)  *LINCOLN-SHIRE/Is all most 60 Miles long/and in some places above 30./miles broad, on the East is/the German Ocean, on the/ North Humber, on the West/Notingham-shire.* [Space for the map, with the Heart symbol followed by *II*]. *on the South parted from/Northampton shire. with y^e/River Welland/ It abounds with fruit and/Cattel, the Shire Town is/ Lincoln, its seated on the/side of a Hill, in it are 630/ Parishes, and many Rivers./* The suit mark has been drawn in outline and filled in solid by stencil.

A pack of playing cards. [1676].[5]

**Copy**   Leeds UL W 39.[6]

(ii) Without stencilled suit marks.

A pack of playing cards. [1677?].[7]

No complete set is recorded.

(iii) With suit mark engraved with vertical bars (to denote a red suit).

A pack of playing cards. [1712?].[8]

**Copy**   The private collection of Mr. C.A. Burden.

(iv)   The border has been increased from a fine line to one 2mm. thick along the top and down the right hand side and 1mm. along the other two.

A pack of cards issued after 1712 and before 1717.

**Copy**   Bod Playing cards. Douce English Geographical.[9]

(v)   A decorated border has replaced that of state (iv). The suit marks have not been stencilled over.

A pack of playing cards, issued by John Lenthall. [1717].[10]

**Copy**   Guildhall Library Phillips Collection 224 and 225.

### References
1    Plomer, 1922, p. 250, cites Nichols, J. Literary anecdotes of the eighteenth century. (London, 1817–1858). Vol. VIII, p. 368; Mann, S. and Kingsley, D. op. cit., pp. 18–20.

2  Plomer, 1907, p. 132; Plomer, 1922, p. 211.
3  Tyacke, p. 99 records an advertisement in *London Gazette* (19–21 Dec., 1710) for a set of cards; however, Kingsley, p. 42 records a set issued by Redmayne (held by Stanley Gibbons group), which he dates to 1712.
4  Hodson, 1984, pp. 71–74; Mann, S. and Kingsley, D. p. 27 *et seq*; Wayland, V. and Wayland, J.H. Lenthall pack no. XIII Historiographical cards. *Journal of the Playing-Card Society,* vol. V, no.2. (Nov. 1976), pp. 14–23.
5  The date of the advertisement referred to above.
6  Hodson, 1984, p. 71 corrects Skelton's entries 96 and 97. The Leeds set of 52 cards consists of 50 from one setting and 2 from the 1677 setting.
7  Only 11 cards (not Lincolnshire) have survived from the edition advertised in *Term catalogue,* Michaelmas, 1677, as 'Geographical, Chronological, and Historiographical, Cards of England and Wales. Sold formerly for 1s.; now newly done, and sold at 6d. a pack. By W. Redmayne at the Crown on Addle Hill'. Kingsley, p. 41 suggests that the reference to 'newly done' need not imply a new edition.
8  The 10 of Spades has a tax stamp in red. Tax on cards came in from 11 June, 1711 (10 Anne c.19). Mann, S. and Kingsley, p. 19 discuss the various types of mark that were used and the periods of their application. From 1714 the mark was usually placed on the ace of spades but this did not become a legal requirement until 1722.
9  Probably issued at the same time as state (iii) above. The Bodleian version only has twenty cards.
10 Advertised in *St. James Evening Post* 19–22 Jan. 1716/7 as: '... there is publish'd ... County Cards, wherein is exactly describ'd the Situation, Climate, Soil, Customs, Manners, and Commodities of every County in England with ... exact Maps of each Place curiously delineated ... Sold by John Lenthall, Stationer, at the Talbot against St. Dunstan's-Church, Fleetstreet, London'.

Skelton nos. 96 and 97; Hodson no. 146.


## 17

**RICHARD BLOME**                                                    **1681**

Size: 24.45cm x 19.55cm.

Blome's work has already been discussed (**14**). It is not known for what purpose he had two sets of plates engraved, but, it is clear from the dates on the plates that they were being prepared simultaneously.[1] The Lincolnshire plate is dated 1671 and reveals particularly Blome's device for raising extra subscriptions; the plate was re-engraved twice as the dedication was changed from, initially, Sir Robert Carr of Sleaford to Sir Edmund Turnor of Stoke Rochford and, finally, to Thomas Drax of Sibsey.[2] The death of Sir Robert Carr in 1682 (only one year after the map appeared in book form) would allow an opportunist like Blome to offer the dedication of his plate in return for a subscription. Blome gave himself the freedom to offer the dedication around each time he produced a new work containing these plates; seventeen of the 38 county plates were altered in this manner in 1685 and 13 in Blome's final publication in 1693. Seven other counties have three different dedications besides Lincolnshire.[3]

The plates are crudely engraved but the name of the engraver is not known to us. They are based, somewhat loosely, on those of Speed. The omission of Spalding on the map is one example of Blome's laxity. Below the plate is a list of the thirty wapentakes, probably taken from Speed's accompanying text. There is evidence that the original plates bore Blome's name in the area occupied by the title but it was deleted before the first book issue, which was, of course, ten years after the Lincolnshire plate was engraved.[4]

**Later history**
After the two titles published by Blome the plates remained unused and their location
unrecorded. Blome died in 1705. In 1715 they had become the possession of Thomas
Taylor who used them for a decade or so. Probably through the successor in the premises
used by Taylor, Timothy Jordan, they ended up in the possession of Jordan's partner,
Thomas Bakewell by about 1731. By 1746 Bakewell had moved his premises and by
1749 the business was in the hands of his widow. No more issues of the plates were made
although copies of the atlas may have been sold by his widow for some time afterwards.[5]

## EDITIONS

(i)   *A Mapp of/the County of/LYNCOLNE./with its Divisions/& Hundreds/* (Ea, in a
horseshoe shaped wreath). *A scale of 10 Miles* [= 4.3cm] (Ae). *To the R*[t]*. Honourable
S*[r]*/Robert Carr of Sleaford K*[t] *and/Bar*[t] *Chancellor of the Dutchy and/County Palatine of
Lancaster and/one of his Matys most Honourable/Privy Councell & ca./The Mapp is
D.D. by RB./* (Ae, above the scale and surmounted by Carr's coat of arms). A single rule
is drawn across the bottom of the map – there is no frame – and below the rule is the
heading: *A Table of the Divisions & Wapontaks or Hundreds in Lyncolne Shire London
Printed for Ric. Blome 1671/* Below the heading are the numbered names of 30 divisions
arranged in six short columns. Compass (Ac).

*Speed's Maps Epitomiz'd: Or The Maps Of The Counties Of England, Alphabetically
placed. London, Printed Ann. Dom. 1681.*[6]

**Copy**   BL 118 b 20.

(ii)   The dedication has been altered to read: *To the R*[t] *Worshipfull S*[r]*/Edmund Turnor of
Stoke/Pochford* (sic) *in Lincoln shire/This Mapp is humbly DD/by RB/* (Ae, above the
scale and with the coat of arms redrawn).

*Speed's Maps Epitomiz'd . . . London, Printed and Sold by Sam. Lownes over against
Exeter Exchange in the Strand. 1685.*

**Copy**   Leeds UL W 40.

(iii)   The dedication now reads: *To the Worshipfull Thomas/Drax of Sibsy Esq./This
Mapp is humbly/dedicated by Rich: Blome.* (Ae, above the scale, with the coat of arms
redrawn).

*Cosmography And Geography, In Two Parts: The First, Containing the General and
Absolute Part . . . Being A Translation From . . . Uarenius . . . The Second Part, Being a
Geographical Description of all the World, Taken from . . . Monsieur Sanson . . . The
Third Impression, Illustrated with Maps. To which is added the County-Maps of England,
drawn from those of Speed. London: Printed by Samuel Roycroft, for Richard Blome,
dwelling near Clare Market in New Weldstreet, at the Green Pallisado-Pales, M DC
XCIII.*

**Copies**   Leeds UL W 41. (Boston RL Map 20; Gainsborough RL Map 1; Grantham RL;
Manchester UL Map C272a).

*Cosmography And Geography . . . M DC XCIII.* [1694?].[7]

**Copies**   CUL Adams 3.69.2; NLS BCL A 4404–5.

Issued as an atlas, without title-page or text.

**Copy**   The private collection of Mr. C.A. Burden.

*England Exactly Described Or a Guide to Travellers In a Compleat Sett of Mapps of all
the County's of England being a map; for each County where every Town & Village in*

*each County is Particulerly Expressed with the Names and Limits of every Hundred &c.
very Usefull for all Gentlemen & Travellers being made fit for the Pockett; Printed
Coloured and Sold by Tho: Taylor at y$^e$ Golden Lyon in Fleetstreet where are Sold all
Sorts of Mapps and fine French Dutch and Italian Prints* [1715].[8]

**Copies**  BL Maps C.25.a.21; CUL Atlas 7.71.3.

(**iv**)  Plate number *23* added (Ea, to the right of the title).

*England Exactly Described . . .* [1715].

**Copies**  BL Maps C.24.a.21, etc.[9]; CUL Atlas 7.71.5; Bod Gough Maps 115; RGS 10
B 1; Leeds UL W 47.

(**v**)  *Spalding* has been added to the map, with the symbol for a market town. Double
lines to indicate roads have been added and mileages have been inserted within circles
along the roads.

*England Exactly Described . . . Counties of England; being a Map for each County,
Wherein every Towne and Village is Particularly Express'd with the Names and Limits
of every Hundred, and the Roads and Distances in Measured Miles according to M$^r$
Ogilby's Survey. Very Usefull . . . being made fit for the Pockett . . .* [1717].[10]

**Copies**  BL Maps C.27.b.6; Bod Allen 13; RGS Fordham 144; Leeds UL W 48a;
Guildhall L. Bay H.24 no.20. (LAO Scorer Map 3; LRL Map 101).

*England Exactly Described . . .* [1718].[11]

**Copies**  BL Maps C.27.b.28; Admiralty L. Vg 8; Hull RL 912.42 B.

*England Exactly described . . . In a Compleat Sett of Most correct Mapps of Counties in
England . . . Sold by Thomas Bakewell Next y$^e$ Horn Tavern in Fleetstreet . . .* [1731].[12]

**Copies**  BL Maps C.24.b.29 and C.18.a.4; CUL Atlas 7.71.4; Bod Allen 20; Leeds UL
W 48b.

### References
1   Skelton, pp. 158–159.
2   Holmes, C. Seventeenth Century Lincolnshire. History of Lincolnshire, Vol. VII. (Lincoln,
    1980) pp. 239–246 for a discussion of the life and political activities of Robert Carr.
3   Harvey and Thorpe, pp. 22–26 discuss Beighton's use of this method to secure patronage. His
    large-scale map of Warwickshire provided spaces for 304 coats of arms round its edge.
    Beighton probably had the idea from Joseph Browne's Staffordshire map (1682). Senex's map
    of Surrey was being subscribed at the same time as Beighton's (1729) but he only found takers
    for 43 out of 152 shields on offer.
4   In the Royal Library, Windsor there are some examples in an earlier (proof?) state. Kingsley,
    p. 37 and Hodson, 1974, p. 33 refer to examples for Sussex and Hertfordshire respectively.
5   Hodson, 1984, pp. 50–60 for details of Bakewell's usage and p. 57 especially for evidence
    that Elizabeth Bakewell carried on her late husband's business and instituted some alterations
    to some plates.
6   Richard Blome signed the dedicatory letter.
7   The CUL and NLS copies have no county maps in the work with this title, but bound in at the
    end each has: *A Geographical Description Of The World . . .* 1680. This has foreign maps
    followed by an English section. Four plates have been altered since the 1693 issue and one
    (Kent) is now re-dedicated to Henry, Viscount Sidney of Sheppey, who was created Earl in
    1694. (Skelton, p. 217). Mr. C.A. Burden's copy also has the four plates in their updated state
    as noted by Skelton.
8   This work was first advertised in *Daily Courant* for 24 Sept. 1715. There were three issues;
    (a) Mr. C.A. Burden has an example that does not have the England map and contains only
    the 40 county maps of the first advertisement; (b) the CUL atlas, which has the England map

but not that of Scotland; and (c) the BL atlas, which has both. Taylor prepared (after the 1715 Jacobite rising) a map of Scotland (advertised first in *St. James's Evening Post* for 29 Oct. – 2 Nov. 1715) that was incorporated at once in this and the later examples, in which the county plates were numbered. Hodson, 1984, pp. 51 and 59 refer to states (b) and (c); state (a) has come to light later.

9  In the BL example quoted (one of three held there) the map of Scotland is undated; all the other examples are dated 1715.

10  A map of Ireland dated 1716 has been added. The work was advertised in a form that mirrors the revised title in *Daily Courant* 21 Feb. 1717/8. See Hodson, 1984, p. 53.

11  These copies are bound with Taylor's maps of the Welsh counties, all of which are dated 1718. Hodson, 1984, p. 59 reports two other examples in NLW and Winchester College (Fellows Library). The Hull copy, of which he was unaware, lacks the main title-page but has the title-page for the Welsh maps bound between the end of the English county maps and the first of the Welsh maps.

12  Bakewell took over the plates from Thomas Jordan, who had taken over Taylor's business. The volume was issued, perhaps, soon after the change of ownership.

Skelton nos. 104, 105 and 114.
Hodson nos. 139, 140 and 141.


# 18

## JOHN SELLER                                                    1694

Size: 12.15cm x 14.95cm.

John Seller (*fl.*1660–1697) is particularly renowned for his sea charts published under the title *The English Pilot* first issued in 1671, ten years after his appointment as Hydrographer To The King.[1] Although this project was not financially rewarding for Seller and he sold his rights in the hydrographic side of his output by 1679–1680 the work was being issued until the early years of the nineteenth century. He began to prepare a county atlas entitled *Atlas Anglicanus* in 1679[2] and was awarded financial support from the Treasury to make a new survey of all the counties. The terms for subscribers are very similar to those offered contemporaneously by Blome. At the beginning of 1680 it was announced in the *Term Catalogue* that three southern counties were ready and work on a fourth (Kent) had begun. Only two more counties were completed and the scheme failed. The plates were sold to Philip Lea by 1693.[3]

Seller turned his attention to a more compact atlas, perhaps as a means of restoring his finances. The maps are crudely based on Speed. The engraver's name is not recorded. The maps prepared for the *Atlas Anglicanus* were the first to adopt the London meridian. In the new work other counties (but not Lincolnshire) show the adoption of London as the meridian, probably measured from St. Paul's.[4] There are no roads and, in place of a compass, the four cardinal points are named in the borders.

### Later history
There were only two issues of the plates before Seller died in 1697.[5] They were used for a few more years in later editions of his *History of England* and another historical work produced by Joseph Wild (1701) and Isaac Cleave (1711). The plates then disappeared from view for over sixty years before being re-worked and adapted for use by the publishers of Sir Francis Grose's *Antiquities of England And Wales,* who issued a variety of editions. Grose (c.1731–1791) was Richmond Herald (1755–1763) and became a captain in the Surrey Militia. His interest was always in antiquities and he became a noted writer of historical works, which used a large array of engraved views, which he had

collected from 1760 onwards. His *Classical Dictionary of the Vulgar Tongue* (1785) was a useful reference work, which reappeared nearly thirty years later. There were companion volumes to *The Antiquities of England and Wales*; in 1789 he met Robert Burns, who noted his corpulence and drinking prowess but also his diligence; his two volumes on Scotland appeared in 1791 and 1795. He died in Ireland in 1791 while working on the work that would have completed his study of all the British Isles.[6] An obituary proposed an epitaph which ended 'Death put an end to His Views and Prospects'.[7]

## EDITIONS

(i)  *LINCOLNE/SHIRE./* (Ea, in a rectangular box). *English Miles 12* [=12.2cm] (Ee). *North, East, South, West* appear between the inner and outer borders (Ca, Ec, Ce and Ac respectively).

*Anglia Contracta, or A Description of the Kingdom of England & Principality of Wales. in Several new Mapps of all the Countyes therein Contained By John Seller Hydrographer to The King.* [1694].[8]

**Copies**   BL Maps C.24.a.9; CUL Atlas 7.69.6; Bod Gough Maps 137; Leeds UL W 43a; William Salt Library (Stafford) bs 1666/1.

*The History of England Giving A True and Impartial Account of the most Considerable Transactions in Church and State, in Peace and War . . . from the coming of Julius Caesar into Britain . . . to the Year 1696 . . . with Exact Maps of each County, By John Seller, Hydrographer to His Majesty. London, Printed by Job and John How, for John Gwillim, against Crossby-Square in Bishopsgate-street, 1696.*

**Copies**   BL 1480 aa.21; CUL R.5.54.

*The History of England . . . Job and John How, and are Sold by H. Newman, at the Grasshopper in the Poultry, 1697.*[9]

*Camden's Britannia Abridg'd; With Improvements, and Continuations, to this present Time. To which are added Exact Lists of the Present Nobility of England, Scotland, and Ireland: Also a Valuation of all Ecclesiastical Preferments At the End of each County. With many other Useful Additions. The whole Carefully Perform'd, and Illustrated with above Sixty Maps Exactly Engraven. Vol. II. London, Printed by J.B.* [=J. Brockwell] *for Joseph Wild, at the Elephant at Charing-Cross. 1701.*

**Copies**   BL 10348 C 9; CUL R4.33–34; Bod Gough Gen Top 164–165; Leeds UL W 44; RGS 262 B 11–12; NLS A 43.e.5.

*The History of England . . .* [some of the wording has been altered from that of the 1696 edition] *. . . With the Maps of all the Counties and Islands belonging to England, being in all about Seventy Copper Cutts; with a Table of Contents. By John Seller . . . The Third Edition. London, Printed for J. Marshall, at the Bible in Grace-Church-street, 1703.*

**Copies**   BL 1326 a 6; RGS 262 a 18.

*Camden's Britannia Abridg'd; With . . . the whole carefully perform'd, and Illustrated . . . London: Printed for Isaac Cleave next to Serjeants-Inn in Chancery Lane. 1711.*

**Copy**   The private collection of Mr. C.A. Burden.

(ii)   The title has been re-engraved and is now enclosed in a double ring oval (Ea). Place-names have been altered: *Waplod* (earlier *Quaplod*); *Market/deeping (Market/ depping)*; *Marketraisen (Marketstrasen)*; *Gainsburgh (Ganesburgh)*. The map appears at the top of a single leaf, with a text on the county occupying the remaining one and a half sides of paper. There are several settings of the text but it is not possible to ally a textual state to a particular edition.

*The Antiquities of England And Wales. By Francis Grose . . . Supplement* [Vol. 1]. *London, Printed for S. Hooper, Nº 25, Ludgate-hill. MDCCLXXVII.* [1787].[10]

**Copies**   BL G 3043–5; Bod G.A. Gough Gen Top. c.30–36; Liverpool Athenaeum 914.2F; Leicester UL F.H942.1 GROS; Edinburgh RL q DA 690. (LAO Scorer Map 8; LRL LA 9 and Map 80; Grantham RL L9; Gainsborough RL L9).

*The Antiquities . . . Vol. II . . . MDCCLXXIIII.* [1787].[11]

**Copy**   BL 2061 i.

*The Antiquities . . . Vol. III. London: Printed for S. Hooper, Nº. 25, Ludgate-hill. MDCCLXXV.* [1795].[12]

**Copy**   Liverpool RL Hornby 24.5.

*The Antiquities . . . Vol. III. New Edition London. Printed for Hooper & Wigstead, Nº 212, High Holborn, facing Southampton Street, Bloomsbury-Square.* [1797].[13]

**Copy**   CUL Li.12.1–8.

*The Antiquities . . . Vol. III . . .* [1797, i.e. 1809].[14]

**Copies**   BL 10348 g 3; CUL S.474.b.78.4–11; Nottingham UL s/DA 620.G7 oversize. (LAO Dixon 21/4/1/2).

*The Antiquities . . .* [1810].[15]

**Copies**   BL 1505/358; John Rylands UL Q 942.006 G912 (R 4823).

*The Antiquities . . .* [1815].[16]

**Copy**   Birmingham RL HGD/094/1797/19Q.

**References**

1  Skelton, pp. 182, 186, and 246 for biographical detail; Tooley, R.V., Maps and mapmakers. (London, 1949), pp. 60–62 for a broad picture of the many volumes and editions of this work which made its final appearance in 1803; DNB. Vol. LI (London, 1897), pp. 227–228.

2  Advertised in *London Gazette* for 24–27 Nov. 1679 and cited by Skelton, 1970, p. 186; Tyacke, pp. 24–25; Rostenberg, L. English Publishers in the Graphic Arts, 1599–1700. (New York, 1963), – the advertisement is illustrated as pl. 32. Another advertisement in *London Gazette* appeared on 19–23 Feb. 1679/80 (Tyacke, p.25). BL Maps 187.l.1(23) is a copy of *Proposals For the Carrying on an Actual Survey of . . . England and . . . Wales . . . By John Seller . . . John Oliver and Richard Palmer. Full cost 50 Shillings – 20 Shillings down payment, 30/- on delivery . . .* [1679].

3  Some of the Seller plates were used in the Saxton/Lea atlas noted at **1** (v) above. Two copies are recorded (by Skelton, pp. 187 & 217–8) of 'mock-ups' for the *Atlas Anglicanus*, the remaining county plates being supplied from stock sheets of mainly Speed maps with Bassett and Chiswell imprints. The example in BL (1 TAB 18) from the Royal Collection of George III may have been presented originally by Seller to Charles II to show his intended plans.

4  Lynam, E. British maps and mapmakers. (London, 1943), p. 31.

5  Kingsley, p. 46 says 1697 and Skelton, p. 246 agrees; Skelton (pp. 186 and 203) also gives the death as 1698. Seller's will was proved on 31 May, 1697.

6  Timperley, Vol. II. pp. 772–773.

7  *St. James Chronicle* May 26, 1791.

8  The engraving of William and Mary (frontispiece) is dated 1688; Queen Mary died in 1694.

9  The work, advertised in *Term Catalogue* as 'Second Edition', does not have maps nor are maps called for in the title. The third edition does have maps. The copy owned by Mr. C.A. Burden of the second edition has maps but with text taken from Grose. Clearly the maps are a later insertion.

10  This work consists of four 'main' volumes dated 1774, 1775, 1775 and 1776 respectively. Two supplementary volumes appeared in 1787, but are frequently found with undated title-pages. The engraving of Grose in Supplement Volume 1 is dated 5$^{th}$ Nov$^r$ 1787. There are numerous varieties of make-up, e.g. the Bodleian copy has 1773 in the imprint of volume

1; the BL copy (2061 i) has the six volumes (including the Supplements) bound as three. The variety stems largely from the work being initially issued in parts. I am grateful to Mr. E. Burden, who drew my attention to an announcement in *The Morning Chronicle, and London Advertiser* for 25 Sept. 1787 that the issue of part XLIII that day completed the publication of the Supplement to *The Antiquities . . .* A further announcement, this time in *The World* for 17 Dec. 1787, referred to the issue of 56 County Maps 'which completes the Supplement to the Antiquities . . . in two volumes . . . 71.13s.'

11  Volume 1 has 'An Index Map' dated 1787.
12  Watermarks in volume III show 795 (i.e. 1795).
13  The new edition was issued originally in 1797 in 8 volumes. Normally the title-pages have no dates but volume 7 in most sets is dated 1797. The *Lincoln, Rutland and Stamford Mercury* for Aug. 3, 1798 carried the following advert: 'Grose's Antiquities. An Entire New Cheap Edition Being A Complete Abridgement . . . of all the most important Subjects . . . in the Original Edition . . . On Saturday next will be published. To be continued weekly and completed in only forty numbers. Price One Shilling. No. 1. Folio. Front . . . 4 views and Map of Beds. London Printed for the Editor and sold by H.D. Symonds, No. 20 Paternoster Row and may be had of R. Newcomb of Stamford . . . 1, 2 or more numbers at a time, the whole complete forty numbers; elegantly bound in 1 vol. Price 21.10s'. A similar advertisement appeared two weeks later when Part 4 was announced.
14  Several maps have a watermark date of 1809 (including Lincolnshire in the Nottingham UL set).
15  The Lincolnshire text (p. 129) has a watermark date of 1810.
16  Several maps have a watermark date of 1815.

Skelton nos. 115, 119, 120, 122 and 124; Chubb no. CCXLVII.

## 19

**ROBERT MORDEN**                                                              **1695**

Size: 36.25cm x 42.3cm.

After the issue of the playing card maps in 1676 (**15**) Morden was concerned with the publication of a number of geographical works. He is perhaps best known among collectors for the series of maps he produced to accompany Gibson's edition of Camden's *Britannia*, discussed below, and, to a lesser degree, for the set of smaller maps, which represent his first efforts for the Camden exercise (**20**).

A new version of Camden was first projected by Dr. Edmund Gibson in 1692. Gibson (1669–1748) was then a Fellow of Queen's College, Oxford, and later Librarian at Lambeth Palace; he was Bishop of Lincoln (1716–1723) and London (1723–1748). The first set of Proposals was probably issued in Michaelmas Term, 1692.[1] A second set is dated 'April the XX[th] 1693'.[2] At this stage Abel Swalle was the sole undertaker.[3] Throughout the same year advertisements were placed[4] and, by July, Gibson had moved to London in order to be nearer the printers, giving Awnsham Churchill's shop as his address.[5] The first set of proposals referred to a subscription price of 26s. in sheets and delivery at Michaelmas Term (presumably 1692). Churchill had begun as a bookseller in 1681 and in 1690 he took his brother John into partnership of what was already a substantial business.[6] He became possibly the most successful bookseller of the time, made a fortune, bought an estate in Dorset and was M.P. for Dorchester, 1705–1710.[7] John died in 1714 and Awnsham in 1728.

Gibson's plan of production was to give the translation work to different authors and pass on sheets for inspection to Dr. Arthur Charlett, Master of University College, Oxford. Gibson was unfamiliar, however, with printing workshop methods, particularly the need to redistribute type once the sheets of a section had been printed off. Delays in the receipt of translations meant the plans for printing went awry since the original plan of Camden's

*Britannia* of starting with Devon and Cornwall, proceeding eastwards and then north-wards, was to be followed. Printing did not begin until January, 1694 but had fallen well behind by June, when Surrey was already set while Gloucestershire, three counties 'in front', lacked a text. By October the county texts were ready but the engraving of the plates was not complete. Only on 19 January, 1695 was Gibson able to report to Charlett that 'All's finisht, and the Present books put into the Binders' hands . . .' Charlett was also told that £2,000 had been put out by the publishers before printing had begun.[8]

Morden seems to have drawn two sets of maps for this project. A set of small maps was prepared (**20**) and Skelton suggests that due to a misunderstanding on Morden's part of what Gibson required the set was rejected.[9] Gibson's preface indicates the method adopted for creating the map plates. Copies of whatever maps were already available for each county were sent out to gentlemen in the counties for their comments and correc-tions. As the rejection of the small maps had not apparently occurred until 1693 and there were bound to be some correspondents who delayed or failed to reply it is not surprising that engraving had been the final problem in the completion of publishing. Some counties had been resurveyed since 1660 and manuscript and printed maps were used in this process. Skelton analyses the sources for each map as they were finally engraved, Lincolnshire being based on Speed.[10] Morden seems only to have added roads (from Ogilby's *Britannia*) or used, for coastal counties, the sea-charts of Greenvile Collins (1693). Whether the urgency at the end in getting the plates engraved contributed to the inaccuracies is not certain, but the maps were soon criticised. Inserted letters in mis-spelt place-names abound and roads are not always in their correct locations. One critic said that Morden was unable to copy accurately from his original models.[11]

The maps have an elegance that appeals to many though there are no decorative features. The title appears in a cartouche that is a distinguishing mark of the series. Sea-banks in the Wash and the Humber are distinctive additions (from Collins). The frames are engraved at one minute intervals to indicate latitude and longitude, the latter marked from St. Paul's Cathedral. Kingsley points out that Sussex shows the meridian from Greenwich[12] and Yorkshire is another example.[13] The top margin is marked with the local time difference from London.

A curious feature is the inclusion of a scale in three measures. It is not clear what is meant as they vary for each county and are not, it seems, based on a local mile. Skelton has calculated the values for many counties and, in the case of Lincolnshire, ten great, middle or small miles equal one degree of latitude.[14] This equates to, using the short mile measure, a scale of five miles to an inch.

Because of the number of corrections on the plates other students have supposed that an earlier unrevised state could exist. It seems likely that sheets in proof state were submitted to authoritative gentlemen in each county (where possible) and that, on their return, the necessary corrections were made on the plates. Such proof sheets have never been found. Hodson[15] draws attention to the Devon map, where all the changes are made within a five mile radius of Hatherley, implying that the Morden/Gibson agent lived in that town. The changes on the Lincolnshire plate are so widely spread that if a single agent were involved he had a very wide knowledge of the whole county. Inserted letters in place-names occur, for instance. at *Kir(k)ton* (in Holland), *Stallin(g)borough, Kir(k)by* [Underwood], and *Lud(d)ington* [north-east of Crowle]. (The inserted letters are given here in brackets). Many examples remain of radically mis-spelt names that one might have expected to be picked up during such an inspection. Incomplete deletions of previously engraved material appear on the 1695 plate in a wide scatter of places.

### Later history
Morden died in 1703. A second edition was printed in 1722 for Awnsham Churchill who owned the plates. Awnsham Churchill died in 1728 and it seems that the plates were acquired (in equal tenth shares) by the booksellers named on the title-page of a new issue

in 1730. Hodson shows that there were a number of sales of one-tenth shares after 1730 and that later the shares were further subdivided and offered in up to 1/160th parts.[16]

That the first edition was printed in a large number of copies is shown by the survival rate of examples and by the fact that, although work on a revised edition may have begun several years before, the second edition did not appear until 1722. Thomas Hearne, the antiquary, suggested that a new edition was in the air as early as 1708[17] and Gibson (preface to the 1722 edition) says [I had] 'about twelve Years since . . . turn'd my Thoughts . . . towards Improvement . . . of this Work'. In 1718 a new edition was announced[18] and Gibson also wrote to Dr. Charlett that '*Britannia* has been ready . . . these 7 years and I have revised it once more this Summer at Bugden . . .'[19] Internal evidence suggests that the work had been completed some time before 1722, e.g. there is no item in the bibliography later than 1719.

Once again a large printing found a ready market in 1722 and there were enough sheets of that printing to make up the volumes for a further issue in 1730. The maps underwent a comprehensive revision in 1722 in an attempt to overcome the criticism to which they had earlier been subjected. In 1753, when the supply of 1722 sheets had been exhausted, a new re-setting of the type took place and the maps were revised again. Fifteen owners of shares were now involved in the printing of an edition, which Hodson shows consisted of 500 copies.[20] In 1772 one more edition was put out by the (by then) thirty share-holders. The text and the plates were not used again since, in 1789, there appeared Gough's fresh translation with newly engraved plates from John Cary (**49**).

## EDITIONS

(i)    *LINCOLN=/SHIRE/by Rob$^t$ Morden/* (Eb). The inner frames (not along the top) are shaded at 1′ intervals and numbered, between the inner and outer frames, at 10′ intervals. The top inner frame is not shaded or numbered but divisions at 10′ intervals mark *Minutes of Time from London* (Ba, between the inner and outer frames) with *Degree* at the 1 degree mark (Be). *Scale of Miles/Great 10* [= 6.4 cm]/*Midle 10* [= 5.8cm]/*Small 10* [= 5.2cm] (Ac). There is no compass.

*Camden's Britannia, Newly Translated into English: with Large Additions And Improvements. Published by Edmund Gibson, of Queens-College in Oxford . . . London, Printed by F. Collins, for A. Swalle at the Unicorn at the West-end of St. Paul's Church-yard, and A. & J. Churchil, at the Black Swan in Pater-noster-Row. 1695.*

**Copies**    BL 1259 d 18 and 7 TAB 51;[21] CUL R.7.1 etc.; Bod Lister D.69 and Gough Gen. Top. 49; Leeds UL W 42; LAO – Law Library; RGS 264 D 18. (LAO Scorer Map 5; LRL Maps 91 and 154; Louth RL Map A 16; Grimsby RL MORDEN 1695 (RM 109); Gainsborough RL Map 9; Grantham RL).

(ii) The names of four wapentakes have been changed: *BRADLEY, LONGOBO=/BY, MANLEY, ASLACKOE* (formerly BRODLEY, BOOTH BY, MANTLE, and ASLACOTE respectively). Many place-names have been changed, e.g. *Scotherne* (previously *Scoghthorn*), Eagle (*Agle*), *Skredington*, (*Sknekington*) and *N Hickham* (*N S$^h$tikham*). Many names are uncorrected, e.g. *Freckingham* (= Threekingham), *Redburn* (= Redbourne), *Aistrop* (= Aisthorpe and south of *Scamton*, and thus repeating Speed's error from Saxton) and *Tharleby* (= Thurlby). *Barkston* still appears twice as do *Ingoldmeles* and *Ingoldmels*. *Moulton Seasend/Second* also suggests that earlier doubts have not been clarified. Retouching in the area around Stamford has led to the town name losing its last two letters, the *Weyland Flu* (= River Welland) being extended into the area between the inner and outer frames, *Vffington* being altered to *Uffington*; the Great North Road is broken for a space of 9 mm.

*Britannia: Or A Chorographical Description Of Great Britain And Ireland, Together with the Adjacent Islands. Written in Latin By William Camden, Clarenceux, King at Arms: And Translated into English, with Additions and Improvements. The Second Edition. Revised, Digested, and Published, with large Additions, By Edmund Gibson, D.D. Rector of Lambeth; and now Bishop of Lincoln, and Dean of His Majesty's Chapel-Royal. Vol. I. London: Printed by Mary Matthews, for Awnsham Churchill, and Sold by William Taylor, in Pater-Noster-Row. MDCCXXII.*

**Copies**  BL 1259 f 20; CUL Atlas 4.72.1–2; Bod Douce C. Subt. 57–58 and S 2 14–15 JUR; Nottingham UL DA 610.C2; London UL [G.L.] 1722. (LAO FL Maps 3; LRL Maps 803 and 803a; Grimsby RL Morden 1695 (RM 116)).

*Britannia . . . Revised . . . By Edmund Gibson . . . now Bishop of London . . . The Second Edition. Vol. 1. London, Printed for James and John Knapton* [and ten others. 1730].[22]

**Copies**  Bod Allen 16; LAO AS 950.2; Birmingham RL 162291; NLS A 426; Sheffield UL \*\*F 914.2(C); John Rylands UL LQ 942.006 C 141 (R17667).

There are no distinguishing features between loose maps from the 1722 and 1730 editions. It is possible, on occasion, to separate the two issues on watermark evidence. The 1722 issue often has a horse or a spread eagle, while the 1730 maps have a large P. LAO AS 950.2 has a clear example of the P watermark.[23]

**(iii)**  The final two letters of *Stamford* have been restored. The road from Spalding to Crowland has been strengthened and consists now of two continuous lines. The road at *Dinnington* [= Donington] has been extended a short distance westwards as two parallel lines. *Christed/Abby* [=Kirkstead Abbey] is now clearly engraved. The engraving of the Great North Road has been deepened, resulting in 'doubling' – most notable near Clipsham.

*Britannia . . . By Edmund Gibson, D.D. Late Lord Bishop of London. The Third Edition. Illustrated With Maps of all the Counties, and Prints of the British, Roman, and Saxon Coins. Vol. I. London: Printed for R. Ware* [and 14 others]. *MDCCLIII.*[24]

**Copies**  BL 10348 l.2; Leeds UL W 43; Sheffield RL 914.2 STF; Canterbury Cathedral L. Elham 4; Derby RL (Local Studies Dept. 9404–9405).

A collection of maps taken from Camden's Britannica. [1730].[25]

**Copy**  Bod Vet A4 b 24.

*Britannia: Or, A Chorographical Description Of Great Britain And Ireland, Together With The Adjacent Islands. Written in Latin By William Camden, Clarenceux King at Arms; And translated into English, with Additions and Improvements; By Edmund Gibson . . . The Fourth Edition is printed from a Copy of 1722, left corrected by the Bishop for the Press. Vol. 1. London: Printed for W. Bowyer* [and 29 others]. M.DCC.LXXII.[26]

**Copies**  BL 2061 k and 192 f 10–11; Bod Vet A4 b.34 etc.; Leicester UL F.SH.T; LAO R 911.42 GIB oversize; Manchester Chethams L. 0021. 5–6; London UL 4a 6–7.

**Reproductions**
(i) The map in state (i) is shown in reduced form and discussed by John Holehouse in *Lincolnshire Life* Vol. 14, no.4 (June, 1974) pp. 44–45. (ii) An earlier reduced reproduction, also of state (i), appeared in *Lincolnshire Life* Vol. 3, no. 4 (Aug./Sept. 1963) pp. 26–27; in the following issue (p. 59) there is a note by D.N. Robinson concerning the dating and features of the map, especially the odd spelling of place-names. (iii) The complete work was issued as: Camden's Britannia 1695 A facsimile . . . With an introduction by Stuart Piggott . . . and a bibliographical note by Gwyn Walters. (Times Newspapers, 1971 – copy in LRL).

## References

1   Bod. 2597.b.1(1).

2   Bod Wood 658.f. 806.

3   Plomer, 1907, p. 174; Skelton, 1970, p.249.

4   *London Gazette* 3–6 April, 1693 and at least eight other occasions up to 7–11 June, 1694. Cited by Tyacke, pp. 58–64.

5   *Term Catalogue,* Vol. 1. p. 460 advertising a law book.

6   John Churchill. Plomer, 1922, p. 70.

7   Awnsham Churchill. Plomer, H.R. 1922, pp. 69–70; Treadwell, M. London Trade Publishers 1675–1750. *The Library,* VI, iv (1982), pp. 99–134; Skelton, p. 249.

8   Walters, G. and Emery, F., Edward Lhuyd, Edmund Gibson and the printing of Camden's Britannia, 1695. *The Library,* V, xxxii (1977), pp. [109]–137.

9   Skelton, p. 195.

10  ibid., pp. 196–198.

11  Rev. F. Brocklesby writing to Thomas Hearne in May, 1711. Reprinted in: The itinerary of John Leland; edited by T. Hearne. Vol. VI. (London, 1711), pp. 91–92.

12  Kingsley, p. 49.

13  Raistrick, A. Yorkshire maps and map-makers. (Clapham via Lancaster, 1969), p. 26.

14  Skelton, p. 197.

15  Hodson, 1984, p. 169, n.3.

16  ibid., pp. 105, 107 and 110–111.

17  Hearne, T. Remarks and collections of Thomas Hearne; ed. by C.E. Doble (Oxford, 1885–1921). Vol. II, p. 146.

18  *Daily Courant* 31 Dec. 1717 stated '. . . a New Edition . . . will speedily be put to Press.' *Evening Post* 7–9 Jan. 1717/8 had a similar notice. Cited by Hodson, 1984, p. 100.

19  Bod. Ballard MSS. vol. 6. f 133, dated 30 Oct. 1718.

20  Hodson, 1984, pp. 107–108.

21  This example has manuscript notes on interleaved pages. Most of the notes refer to Knights of the Shire, e.g. those for Cornwall refer to those elected in 1713 but space prepared for the results of 1714–5 is left blank. The notes for Lincolnshire give the result of the 1714–1715 election 'Sir John Brownlow and Sir W. Hickman as before.' One must presume that this was a private compilation; CUL Atlas 4.69.1 is also a privately made up copy with an interpolated map dated 1715.

22  Two of the others were Thomas Longman and Charles Rivington. Longman was the son-in-law of John Osborne, who acted as executor for William and John Taylor (best remembered for being publishers of *Robinson Crusoe*). Longman was persuaded by Osborne to buy the Taylors' business and thus founded the firm that has survived to the present day and is especially noted for the production of dictionaries, having begun with Dr. Johnson's. Plomer, 1922, pp. 158–159 and 285; *The Critic*, New Series, xx (1860), pp. 366 et seq; Cox, H. The House of Longman, 1724–1924. (London, 1925); Briggs, A. (ed.) Essays in the history of publishing . . . the House of Longman, 1724–1972. (London, 1974); Longman, C.J. The House of Longman, 1724–1800 . . . (London, 1936). A group of 168 catalogues from Longmans, Green, spread over the period 1704–1768 provides much valuable detail on the workings of the printing industry and the disposal of shares in the rights for this version of *Britannia*. (BL C.170.aa).

    Charles Rivington bought Richard Chiswell's business and later published for the Wesleys and shared the publication of Richardson's *Pamela*. John and James succeeded their father on his death in 1742 and were two of the partners in the 1753 edition of *Britannia*. The firm continued until it was finally sold to Longmans in 1890. Plomer, H.R. 1922, p. 254; Rivington, S. The Publishing Family of Rivington. (London, 1919).

    The date of publication is based on two advertisements in January, 1730, which closely follow the text of the title-page. (Hodson, 1984, p. 105).

23  Heawood, E. Watermarks mainly of the seventeenth and eighteenth centuries. (Hilversum, 1950), item 2803.

24  The business records of William Strahan whose firm helped to print this edition (BL Add. MS. 48803 A f.16v) show that he produced 500 copies of the sections he was contracted for. This

clearly suggests that the total printed was 500 sets. Cited in: Hodson, 1984, p. 105; Plomer, 1922, pp. 239–240 provides biographical notes on Strahan.

25  This example lacks a title page, all preliminary material, etc. The date is that of the Bodleian's catalogue.

26  Hodson, 1984, p. 111 suggests that the work was published in December, 1772, quoting an advertisement in *Public Advertiser* for 19 Dec. 1772 'This Day is published in Two Volumes, Folio, Price Four guineas bound . . . the 4th Edition . . . of Britannia . . .' One of the Bodleian copies (Arch. A b.12 (1–2)) bears the plate of George Scott of Woolston Hall in Essex and on the flyleaf is pasted a letter recording the donation to the university of a set of the fourth edition and which is dated Feb: 19, 1772. It seems likely that this should read 1773.

Skelton no. 117. Hodson nos. 169–172.


# 20

## ROBERT MORDEN                                                        1701

Size: 22.1cm x 17.4cm.

The preparation of these plates and their rejection by Gibson has already been discussed (**19**). Six of these plates are signed, Lincolnshire included, by R. Spofforth. Like the larger maps these have degrees and minutes of longitude along the bottom, with London as the prime meridian and the time differences from London along the top border. The spelling of the names of the hundreds follows that of the first state of the larger plates, which suggests the two sets were engraved from the same erroneous source material. Like the larger map there is a scale in three forms. The author of the text is unknown.

### Later history
Morden died in 1703, but it is not known how his plates came into the hands of the next owners. There were several editions in 1704 and there were, by then, six partners in the project. In 1708 a new issue of the plates appeared with the assistance of Herman Moll (**24**). In this item there was no text, a characteristic of several of Moll's own volumes. Finally the plates were used in the part work, which commenced in 1714: *Magna Britannia et Hibernia.*

*Magna Britannia* is one of the first works to be issued in serial form.[1] Originally conceived as part of *Atlas geographus: or, a compleat system of geography, ancient and modern* it began in monthly parts at one shilling each in June, 1708.[2] By the end of 1710 the whole of Europe except Great Britain and Ireland was completed and was to be bound up in two volumes.[3] From January in the following year fifteen monthly parts comprised Asia, immediately followed by a further fifteen parts, which dealt with Africa.[4] This took until December, 1713. Beginning again in March, 1714 America took up one volume made up of seventeen parts, spread over two years. The chief undertakers of this project seem to have been John Morphew and John Nutt and, after the death of Nutt in 1716, his widow Elizabeth. Others whose names appear on the title-pages of the volumes or issues in parts may have been vendors only and never shareholders. John Nicholson, who may have been involved at the start had, by the time of his death in 1717, become owner of all the shares, since his trustees offered them in a trade sale in 1740.[5]

So much information had been gathered by 1712 that it had been decided to issue the material for Great Britain and Ireland separately. Morphew, who was one of the principal booksellers of his time was again involved.[6] Nicholson, although his name does not appear on any part, seems also to have been involved since letters survive from him to Sir Hans Sloane in which his help is requested for the loan of maps that must have been intended for this project.[7] Nicholson died in 1717 and Morphew in 1720 and sole ownership passed to Elizabeth Bell who possibly purchased the work in 1721.

Various announcements that issues were about to appear occurred during 1710–1713, but the first number did not appear until 4 January, 1713/4.[8] With a number of breaks, particularly between February, 1720 and March, 1722 the complete work took over seventeen years before part 92 came out in April, 1731. The intention to produce numbers in monthly parts was rarely adhered to for very many months in succession. Beginning with an introduction spread over two and a half parts the counties were dealt with in alphabetical order and, at the beginning (at least as far as Kent), the first part of a county's treatment appeared as the second half of a part, with the second, concluding part, forming the first half of the following issue. With Kent the totality of material demanded that several issues should be devoted to the county and. with a few exceptions, that remained the final pattern. One oddity was that Lancashire still only appeared in two half parts. Lincolnshire formed the second half of part 28 (the first half being the concluding portion of Leicestershire), which appeared in October, 1719 plus the whole of parts 29 and 30, which were issued in December, 1719 and February, 1720 respectively – each a shilling apiece. At the appearance of part 30 the 'Undertakers . . . have thought fit to end Two Vols. and have in this No. added a general Title and Preface.'[9] When completed in 1731 the whole set comprised six volumes, published by E. and R. Nutt and sold by T. Cox and it is in this form that the work is now normally found. In spite of the title of the parts and in the final volume form only the English counties were included.

The single issues are very rare and those of Lincolnshire have not yet been found. The map was usually issued with the first part issue relating to that county, so that the map of Lincolnshire would have been found in part 28. It was usual to end each county part with a mileage chart and this explains why, when there was a later issue of each county gathered up as a separate entity, the first page of Lincolnshire is on the verso of a leaf with the mileage chart of Leicestershire.

Elizabeth Bell, the owner of the material, died in 1724 and Thomas Cox became the new proprietor and saw the work to completion; individual counties were sold as separates. What remained unsold came into the hands of Ward and Chandler at a sale held in Oct. 1738 and they lost no time in trying to recoup their £200 outlay.[10] Apart from part one (which was out of print) they had enough copies to make up 108 sets to supplement the 36 complete sets in volume form that came with this material. Ward and Chandler, therefore, had new title-pages printed and the first part reprinted so that they could make up more sets as needed. The plates were sold in 1746 to Henry Lintot, after Chandler's suicide in 1744 over his debts, which, in turn, led to Ward's financial downfall. The resultant sale included copies of the texts and the plates for the maps. By 1751 Joseph Marshall held the stock, which was now reduced to individual counties only. The plates were probably sold as scrap metal after 1746 but sheets turned up around the end of the eighteenth century with a number of new and spurious title-pages provided. In Lincolnshire's case six different forms of title-page have been found, but who produced them is still unknown.

The supposed author of the text is Thomas Cox and it is based on Camden. There was a range of additional information included as a result of advertisements for assistance in the total undertaking, beginning with *Atlas geographus*.[11] There are a pair of candidates, who may be the Thomas Cox, the general editor of the text. The British Museum catalogue awards the palm to Thomas Cox, Rector of Stock-Harvard. Richard Gough believed the author to be Thomas Cox, Vicar of Bromfield, Essex.[12] Hodson thinks Gough confused Cox the author with Thomas Cox, the final publisher of the work. In fact, there is good evidence to suggest that much might be the work of Rev. Anthony Hall (1679–1723), Fellow of Queen's College, Oxford, whose assistance in writing the introduction Cox acknowledged later; but Hall probably compiled some material for the county sections also.[13]

## EDITIONS

(i) *LINCOLN/SHIRE/By/Rob' Morden/* (Ae, in a leafy cartouche). *Scales of Miles 10 Great* [= 3.8cm]/*Midle 10* [=3.6cm]/*Small 10* [= 3.15cm] (Ae, below title). *Minutes of Time from London* (Aa,between the inner and outer border). *R' Spofforth Sculp* (Ee). *Hundreds &/ Wapontakes/* above a numbered list of 30 names (Ac–Ad), with the names of the three divisions reading upwards alongside the groups of names (Ac–Ad, between the inner and outer frames). The wapentakes are numbered on the map. Marks between the inner and outer frames are numbered in 5s to indicate latitude and longitude. *Degree* (Ae, between the inner and outer frames next to the 1 degree mark). There is no compass.

*The New Description and State Of England, Containing the Mapps Of The Counties of England and Wales In Fifty Three Copper-Plates, Newly Design'd, Exactly Drawn and Engraven by the Best Artists . . . London, Printed for Robert Morden in Cornhill; Thomas Cockerill, at the Three Leggs and Bible against Grocers-Hill in the Poultry; and, Ralph Smith, at the Bible under the Piazza of the Royal Exchange, 1701.*

**Copies**  BL 579 d 28; Bod Allen 11, etc.; RGS 262 B 3; Birmingham UL r GB 181; Edinburgh UL P* 22.12; Manchester RL BR 912.42 Mo 1. (LAO Scorer Map 6; LRL Map 72a; LRL A9 (UP 1675) consists of the title-page, the 19 pages of preliminary matter, pp. 89–92 (the county text), the county map and pp. [213–236]).

*The New Description and State Of England, Containing The Maps Of The Counties of England and Wales, In Fifty Three Copper-plates, Newly Design'd, by Mr. Robert Morden, Exactly Drawn and Engraven by the Best Artists . . . The Second Edition . . . London: Printed for S. and J. Sprint, J. Nicholson, and S. Burroughs in Little Britain, A. Bell, and R. Smith in Cornhill, 1704. Price Bound 8s.*

**Copies**  BL 577 f 3; CUL Atlas 7.70.3.

*The New Description and State of England . . . Newly Design'd by Mr. Robert Morden . . . Printed . . . in Little-Britain, and A. Bell and R. Smith in Cornhill. 1704.*[14]

**Copies**  CUL Atlas 6.70.6;[15] Bod Gough A Gen Top 358.

*The New Description . . .exactly drawn and engraven by Mr. Hermann Moll, and the best Artists . . .Printed for Ralph Smith at the Bible, under the Royal Exchange, Cornhill. Pr. 8s.* [1704].[16]

(ii)    The following places have been added to the plate: *Heckington, Morton, Briggeslay, Corobet* [= Cowbit), *Wranbye* [= Worlaby], *Kirby* (Laythorpe), *Boultham, N. Hikeham.* Roads have been upgraded (two parallel lines): from Deeping to Barton; Crowland to Boston; Sleaford to Boston, in a curve just avoiding Swaton and *Dinnington* [Donington]; Grimsby to Lincoln, and now, newly drawn, continued to Newark. The Great North Road has been added and directions where some roads leave the county, e.g. *to London.* The county's three divisions are named; previously only Holland had been included. *Witham R* is named near Grantham; (it is already named west of Boston). The *Nine R* [River Nene] has been continued parallel to the Cambridgeshire border (which has now been added as broken lines, instead of dots) and continues into the Wash. Yorkshire, Nottinghamshire and Leicestershire have now also been named as adjoining counties. The boundaries between the adjoining counties are now shown by dotted lines. A small compass has been added (Da). The circle marking [Sutton] *S Maryes* has been moved northwards.

*Fifty Six new and Acurate* (sic) *Maps of Great Britain, Ireland and Wales; With All the Direct and Cross Roads exactly Tracted in the Maps, which are more Full and Exact than any Extant, having all the Cities, Parliament and Market Towns, Villages, Parks, Hundreds, Wapentacks, &c. distinguish'd. Begun by Mr. Morden: Perfected, Corrected, and Enlarg'd by Mr. Moll . . . London: Printed for John Nicholson at the King's-Arms,*

*and John Sprent at the Bell in Little-Britain; Andrew Bell . . . and Ralph Smith . . . 1708.*
*Price Plain 7s. Colour'd 8s . . .*

**Copies**   BL Maps C.24.b.25; CUL Atlas 6.70.3; Bod Allen 12; RGS 4 B 16; Leeds UL
W 45. (BL Map 3355 (8); LRL Map 72).

*Magna Britannia Et Hibernia, Antiqua & Nova. Containing, 1. The Geographical*
*Description of each County in Alphabetical Order. 2. The Ecclesiastical History. 3. The*
*Civil History. 4. The Natural History. 5. Literary History. 6. Antiquities. 7. The best Map*
*of each County yet extant, and opposite to it the Names of each City, Town, Village,&c.*
*with the Value of the Living: Also a Scheme of all the Market-Towns, &c. their Distance*
*from London, and from one another, &c. [Nᵒ 28.] This finishes Leicestershire, with a*
*Table thereof, and contains part of Lincolnshire, with a Map of it, &c. In the Savoy,*
*Printed by Eliz. Nutt and sold by M. Nutt and J. Morphew.* [1719].[17]

*Magna Britannia Et Hibernia, Antiqua & Nova. Or, A New Survey of Great Britain,*
*wherein to the Topographical Account given by Mr. Cambden, and the late Editors of his*
*Britannia, is added a more large History, not only of the Cities, Boroughs, Towns, and*
*Parishes mentioned by them, but also of many other Places of Note, and Antiquities since*
*discovered . . . Collected and Composed by an impartial Hand. Vol. II. Containing the*
*Counties of Gloucester, Southampton, Hereford, Hertford, Huntingdon, Kent, Lancaster,*
*Leicester, and Lincoln. In the Savoy: Printed by Eliz. Nutt; and Sold by M. Nutt in*
*Exeter-Exchange in the Strand, and J. Morphew near Stationers-Hall. MDCCXX.*

**Copies**   BL 289 i 25–30, etc.; CUL VIII.3.1–6; Bod Tanner 355, etc.; Nottingham UL
s/DA 600.C6; Birmingham RL HGD/094/1720/1; NLS NE 54 d.12. (LAO Sc 942.52 and
LRL L9 are examples of the Lincolnshire text (pp. 1404–1516), with map and mileage
chart).

*A Compleat History of Lincolnshire. Containing 1. The Geographical Description . . . 7.*
*A Map of the County. 8. A Table of the Names of all the Towns and Villages,&c. with the*
*Value of the Livings, the Patrons, Incumbents, and Gentlemens Seats: Also a Scheme of*
*all the Market-Towns, &c. their Distance from London, and from one another, &c. In the*
*Savoy: Printed by E. and R. Nutt; and Sold by T. Cox, at the Lamb under the Royal*
*Exchange, Cornhill. M.DCC.XXX . . .* [18]

*Magna Britannia Antiqua & Nova: Or, a New, Exact, and Comprehensive Survey Of The*
*Ancient and Present State Of Great-Britain . . . the whole being more Comprehensive*
*and Instructive than Camden, or any other Author on this Subject. And Illustrated not*
*only with General Mapps, and also particular ones of each County, and other Plates . . .*
*Collected and Compiled . . . and communicated by several Judicious Hands. Vol. II.*
*London, Printed for and sold by Cæsar Ward and Richard Chandler, Booksellers, at the*
*Ship without Temple Bar, and at their Shops in Coney Street, York, and at Scarborough*
*Spaw. MDCCXXXVIII. Price three Guineas bound and gilt. In Six Volumes. N.B.*
*Particular Counties may be had separate* [1739].[19]

**Copies**   BL 1484 k 3;[20] BL 289 i 25–30; Manchester Chethams L. 942 (2 H3. 32–37).

*A Compleat History Of Lincolnshire . . . London: Printed for C. Ward and R. Chandler,*
*Booksellers, at the Ship without Temple-Bar, and sold at their Shops in Coney-Street,*
*York, and at Scarborough-Spaw.* [1739] . . .[21]

**Copies**   BL C.115.f.1; LAO Sc 942.52; Louth RL L9.BL.[22]

The following items consist of the text for Lincolnshire (i.e. pp. 1404–1516, with the
mileage chart facing the final page, the map, (facing p. 1404), and a spurious title-page.[23]

*A Topographical, Ecclesiastical, And Natural History Of Lincolnshire. With pedigrees of*
*all the Noble Families and gentry, both Ancient and Modern, Biographical Notices of*
*eminent and learned Men to whom this County has given Birth; also an Alphabetical*

*eminent and learned Men to whom this County has given Birth; also an Alphabetical Table of the Towns, Villages, and Hamlets, with the several Hundreds and deaneries in which they stand, together with the value of the Churches in the King's Books, collected and composed according to the best relations extant. By The Rev. Thomas Cox. In the Savoy: Printed by Eliz. Nutt; and Sold by M. Nutt, in Exeter-Exchange in the Strand and J. Morphew near Stationers-Hall. MDCC.*

**Copy**   A private Lincolnshire Library.[24]

*A Topographical . . . History Of* [blank space, with county name to be filled in by hand] *. . . By Thomas Cox. In the Savoy: Sold by M. Nutt, in Exeter-Exchange. 1700.* (a) and (b).[25]

**Copies**   (a) Grimsby RL X 000:942 COX. (b) A private Lincolnshire Library.

*Magna Britannia Et Hibernia, Antiqua & Nova. Or, A New Survey of Great Britain . . . Collected and Compiled by an impartial Hand. [The Rev. Thomas Cox.]* [space for county name to be filled in by hand]. *In the Savoy: Printed by E. and R. Nutt; and Sold by T. Cox at the Lamb, under the Royal Exchange, Cornhill. MDCCXX–XXXI.*

**Copies**   LRL L9. Grantham RL L9 is another example of this type but lacks the map.

*Magna Britannia; Or Topographical, Historical, Ecclesiastical, And Natural History Of Lincolnshire, With Pedigrees of all the noble Families . . . composed to the best relations extant. By The Rev. Thomas Cox. In the Savoy: Printed by Eliz. Nutt; and Sold by M. Nutt, in Exeter-Exchange in the Strand, and J. Morphew near Stationers-Hall.*

**Copies**   LRL L9 and (a second copy) A9; Louth RL GL 9.

*Magna Britannia; Or Topographical . . . History Of Lincolnshire . . . composed to the best relations extant. With Map of the County by Robert Morden. By The Rev. Thomas Cox. In the Savoy: Printed by Eliz. Nutt, in Exeter-Exchange in the Strand, and J. Morphew near Stationers-Hall. MDCCXX.*

**Copy**   LAO Sc 942.52.

### References

1   Wiles, R.M., Serial Publications in England before 1750. (Cambridge,1957), pp. 270–274.
2   Tyacke, p. 92 cites *London Gazette* for 24–28 June, 1708; Hodson, 1984, p. 12 cites *Daily Courant* for 18 June, 1708.
3   *Tatler* 7–9, Dec. 1710; Tyacke, p. 98 cites *London Gazette* 14–16 Dec. 1710.
4   There were a number of advertisements cited by Hodson and by Tyacke (e.g. *London Gazette* 30 Jan–1 Feb. 1710/1).
5   Sale catalogues owned and annotated by Aaron Ward, 1718–52, 1792 and 1797; no. 79 now preserved in Johnson Collection in the Bodleian. Cited by Hodson, 1984, pp. xvi and 13.
6   Plomer, 1922, pp. 210–211; Hodson, 1984, pp. 12–18 *passim*.
7   BL Sloane MS 4044 f. 54.
8   Tyacke, p. 102.
9   *Daily Courant* 8 Feb. 1719/20; cited by Hodson, 1984, p. 18.
10  Hodson, 1984, p. 26 illustrates the leaf from the sale catalogue, which lists how many copies were available of each part.
11  At the end of the separate parts requests for information were printed and receipt of information was acknowledged. General requests were also placed in the press; Hodson, 1984, p. 17 cites *Evening Post* 6–8 Sept. 1716.
12  Gough, 1768, p.21; Gough, 1780, p. 33.
13  Hodson, 1984, p. 21 discusses Cox's identity and the evidence for Anthony Hall's involvement.
14  This version is quarto sized. Formerly octavo the text has been reimposed by printing two of the small pages side by side divided by a vertical rule. The larger size permits the maps to be bound in unfolded.

15  Although the CUL copy lacks a title-page it conforms to the new format.
16  Described by Burne, S.A.H. Early Staffordshire maps, Addenda. *The North Staffordshire Field Club Transactions,* LX (1926), p. 77. Advertised by Smith in *Post-man* 12–14 Oct. 1704. No copy has been found. Smith had advertised the octavo edition (*Daily Courant* 9 March, 1703/4) and it is assumed that the two editions were on sale concurrently at the end of 1704.
17  The title is assumed from other known parts. The imprint and date are supplied by Hodson, 1984, p. 18, supported by an advertisement (in *Post Boy* 1–3 Oct. 1719). As stated above no copy has been found.
18  Thomas Cox had assumed control in April, 1724 and advertised in some issues that some counties were available as separate items. Prices ranged from 1s. for Berkshire (two half parts) to 7s. for Warwickshire (5 whole and 2 half parts). The title is taken from the reissue in [1739] by Ward and Chandler – see note 21. Normally the first leaf of the county text appears on the verso of the previous county's mileage chart; in these separate issues of individual counties the first leaf was cancelled and replaced by a new title-page with the first page of text reset word for word on its verso. County 'booklets' were not advertised by Cox after April, 1731. It is assumed, therefore, that if a Lincolnshire example exists it would be dated 1730, as in the Dorset example owned by Mr. Donald Hodson.
19  The new title-pages, replacement text and imprint labels did not arrive until January, 1739 and the title-page date is, therefore, inaccurate. (Hodson, 1984, p. 28).
20  This BL copy is one of the second versions, i.e. it includes the reset version of Part 1.
21  As the new title-page of 1738 reports, separate counties were also available from Ward and Chandler, who used up already prepared separates from Cox's time and stuck a new label with their own imprint over Cox's imprint, early in 1739.
22  Pasted inside the set at CUL (VIII.3.1–6) is a copy of Ward and Chandler's price-list for the individual counties. Lincolnshire was offered at 2s. 'sew'd up in blue Paper.'
23  No-one knows who produced these late examples, why so much trouble was taken to produce so many variant title-pages or why two of these examples are dated 1700, when the plates did not appear until 1701. The date 1720 in the other cases relates to the first appearance of Lincolnshire in bound form, but that date is also found on the spurious titles for counties which were issued after 1720 in part and volume form.
24  The title-page of this copy has a watermark GR in a large italic script above the date 1797. A search of the *Lincoln, Rutland and Stamford Mercury* yielded no advertisement from the county book trade for this work between 1796 and 1801.
25  In (a) the setting is in Baskerville type and (b) a Bodoni-style type face. In addition the setting of the title differs; in (a) the last line of the title begins 'composed'.

Skelton no. 123. Hodson nos. 125–129.

# 21

**SUTTON NICHOLLS**                                                                    **1712**

Size: 39.1cm x 48.2cm.

Some discussion has taken place already (**6** and **13**) of the atlases put together by John Overton from c.1670 onwards and the continuation by his second son, Henry, to whom he sold the business in 1707 on the latter's marriage.[1] John Overton acquired twelve county plates (those engraved by William Smith in c.1602–3) after the death of a later owner, Peter Stent, in 1665 plus one engraved by Norden. Between 1666 and 1685 five more county plates were engraved to his order; he was thus enabled to make up complete atlases mainly comprising sheets printed by Blaeu and Jansson. In c.1711 Henry began to use Sutton Nicholls to engrave plates for those counties for which he had no plates (and, presumably, no supply of ready printed sheets). He had no further need of this approach when he bought the Speed plates, which had belonged to Christopher Browne, in 1713 or 1714. Of the six plates prepared by Nicholls Lincolnshire was one. By 1715

Henry Overton was in a position to issue atlases made up from plates in his sole ownership.

Five of the plates engraved by Nicholls are based on Speed via Jansson, the other (Nottinghamshire) on Morden. Roads have been added and stars are included to denote towns returning members to parliament. The roads have been taken from Morden's map for Camden's *Britannia*, with their inaccuracies uncorrected and one major one added for good measure.[2] There is a good deal more of Morden also; e.g. in adding the names of the wapentakes the four mis-spelt names of the 1695 Morden are followed.

Sutton Nicholls is first noted as an engraver for the new map of England and Wales for the Saxton atlas issued in the version revised by Philip Lea in c.1689 (**1 (iv)**). His name appears on five of the maps engraved for Camden's *Britannia* in 1695.[3] Although his name does not appear on the Nottinghamshire map of 1714 prepared for Overton it seems probable that he did engrave it and that it was his last identifiable work. In the assessment for the parish of St. Faiths-under-St. Pauls for 1695 he was described as a printseller.[4] In 1692 when he advertised a pair of globes they were 'Made and Sold' by him.[5] He died in the period between 1721 and 1731.[6]

**Later history**

Overton issued his first atlas with the complete assembly of plates in his possession in c.1720. Hodson puts the atlas as 'post-1716' as the Northumberland plate copies details from Warburton's map of that county that had not been added until late 1716, but Hodson also provides good reasons for a later date. Overton altered the northern counties of his Speed plates to indicate his ownership, probably in 1715 or 1716, perhaps in an attempt to catch a public then interested in the Jacobite rebellion. It then seemed natural to treat the plates for the other counties similarly and, allowing time for the work, Hodson suggested that 1720 would be as near as we could now ascertain.[7] By the time the atlas form appeared the Lincolnshire map was, however, in a second state with a tide table calculated by Stukeley, which was used also on the plate for Lincolnshire by Stukeley's friend, Herman Moll, in 1724 (**24**). We do not know when Stukeley prepared his chart but, as the date in the example in the tide table is Nov.7.1723 that might imply its preparation in 1722 in time for the 1723 almanacs or, perhaps more likely, around the date in the chart in readiness for almanacs issued for use in 1724. It is worth noting that the wording in the instructions for the use of Stukeley's chart is (excepting one minor difference) identical in both Moll and Nicholls. Since there is a notable spelling error (peice, instead of piece) in both, clearly one has been copied faithfully from the other or, less likely, they were copied simultaneously from an erroneous source. We know Moll used the chart no earlier than 1724 so it seems probable that Nicholls used the chart first, but not perhaps four years earlier as present knowledge suggests.

Overton issued atlases until his death (1751) but only at widely spaced intervals. After his death his shop was taken over by his nephew, also Henry, but by 1754 the plates were owned by W. and C. Dicey, who listed them in their catalogue of that year.[8] Only four examples of an untitled atlas have been found and they are all different. The latest was probably put together in about 1782 and, after that, the plates disappear.

**EDITIONS**

(**i**)  *A New Mapp of/LINCOLN SHIRE/with the Post & Cross Roads, & other/remarks, according to the latest & best/observations 1712/ Printed & Sold by Henry Overton at the/white Horse without Newgate LONDON./* (Ae, in a rectangle around which are *putti*, a hunter and two fishermen). *Scale of English Miles/8* [= 5cm] (Ee). *Sutton Nicholls sculp* (Ee, below the scale and between the inner and outer borders). Eleven shields, one blank (Aa–Ac). *LINCOLN SHIRE is 180 miles in circumference, con/taining, seventeen hundred & fortey thousand Acres of land:/is divided into 30 Wapontakes or Hundreds:*

*has 30 Markett/towns; 161* (sic) *Parishes; and by computation has in it 40590/houses; sends 12 members to Parliament./The Air is thick & good, it's Soil fruitfull, the most noted/Rivers are the Trent, Humber, & Wytham. It's commoditys/are fat Cattle, Pikes, Mallards, Horses, Wool, & Pippins./ It's chief City is Lincoln, which has in it 14 Parish Churches,/is Scituated North-West* (sic) *from London, distant 102 miles./the market day is Friday./* (Ea). *Note. This starr * Shows the Towns which send/Members to set in Parliament./The double lines thus* ══════ *are Post Roads/The single lines thus* _____ *are Cross Roads/* (Ab). The four cardinal points are named in the centre of the respective borders between the inner and outer frames. The inner frames are marked off in minutes at 2′ intervals, with (between the frames) numbers at 10′ gaps. The figures for longitude use the Azores meridian, indicated by the number *17* (Ce).[9]

Issued as a separate sheet. [1712].

**Copy**   Bod Gough Maps. Lincolnshire 3.[10]

(**ii**)   The descriptive text has been replaced by a tide table in one panel (Ea), with an explanatory text of 17 lines in a separate panel (Da).[11]

*England Fully Described in a Compleat Sett of Mapps of y[e] County's of England and Wales, with their Islands, Containing in all 58 Mapps Printed & Sold by Henry Overton at y[e] white Horse without Newgate London* [1720].[12]

**Copies**   BL C.29.f.2;[13] Leeds UL W 13. (Bod Gough Maps Lincolnshire 2; LRL Map 84a; Grimsby RL NICHOLLS 1712).

Atlas without title-page. [c.1738].[14]

**Copy**   The private collection of Mr. C.A. Burden.

(**iii**)   The imprint date has been scratched out.

Issued as a separate sheet. [After 1743].[15]

**Copy**   (LRL Map 84).

(**iv**)   Overton's name and address have been deleted; the imprint now reads: *Printed & Sold by C. Dicey & Co. in Aldermary/Church-Yard LONDON./*

An untitled atlas, issued by C. Dicey & Co. [c.1770].[16]

**Copies**   CUL Atlas 4.77.2. (LAO Scorer Map 19; LRL Map 93; Manchester UL Map C273).

An atlas without title-page issued by C. Dicey & Co. [1782?].

**Copy**   Pembroke College, Cambridge. e 3.1.[17]

### References

1   Plomer, 1922, p. 225; Hodson, 1984, pp. 60–63.
2   The Great North Road at Grantham continues (with double lines) directly to Lincoln. The continuation beyond Grantham to Newark is marked by a single line denoting a cross road.
3   Skelton, pp. 173 and 194.
4   Tyacke, p. 128.
5   *London Gazette,* 1–5 Sept. 1692.
6   Vertue, G. Vertue Note Books; edited for the Walpole Society. (London, 1937–1947). Vol. 2, p. 11. Vertue marked his list of practising engravers 'ob' [=obit.] between those dates. Cited in the biographical note on Nicholls in Tyacke, pp. 128–129.
7   Hodson, 1984, pp. 44–47 and 63–64.
8   Bod 258.c.109 (p. 37).
9   Further proof that Nicholls had used an early map to make his own copy. The meridian based

on Greenwich was generally used from 1694. (See Fordham, H.G. Studies in carto-bibliography. (Oxford, 1914), p. 15.).

10 This example was sold, presumably from 1712 or shortly afterwards. No atlas form has been found containing it; by the time Overton produced an atlas the Lincolnshire plate had been reworked.

11 The text begins by acknowledging that it was 'communicated by D$^r$ Stukeley.' The date chosen for calculating a passage 'over the Washes 7.Nov.1723.' may indicate that of compilation.

12 As discussed above Hodson's view is that the atlas appeared in c.1720. If the date in the tide table has any significance it perhaps favours an argument that the plate was prepared some time after 1720. What remains a mystery is why Overton used his Speed map in favour of the new plate he had just had engraved and why the two plates were used interchangeably for twenty or more years. (See 1 (vi–viii) above).

13 The BL copy lacks the title-page.

14 The Huntingdonshire map is dated 1738.

15 In the issue of an atlas by Overton (BL Maps 145.c.9 – see 6 (viii) above) there were only two plates of Nicholls (Derbyshire and Yorkshire), Lincolnshire being a revised Speed. Hodson (1984, p. 62) had wondered whether the other Nicholls maps would turn up in a similar state, i.e. with date scratched off. LRL Map 84 bears out his hypothesis. No atlas has been found containing this version.

16 One of the general maps in the CUL copy is dated 1770.

17 Bound in, often on the backs of the Overton/Dicey plates, are other maps. One that might have been included when the work was first assembled is dated 20 May, 1782 (Bowles's New Pocket Map of Scotland).

Hodson nos. 142–145. Hodson, 1989, pp. xiii–xiv amend or add to earlier details.

## 22

## JOHN LENTHALL                                                1717

Size: 5.85cm x 5.6cm (the card is 8.9cm x 5.6cm.)

The career of Lenthall has been discussed in connection with a re-issue of Redmayne's playing cards (16). In 1717 he produced a set of county map cards, which are closely modelled on the third state of Morden's cards (15 (iii)). Some place-names have been changed, e.g. *Spon Head* is now *Spurn Head*, the missing letter has been inserted in *Splding* and other errors rectified. Not all is better, however; *Shelford, Cowland* and *Quapled* survive instead of Sleaford, Crowland and Whaplode respectively. As in Morden Lincolnshire is the eight of clubs; some other counties differed from Morden.

The cards were published in October 1717 and he advertised them simultaneously with the Redmayne set and many other playing card sets.[1] Originally sold for 2s. per pack, the price was increased to 5s. by 1733, reduced to 2s. in 1741 and offered at 2s.6d. in 1750. Lenthall had died or retired by 1734 and these offers were made by quite a large number of retailers until the mid-1750s.[2]

Perhaps surprisingly there have been no new examples of the fitting of fifty-two county maps for England and Wales on to a set of fifty-two playing cards after this issue; after the 1974 changes to local government boundaries such an issue became anachronistic.

## EDITIONS

(i)    *Lincoln Shire*. (In top panel, with club sign to the left and *VIII* to the right). Compass (Eb in map panel). In the lower panel the following details are given: *Length. .60/Breadth. .44/Circumference. .170/ Lincoln. { Distance from London: 103: 128/Latitude 53/* [Scale] *10/Miles* [= 1cm] (Ae, in the map panel). The border of the card is plain.

**Copy**   No set has been found that contained Lincolnshire.[3]

(**ii**)   As in state (i) but with a decorated border of alternating black and white strips.

**Copy**   The private collection of Miss Sylvia Mann.[4]

### References

1   *Evening Post* 22–24 Oct.1717 is cited by Hodson, 1984, p. 75 as the first advertisement for the new set of cards.
2   Hodson, D. 1984, pp. 71–76; Mann, S. and Kingsley, D. 1972 op. cit., p. 21; Wayland, H. and Wayland, V. Lenthall pack no. 4 Map Cards. *Journal of the Playing-card Society,* Vol. II, no. 1 (Aug. 1973), pp. 7–11.
3   Bod. Douce Collection has two part sets of 18 cards each, which (remarkably) are identical; they comprise all the Diamond suit, the king, queen and knave of Hearts and the two explanation cards. The Cary Collection in the Beinecke Library of Yale University has 17 cards – exactly as those in Bodleian except that it lacks the 5 of Diamonds. The Bod. Douce collection also has a third example of the 3 of Diamonds.
4   Miss Mann has 40 cards and very kindly sent me a photocopy of the Lincolnshire card. A full set is part of the collection of Mr. Field of New York. There is one further card known – in the Bod. Douce Collection – 5 of Clubs (Derbyshire).

Hodson no. 147.


## 23

**EMANUEL BOWEN**                                                           **1720**

Size: 11.5cm x 11.5cm.

Emanuel Bowen was a prolific producer of maps over a long career spanning almost fifty years. At one time Bowen was engraver to George II and held a similar appointment to Louis XV. The work discussed here had many editions between its first appearance in 1720 and the last in about 1767, the year he died, and the number of surviving copies is testimony to its sales and continuing usefulness.

*Britannia Depicta* is a road-book with county maps inserted in a random manner; there is no geographical or alphabetical scheme for their placing. After Ogilby's *Britannia* (1675; second edition, 1698) no rival work had been produced for many years. As a large folio volume the work was not easily portable and the need for something more pocketable was recognised. In 1717 or 1718 Moll began a work that he left incomplete (Appendix, **2**) but a pair of such works produced by Gardner (Appendix, **3**) and Senex (Appendix, **4**) came out in 1719 and, in the following year Bowen's *Britannia Depicta*. All four were copies of Ogilby and showed no new material based on fresh surveys. Bowen only differed by the inclusion of very cramped county maps which have no claim to beauty or originality and in the inclusion of notes on the historical sites the traveller might encounter.

In this project Bowen had the help of Thomas Bowles, who had a hand in Moll's abortive scheme. Bowles advertised in 1718 his intention to produce a version of Ogilby that added interesting information to the strip maps.[1] When Bowen was brought in is not clear and he is only referred to in the announcements near the completion of the project. It is possible that Bowen was made a junior partner in the enterprise as whole or part payment for the work of engraving.[2] The work was rushed out early in 1720[3] to thwart the rivals; early copies are 'unstable' with corrections being added to plates as quickly as possible and each volume having individual mixtures of sheets in various stages of correctness. Four basic states for 1720 have been identified and two further states, both dated 1720, probably published in 1722 and 1723.[4] Only those editions that show a variation of title-page or text on the map page are recorded here since the map never varied throughout its long use. The strip maps showing Lincolnshire roads are listed below (Appendix, **5**).

The text was written by John Owen, an antiquarian who had been admitted to the Middle Temple in 1718. Bowles made reference to an involvement in the project by Henry Overton, who acted as a distribution agent perhaps, but never to Owen or Bowen; an irony now that Bowles' work here is usually referred to as Owen/Bowen.

### Later history
After Bowen's death in 1767 his other works continued to be put out by his son, Thomas. *Britannia Depicta* was not included in the inheritance, since it had been published at first by Thomas Bowles with Bowen and then, from 1730, by Bowles alone. The final two editions were issued by Carington Bowles to whom the plates had passed.

### EDITIONS

(i)   The map appears on page 206 (given in the top left hand corner of the sheet). Above the map, with the heading: *The Road from/ NOTTINGHAM TO GRIMSBY IN/LINCOLN-SHIRE. Containing* . . . are the places and mileages along that road. The map is on the verso of a strip map showing the final section of the road from *Breakneck* to *Lanbedor*.

*A MAP of/LINCOLN/SHIRE-/* (Be). *English Miles/ 10* [= 2cm]. (Ad). *The Hundreds & Wapontakes* appears above a numbered list with four names in the first column, seven in the second and twelve in the third. (Da–Ea). *The rest of $y^e$ Wapontakes &c* above the remaining seven names (Ae). A panel of text beginning: *The County of/LINCOLN-/is 180 M. in/Circumference,/contains abt/1740000 A=/cres* . . . (Ab–Ac). There is no compass. A decorative border surrounds all but the bottom edge.

*Britannia Depicta Or Ogilby Improv'd; Being a Correct Coppy* (sic) *of $M^R$ Ogilby's Actual Survey of all $y^e$ Direct & Principal Cross Roads in England and Wales: Wherein are exactly Delineated & Engraven, All $y^e$ Cities, Towns, Villages, Churches, Seats &c scituate on or near the Roads, with their respective Distances in Measured and Computed Miles. And to render this Work universally Usefull & agreeable, [beyond any of it's kind] are added in clear & most Compendious Method 1, A full . . . Account of all the cities . . . By $In^o$ Owen of the Midd: Temple Gent. 2, The Arms of the Peers . . . 3, The Arms of all $y^e$ Bishopricks & Deanaries . . . 4, The Arms, & a succint* (sic) *Account of both Universities . . . Lastly Particular & Correct Maps of all $y^e$ Counties of South Britain . . . By Eman: Bowen Engraver London Printed for, & Sold by Tho: Bowles Print & Map Seller Next $y^e$ Chapter House in $S^t$ Pauls Church-Yard & Em. Bowen next $y^e$ King of Spain in S: Katherines. 1720.*

**Copies**   BL Maps C.27.a.12; CUL Atlas 7.72.8; Bod G.A. Eng. Roads 8° 92; Leeds UL W 244; Nottingham UL s/G 5514; Birmingham RL A 912.42.

*Britannia Depicta . . . & E. Bowen Engraver & Print Seller near $y^e$ Stairs in $S^t$ Katherines, 1720*

**Copies**   BL Maps C.27.a.15 etc.;[5] CUL Atlas 7.72.6; Bod Vet A4 e 942; RGS 10 B 33; Manchester RL BR 912.42 B5; Leicester RL L912.[6]

*Britannia Depicta . . . [$Y^e$ $4^{th}$ Edition.] . . . Sold by Tho: Bowles . . . & I. Bowles Print & Map Seller over-against Stocks Market 1724*

**Copies**   BL Maps C.27.a.17; CUL Atlas 7.72.5; Bod Johnson Maps 241 and GA Eng Rds 8° 87; Leeds UL W 245; Northampton AO NRO 53; LAO Foster L. 950.2.[7]

*Britannia Depicta . . . Bowen Engraver The fourth Edition. Printed & Sold by Tho:
Bowles Print & Map Seller in S<sup>t</sup>. Pauls Ch Yard 1730*

**Copies**  Bod Allen 19 and Vet A4 e 1295; NLS Newman 843.

*Britannia Depicta . . . 1731*

**Copies**  BL Maps C27.a.18; CUL Atlas 7.73.2; Bod Gough Maps 116.

*Britannia Depicta . . . 1734* [8]

*Britannia Depicta . . . 1736*

**Copies**  BL Maps C.27.a.19 and C.27.a.20; CUL Atlas 7.73.1; Leeds UL W 246;
Birmingham RL A 912.42; Liverpool Athenaeum 910.3; NMM 912.44 (42) "1736": 094
[E 8671].

*Britannia Depicta . . . 1749*

**Copies**  Manchester RL BR 912.42 B6; John Rylands UL 942.006 Og (R14856).

*Britannia Depicta . . . 1751*

**Copies**  BL Maps C.27.a.21; Bod Allen 28 and Map C.17.e.23.

*Britannia Depicta . . . 1753*

**Copies**  BL Maps C.27.a.22; CUL Atlas 7.75.5; Leeds RL R912.42 OG4.

*Britannia Depicta . . . 1759*

**Copies**  BL Maps C.27.a.23; Bod Vet A5 e 2499.

(**ii**)  The map remains unaltered, but the mileage in the heading has been changed to 67
(formerly 67′1); the first column of figures below is unchanged but those in the second
column read: *17, 31′6, 38′6, 48′4* and, in the fourth column, the last figure is now 67.

*Britannia Depicta: Or, Ogilby Improved. Being an Actual Survey of all the Direct and
Principal Cross Roads Of England and Wales; Shewing All the Cities, Towns, Villages,
Churches, Gentlemen's Seats, &c. situated on, or very near any of the Roads; With
Distances laid down in Measured Miles, through each Road. Engraved By Emanuel
Bowen, Geographer. To which is added, An accurate Historical and Topographical
Description of all the Cities, Boroughs, Towns Corporate, and other Places of Note.
Compiled . . . By John Owen, Gent. The Whole illustrated with Maps of all the Counties
of South Britain, and a summary Description of each. London: Printed for Carington
Bowles, in St. Paul's Church Yard. MDCCLXIV.*

**Copies**  BL G 4697; CUL Atlas 7.76.4; Bod Vet A5 e 5417; Leeds RL R912.42 OG4;
John Rylands UL 942.006 Og 4 (R85119); Grimsby RL 914.2 OWE.

*Bowles's Britannia Depicta . . . London: Printed for the Proprietor Carington Bowles,
No. 69, St. Paul's Church Yard.* [1767].[9]

**Copy**  BL Maps C.27.b.68.

## Reproductions
(i) Britannia Depicta or Ogilby Improved By Emanuel Bowen 1720 Facsimile Reprint.
(Newcastle-upon-Tyne, 1970). The very useful introduction is by J.B. Harley. (ii) A
facsimile edition of the 1731 edition has also been published (Tiverton, 1969).

## References
1  *Post Boy* 22–24 Jan. 1718/9; his first proposals were given in the *Daily Courant* 29 Dec.1718,
   the day that Gardner's road-book was published and four days before the version by Senex.
   Hodson, 1984, pp. 80–81.
2  Hodson, 1984, p. 82.

3   *Post Boy* 9–11 Feb. 1719/20 and *Courant* 12 Feb. 1719/20; both announced its issue after
    other promised publishing dates had passed. Hodson, 1984, p. 82.
4   Hodson, 1984, pp. 78–87. The point is made that more variants will exist because of the work's
    basic instability bibliographically. Indeed, he cites several further variants from his six basic
    types and others have been found since.
5   The BL has several examples, which are all differently made up, but they all have in common
    the title-page as quoted and the map.
6   Leicester City Library's copy lacks pp. 85–88, but it is in all other respects an example of
    Hodson's sixth state ascribed to 1723.
7   This copy came to Lindsey County Council from the library of Canon Foster. From signatures
    on the leaf facing the title-page it had apparently been in family ownership since 1726.
8   Chubb, No. CLIIa cites, for an edition in 1734, a reference in: Fordham, F.G. John Ogilby
    (1600–1676) His Britannia, and the British Itineraries of the eighteenth century. *The Library.*
    IV,vi (1926), p. 175. In 1730 Thomas Bowles took over production; every spring he advertised
    'This day is published . . .' However, no copy of the work (or an advertisement) dated 1734
    has been found. Mr. C.A. Burden owns a copy, which, in the make-up of its preliminary pages,
    matches the 1736 edition but with 1731 on the title-page, although the final digit shows signs
    perhaps of an intended alteration.
9   The numbering of houses in London began in the 1760s and by 1767 Bowles was quoting his
    address as shown. The work was for sale still in 1795.

Hodson nos. 149–165.


## 24

**HERMAN MOLL**                                                    **1724**

Size: 25.75cm x 19.1cm.

Moll was probably German[1] and his date of birth has been put at 1654[2] in Bremen[3]
although the evidence has not been found.[4] He is initially recorded as an engraver and
probably in that capacity he worked on Moses Pitt's abortive *English Atlas* in 1678. In
1688 he was advertising a general map of England and Wales, which he seems to have
engraved and was also, with Christopher Brown[e], selling from his house in Blackfriars.
For a year or two he was a map-seller as well as engraver, but after 1691 he seems only
to have worked at making plates for others.[5]

He was much involved with the engraving of plates for the *Atlas Geographus* (**20**)
and, in 1708, he revised Morden's smaller plates; the additional place-names are in his
neat and distinctive lettering (**20 (ii)**). In 1723 he engraved *A MAP of the/LEVELS/
in/LINCOLN SHIRE/Commonly called/HOLLAND/ Described by W^m Stukeley 1723/*;
although Moll's name does not appear there can be no doubt from the style of the lettering.
We know that Stukeley and Moll were on good terms, as Stukeley notes in his Commonplace book (September, 1732) 'My old acquaintance Mr. Moll, the geographer, dy'd.' [6]
Moll had signed his will on 3 August, 1732 and it was proved three days later (note 1
below).

Moll's atlas of 1724 is his only work of this type and, if his date of birth was 1654 he
was already 70 years old. The maps are often based on Morden but, if more recent
examples were available, he sometimes went to much trouble to incorporate new
information on his plate.[7] He seems to have preferred to use Ogilby for details of roads
rather than follow the use Morden had made of the same source material. The mileages
shown along the roads on the map are taken from Ogilby's computed miles. The maps
are embellished with engravings of items of an antique nature – in Lincolnshire's case
there are engravings of two Saxon coins struck at Lincoln. Further information comes in
the form of the tide table drawn up by Stukeley and also used by Sutton Nicholls (**21**

(ii)). Moll was engaged in the engraving of the charts issued by Capt. Greenvile Collins during 1693 as *Great Britain's Coasting Pilot* but such details as the sandbanks in the Wash and Humber estuary, which Morden used on his maps, do not appear on this map. The shape of the coastline of the Wash is taken from Stukeley's fen map, also used for details on the Cambridgeshire plate.

Stukeley was involved further with this atlas. The first map in the volume is: *Ingratiam Itinerantium Curiosorum, ANTONINI aug. ITINERARIUM per BRITANNIAM, tentavit W. Stukeley 1723*. According to the preface (p.iij) Stukeley gave this map, which shows the supposed route of Antoninus through Britain. The map also appears in Stukeley's own *Itinerarium Curiosum* . . . (London, 1724).

The three partners involved with Moll and whose names appear in the imprint of the first editions were active in the map world over many years. Thomas Bowles, the founder of a dynasty of sellers of maps and prints and publishers, flourished from the 1680s and died in 1721; he has already been encountered in connection with *Britannia Depicta*. Bowles' son, also Thomas, was the joint publisher of this work with his brother John and Charles Rivington.[8]

### Later history

After Moll's death in 1732 later editions were issued by the Bowles brothers until 1753. Rivington (1688–1742) founded a firm with a long history in publishing, only finally being sold to Longmans in 1890.[9] Rivington's interest seems to have been sold to the Bowles by 1739. His sons were share-holders later in some editions of Camden's *Britannia* with Morden maps.[10]

Moll's plates in a later reworked state came to light recently; it had been issued by a later generation of the Bowles family about twenty years after what was previously thought to be the last use of the plates in 1753.

### EDITIONS

(i)  *LINCOLN/SHIRE./By H. Moll Geographer./* (Ee, in a rectangular box). *10/English Miles* [=3.3cm] (Da). The wapentakes are given upper and lower case letters on the map; keys for those with reference letters A–Z (less J and U) and a–e are given (Ea and Ae, respectively). The inner frame is shaded at 5′ intervals, while numbers appear at 15′ intervals between the inner and outer frames. *The first Meridian from London* (Da, between the inner and outer borders). Compass (Ed). Above the map are, at left, two engravings of *Saxon Coin struck at Lincoln* and, to their right, an introduction to the Stukeley tide table that fills the plate below the map.

*A New Description Of England and Wales, With the Adjacent Islands. Wherein are contained, Diverse useful Observations and Discoveries In respect to Natural History, Antiquities, Customs, Honours, Privileges, &c. With A Particular Account of the Products, Trade, and Manufactures of the respective Places in every County . . . Also, Several Errors of different kinds are Rectified. The Whole illustrated with many Historical and Critical Remarks. To which is added, A new and correct Set of Maps of each County, their Roads and Distances; and, to render 'em the more acceptable to the Curious, their Margins are adorn'd with great Variety of very remarkable Antiquities, &c. By Herman Moll, Geographer. London: Printed for H. Moll over-against Devereux-Court in the Strand, T. Bowles Printseller near the Chapter-House, and C. Rivington Bookseller at the Bible and Crown, in St. Paul's Church-yard, and J. Bowles Printseller over-against Stocks-Market. M.CC.XXIV.*

**Copies**  BL G 1290 and 191 e 9;[11] Bod Gough Gen Top 220; Leeds UL W 49; NLS EME b.1.19; Manchester UL AMGSB1; Manchester Chethams L. 2 H 6. 37. (LRL Map 49a).

*A Set of Fifty New and Correct Maps of England and Wales. &c. With The Great Roads and Principal Cross-Roads, &c. Shewing the Computed Miles from Town to Town. A Work long wanted, and very useful for all Gentlemen that Travel to any Part of England. All, except two, composed and done by Herman Moll, Geographer.* [Numbered list of maps]. *And, to render this Work more acceptable to the Curious, the Margins of each Map are adorn'd with great Variety of very remarkable Antiquities. London: Sold by H. Moll over-against Devereux-Court in the Strand; Tho. Bowles, Print and Map-Seller near the Chapter-House in St. Paul's Church-Yard, and J. Bowles Print and Map-Seller over-against Stocks-Market. 1724.*

**Copies**   CUL Atlas 7.72.4; Bod Allen 18[12] and M.S. Top. Gen. b.73;[13] NMM 912.44 (42) "17": 094 [B 5642].

(ii)   A plate number has been added: *(28)* (Aa, between the inner and outer border).

*A New Description of England and Wales . . . M.DCC.XXIV.*

**Copies**   CUL Atlas 5.72.1; Bod Allen 17; RGS 265 E 3; Leeds UL W 49; Birmingham RL HGD/094/1724/12F; West Suffolk CRO (Cullum 942). (LAO Scorer Map 7 and Foster 912; LRL Maps 798 and 49; Louth RL Map A 18; Grantham RL).[14]

*A Set of Fifty New and Correct Maps . . . 1724.*

**Copies**   BL 118 c 19 and B264 (3);[15] Bod Gough Maps 108, etc.; Leeds UL W 50; Birmingham RL A912.42 (incomplete); RGS 8 B 5 and 8 B 37; Sheffield RL 912.42 STQ (ST 6).

*A New Description of England and Wales . . . 1728.*[16]

*A new description of England and Wales . . . By Herman Moll . . . London: Printed for J. Wilford, behind the Chapter-house; T. Bowles printseller, and C. Rivington bookseller, in St. Paul's Church-yard; and J. Bowles print-seller, in Cornhill. MDCCXXXIII.*[17]

*A Set of Fifty New and Correct Maps Of the Counties of England and Wales,&c . . . London: Printed for, and Sold by Tho. Bowles, Print and Map Seller, near the Chapter-House in St. Paul's Church-Yard; and J. Bowles, Print and Map-Seller, at the Black Horse, Cornhill. 1739. Where may be had . . . Thirty-six Maps of . . . Scotland . . . Twenty Maps of . . . Ireland. Atlas Minor, or . . . Sixty-two Maps of . . . the World. All by H. Moll . . .*

**Copies**   BL Maps C.10.a.9; CUL Atlas 6.73.1; Liverpool RL G 229; Leeds RL SRQ 912.42 M735; Birmingham UL r q.G1854; NLS EME b.1.5.

(iii) The plate number has been changed to *(25)* (Aa, between borders). The maps are now arranged in alphabetical order, the new numbers facilitating binding.

*The Geography of England and Wales; Or, A Set of Maps of all the Counties In England and Wales, With The Great Roads . . . the Distances from Town to Town. A Work long wanted . . . 1747. Where may be had . . .*

**Copies**   CUL Atlas 5.74.1. (BL Map 3355 (12); RGS 1.C.88 (17 A & B – i.e. two exx.); Boston RL Map 4).

(iv)   All the matter above and below the map has been removed. Roads have been added: *Eston* [= Eastoft, north-east of Crowle] to Barton and Grimsby; Lincoln to Bawtry through *Ganesborough.* Over eighty place-names have been added, e.g. Baston, Utterby, Hundleby and Scotherne and including the term: *Salt Marshes* (north of Holbeach) and in Norfolk. All mileages have been changed. More than twenty place-names have been corrected, e.g. *Knaith* (previously *Knath*), *Potter* added above *Hanworth, Carlton/Scroop* (previously *Carliton*) and *Belvoir* (previously *Bever*). Other changes on the plate have also been made, e.g. in order that *Timberland* could be entered *Thorpe Tilney* has been removed; similarly *Goultho* provides space for *Wragby* to be shown; and, *Glanford/ Briggs* has been re-engraved as *Glanford/Bridge*. Many other changes of this type occur.

*H. Moll's British Atlas: Or, Pocket Maps of all the Counties. In which are carefully laid down All the Great Roads, and the Principal Cross Roads in each County, with the Distances from Town to Town. Composed and Engraved by Herman Moll, Geographer; And lately revised and improved, with the Addition of many Hundred Places, by Emanuel Bowen, Geographer to His Majesty. The Maps are placed in Alphabetical Order . . . London: Printed for Tho. Bowles, near the Chapter House in St. Paul's Church Yard; and J. Bowles and Son, at the Black Horse in Cornhill. MDCCLIII. Where may be had . . .*

**Copy**  Admiralty Library Vd 7.

*The Traveller's Companion; or, A Complete Set of Maps of All the Counties in England and Wales: Laid down upon the latest Surveys, and other best Authorities. Distinctly shewing the Cities, Towns and chief Villages; the Rivers, Sea-Coasts, and Harbours. And being designed for Use in Travelling, all the Great Roads, and principal Cross Roads, are particularly described. Composed by Herman Moll, Geographer. Revised and improved by Emanuel Bowen, Geographer to his Majesty . . . Printed for John Bowles at No. 13 in Cornhill.* [1775?].[18]

## Reproductions
(i) The Lincolnshire map in state (i) is illustrated in colour in: Beresiner, p. 150. (ii) Reduced facsimiles were used to illustrate an edition of *A tour thro' the whole island of Great Britain . . . by Daniel Defoe, Gent.* (London, 1927). The Moll maps were thought the most appropriate as the atlas and Defoe's book came out in the same year. (ii) *A Set of Fifty New And Correct Maps of England And Wales* (1724) has been re-issued in a facsimile edition (Old Hall Press, Leeds, 1993).

## References
1  Tyacke, pp. 122–123, on the basis of his will (PRO PROB 11/654/251) in which his goods in Great Britain and Germany are left to his daughter, Henderina Amelia Moll.
2  J.N.L. Baker. The earliest maps of H. Moll. *Imago Mundi*, Vol. 2 (1937), p. 16.
3  Bonacker, W. Kartenmacher aller Länder und Zeiten. (Stuttgart, 1966), p. 162. He refers to a memory of finding such a reference but its origin is not now to be found ('deren Herkunft nicht mehr auszumachen ist'). Since he gets Moll's date of death badly wrong his dictionary's note is quite dubious.
4  Reinharts, D. New information on Herman Moll. *Imago Mundi*, Vol. 40 (1988), pp. 113–114. Reinharts failed to find any worth-while details to prove Moll came from Bremen, largely because Moll(e) was a common German name and the Russians had removed the bulk of the necessary documents in 1945.
5  Tyacke, pp. 45–49 *passim* for announcements in *London Gazette* in which he also sold maps; the last appeared on 13–16 April, 1691.
6  The Family Memoirs of the Rev. William Stukeley and other correspondence . . . Surtees Society. Vol. LXXIII (1880). This is the first of a three volume set completed in 1887.
7  Kingsley, p. 64 shows that Moll carefully reduced the first large-scale map of Sussex, i.e. Budgen (1723) 'a map drawn on three sheets and six times the scale'.
8  Hodson, 1984, pp. 186–191 provides a family tree and a history of the Bowles family, its works and addresses.
9  Plomer, 1922, p. 254 says he was the son of a Chesterfield bookseller, who set up in 1711. For a history of the firm: Rivington, S. The publishing family of Rivington. (London, 1919). Rivington is also noticed above (**19** note 23).
10  They are two of '11 others' who published the 1753 edition (see **19** (**iii**) above).
11  Copies occur in intermediate stages. Plate numbers must have been added quite soon after publication in July, 1724. In BL 191 e 9 only one plate has had a plate number added – Northumberland.
12  Although lacking a title-page the Bod copy conforms in every other detail.

13 This consists of two sets of the maps in state (i), which have been cut down to the inner frame line; there are no preliminary leaves or title-pages. On the Lincolnshire examples Stukeley has drawn the roman roads and roman camp sites.

14 Single sheets with plate number 28 can not be ascribed to a precise edition between 1724 and 1739 (but see note 16 below).

15 BL B 264(3) appears to be Sir Joseph Banks' copy. On the verso of the title appears: Jos: Banks, applied by rubber stamp.

16 Hodson, 1984, p. 115: 'It has been suggested that an edition was published in 1728. This is based on an entry in John Worral's *Bibliotheca Topographica Anglicana* (London, 1736).' Whitaker, H. The printed maps of Northamptonshire. (Northampton, 1948), p. 59 first picked up the reference in Worral's work, but no copy of the item has been found. It is now thought that Worral erred in thinking that John Bowles' catalogue of 1728, which listed Moll's *New Description . . .* was referring to a new work.

17 Hodson, 1984, pp. 119–122 gives a detailed account of the reissue of the text of 1724 in 19 monthly parts, between March and August, 1733. No copy of the text bound as a volume has been found. The title quoted is that given in: Whitaker, H. A descriptive list of the maps of Northumberland, 1576–1900. (Newcastle-upon-Tyne, 1949), no. 184a.

   Several examples exist of two maps with consecutive plate numbers appearing back to back; it is believed that such plates could come from an edition of this date. Two are known in Bedfordshire locations of Bedfordshire and Hertfordshire together; NLW has a copy of S. Wales with Monmouth; a private example of Warwickshire/Worcestershire is known, and Mr. E. Burden of Ascot has a copy of Berkshire with the Isle of Wight. If Lincolnshire were found thus it would have Rutland on the reverse. Mr. M. Goldmark of Uppingham and part author of *Maps of Rutland* (Stamford, 1985) has never seen an example.

18 An example of this title was offered by Beeleigh Abbey Books (Catalogue BA/36, item 240) in 1983, where the suggested date of issue was c. 1780. At the end of 1991 the present owner allowed Mr. D. Hodson to examine the volume and he suggests the date 1775. I am grateful to Mr. E. Burden for a full transcript of the title.

Hodson nos. 173–178.


## 25

**WILLIAM HENRY TOMS**                                                    **1741**

Size: 14.35cm x 14.85cm.

W.H. Toms claimed to be the sole proprietor of this very successful pocket atlas. The first title-page and the imprints of the early states of the map indicate that Badeslade was from the first involved in the project, preparing the maps, which Toms then engraved. The partnership quickly came apart, if one is to judge from the evidence in the imprints of the maps. The claim on the title-page that the maps were prepared for George I would mean that they were drawn by 1727 but not before 1724, since they are modelled on those of Moll (**24**). The plates were engraved a year before the atlas came out and Badeslade left the partnership during the interim, it is assumed.[1]

Thomas Badeslade was active from 1712.[2] He not only surveyed the fen area in the Bedford Level and around Kings Lynn but also acted as the undertaker of *The History Of The Ancient and present State Of The Navigation Of The Port of King's-Lyn* which appeared in 1725.[3] The anonymous author was John Armstrong[4] but the dedication to Sir Robert Walpole was made by Badeslade. The map of the Bedford Level is dated 1723 and the copy of the fen map of 1604 by Wm. Hayward is dated 1724. Badeslade also drew *A Map of the NORTH LEVEL . . .*[5] (which includes the area east of Spalding), ascribed to 1744 and engraved by S. Parker.

Toms the engraver was also a print-seller in London and active between 1723 and

1758. As an engraver he was responsible for the well known map of Boston, 1741/2, by William Nichols and Robert Hall.[6] Other Lincolnshire work included the 1743 Grundy map of the River Witham[7] and *A MAP of the great LEVEL of the FENS . . .* which is included in *The Ancient and present State Of The Navigation Of The Towns Of Lyn, Wisbeach. Spalding and Boston . . . 1751* [by Nathaniel Kinderley].[8]

### Later history
There were several editions in the 1740s by Toms; the last was probably issued in 1749. Hodson shows that shares in the work were still being traded until 1765[9] when the total valuation of whole work may have been the worth of the copper at scrap value prices.

## EDITIONS

(i)  *A Map of LINCOLN SHIRE North from London* (Ba–Da, OS). 25 (Ea, OS). *Published by the Proprietors T. Badeslade & W.H. Toms Sept*<sup>r</sup>. *29*<sup>th</sup>. *1741.* (Be–De, OS). *T. Badeslade delin* (Ae, OS). *W. H Toms Sculpt* (Ee, OS). *English Miles/15* [= 2.75cm] (Ee). Compass (Eb). Down the left hand side in a separate panel is information about the county; Lincoln, Boston, Grantham and Stamford are followed by other towns in alphabetical order. There is no information shown against five places in the list: Helby, Beckingham, Binbrook, Epworth and Stainton.

*Chorographia Britanniæ. Or A Set of Maps of all the Counties in England and Wales: to which are prefix'd the following general maps, viz. 1 . . . the Sea Coast . . . II . . . England and Wales . . . III . . . Roads from London . . . IV . . . Cross Roads . . . With the particular Map of each County, is an Account of all the Cities, Boroughs, Market Towns, Parishes, and Rivers therein: the Number of Members it sends to Parliament, the Market and Fair days . . . This Collection . . . was first Drawn and compiled into a Pocket Book, by Order and for the use of his late Majesty King George I. By Thomas Badeslade Surveyor and Engineer, and now neatly Engrav'd by Will: Henry Toms. Printed and Sold by W H Toms Engraver, in Union Court, near Hatton Garden, Holbourn. 1742.*

**Copies**  BL Maps C.24.aa.23; CUL Atlas 7.74.5 and 6.74.2; Bod Allen 21;[10] Sheffield RL 912.42 ST.[11]

*Chorographia Britanniæ . . . 1742. Price Bound 5*<sup>s</sup>

**Copy**  RGS Fordham 146.

*Chorographia Britanniæ . . . 1742. Price in Sheets 5*<sup>s</sup>. *Bound 6.*<sup>s</sup>

**Copies**  Bod Allen 22; Leeds RL 912.42 B141; NLS Blk 156.

(ii)  The text at the left has been altered: the second line of Boston text reads *& Sat: Fairs July 25. & Nov: 29* (previously the dates for the fairs were *April 23. & July 25.*).

*Chorographia Britanniæ . . . 1742.*

**Copies**  Leeds UL W 51; Manchester RL BR 912.42 B 10.

(iii)  The text at the left has been altered further. The Boston text now occupies three lines; the original fair days have been reinstated with lines 2 and 3 continuing: *& a Mart which/begins the 30 Novomb.* (sic) *& holds 9 days./* Alford now reads *Market Tu Fair Whitsun Tues.* Beckingham has been removed and replaced by *Burgh Mark*<sup>t</sup> *Thurs. Fair May 1. Dunnington* [= Donington] now reads *Mark Sat. Fair May 15.* Under Horncastle *Fair Aug:10* now reads *Fairs June 11 Aug 10* and, under Tattershall, the text is now *Market Frid Fair May 13.*

*Chorographia Britanniæ . . . 1742. Price in Sheets 5ˢ. Bound 6.ˢ*

**Copy**   The private collection of Mr. D. Hodson.

(**iv**)   Many place names have been entered, e.g. Aserby [= Aswardby], *Mablethorp, Leverton, Hiberstow* [=Hibaldstow] and *Hemingsby. P OF RUTLAND* and *Steping R.* have been added. The road from Grimsby to Newark has been drawn in.

*Chorographia Britanniæ . . . 1742 . . .*

**Copies**   BL Maps C.27.b.15; CUL Atlas 7.74.6 and 7.74.15;[12] RGS 9 A 1 and 262 A 21. (LRL Map 50).[13]

(**v**)   More place names have been added, e.g. *Aslackby, Gateburton, Sutton* [= Sutton on Sea], Wilsthorp and *Munbye* [= Brumby, now part of Scunthorpe, or, possibly, Manby (west of Broughton)]. The imprint now reads: *Publish'd by the Proprietor W H Toms Sept. 29. 1742.*

*Chorographia Britanniæ . . . 1742 . . .*

**Copies**   CUL Atlas 7.74.7. (LAO Scorer Map 10; Grantham RL).

*Chorographia Britanniæ . . . Printed for and Sold by W H Toms Engraver, in Union Court, · near Hatton Garden, Holbourn, 1742. Price in Sheets 5ˢ. Bound 6ˢ.& Colour'd 12ˢ. Where is made & Sold the best Transparent Paper for taking off Drawings or Prints . . . No less Quantity than One Quire at 5ˢ. Likewise the best Ground for Etching at 20ˢ Pʳ. Pound*

**Copy**   The private collection of Mr. C.A. Burden.[14]

*Chorographia Britanniæ. Or A New Set of Maps of all the Counties in England and Wales: . . . This Collection (conveying a more comprehensive Idea of South Britain than any thing hitherto publish'd) was first Drawn and compiled into a Book by Order and for the Use of his late Majesty King George I. For his intended Tour thrô England and Wales. By Thoˢ. Badeslade Surveyor. & Engrav'd by W.H. Toms. The Second Edition to which is added the Rates for Hackney Coaches . . . Printed for C. Hitch in Paternoster Row, and W.H. Toms Engraver in Union Court Holbourn 1745. Price in Sheets 5ˢ. Bound 6ˢ. Bᵈ. & Colour'd 12ˢ.*

**Copies**   BL Maps C.27.b.8; CUL Atlas 7.74.8; Bod Maps C.17.f.1; Leeds UL W 53; RGS 262 A 22: Glasgow Mitchell L. S912.42 B 93. CUL Atlas 7.74.9 and Atlas 7.74.7 and Manchester RL BR 912.42 B11 are further examples but are made up of old sheets or an earlier state of title-page and do not conform to the 'standard'; e.g. the Lincolnshire maps in CUL Atlas 7.74.9 and the Manchester RL copy are state (iii).

*Chorographia Britanniæ . . . Toms. Published According to Act of Parliament. Sold by J. Clark Bookseller at yᵉ Golden Ball, in Sᵗ Paul's Church Yard, C. Hitch Bookseller in Pater Noster Row, & W. H. Toms Engraver, in Union Court Holbourn, Price in Sheets 5.ˢ Bound 6.ˢ Bound & Coloured 12.ˢ [1746].*[15]

**Copies**   BL Maps C.27.b.33; Bod Maps C.17.f.4; John Rylands UL 912 B 141 (R61887); Leeds UL Case K-30; NMM 912.44 (42) "17": 094 [C4606]. Leeds UL W 52 is another example, but the right hand half of the Lincolnshire map is missing.

*Chorographia Britanniæ . . . Sold by C. Hitch Bookseller in Paternoster Row, W. Johnston Bookseller at the Golden Ball in Sᵗ. Pauls Church Yard, & by W:H: Toms Engraver & Printseller at yᵉ Golden Head over against Surgeons Hall near Ludgate Hill. Price in Sheets 5ˢ Bound 6ˢ. Bound and Colourd (sic) 12ˢ. [1749].*[16]

**Copies**   BL Maps C.26.a.30; CUL Atlas 7.74.1; Bod Maps C.17.f.3; Leeds UL W 54; Wisbech Museum TB 8.37; Nottingham UL G 5512.

**Reproduction**
(i) The county plate in state (i) appears in Beresiner, p. 50.

**References**
1   Hodson, 1984, p. 155.
2   Eden, P. Land Surveyors in Norfolk, 1550–1850. Part 1. Estate Surveyors . . . *Norfolk Archaeology.* Vol. 35, part 4 (1973), pp. 474–482; the references to Badeslade (pp. 480–481) show he was working in Kent in 1712.
3   Copies in BL 578 1 8; CUL Cam a.725.2; Bod G A Eng Rivers C17; LRL A 386; Wisbech Museum BB 7.39.
4   In the second edition (1766) Armstrong claims authorship on the title-page.
5   LAO Monson 7/16.
6   BL K19.22; Bod Gough Maps Lincs. 8; Nottingham UL Dept. of Geography F 133 (Drawer 121); LAO FL Maps 24.
7   LAO 2 Anc 5/22/1; LRL Map 661; RGS 1 C 88 (44).
8   CUL 7400.c.5; Bod Gough Maps Lincolnshire 12; LAO Sc P2/28; LRL L. A 386; Boston RL L 386; Grimsby RL F301:627.5 KIN.
9   Hodson, 1984, p. 163 quotes the Longman trade catalogue archive, item 139. (BL C.170.aa.1).
10  Bod Allen 21 has the four preliminary pages all dated 1742, implying a later date of assembly than the others quoted here. A large-paper example of this first edition is owned by Mr. C.A. Burden.
11  The Sheffield copy lacks many plates.
12  CUL Atlas 7.74.15 is an intermediate copy, with some plates (including Lincolnshire) dated 1741 and others dated 1742.
13  The date has been scratched off the copy in LRL.
14  Hodson, 1984, pp. 158–159 and 165 (example H) records a copy in Naval Historical Library (Vd 15).
15  Clark joined Hitch and Toms as a joint owner of the plates in 1745 as evidenced by adverts in *General Evening Post (London)* for 26–28 Nov. 1745 and repeated several times later. The re-engraving of the title-page imprint was presumably completed soon after, perhaps in early 1746 but before Clark died in April, 1746.
16  Clark's share had been acquired by William Johnston before December, 1748 when he placed advertisements in *General Evening Post (London).* At that time Toms was still in Holborn, but he was at the new address quoted on the title-page by July, 1750, thus suggesting that 1749 may have been the date of issue. Toms seems to have disposed of his shareholding by 1751. (Hodson, 1984, pp. 162–163 give evidence for the ascription).

Hodson nos. 188–193.

## 26

**JOHN COWLEY**                                    **1743**

Size: 18.3cm x 13.2cm

Robert Dodsley, the publisher of the *Geography of England*, was born in Mansfield in 1703, the son of a schoolmaster. Initially a footman he began, while still young, to write. Pope persuaded Rich (who had made a fortune with *The Beggar's Opera*) to put on *The Toy-shop* and after its success lent Dodsley £100 to set up a bookshop.[1] From 1735 until his death in 1764 he published the work of many of the notable literary people of the age – his friend Dr. Johnson (from 1738 onwards), Shenstone and Mark Akenside *inter alia*. He is more well-known now as the founder of the *Annual Register* in 1744.[2]

In January, 1741 he began publication of a serial *The publick register: or, the weekly magazine.* Only twenty-four numbers appeared before it closed in June, 1741, largely because it had fallen foul of the law relating to stamp duty on newspapers. In March he had decided to avoid printing news (and the consequent tax) and include topographical descriptions with maps. In the event only six counties were dealt with before closure but the seed of a new publishing venture had been sown. The result was the *The Geography of England . . .* In spite of the date on the title-page being 1744 it was published in November, 1743.[3] As Nichols says: 'The Rule in general observed among Printers is, that when a Book happens not to be ready for publication in November, the date of the ensuing year is used'.[4]

The map is based loosely on Morden's small map (**20**); this is especially marked in the delineation of the sandbanks in the Wash and Humber estuary. Extra roads are added, which have not appeared before on any county map; e.g. *Burton* [on Stather] – *Glanford/ Bridge,* then by way of *Castor* and *Binbrook* to *Salt/fleet,* where a road from Horncastle via Louth meets. Apart from these innovatory features one further detail calls for attention. To the west of Boston on the Witham are drawn *N. Canal* and *A New Canal proposed/but not Perfected/.* This probably refers to the commission undertaken by John Grundy, Snr. in 1742[5] which led to the production of three proposals for the improvement of the flow of the river and the avoidance of flooding.[6] Such awareness of recent events had been in very short supply among previous county map makers, who were generally content to copy from Saxton, Ogilby and Greenvile Collins.

John Cowley (*fl.*1733–44)[7] was a political writer and geographer who is not recorded in any other map production. The maps were re-used in a little atlas in 1745 and are not heard of again. A year after Dodsley's death a new edition of *The Geography . . .* appeared, but without maps.

**EDITION**

(**i**)   *An Improved MAP/of/LINCOLN-SHIRE/containing yᵉ Borough/and Market Towns, with/those adjoyning; also/it's Principal Roads and/Rivers. by I. Cowley,/Geographer to his MAJESTY/* (Ae). *Min. E. Long from London* (Ea, between the inner and outer frames). The inner frames are shaded at 5 minute intervals and numbered every 10. *English Miles 10* [= 22.5cm] (Be). *Explanation/Cities/Boroughs/ Market Towns/Abbys/* (with symbols) (Ee). There is no compass.

*The Geography Of England: Done in the Manner of Gordon's Geographical Grammar, Each County being consider'd under the following Heads: viz. The Name, Situation, Air, Soil, Commodities, Rivers, Chief Towns, Noblemen's Seats, Curiosities natural and Artificial, Remarkable persons, Various Particulars. To each County is Prefix'd A compleat Mapp from the Latest and Best Observations, shewing the Chief Towns, Parks, Rivers and Roads, both direct and across. Also a Separate Mapp of England, of the Roads, Of the Channel, and a Plan of London. Likewise . . . A Clear . . . View of our Constitution, And . . . Legislature. London: Printed for R. Dodsley, at Tully's Head in Pall Mall. M,DCC,XLIV.* [1743].

**Copies**   BL 796 e 56; Bod Gough Gen Top 321; Leeds UL W 55. (Nottingham UL Li C17.D 45).

*A New Sett of Pocket Mapps Of all the Counties of England and Wales. Shewing, The Situation of all the Cities, Boroughs, Market-Towns, and most considerable Villages, with the Distances between each. Also the Rivers and Roads both direct and across. Together with A Separate Mapp of England, a Plan of the Roads, and a Chart of the Channel. London: Printed for R. Dodsley in Pall-Mall, and M. Cooper in Pater-noster-Row. 1745. [Price 10s.]*

**Copy**   CUL Atlas 7.74.10.[8]

*A New Sett of Pocket Mapps Of all the Counties of England and Wales. In Which Particular Regard has been made to the Rivers and Roads both direct and across. Shewing also, The Situation . . . Together with A Separate Mapp . . . 1745. [Price 4s.]*

Copy   Guildhall L. s 912/42.

### References

1   Plomer, 1932, pp. 76–77; Knight, C. Shadows of the old booksellers. (London, 1865), pp. 189–213.
2   Strauss, R. Robert Dodsley, poet, publisher and playwright. (London, 1910), pp. 11–13; Knight, C. 1865, op. cit., pp. 11–13.
3   Hodson, 1989, pp. 6–7; the advert in *Daily Gazetteer* for 24 November, 1743 is reproduced on p. 7.
4   Nichols, J. 1812–5 op. cit., Vol. III, p. 249n. Examples of this practice include Tennyson's *Poems, 1833,* which appeared in December, 1832 and *The Holy Grail . . . 1870,* which came out in December, 1869. The practice still continues.
5   Wright, N.R. John Grundy of Spalding, Engineer, 1719–1783 . . . (cover title) (Lincoln, [1983?]), p. 11; Skempton, A.W. The Engineering Works of John Grundy (1719–1783). *Lincolnshire History and Archaeology,* vol. 19. (Lincoln, 1984), pp. 65–82.
6   Grundy's map is dated 1743. (LAO 2 Anc 5/22/1; RGS 1 C 88 (44); LRL Map 661). The report was not published until 1744. (Wright, N.R. [1983?], op. cit., p. 22).
7   Moreland and Bannister, p. 169.
8   The CUL copy has the price of 10s. but an attempt has been made to erase what was an erroneous price. A privately owned copy (Mr. C.A. Burden's) has the price 4s. inserted by hand but it is impossible to decipher what was formerly printed.

Hodson nos. 194–196.


**27**

**THOMAS READ**                                                                              **1746**

Size: 18.9cm x 15.78cm.

The maps discussed below first appeared in volume form in 1746, but were first issued, with an accompanying text in a part work, which has some similarities to *Magna Britannia et Hibernia . . .* (**20**). The latter was the text's main source, though it has been shown that other sources were also used.[1] Thomas Read, the printer, began the issue of the parts in November, 1743.[2] Each part was to appear at weekly intervals on Saturdays 'stitch'd in blue covers' at threepence a copy. The promise to complete the work in twenty-five parts was obviously an impossibility unless each number contained more than one county. The problem was only exacerbated by the promise made in the announcements to provide a county map in each number, since the plates would have to be prepared and engraved very quickly for a publisher who had no access to any already existing plates.

In the event Read kept to his schedule for the maps for nearly a year. Forty parts were issued by September, 1744 at 3d. per part, but the text was only complete up to Gloucestershire; following the model noted above the text for all but the first counties was spread over several issues, while, unlike *Magna Britannia . . .* maps were included with each part. The Lincolnshire map was included with part 22, which contained the second part of the text for Dorset and seems to have appeared on 28 April, 1744, being advertised in the *Daily Post* that day. Part 35 concluded the first intended volume as the *Daily Post* announced on 28 July.[3] After Part 40 (Sept., 1744) no further announcements appeared, although, when the work came out in book form, there were three volumes of text, extra maps and Wales had been included in the project (in spite of the title). Hodson has estimated that sixty-six more parts came out and, if weekly schedules were maintained, it was completed in November, 1745. When the book form came out the maps were united to their county text. However, the three volumes as published in 1746 do not contain counties or maps after Shropshire, although maps as far (alphabetically) as Worcestershire had been announced in the weekly parts in 1744. It now seems clear that a full set of the work should have included a fourth volume, which for some reason never reached completion.

It is not known who engraved the maps and it seems likely that a number of engravers were involved. The plates are closely based on those of Moll (**24**). In Lincolnshire's case the sandbanks in the Wash and Humber are taken from the little Morden map (**20**). Curiously, because Moll had placed his list of wapentakes in the top right hand corner Moll had no space to show the eastwards extent of the north side of the Humber; Read's engraver did not know, it seems, how the shore should appear and continued in a straight line off the plate so there is no Spurn Head. By following the first state of Moll's map Rutland is not named, although the division between Rutland and Leicestershire is marked.

Thomas Read was a printer[4] who had two addresses in the Fleet Street area from c.1726 to c.1753. He first advertised in May, 1726 but the office may have been opened earlier. The apparent last work was advertised in *Public Advertiser* for 12 Jan. 1753.

**Later history**
The remaindered sheets were, it seems, purchased by James Corbett for, in 1749, he issued volumes one and three only, but without the maps. This would suggest that the text part of Read's work was never completed. One county section is known using remainder sheets in 1757 but once more there was no map.[5] The map plates had passed into the hands of John Rocque (c.1704–1762), who published three editions in 1753 and another in the year he died. His business was continued by his widow Mary Ann, who must have issued the edition of 1764 as well as seeing through the press his nine-sheet map of Surrey (c.1768). Another piece of Rocque's work, seen through the press after his death (as noted on the title page) is *A Collection of Plans of the Principal Cities Of Great Britain and Ireland . . .* Among this series is *A PLAN/of/BOSTON/in the County of/LINCOLN./*[6] Rocque had already won recognition for his map work with the large-scale plans of London in twenty-four and sixteen sheets, published in 1746 and 1747 respectively. Rocque was probably a Huguenot whose family arrived in England possibly by way of Switzerland[7] by 1734 the date of his earliest work.[8] His early training as an estate surveyor was, no doubt, the inspiration for these large mapping exercises. He pioneered new ways of delineating land use, distinguishing between orchards and woods, hedges, arable lands and many other features in the course of engraving maps of the counties of Middlesex, Shropshire, Berkshire and Dublin, each in four sheets. The map of Dublin was still 'in print' in 1821 and he was among those called the 'French school of Dublin Land Surveyors'. He also projected similar mapping ventures for Buckinghamshire and Oxfordshire.[9] Lynam says he was ahead of his time;[10] he also avers that these new ways of delineating features were 'not previously on any English maps except estate plans . . .'[11] From 1751 he styled himself Chorographer or Topographer to the Prince of Wales and, from 1760, to the King.[12]

The plate for Lincolnshire made one final appearance in *England Displayed . . .* issued in parts and volume form in 1769 [–1770]. The maps include the work of Kitchin (nine or ten examples depending on the copy of the work being examined) and Rollos with five maps, the remainder being the Read/Rocque plates. To many collectors and dealers these maps are known as 'Rocques'.

**EDITIONS**

(i)  *LINCOLN SHIRE* (Be–De, OS). Compass (Ee). *10/English Miles* [= 2.6cm] (Ea).

*The English Traveller: Giving a particular Description, both Geographical and Histori-cal, of every County in England . . . with a Map of every County . . . after the design of Herman Moll. Sold by T. Read, in Dogwell-Court, White-Fryers, Fleet-Street; and by all the Booksellers . . .* [April, 1744].[13]

*The English Traveller: Giving A Description Of those Parts of Great-Britain Called England and Wales. Containing I. A particular Survey, both Geographical and Historical of every County in the same. II. An . . . Account of all the Cities, Boroughs, Market-Towns, Villages, Rivers, Royal-Palaces, Noble-men's and Gentlemen's Seats. III. The Fairs, Trade, Commerce, and Product of each County . . . IV. A Map of every County, from the best and latest Observations; wherein are mark'd the Number of measured Miles, and the usual Roads from one Post-Town to another after the designs of Herman Moll. V. Of the Isles of Wight, Man, Jersey, Guernsey, and other British Isles, with the Maps of the same. Vol. III London: Printed for T. Read, in Dogwell-Court, White-Fryers, Fleet-street. MDCCXLVI.*

**Copies**  BL G 3154–6; Guildhall L. AN 18.4.4.

(ii)  Plate number 7 added (Ea, above the scale). The scale has been altered slightly; there were originally three figures (*0,5* [and] *10*); now only the two final figures remain. Main roads have been retouched, with one of the two parallel lines used in their delineation now cut more deeply.

*The Small British Atlas: Being a new Set of Maps of all the Counties in England and Wales; to which is added a general Map, with Tables of Length, Breadth, Area, Cities, Boroughs and parishes in each County. Likewise a Parliamentary Map of England, with Tables of the Produce of the Land-tax. A Scheme of the Proportion the several Counties paid to the three Shilling Aid, 1699, compared with the Number of Members they send to Parliament. Publish'd according to Act of Parliament, by John Rocque, Chorographer to His Royal Highness the Prince of Wales, near Old Round Court in the Strand.*

*Le Petit Atlas Britannique: Ou recueil des Provinces d' Angleterre & de la principauté de Galles . . . Publié par Jean Rocque . . . dans le Strand à Londres. [Price Seven Shillings and Six-pence.]* [1753].[14]

**Copy**  Private collection.

(iii)  An attempt to erase the plate number has been made but the upper (horizontal) stroke of the 7 is still visible.

*The Small British Atlas: Being a New Set of Maps . . . To which is added A General Map . . . Tables . . . of the Land-Tax; Also A Scheme of the Proportion . . . paid to the Three Shilling Aid . . . London Published according to Act of Parliamᵗ. 1753. By John Rocque . . . Strand. (Price 7ˢ. 6ᵈ.)*

*Le Petit Atlas Britannique . . . Et de la Principauté de Galles . . . dans le Strand, à Londres.*[15]

**Copy**  CUL Atlas 7.75.1.

*The Small British Atlas . . . To which is added, A General Map . . . London: Published according to Act of Parliament 1753. By John Rocque . . . in the Strand, and Robert Sayer, Map and Printseller, at the Golden Buck, opposite Fetter Lane, Fleet-Street. [Price 7s. 6d.]*

**Copies**  BL C.24.aa.6 and 118 b 21; Leeds UL W 65; RGS 8 A 1. (LAO Scorer Map 14; LRL Maps 806).[16]

(**iv**)  Plate no. 20 has been added (Ea, OS).[17]

*The Small British Atlas . . . Published according to Act of Parliam$^t$ By John Rocque, Chorographer to his Majesty. near Old Round Court in the Strand 1762 (Price 7$^s$. 6$^d$.)*

*Le Petit Atlas Britannique . . . Publié par Jean Rocque, Chorographe de Sa majesté Britannique. dans le Strand, à Londres.*

**Copies**  BL Maps C.24.b.14; CUL Atlas 6.76.7;[18] Bod Allen 33.

(**v**)  The plate number has been changed to *21* (Ea, OS).[19]

*The Small British Atlas . . . Published according to Act of Parliam$^t$. 1753. By John Rocque . . . 1764 . . .*[20]

*Le Petit Atlas Britannique . . .*

**Copies**  CUL Atlas 7.76.5; Bod Vet A5 e 1906; Leeds UL W 66.

(**vi**)  The plate number has been removed but clear traces remain.

*England Displayed. Being A New, Complete, And Accurate Survey and Description Of The Kingdom of England, And Principality of Wales . . . By a Society of Gentlemen: Each of whom has undertaken that Part for which his Study and Inclination has more immediately qualified him. The Particulars respecting England, revised, corrected, and improved, By P. Russell, Esq: And those relating to Wales, By Mr. Owen Price . . . Vol. I. By the King's Authority. London: Printed for the Authors, by Adlard and Browne, Fleet-Street: And sold by S. Bladon, No. 28, T. Evans, No. 54, and J. Coote, No. 16, in Pater-Noster Row; W. Domville, and F. Blythe, at the Royal Exchange. MDCCLXIX.* [1770].[21]

**Copies**  BL Maps C.25.c.12 and 1034811; CUL S696.a.76.1; Bod Gen Top c 5–6; Leeds UL W 78; Nottingham UL s/DA 620.R8. (LRL Map 127; Manchester UL Map C273.2).

### References

1  Hodson, 1989, pp. 14–15.
2  *Daily Advertiser* 26 Nov. 1743, and two other papers; cited by Hodson, 1989, p. 15.
3  Hodson, 1989, p. 18 for both advertisements.
4  Plomer, 1932, p. 208.
5  Gloucestershire. In: *County Curiosities, or, a new description of Gloucestershire.* Noted by Hyett, F.A. and Bazeley, W. The bibliographer's manual of Gloucestershire literature. (Gloucester, 1895), Vol. I, pp. 20 and 123.
6  CUL Atlas 7.76.2; Bod Maps C.17.f.1.
7  Kingsley, p. 84.
8  Varley, J. John Rocque, Engraver, Surveyor, Cartographer and Map-seller. *Imago Mundi*, V, (1948), pp. 83–91. The first map is of the House, Gardens and Hermitage of Their Majesties at Richmond; Phillips, H. John Rocque's career. *London Topographical Record.* Vol. XX. No. 85, (1953), pp. [9]–25.
9  Hodson, 1989, p. 29 cites *London Evening Post* for 1751–1752.
10  Lynam, E. op. cit., p. 32.
11  Lynam, E. Period ornament, writing and symbols on maps, 1250–1800. *Geographical Magazine*, Dec. 1945, p. 326.
12  Dictionary of National Biography Missing Persons. (Oxford, 1993), p. 564; Andrews, J.H. French School of Dublin Land Surveyors. *Irish Geography*, Vol. V, no. 4. 1967, pp. 275–292; Phillips, H. op. cit., pp. 21 *et seq.*
13  The title is taken from the advertisement in *Daily Post* 3 March, 1743/4 (illustrated in Hodson, 1989, p. 16) for Part XIV (text of Derbyshire, map of Gloucestershire). Lincolnshire was

similarly announced on 28 April, 1744 with the second part of the Dorset text. The date of issue of the Lincolnshire text is unknown but if weekly issue were kept up it should have appeared in the Spring of 1745.

14  The description is based on Hodson, 1989, pp. 23–27, who had access to the only known copy in private ownership. The Parliamentary map is dated 1753. A fire in 1751 destroyed Rocque's shop and stock; he moved to the imprint's address in 1753.

15  Rocque, commonly, had title-pages in English and French and it is not unknown in other productions of the period. Laxton, P. Introductory notes to a facsimile of Rocque's map of Berkshire, 1761 (Lympne, 1973) gives details of Rocque's problems after the fire. Read last advertised his work in January, 1753 and perhaps Rocque bought this material to replace his lost stocks.

16  The LRL copy differs slightly from all other examples seen; the names of adjoining counties have a final abbreviation *SH*; here a line through the top part of the S after *NORTHAMPTON* almost converts it into a dollar sign.

17  The plate numbers in state (ii) started with the lowest numbers allocated to the northern counties and moving in bands across the country from east to west, ending with Cornwall. The early removal arose, presumably, because the numbers did not coincide with an alphabetical order (or the reverse alphabetical of the first issue of 1753). The new numbers were meant to relate to an alphabetical order, but four numbers were left out and several counties were misplaced within the sequence of Welsh counties, which followed the English.

18  The CUL copy lacks a title-page but the plates agree in all essentials except that Rutland is not numbered.

19  See note 17. The new numbers rectify some of the anomalies in the 1762 edition.

20  John Rocque died in 1762 and his widow, Mary Ann Rocque, saw this edition through the press.

21  The work was initially issued in weekly numbers from 8 April, 1769. A copy of the prospectus dated '28th Day of February, 1769' (CUL 7850 c.54[6]) announces the date. Hodson, 1989, p. 37 illustrates an advertisement for 6–8 April, 1769 in *Whitehall Evening-Post* which confirms the start. If weekly publication were maintained the complete work would have been completed on 13 Oct. 1770. It was only advertised twice as a two volume work in January, 1771; copies are found bound as one or two volumes, with or without two title-pages.

Hodson nos. 197–203.

## 28

**ROBERT WALKER**                                                                    **1746**

Size: 17.9cm x 15cm.

Walker was a prolific printer with a well-developed network for the distribution of his wares. His earliest work is dated 1729 but between 1744 and his retirement (before 1764) he had three shops open at the same time in London and operated in Cambridge as well.[1] In Cambridge he was responsible, with Thomas James, for the town's first newspaper – *Cambridge Journal and Flying Post*. Under Letters Patent dated 20 July, 1534, the University Printers had had a monopoly and the newspaper's publishers were the first in the town to publish without the authority of the University.[2]

Like the work of Read *The Agreeable Historian* was issued in parts and in its final form was published in three volumes. The text is so closely related to that of Read's work that it seems that the efforts of the same writer were shared.[3] In some areas Walker relied on the text of *Magna Britannia* . . . for lack of a suitable text. The compiler, Samuel Simpson, has not been identified.

The map is almost identical to that in Read (**27**) and must have been based on state (i) of that map, as the divisions on the scale bar bear three numbers. Two maps only

(Huntingdon and Northumberland) are signed, by H. Burgh; there is no clear evidence that he engraved the Lincolnshire plate, nor who did.

## EDITION

(i)  *LINCOLN SHIRE* (Ba–Da, OS). *The Arms of Lincoln-shire* (Ae, below the arms of the City of Lincoln).[4]*10/English Miles* [= 2.7cm] (Ea). Compass (Ee).

*The Agreeable Historian, Or the Compleat English Traveller: Giving A Geographical Description of every County in . . . England . . . With a Map of every Country* (sic) *prefix'd to each . . . after the Designs of Herman Moll, and others. Vol. II . . . Compiled from Camden, Leland, Dugdale, Ogilby, Morgan, and other Authors, By Samuel Simpson, Gent. London: Printed by R. Walker, in Fleet Lane; and Sold by the Booksellers in Town and Country, 1746.*

**Copies**   BL 10348 b 20; Bod Gen Top 8° 569–571; Manchester RL 942 S 90. (LRL L9 consists of the complete Lincolnshire text (pp. 537–568) and the map).

**References**
1   Plomer, 1932, pp. 252–3.
2   Bowes, R. On the First and other early Cambridge Newspapers. *Cambridge Antiquarian Society's Communications. Vol. VIII.* 1894, p. 348; Cranfield, G.A. A hand-list of English Provincial Newspapers & Periodicals, 1700–1760. *Cambridge Bibliographical Society Monograph*, No. 2. (London, 1961), p. 4.
3   Hodson, 1989, pp. 42–43. The chief reason for thinking that they used the services of the same text editor is that Part 1 of Walker's work appeared only nine days after Read's first number.
4   Most plates have a coat of arms and the mistaken attribution on the Lincolnshire plate is not the only example; Hodson, 1989, pp. 44–45 notes that the arms of St. Albans appear on the Herefordshire map, while the Hertford map has the arms of the city of Hereford.

Hodson no. 204.

**29**

**THOMAS HUTCHINSON**                                            **1748**

Size: 14.6cm x 17.1cm.

Although Hutchinson only signed the general map of England and Wales and that of Gloucestershire he may have engraved all or part of the other plates. This little atlas was first publicised as being in preparation in July, 1747 and the intention was that Scotland would be included. An atlas of Scotland was prepared and sold separately, although the two works are sometimes found bound together with two title-pages. The maps loosely ape Morden's map of 1695 (**19**), the resemblance being especially strong in the use of *Minutes of Time* for the longitudinal marks. A re-issue occurred in 1756, almost certainly when the Scottish atlas was also re-issued. It was last advertised in 1761.[1] In spite of the spelling of his name in the 1748 edition, Andrew Millar was a noted publisher in his own right and published James Thomson's *The Seasons* in 1730. He is noted for the prices he paid for copyrights and this moved Samuel Johnson to say 'I respect Millar, Sir. He has raised the price of literature'. [2]

## EDITIONS

(i)  *A/Correct MAP/of/LINCOLN/SHIRE./* (Ea, in a rectangular frame). *21* (Ea, OS). *Miles/15* [= 3.45cm] (Ae). The inner frames are marked at 5′ intervals and numbered every 10 minutes. *Minutes of Time* (Aa, between the inner and outer frames). *1° West from London* (Be, between the inner and outer frames). Compass (Ec).

*Geographia Magnae Britanniae. Or, Correct Maps of all the Counties in England, Scotland, and Wales; with General ones of both Kingdoms, and of the several Adjacent Islands: Each Map expressing the Cities, Boroughs, Market and Presbytery Towns, Villages, Roads and Rivers; with the N°. of Members sent to Parliament; together with Tables of the high and cross Roads, market Days, &c. Printed for S. Birt. T. Osborne. D. Browne. I. Hodges. I. Osborne. A. Miller. I. Robinson. Published according to Act of Parliament October 12th 1748.*

**Copies**    CUL Atlas 7.74.4; Bod Allen 24; Leeds UL W56; RGS 262 a 19; Sheffield UL *912.42 (G); Guildhall L S388/1 GEO.

*Geographia Magnae Britanniae . . . Printed for T. Osborne, D. Browne, J. Hodges, A. Millar, J. Robinson, W. Johnston, P. Davey & B. Law . . . 1748.* [1756].[3]

**Copies**    BL Maps C.24.aa.29; Bod Vet A4 f 837 and Allen 25; RGS 26 A 141; Birmingham RL A912.42; Hull RL 080.91242; NLS Grindlay 117.[4]

### References

1    Hodson, 1989, p. 52, n. 1 cites *Public Advertiser* for 9 Nov. 1761.
2    Knight, C. op. cit., pp. 216–220.
3    Osborne gave up publishing in 1751 and Birt died in Nov. 1755; most copies are bound with *Geographia Scotiæ*, dated 1756.
4    CUL Atlas 7.74.3 is another example, but six maps are missing, including Lincolnshire.

Hodson nos. 205–206.

## 30

**GEORGE BICKHAM**                                                    **1749**

Size: 22.6cm x 14.4cm (26cm x 15.6cm for the total plate area).

Although not strictly a geographical map it is included here as convenient for collectors. The plates for the text and maps are all engraved; the maps show bird's-eye perspective views and Lincolnshire is drawn from a point south-west of Stamford. The plate shows places dotted around with a very inaccurate idea of correct geographical relationships and perspective. A few examples will suffice: *Stamford* is pictured 6cm. above the lower frame with the *Humber* only 5cm. further up the plate; all the county is, therefore, squashed into little more than a quarter of the available plate area. *Boston* is marked due north of Stamford, with *LINCOLN* due west of the former; *Alford* is placed half-way between *Castor* and *Barton*; Louth and Grimsby are not named. The foreground consists of a rural scene with a lady on a horse and a shepherd boy with sheep, looking towards a stylised view of Stamford; a large tree frames and dominates the county view.

George Bickham senior (1684–1758) and his son, also George (*fl.*1735–1767, died 1771) were renowned as engravers and their fame rests on their work on penmanship *The Universal Penman*.[1] The item under discussion here also provides another well-known example of their combination of calligraphy and engraving.

*The British Monarchy* was published in weekly parts, commencing on 8 October, 1743[2] and did not contain maps. This version was completed in 1748 on 190 numbered plates and the weekly plan had been a failure as only forty-two parts appeared over the five years. In November, 1748 the Bickhams announced a re-issue in two volumes or in parts with 'Maps . . . engrav'd in a quite new Taste'.[3] Volume one was completed by

January, 1752[4] but what was exactly contained in it, as envisaged by the publishers, is not now clear.[5] Since most of the maps are dated one assumes the map engraving was completed in 1754 and the complete work was ready in 1755.[6]

Eleven sets (so far) of *The British Monarchy* have been found, nine bound as a single volume. Four[7] of them have no maps, but three of them are the only examples containing the final two plates (189 and 190). Of the six that have maps two have a 1743 title-page, which is anomalous; and the other four have title-pages dated 1749. A second undated title-page occurs in those examples bound in two volumes. At least one other copy must exist (see the facsimile edition noted below).

**Later history**

The only other issue of the plates in a volume was produced by Laurie and Whittle in 1796. Although the title-page refers to their being successors to Robert Sayer there is no evidence that these items were in Sayer's stock. How they acquired these plates is not known, but as they do not appear in their catalogues of 1794 and 1795 they were probably bought in the year they published them. It is assumed that if they had bought the text plates they would have issued them as well as the maps. Three examples of the plates in their cut down state but lacking Laurie and Whittle's additions have come to light. They may be proof states for the 1796 edition, but they could have been produced by whoever purchased the Bickham's stock after the son died (1771).[8]

**EDITIONS**

(i)    *A Map of LINCOLNSHIRE North from London/Humbly Inscribed to his Grace y^e Duke of Ancaster Lord Lieut^t. of the County/* (Aa–Ea, above the top frame). *From Stamford to Deeping 10.6. to Bourn 6.4. to Moreton 2.3. to Sleaford 14:5/to Lincoln 18, & 128 from London Has 3 Divisions Lindsey Kesteven and/Holland in which are 30 Hundreds 35 Market Towns 12 members of Parliament/ According to Act of Parliament by G Bickham 1753.* (Ae–Ee, below the lower frame).

*The British Monarchy: Or, a New Chorographical Description of all the Dominions Subject to the King of Great Britain. Comprehending The British Isles, The Electoral States, The American Colonies, The African & Indian Settlem.^ts And enlarging more particularly on The respective Counties of England and Wales . . . The Whole Illustrated with suitable Maps . . . Engrav'd by George Bickham. GB. fecit. Publish'd according to Act of Parliament, October 1^st 1743 and Sold by G. Bickham in James Street, Bunhill-Fields, & by the Booksellers . . .* [1755].

**Copies**    BL 800 cc 127; Guildhall L Bay H.1.5 no. 6.

*The British Monarchy: Or, a New Chorographical Description Of . . . England and Wales. With Maps of each County in a New Taste . . . Publish'd according to Act of Parliament, Decem^r. 2^d: 1749, by G. Bickham Jun. in May's-Buildings, Covent Garden London, & by y^e Stationers, Book & Printsellers, in Town & Country.* [1755].

**Copies**    Leeds UL W 62; Sheffield UL \*\*F914.2(B); Birmingham RL F942.072; Manchester RL BR f 942 B 66.

(ii)    The text above and below the plate has been removed.[9]

(iii)    *LINCOLNSHIRE.* has been added (Ca, OS). *21* has been added (Ea, OS).

*A Curious Antique Collection Of Birds-Eye Views Of The Several Counties In England & Wales; Exhibiting A Pleasing Landscape Of Each County; With A Variety Of Rustic Figures, Ruins, &c. &c. And The Names Of The Principal Towns And Villages, Interspersed According To Their Apparent Situation. Finely Engraved on Forty Six Plates. By*

*George Bickham, Junior. London: Published By Robert Laurie And James Whittle, Map, Chart, And Print Sellers, No. 53, Fleet Street, (Successors To The Late Robert Sayer.) 1796.*

**Copies**   BL 191 b 9; CUL Atlas 6.79.1; Liverpool RL FQ 295; Leeds UL W 63.

**Reproductions**
(i) A facsimile edition: The British Monarchy or A New Chorographical Description Of . . . Great Britain by George Bickham. (Newcastle-upon-Tyne, 1967). This edition differs in make-up from all other known copies, but the provenance of the copy upon which the facsimile is based is not recorded. (ii) Holehouse, J.E. Famous Early Maps in Lincolnshire: G. Bickham 1753. *Lincolnshire Life* Vol. 13, no. 9 (Nov. 1973) pp. 48–51. A copy of the map appears on p. 48 and the three pages of county text (plates 114–116, in the original) are also fitted in.

**References**
1   Lyon, D. A Bird's-eye View of the Bickhams. *Map Collector.* No. 2 (March, 1978), pp. 30–33; Heal, A. English writing masters. (London, 1931), pp. 14–16; Muir, P. H. The Bickhams and their Universal Penman. *The Library.* Fourth Series, Vol. 25 (1944), pp. 162–184.
2   Hodson, 1989, p. 83 shows the *Westminster Journal* . . . advertisement for that date.
3   ibid., p. 85 notes the *General Advertiser* for 4 Nov. 1749.
4   as note 3 – date 21 Jan. 1751/2.
5   Two copies are known, bound as two volumes, one of which (Mr. C.A. Burden's) is misbound in that the first four pages of Middlesex text are in the first and the remainder of that county's text and the map in the second volume. The Manchester RL copy is bound with the first volume ending with the last page of Middlesex text and map. The second title-page states that there were 25 numbers altogether, each 6d. and the 'first volume neatly Bound [is] 15ˢ'.
6   *Public Advertiser* 10 Nov. 1755 announced the completion but 'any Person may begin with No. 1'. (Hodson, 1989, pp. 86–87). No title-page dated later than 1749 has been found.
7   Bod GA Gen Top c 29; BL 190 d 5 and 796 i 1; West Suffolk CRO 942T; the Bodleian copy alone of these has a title-page of 1743; BL 796 i 1 has the 1749 version and the other two have the 1748 version of the title-page.
8   George, junior, died 3 July, 1771. (Lyon, D. op. cit., p. 30). Hodson, 1989, p. 87 notes the sale notice for the remaining stock of the Bickhams in *Public Advertiser* for 18 Dec. 1772.
9   I am grateful to Mr. E. Burden who told me that examples of the maps of Cornwall, Devon and Surrey in this state were offered for sale in September 1991.

Hodson nos. 217–218.

## 31

### THOMAS JEFFERYS and THOMAS KITCHIN                              1749

Size: 12.95cm x 13.35cm.

Jefferys and Kitchin (sometimes spelt Kitchen) were together and separately prolific engravers and map publishers for almost forty years. Jefferys flourished from 1732 until his death in 1771 and was in the forefront of those surveyors who did so much to undertake large-scale surveys of the counties in the second half of the eighteenth century. He went bankrupt in 1766 and, yet, with the help of friends, he was still able to finish a number of his large surveying tasks.[1] He began with a plan, a reworking of an older map, which covered the City, Westminster and Southwark, which was issued in 1732.[2] In 1749 he engraved a map entitled *INCLOSED/LANDS IN/LINCOLN SHIRE*; this appeared in *Observations On The Decay of the Outfalls . . . of divers weak Rivers, Particularly the*

*River Neen . . . And Shire Drain . . . By Richard Edwards.*[3] Much later he became involved in county surveying, firstly with Donn in the survey of Devonshire (1765); that was followed by the surveying of Bedfordshire, Huntingdon, Oxfordshire and Buckingham. Gough says that he also published proposals for a survey of Nottinghamshire.[4] In 1747 he engraved a map of Staffordshire, after the first large-scale map of the county by Robert Plot (1682).[5] At that time he was regarded as an engraver only, whereas in his later county work he acted as the surveyor and chief undertaker, until his financial difficulties became so serious, largely as a result of the high costs involved in county surveys.

Kitchin (*fl.*1732–1776, d.1784[6]) was equally prolific but not involved in very original work or in the larger scale projects that eventually led to Jefferys' downfall. He was apprenticed to Emanuel Bowen in 1732 and came out of his time in 1739, marrying Bowen's daughter, Sarah, two weeks later.[7] By 1746 he was taking apprentices of his own and had established himself fully with his engraving of the map of Scotland for John Elphinstone. In 1748 he engraved many plates in *Geographia Scotiæ,* the companion piece to **29** above. His work is met again in his maps for the London Magazine and, with Emanuel Bowen, in *The Large English Atlas* (**32** and **34**).

Both men were honoured for their work. Jefferys was appointed as Geographer to Frederick, Prince of Wales, in 1746. On his map *A PLAN of/the/HIGH HUNTRE,/ OR/HOLLAND FEN,/Engraved by/ Thomas Jefferys Geographer to His Majesty/1767.*[8] the wording makes clear his final title. Kitchin had the title of Hydrographer to the King and also engraver to the Duke of York from 1758 to 1763.[9]

The maps for *The Small English Atlas* were issued in weekly parts from 19 November, 1748 and there seems to have been four maps to each 6d. part. Lincolnshire appeared in this first part, together with the maps of Herefordshire, Northamptonshire and Westmoreland.[10] The plan was to complete the whole work in twelve parts and that implies completion on 18 February, 1749. The plan could not have been maintained quite strictly as there are fifty-two plates, which indicates issue in thirteen parts. Hodson has shown that the maps were engraved four up, with two inverted in relationship to the others and then separated at the binding stage.[11] The maps are loosely based on the Morden maps produced in 1695, but there are additions taken from later sources. Badeslade's idea of a panel of information has been adopted and the information on the county's towns forms the lower part of the plate.

### Later history
There was no further edition in Jefferys' lifetime, but after his death in 1771 his materials passed to Robert Sayer, who had been involved in re-establishing him after his bankruptcy. By the time of Sayer's death in 1794 four more issues had been published. The plates passed to Laurie and Whittle and they put out the final version; sales must have been poor by then since copies of the last three editions are rare.

### EDITIONS
(i)   *A Map of LINCOLN-SHIRE* (Ba-Da, OS). *14/English Miles* [= 2.15cm] (Ea). Compass (Eb). The inner frame is shaded at 5′ intervals and, between the inner and outer frames, are minutes of latitude and longitude at 30′ intervals. *1° Longit. W. from London* (Be, between the inner and outer frames). In an unframed panel below the map are lines of text: *Lincolnshire Contains 1 City, 4 Parliamentary Boroughs, & 26 other Market Towns, & sends twelve/Members to Parliament viz. 2 for the County, 2 for the City of Lincoln, & 8 for the 4 Boroughs./* Below are three columns (separated by double vertical rules), in which the county's towns are listed, the first column beginning with Lincoln, followed by Stamford, Grantham, Boston, Grimsby and then by the market towns in no special order. Compass (Eb).

*The Small English Atlas being A New and Accurate Sett Of Maps of all the Counties in*

*England and Wales London, Publish'd according to Act of Parliam*. by Mess*rs*. Kitchin
& Jefferys 1749. and Sold by M Payne at the White Hart and M. Cooper at the Globe in
Paternoster Row.*

**Copies**  BL Maps C.24.aa.6; CUL Atlas 7.74.17; Bod Allen 26; Leeds UL W 57; RGS
10 B 2. (LRL Map 947).

*The Small English Atlas . . . 1751. and Sold by Tho*s*. Jefferys Geographer to his Royal
Highness the Prince of Wales at the Corner of S*t*. Martin's Lane Charing Cross, by M*r*.
George Faulkner in Essex Street Dublin. A Paris chez le S*r*. le Rouge Ingenieur
Geographe du Roy de France.*

**Copies**  CUL Atlas 7.75.2; Leeds UL W 58; RGS 262 a 31; NLS EME b.1.26.

(ii)   plate number 25 added (Ea, OS).

*The Small English Atlas . . . 1751 . . .*

**Copies**  RGS 10 B 3. (BL 3355 (11)).[12]

(iii)   Roads have been added, but not with any degree of accuracy. A road from
Horncastle is drawn eastwards to Alford and, with a slight bend, continues south of Alford
where it makes a junction with a road from Tattershall which passes through *Bullinbrook*
and Spilsby and then together go on to the coast; the road from Saltfleet to *Castor* misses
Louth but crosses the Lincoln-Grimsby road half way between Binbrook and Market
Rasen. The panel of text below the map has been re-engraved: the first line is now: *This
COUNTY contains 1 City, 4 Boroughs and 27 Market Towns./*; the list of towns is now
in alphabetical order, the one minor exception being that *Market Deeping* appears
between Crowle and *Dunnington*. The boundaries of the wapentakes, etc. are marked by
dotted lines.

*The Small English Atlas . . . By T. Jefferys, Geo: to the King, and Tho*s*. Kitchin Sen*r*.
London: Printed for Robert Sayer and John Bennett, N*o*. 53 in Fleet Street, John Bowles,
N*o*. 13 in Cornhill, and Carington Bowles, N*o*. 69 in S*t*. Pauls Church Yard.* [1775].[13]

**Copies**  CUL Atlas 7.77.3; Bod Allen 40 and Maps C.17.e.26; Leeds UL W 59 and W
60; RGS 10 B 4. (LRL Map 187; Boston RL Map 21).

(iv)  A road has been added from Lincoln in a straight line to Horncastle.

*An English Atlas Or A Concise View of England And Wales; Divided into Counties, and
its Subdivisions into Hundreds &c. Describing Their Situation, Extent, Boundaries,
Circumference, Soil, Product, Chief Rivers, and the Principal Great and Bye-Roads; with
a Chart of the Distances . . . with a Description of Antiquities . . . Castles, Palaces, or
Monasteries . . . On Fifty Two Copper Plates. Published as the Act directs 1 Aug*t*.1776.
Printed for Rob*t*. Sayer, and Jn*o*. Bennett, Map and Printsellers, N*o*. 53, in Fleet Street.*[14]

**Copy**  The private collection of Mr. C.A. Burden.

*The small English atlas, London, Robert Sayer.* [c.1787].[15]

*An English Atlas Or A Concise View Of England and Wales . . . Published as the Act
directs 1 Aug*t*. 1787. London. Printed for Rob*t*. Sayer No. 53, in Fleet Street.*

**Copies**  BL Maps C.25.a.3; CUL Atlas 6.78.1.

*An English Atlas . . . 1787 . . .* [1794].[16]

**Copy**  Leeds UL W 61.

*An English Atlas . . . 1787 . . .* [1796].[17]

**Copy**  Manchester RL BR 912.42 S 14.

## References

1   Harley, J.B. The bankruptcy of Thomas Jefferys. *Imago Mundi*, Vol. XX (1966), pp. 27–48.

2   Chambers, B. Thomas Jefferys and the map of Bedfordshire. *Bedfordshire Historical Record Society.* (Bedford, 1983): issued with that Society's facsimile reprint of Jefferys map.

3   Bod Gough Lincolnshire 12.

4   Gough, 1780, Vol. II, p. 78.

5   Harley, J.B. 1966, op. cit., p. 30; King, G.L. The printed maps of Staffordshire, 1577–1850; 2nd ed. (Stafford, 1988), pp. 25 and 31.

6   Kingsley, p. 74; Maxted, p. 130 says 1783; Worms, L. Thomas Kitchin's 'Journey of life' . . . Part One. *Map Collector.* No. 62 (1993), pp. 2–8. Worms illustrates Kitchin's apprenticeship indenture (City of London Record Office CF 1/697/18); he notes that he died in 1784 aged 66 but plumps for 1719 as the year of his birth; his search of the records of St. Olave, Southwark produced the baptismal records of two Thomas Kitchins, one born in 1719 and the other in 1717. Worms also contributes the life of Kitchin to *Dictionary of National Biography Missing Persons* (Oxford, 1993), pp. 378–379.

7   Maxted, p. 130.

8   CUL Maps bb.44.76.1; Bod Gough Maps 16, Fol. 3.

9   Worms, L. Thomas Kitchin's 'Journey of life' . . . Part Two. *Map Collector.* No. 63 (Autumn, 1993), pp. 14–19, notes that Kitchin was so described in *Royal Kalendar* from 1773 jointly with his son, Thomas Bowen Kitchin.

10  Hodson, 1989, p. 58 reproduces the text of the full announcement in *London Evening Post* 17–19 Nov. 1748.

11  Hodson, 1989, pp. 61–63 for an elegant bibliographical analysis, showing how the maps were engraved together.

12  Plate numbers were added piecemeal to the 1751 edition and all the extant copies are unstable. A full analysis appears in Hodson, 1989, pp. 65–71.

13  John Bennett joined Sayer in 1774 and it is likely that this edition is the one advertised in their list dated 1775.

14  This edition undoubtedly appeared after the [1775] edition. While this copy contains a general map dated 1777 it was probably inserted by a later owner as the map is not known in Sayer's stock. *An English Atlas* differed from the *Small English Atlas* by the inclusion of letter-press on each county.

15  Chubb, No. CXCIV notes a copy in Sir George Fordham's possession; the abbreviated details are quoted here but the copy has since disappeared. Chubb suggested 1785 but it now seems likely that it appeared after Bennett had left the partnership (in April, 1786); he died in Dec., 1787. Hodson, 1989, pp. 77–78 and 187; *Gentleman's Magazine* (Dec. 1787). Leeds UL W 60 has no title-page; Whitaker thought it was an example of the 'missing' work; however its contents are indistinguishable from the edition of [1775].

16  A new 'Longimetric Table of Distances . . .' has been added; it bears the imprint of Laurie & Whittle and is dated 12th May, 1794.

17  A new issue; the general map bears the imprint of Laurie & Whittle and is dated 1796; Hodson, 1989, pp. 78–80 notes that the atlas with both versions of the title was advertised until the mid-1820s; no copy later than 1796 has yet been found.

Hodson nos. 209–216.


# 32

## THOMAS KITCHIN                                                        1751

Size: 21.3cm x 16.5cm.

*The London Magazine* was established in 1732 and ran at monthly intervals until July, 1783. From 1747 to 1763 English and Welsh county maps were inserted at irregular intervals and thereafter maps of the Scottish and Irish counties appeared until 1781.

Lincolnshire appeared in volume 20 (January, 1751), together with three pages of letterpress (pp. 6–8).

Kitchin followed Moll for the basic form of the map but added roads. The road from Lincoln to Horncastle is drawn as a curve but south of Wragby and there are other general inaccuracies in delineating roads.

Robert Baldwin was a well-known publisher of political pamphlets and his 'industry and integrity were almost proverbial'.[1] The *London Magazine* had been set up as a rival to *Gentleman's Magazine* and Baldwin became associated with it in 1746. Gradually it seems that he assumed control as the original partners died or sold their shares. The issue of maps probably resulted from the rival *Universal Magazine of knowledge and pleasure* having started to produce county maps in the issues of 1747 (**35**). The circulation grew from 4000 copies per month to 8000 at its peak (Aug. 1739) but was still 7500 in February, 1747.[2]

The plates were not used again in Baldwin's lifetime. Some maps, but not Lincolnshire, appeared in *England Displayed* (1769) (**27**) – but the next appearance of all the plates occurs in the works of Alexander Hogg from 1786. How the plates reached him is not clear. Hogg was active from c.1778 until 1824 and was known for the issue of part-works (referred to, then, as Paternoster numbers).[3] Little of the contents of his works was original. Hogg's unashamed copying is clearly indicated in the title of the volume issued in 1795. The final use of the plates came in an abridged version of Grose's *Antiquities* . . . (see also **18(ii)** above).

## EDITIONS

(i)  *LINCOLN/SHIRE/Drawn from the best/Surveys & Maps/Corrected from /Astron[l]: Observ[ns]:/By T. Kitchin/Geog:[r]/ (Ee)*. British Statute Miles/12 [= 2.85cm] (Ad). *Printed for R. Baldwin Jun[r]. at the Rose in Pater Noster Row.* (Be\–De, OS). *Explanation/8 lines of text and symbols* (Ae). The inner frame is shaded to show latitude or longitude at 1' intervals; numbers (between the inner and outer frames) record the same at 15' gaps. *Longitude W. from London* (Ae, between the inner and outer borders). Compass (Ec).

**Copy**  (The author's collection).[4]

(ii)  *for the London Magazine* added (Ca, OS).

*The London Magazine: Or, Gentleman's Monthly Intelligencer. Vol. XX For the Year MDCCLI Printed for R. Baldwin at the Rose in Pater-Noster Row. T. Kitchin sculp.*

**Copies**  CUL 7900.1.20; Bod Hope Adds 406; NLS Q 6. (LAO Scorer Map 13 consists of the map and the county text). BL Maps C.24.d.20 is an untitled collection of maps, mostly by Kitchin and mostly from the London Magazine; the Lincolnshire map is included but there is no text; there is a set of the periodical in the BL (PP 5437) but most of the maps are missing, Lincolnshire included.

(iii)  The imprint and the heading have been removed.

*Historical Descriptions of New and Elegant Picturesque Views of The Antiquities Of England and Wales: Containing A . . . Collection of Superb Views . . . Accompanied by . . . Descriptions of the several Places Delineated . . . Picturesque Views of the principal Seats of the Nobility . . . Published Under The Inspection Of Henry Boswell . . . Assisted by Robert Hamilton . . . London: Printed for Alex. Hogg, at the King's-Arms, No. 16, Paternoster-Row . . . [1786].[5]*

**Copies**  BL 1700 a 26; CUL S.474.a.78.1; Bod Vet A5 b.75; Leeds UL W 82 and Special Collections. Architecture B – O.12 Fol. BOS; Leicester UL F H942.1 BOS; Birmingham RL HGD/094/1786/11F. (LRL Maps 812 and 812a; Grimsby RL – Kitchin RM 92).[6]

*Complete Historical Descriptions of a New and Elegant Collection of Picturesque Views and Representations of the Antiquities Of England and Wales . . . Containing . . . By Henry Boswell . . . Assisted by many Antiquarians . . . London: Printed for Alex. Hogg, at the King's-Arms, (No. 16) Paternoster-Row, And Sold . . .[1790].* [7]

**Copies**   Nottingham UL s/DA620.B6 (oversize X); West Suffolk CRO Cullum Q 942T.

*The Antiquities of England and Wales Displayed; Being A Grand Repository . . . Comprising Every Thing of Importance in the Works of Leland, Maitland, Gibson, Dugdale, Buck, Speed, Pennant; &c. And including particularly every Article worthy of Notice in that Voluminous and very Expensive Work published by Captain Grose . . . By Henry Boswell . . . Assisted by Many Antiquarians . . . A New Edition, With Many Alterations And Improvements. London: Printed For Alex. Hogg, At No. 16, Paternoster-Row. 1795.*

**Copies**   Bod G.A. Folio A 359; Manchester RL BR f 942 B71.

*A New and Complete Abridgement or Selection of the most Interesting and Important Subjects in The Antiquities Of England And Wales: Being A Complete Collection Of Beautiful Views Of The Most Remarkable And Ancient Ruins . . . By Francis Grose . . . A Complete Set of County Maps, By The Best Artists . . . London: Printed for the Editor, and sold by H.D. Symonds, No. 20, and Alex. Hogg, No. 16, Paternoster-Row . . . MDCCXCVIII.*

**Copy**   Leeds UL W 104.

**References**

1   Mackenzie, D.F. and Ross, J.L. (eds.). A ledger of Charles Acker, printer of the London Magazine. *Oxford Bibliographical Society Publications*, new series, Vol. 15. 1968, p. 68 *et seq.* The ledger deals with the period 1732 to 1748.
2   Hodson, 1989, p. 149.
3   Maxted, p. 112; Timperley, p. 838.
4   Hodson, 1989, p. 151 notes that the maps were advertised separately at 3d. each (*London Evening-Post*, 20–22 Sept. 1750) and repeatedly until July, 1751, by which time the Lincolnshire map would have appeared. This loose copy could be a proof or simply an error (soon corrected) in a print-run of 7–8000 copies.
5   The frontispiece has the engraved date Nov. 25, 1786. In the CUL copy a note on a preliminary page states: 100 Numbers £3.0.0; binding 10.0; [Total] 3.10.0.
6   LAO has a copy (AS.A7) but the Lincs. map is missing.
7   The text is as in the previous and later Hogg works, only the preliminaries being varied. It is assumed that the work appeared after the [1786] edition but before that of 1795. The make-up of all these differing editions/copies varies. The text pages are unnumbered, but Lincolnshire's has *(40.)* in the lower left corner of the first text page.

Chubb no. CCLVII (1786 edition only). Hodson no. 229.


## 33

**SARAH and JOHN EXSHAW**                                                    **1751**

Size: 21.2cm x 16.5cm.

The *London Magazine* was also published in an edition put out in Dublin by Edward Exshaw.[1] It was intended as a reprint of the London edition and began in May, 1741. According to the Preface of the 1741 volume there was also an Edinburgh edition, but it has not been identified as a separate and locally issued item. John Exshaw joined the Irish enterprise in 1745 and when Edward died in 1748 his stepmother Sarah joined.[2] Until 1751 the text in Dublin had been closely copied from sheets of the London text. From that year a more independent editorial line was pursued and extracts copied from contemporary journals were introduced. In 1755 the title was changed to *The Gentleman's And London Magazine: And Monthly Chronology*. By then Sarah had her own lace shop and John carried on the business until he retired in 1776; he died in 1777.

Between 1749 and 1752 maps for twenty-eight counties had been issued; the same order was pursued as with the London edition. The only exceptions to this were that the London maps of 1747 and 1748 were not issued at all and the map of the Shetland Islands, which had slipped in to the London sequence, was omitted. The maps were re-engraved from copies of the London maps sent to Ireland for the purpose. The maps are quite faithful copies but the title panel is distinctive, and the heading and imprint are changed. Two maps each are signed by D. Pomarede[3] and I. Ridge[4] but Lincolnshire's plate is unsigned.

## EDITION

(i)  *LINCOLN SH./Drawn from the best/Surveys & maps/Corrected from/ Astron[l]. Observations/By T. Kitchin Geograph[r]./* (Ee). *Dublin Printed for S. & J Exshaw.* (Be–Ce, OS). *British Statute Miles/ 12* [= 2.85cm] (Ad). *Explanation/8 lines of text and symbols* (Ae). *Lon: Mag: 1751* Ca, OS). *Page. 8.* (Ea, OS, the figure being engraved upside down). The inner frame is shaded to mark degrees of latitude and longitude at 1′ intervals, while numbers between the inner and outer frames record the same at 15′ intervals.

*The London Magazine: And Monthly Chronologer MDCCLI . . . Dublin: Printed for Sarah and John Exshaw, at the Bible on Cork-Hill. Of whom may be had compleat Sets from May, 1741.*

**Copy**  Bod Vet A5 e 3474.[5]

### References
1  Munter, R. A Dictionary Of The Print Trade In Ireland, 1550–1775. (New York, 1988), p. 92; Plomer, 1932, p. 383.
2  Munter, op. cit., pp. 92–93; Pollard, M. Dublin's Trade in Books, 1550–1800 . . . (Oxford, 1989), *passim*.
3  Oxfordshire and Berkshire.
4  Cheshire and Lancashire.
5  Hodson, 1989, pp. 176–179 notes a set in Trinity College, Dublin (shelf no. 00.rr).

Hodson, 1989, Appendix I, no. 2.

34

**EMANUEL BOWEN**                                      **1751**

Size: 70.55cm x 52.8cm.

The plates that eventually formed *The Large English Atlas* were a long time in preparation and their ownership passed through the hands of several publishers before the work was completed. There had been no new large scale atlas of the English counties since Blaeu and Jansson just over a century before; where large plate maps had been published they were either the accompaniment to a work of history, e.g. *Britannia* or they completed an atlas made up of a mixture of plates of varying provenance. The bulk of large map plates still being offered went back to Speed or Saxton for their data; it should be remembered, indeed, that the plates of both of these pioneers were still in use (**1 (vii)** and **6 (x)**) at the same time as this new venture.

John Hinton approached Thomas Kitchin and Emanuel Bowen with a view to their engraving large maps of the counties all designed to a uniform plan, probably in 1748. They were to use the best of what newer large-scale county surveys were to hand. Hinton was active from c.1739 until his death in 1781 as a bookseller and publisher.[1] He used Kitchin and Bowen to engrave maps for his successful *Universal Magazine of Knowledge and Pleasure* (**35**). In 1749 the first map (Sussex) was advertised.[2] Monthly issue of maps was promised and five maps at a shilling each had arrived by September. Under a new scheme of production, groups at 1s. 6d. each were announced; six appeared together in March, 1750; further promises were made but were slow to be honoured.

In 1752 Hinton left the map publishing business when the costs and the difficulty of keeping to publication dates threatened the success of his other ventures. By then twenty eight plates had been issued and they were sold to John Tinney who put his name on a number of them. Though Tinney had been an apprentice engraver he had been a print-seller from 1734. By 1761 when he died only thirteen more plates had been engraved and, in three cases, pairs of Welsh counties had been placed together on the same plate, in an attempt to economise. Already in 1756, Thomas Bowles, John Bowles And Son, and Robert Sayer had joined him in the venture. In fact, only one plate (in 1758) came out in three years. In 1760 the last three plates were produced – two pairs of English counties and North Wales on one plate. The total coverage of the counties in England and Wales was thus complete on forty-five plates. During this period individual sheets were on sale. The first 'official' issue of all the maps together was announced in May, 1760.[3]

Lincolnshire was one of a group of six counties announced to appear in May, 1751 but there are no further advertisements in this period to confirm their appearance as announced. The maps are based on the Morden 1695 plates and, as such, continue to perpetuate the errors and oddities stemming from Saxton, without adding anything very original. The roads also come from Morden and still show many misplaced routes. The decorative feature that adds to their appeal is the inclusion of snippets of information about the county, which are placed around the outside of the map area. In the upper left corner appears the Stukeley tide-table, probably taken from Sutton Nichols' plate, but with the table redrawn and the text in reported speech (as though from Stukeley direct); the date in the example given to show its working is still that of 1723, as in the original. For the first time a map of the county shows mileages based on a degree of 69 miles, although its actual value is, in fact, 69.15 miles. As the title-page indicates, letters are placed by the place-names on the map to indicate whether they had rectors or vicars; an interesting novelty not taken up by any rivals.

### Later history

The various partners (as from 1756) and their successors issued atlases until the 1790s by which time Cary's *New and Correct English Atlas* (**47**) must have begun to have a severe effect on the sales of the *Large English Atlas*. The plates went to Laurie and Whittle, Sayer's successors, and although entered in their catalogue no fresh title-page was issued to go with the plates to form a new atlas. In 1825 R.H. Laurie sold the plates as old copper.[4]

### EDITIONS

(i)    *An ACCURATE MAP of/LINCOLNSHIRE,/Divided into its/ WAPONTAKES./Laid down from the best Authorities, and/most approved Maps & Charts,/with various additional Improvements./ Illustrated with HISTORICAL EXTRACTS/relative to the Air, Soil, Natural Produce,/Manufactures, Trade & present State/ of its principal Towns./By Eman. Bowen GEOG<sup>R</sup>. to HIS MAJESTY./* (Ea). *British Statute Miles/15* [= 12.25cm] (De). *Explanation/8 lines of notes and symbols* (Ee). *To the Most Noble/Peregrine*

*Bertie,/Duke of Ancaster,/Lord Great Chamberlain of ENGLAND,/Custos Rotulorum &
Lord Lieutenant/for the County of LINCOLN:/This MAP is humbly Dedicated by
His/Grace's most Obed<sup>t</sup>. Serv<sup>t</sup>/Eman Bowen./* (Ae). Compass (Db). *A PERPETUAL TIDE
TABLE for Fosdike & Cross Keys* . . . (Aa). The inner border is shaded to indicate latitude
and longitude at 1′ intervals; numbers mark the same at 5′ gaps with lines between the
borders, which are carried across the map to provide a graticule. *Longitude from London*
(De, between the inner and outer frames).[5]

(ii)   Imprint added: *Sold by J. Hinton at the Kings Arms in S<sup>t</sup> Pauls Church Yard London
1751.* (Be–Ce, OS).

Untitled collection of maps put together in c.1755.[6]

**Copies**   CUL Atlas 1.75.1. (BL K.XIX.18).

Untitled atlas put together in c.1760.[7]

**Copy**   The private collection of Mr. C.A. Burden.

(iii)   New imprint: *Sold by J. Tinney at the Golden Lyon in Fleetstreet London 1753.*

Untitled collection of maps put together in c.1755.[8]

**Copies**   BL Maps C.10.d.18; CUL Atlas 1.75.1. (Bod (E) C17:39 (2); LAO Misc Don.
81/3; LRL Maps 87 and 180).

Untitled collection of plates put together in c. 1763.[9]

**Copy**   Bod Dep. a.19.

(iv)   New imprint: *Printed for J. Tinney, at the Golden Lyon in Fleetstreet T. Bowles in
S<sup>t</sup>. Pauls Church Yard, John Bowles & Son, in Cornhil, & Rob<sup>t</sup>. Sayer in Fleet Street.*

*The Large English Atlas: Or, A New Set Of Maps Of All The Counties In England And
Wales, Drawn From The Several Surveys which have been hitherto published:* [numbered
list of 45 maps] *Laid Down On A Large Scale, And Containing all the Cities, Towns,
Villages, and Churches, whether Rectories or Vicarages, Chapels, many Noblemen's and
Gentlemen's Seats,&c.&c. Each Map Is Illustrated With a General Description of the
County, its Cities, Borough and Market Towns, the Number of Members returned to
Parliament, of Parishes, Houses, Acres of Land,&c. And Historical Extracts . . . By
Emanuel Bowen, Geographer to His Majesty, Thomas Kitchin, and Others. London:
Printed and Sold by T. Bowles, in St. Paul's Church-Yard; John Bowles and Son, at the
Black House, in Cornhill; John Tinney, at the Golden Lion, and Robert Sayer, at the Buck,
both in Fleet-Street.* [1760].[10]

**Copies**   BL Maps C.10.d.10 and 1 TAB 20; Leeds UL W 68.[11] (Bod Gough Maps
Lincolnshire 10; LRL Map 87a; Grimsby RL Bowen).

*The Large English Atlas* . . . [1761].[12]

**Copy**   Bod Allen 30.

(v)   The imprint now reads: *Printed for R. Sayer at the Golden Buck in Fleetstreet, T.
Bowles in S<sup>t</sup>. Pauls Church yard, John Bowles & Son in Cornhil.* Plate numbers have
been added: *21* (Ea, OS and Ee, OS).

*The Large English Atlas . . . With three general Maps of England, Scotland, and Ireland,
from the latest and best Authorities . . . London: Printed and Sold by T. Bowles, in St.
Paul's Church-Yard; John Bowles, at the Black Horse, in Cornhill; and Robert Sayer, at
the Buck, in Fleet-Street.* [1763].[13]

**Copies**   BL Maps C.10.d.8. (LRL Map 92; Grimsby RL Bowen 1763).

Issued as a loose sheet in a cardboard holder. [1763].

**Copy**   Nottingham UL Li 1.C17.DO5.

*The Large English Atlas . . . London: Printed and Sold by John Bowles, at the Black Horse, in Cornhil; Carington Bowles, next the Chapter-House in St. Paul's Church Yard; and Robert Sayer, at the Buck, in Fleet-Street.* [1764].[14]

**Copy**   CUL Atlas 2.76.2.

(**vi**)   The imprint now reads: *Printed for R. Sayer at the Golden Buck in Fleetstreet, Carington Bowles in S$^t$. Pauls Church Yard, and John Bowles in Cornhil.* Plate number changed from *21* to *23*.

*The Large English Atlas . . . London . . . T. Bowles . . . John Bowles . . . Robert Sayer . . .* [1764].[15]

**Copies**   Nottingham UL s/G 5512. (LAO Yarb 4/27/1; LRL Map 160; Scunthorpe Museum[16]).

(**vii**)   The imprint now reads: *Printed for R. Sayer at the Golden Buck in Fleetstreet, Carington Bowles in S$^t$. Pauls Church Yard.*

*The Large English Atlas . . . T. Bowles . . . John Bowles . . . Robert Sayer . . .* [1764].

**Copies**   Birmingham RL AE912.42 MR; Manchester Chethams L. G 10.1 (2269).

*The Large English Atlas . . . London: Printed for and sold by Carington Bowles, Map and Print Seller, at No. 69, in St. Paul's Church-Yard. MDCCLXVII.*

**Copies**   Leeds UL W 69; NMM 912.44(42) ″17″: 094 (D5439); PRO M21 – Press 21; Nottingham UL G 5512; Nottingham RL R912.42. (CUL Maps AA.70.76.1; Grantham RL; Louth RL Map A5).

Issued folded in a cardboard holder: the label on the front reads: *LINCOLN/SHIRE/Printed for Carington Bowles in S$^t$. Pauls Church Yard./* [1767].

**Copy**   Nottingham UL Li 1.B8 D85.

*The Large English Atlas . . . London: Printed and Sold by Robert Sayer, Map and Print Seller, at no 53, in Fleet-Street.*

*Le Grand Atlas Anglois: Ou, Nouveau Recueil Des Cartes De Toutes Les Provinces D'Angleterre Et De La Principauté De Galles . . . Chez Robert Sayer, Marchand Des Cartes & d'Estampes dans Fleet-Street, No. 53.* [1767].[17]

**Copies**   NMM 912.44(42) ″17″: 094 [D5441]; Chetham's L. G.10.1; NLS EU.15.B.

*The Large English Atlas . . . London: Printed and Sold by Robert Wilkinson, at No. 58, in Cornhill, Successor to Mr. John Bowles, deceased.* [1779].[18]

**Copy**   BL Maps C.10.d.11.

*The Large English Atlas . . . Drawn From The Several Surveys Hitherto Published. To which are added, a Map of the Country 35 Miles round London, a Plan of London and Westminster, and general Maps of Scotland and Ireland, From the latest and best authorities. The Whole Engraved on 50 Copper Plates. Laid Down On A Large Scale . . . By Emanuel Bowen, Thomas Kitchen, Captain Andrew Armstrong, and Others. London: Printed for Robert Sayer, Map, Chart, And Printseller, No. 53, Fleet-Street. MDCCLXXXVII.*

**Copy**   RGS 1 C 73.[19]

(**viii**)   The imprint now reads: *Printed for R. Sayer at the Golden Buck in Fleetstreet, Carington Bowles in S$^t$. Pauls Church Yard, and Robert Wilkinson in Cornhill.*

Issued in a cardboard cover, the label on the cover reads: *LINCOLN/ SHIRE/Printed for Carington Bowles in S<sup>t</sup>. Paul's Church Yard/* [1787].

**Copy**   Grimsby RL X000:912 BOW.

*The Large English Atlas . . . MDCCLXVII.* [c.1794].[20]

**Copy**   BL Maps C.10.d.19.


**(ix)**   The imprint now reads: *London. Printed for Laurie & Whittle, 53 Fleet Street R. Wilkinson 58 Cornhill, Bowles & Carver 69 S<sup>t</sup>. Pauls Church Yard.* [1794 or later].[21]

**Copies**   (LRL Map 87b; Manchester UL Map C273.1).

### References

1   Plomer, 1932, p. 126.
2   11 May, 1749 in *General Advertiser* for publication on 24 May. Hodson, 1989, p. 99 illustrates the announcement.
3   *Public Advertiser* 23 May, 1760; Hodson, 1989, p. 107 reproduces the original announcement.
4   ibid., p. 123 records Sotheby's auction of 29 Sept. 1825. It was calculated that there were 569 lbs. of copper, which fetched £28.9.0 at 1s. per lb.
5   Examples of other counties with the map without imprint are recorded: Warwickshire is noted by Harvey and Thorpe, p. 99 and there is an example in BL (K.Top. XLII.75); Hertfordshire is noted by Hodson, 1974, p. 50 (example in Bod – Gough Maps Hertfordshire 8A); Middlesex is recorded in this state and appears thus in some of the early atlas compilations. These are probably proofs and Lincolnshire may exist in a similar state.
6   This collection contains none of the maps issued in 1756 or later. The CUL copy has two different copies of Lincolnshire – state (ii) and state (iii).
7   This includes all 43 plates, which are bound in the order of publication. This suggests that it was put together in 1760. The binding is, however, modern.
8   The two atlases each have two maps of Lincolnshire; both have the map with the 1751 imprint and also the 1753.
9   This copy, formerly in Oriel College, contains a mixture of plates, including maps of Scotland and Ireland, which appeared in May, 1763, when the set was advertised as having 47 plates.
10  The atlas and the title-page must have been issued in 1760 or soon after when the plates were all ready.
11  The Leeds copy has no title-page but was perhaps assembled in 1760 just prior to the title-pages becoming available.
12  Tinney died in early 1761; his name was soon removed from the imprint of some plates. This example was probably prepared later that year, but before state (v) was introduced.
13  The general maps were first advertised in *Public Advertiser* on 16 May, 1763. William Strahan was paid £1.12.0 to print 200 title-pages in black and red in October, 1763. (BL Add. MS. 48803A.f.69<sup>v</sup> – cited by Hodson, 1989, p. 116, n. 1).
14  Carington Bowles began to trade under his own name in early 1764; this atlas probably appeared in that year. It was listed in Robert Sayer's catalogue for 1766.
15  In the Nottingham and NLS copies Norfolk also has plate numbers 23, the original number assigned to that county; the title-page is from older stock and several plates show two different numbers on the same plate, all indications of the instability of the work.
16  The Scunthorpe map lacks the plate numbers; they were lost probably when the margins were cropped.
17  Hodson, 1989, pp. 119–120, 141 and 147 noted two copies with the title-page in French also; the Birmingham copy noted above and the Chetham's L. copy. The NLS copy is a third example. The NMM copy only has the English title-page. Sayer probably issued his own title soon after Carington Bowles did the same in 1767.
18  Bowles died in Summer, 1779. Wilkinson had 150 copies of a new title-page printed by Strahan in December, 1779. (BL Add. MS 48815.f.44<sup>v</sup>).
19  Capt. Armstrong's name on the title-page recognises the use made of his reduced version of Lincolnshire (**45**) and the single sheet map of Rutland (1780).
20  In spite of having a title-page for 1767 the BL example was compiled in 1794 or just after,

since many plates have Laurie & Whittle's names in the imprint; they did not own the plates until 1794. Several plates, including Lincolnshire, are in this earlier state.
21  County maps with similar imprints are recorded in BL Maps C.10.d.19 and RGS 1 C, but those atlases have Lincolnshire in states (vii) and (vi) respectively. With the names of Laurie & Whittle in the imprint these plates must have been issued in or after 1794. No atlas has yet been found containing a Lincolnshire map similar to the loose sheets noted here.

Hodson nos. 221–228.

## 35

**EMANUEL BOWEN**                                                            **1752**

Size: 19.4cm x 17.9cm.

From its first issue (June, 1747) the *Universal Magazine of Knowledge and Pleasure* had included topographical pages on the counties, each illustrated with a map. The first was Berkshire[1] but production was irregular and slow; Radnorshire (the last) did not appear until 1766. Among those employed as engravers were: Kitchin, Seale and the engraver of the Lincolnshire plate, Emanuel Bowen. It is clear that Hinton maintained only an irregular schedule of map production, perhaps due to problems in acquiring the services of engravers. Three maps appeared in 1747, five in 1748, only two in 1749, four in 1750, five dated 1751[2], three dated 1752[2] and one in 1753. Only the first four are signed by Kitchin; sixteen are by Bowen (1751–1759) and nine by Seale (1759–1764); in addition there were several periods when plates had no signatures, including all of the Welsh counties and the general maps which were the last to be produced. Six Welsh counties had no maps. The final series consisted of fifty-one plates spread over twenty years.[3]

    The Lincolnshire map is a fairly exact copy of that by Kitchin for *The London Magazine*, which had appeared a year before (**32**) with the arms of Lincoln (named correctly this time) taken from Walker's plate (**28**). It appeared in the April, 1752 issue with a text occupying pages 145–151, continued in May (pp. 204–210) and concluded in June (pp. 241–244). After this single use a fresh set of plates was engraved when the same journal issued county plates in the 1790s (**53**).

## EDITION

(i)   *LINCOLNSHIRE/Drawn from the best/Authorities/By Eman: Bowen/ Geograph^r. to His Majesty./* (Ac). Compass (Eb). *The Arms of Lincoln* (Ad). *Explanation./8* lines and symbols/ (Ae). *English Miles/12* [= 2.85cm] (Ee). *Printed for I. Hinton at the Kings Arms in Newgate Street: 1752.* (Be–De, OS). *Engraved for the Universal Magazine.* (Ca, OS). The inner border is marked with degrees of latitude and longitude at 2′ intervals with figures every 30 minutes between the borders. *1° Longit: West from London* (Be, between the inner and outer frames).

*The Universal Magazine of Knowledge and Pleasure . . . Vol. X. Published Monthly according to Act of Parliament For John Hinton at the King's-Arms in Newgate Street, London. price Sixpence.* [1752].[4]

**Copies**   CUL T.900.d.75.1; Bod Per 2705.e 552/1. (LAO Scorer Map 12; Grantham RL).

Atlas without title-page. [1766].[5]

**Copy**   BL Maps C.24.d.20.

Atlas without title-page. [1769?].[6]

**Copy**   CUL Atlas 7.74.13.

**References**

1   Burden, item no. 29.
2   One so dated was issued in the following year.
3   Chubb, pp. 146–149.
4   The title is taken from the bound set in BL (PP 5439), but this lacks the Lincolnshire map; the first leaf of the April text refers to the inclusion of the Lincolnshire map.
5   A miscellaneous collection of maps many from *The Universal Magazine*; there is no map issued after 1766 and it was probably assembled at that time.
6   A miscellany, which includes two maps of London, one dated 1765 and a second, from *The Universal Magazine*, issued in May, 1769.

Chubb no. CLXXXVIII.

**36**

**JOHN GIBSON**                                                                      **1759**

Size: 11.2cm x 6.35cm.

John Gibson, the engraver of this group of little pocket-book maps, flourished from c.1750 until 1792.[1] In 1758 he had been involved with Bowen in the production of *Atlas Minimus* . . . issued by John Newbery and intended for children. Newbery was the first publisher seriously to cater for children and their reading by producing a wide variety of story books and factual works designed to inform the young. His name has been honoured since 1922 in the annual award of a medal in U.S.A. for the most distinguished contribution to children's literature by an American resident.[2] It is not clear whether the atlas below was intended only for children, although Hodson notes that the atlas appears in a list of twelve books for 'the little Gentlemen and Ladies of these Kingdoms' from 'their old Friend, Mr. Newbery's'.[3] Newbery (1713–1767) married the widow of William Carnan whose business in Reading he continued before setting up in London in 1744.[4] He was a friend of Goldsmith, who may have written for him *Goody Two Shoes* and who depicted him as the philanthropic bookseller of St. Paul's Church Yard in *The Vicar of Wakefield*, published by Newbery's nephew, Francis Newbery, in 1766. The atlas appeared in May, 1759. The Lincolnshire map is based on, and much reduced from, the Kitchin map for *London Magazine*.

**Later history**
The map appeared only once more – in an atlas printed by Thomas Carnan, who was the son of the Carnan, whose widow Newbery had married. After Newbery died in 1767 Carnan was in partnership with his half-brother, Francis Newbery.[5] The atlas was issued after they went their separate ways in 1779. Cannan appears to have lost a case involving the printing by Paterson and Bowles of Paterson's *British Itinerary*, which infringed, he claimed, the copyright he had purchased in Paterson's *Roads*.[6] A further challenge, in which he succeeded, was against the Stationers' Company and broke the monopoly for printing almanacks.

**EDITION**

(i)   *Lincoln/Shire* (Ee, in cartouche). *English Miles/20* [= 1.7cm] (Be). 22 (Ea, OS). *Lincolnshire is in the Diocese of Lincoln and is 60 Miles long/ & 55 broad it contains 4590 Houses, 243540 Inhabitants,/631 Parishes, 35 market Towns, & sends 12 members to/Parl*. *The Air in the South & East parts is thick & foggy,/& the Soil Fenny & barren, but the West & Northern parts/are pleasant & fruitful It abounds in fat Cattle, Horses,/Sheep, Wool, Fowls, & Fish, Lincoln is a City, a Bishops/See, & a County of itself./* (above the map). Compass (Eb). The inner frame is shaded to indicate latitude and longitude at 5′ intervals numbered every 15 minutes. *1° Long. West from Lond^n*. (Ae–Ce, between the inner and outer frames).

*New and Accurate Maps of all the Counties of England and Wales Drawn from the latest Surveys By J. Gibson London Printed for J. Newbery at the Bible and Sun in S^t Pauls Church Yard* [7 lines of symbols used in the atlas]. [1759].[7]

**Copies**   BL Maps C.24.a.22; CUL Atlas 7.75.9; Bod Gough Maps 140. (LAO Scorer Map 9).

*New and Accurate Maps . . . London. Printed for T. Carnan, in S^t. Paul's Church Yard . . .* [Between 1780 and 1788].[8]

**Copies**   Bod Allen 37; Leeds UL W 79; NLS EME b.2.8.

**References**

1   Moreland and Bannister, p. 171.
2   Campbell, A.K.D. Outstanding Children's Books. (Swansea, 1990), pp. 10–13.
3   Hodson, 1989, p. 94.
4   Plomer, 1932, p. 179; Tooley, R.V. Dictionary of mapmakers. (Tring, 1979), p. 463; Roscoe, S. John Newbery And His Successors 1740–1814: A Bibliography. (Wormley, 1973), pp. 5–6.
5   Plomer, 1932, p. 43.
6   Roscoe, S. op. cit., pp. 22–23.
7   Announced in *Lloyds Evening Post* . . . 22–24 May, 1759 – cited by Chubb, p. 175.
8   Carnan was working alone by 1780 and died in 1788. Roscoe, op. cit., p. 328 notes that Carnan lists it in a book dated 1788 as on sale for 4s.

Hodson nos. 219–220.

# 37

## EMANUEL BOWEN                                                              1761

Size: 18.8cm x 16.55cm.

These small maps by Bowen were issued in *The Natural History Of England*, which formed part of *The General Magazine of Arts and Sciences*, a larger work issued in parts from 1755 until 1764. The first issue was announced for 1 Feb., 1755[1] and the series continued somewhat haphazardly until completed in 1762. The plan of issue was similar to that adopted in Camden's *Britannia*; the south-west counties were followed by a sequence that moved eastwards and then northwards. Benjamin Martin (1704/5–1782) wrote most of the text; he had started life as a plough-boy, became a teacher and, on receiving a legacy, took up writing, ran a publishing business and a school. He also became a notable maker of optical instruments before being bankrupted in 1781.[2] William Owen, the publisher of the part work and a well respected book-man, first set up in c.1748. In 1781 he was Master of the Stationers Company; joined by his son in the business in 1783 he remained active until at least 1793.[3]

   The Lincolnshire map is dated 1761 but it is not clear in which monthly part it made its first appearance. The plate is a small-scale copy of that used in *The Large English Atlas* (**34**). It was not issued again.

## EDITION

(i)  *LINCOLN SHIRE/Divided into its/WAPONTAKES./containing the/City Borough &*
*Market Towns/Rivers, Roads, Distances &ct/By Eman: Bowen/Geog$^r$. to His Majesty./*
(Ae). *Engrav'd for the General Magazine, of Arts & Sciences for W. Owen at Temple Bar*
*1761.* (Ae–De, OS). *Explanation* followed by three lines, with symbols, for Borough
Towns, Market Towns and Post Stages (Ea). Compass (Ed). The inner frame is shaded
to mark latitude and longitude at 1' intervals, with the numbers (between frames) at 10'
gaps. *Long. from London* (De, between the inner and outer frames). *Wapontakes* followed
by a list, lettered A–Z, a–g (Ab–Ad).

*The General Magazine Of Arts and Sciences, Philosophical, Philological, Mathematical,*
*and Mechanical . . . By Benjamin Martin. London: Printed for W. Owen, at Homer's*
*Head, in Fleet-street.* [1761].

**Copy**  BL 250 k 8–21.

*The Natural History Of England; Or, A Description of each particular county, In regard*
*to the curious Productions of Nature and Art. Illustrated by a Map of each County . . .*
*Vol. II . . . By Benjamin Martin. London: Printed by W, Owen, Temple-Bar, and by the*
*Author, at his House in Fleet-street. MDCCLXIII.*

**Copies**  BL 250 k 10–11; CUL Lib.6.75.2–3; Bod Gough Nat Hist 48–49; Leeds UL W
67; Leeds RL 942 M363; RGS 262 b 7–8. (BL 3355 (14); LRL Maps 813 and 813a; LRL
has a booklet (shelf no. A 57) containing the map and the text pages 207–219; p. 207 has
the heading: The Natural History Of Lincolnshire. In Grantham RL a copy of the map
has been bound with the county section of *Magna Britannia et Hibernia* (which lacks
the correct Morden map). LAO Scorer Map 15 appears to be a copy of the map without
the imprint; the sheet has been trimmed at the bottom but 3mm. remain below the neat
line – the top of the capital letters in the imprint are normally 2mm. below the lower
frame.

### References

1  *The Public Advertiser* for 31 January, 1755 said 'Tomorrow will be published . . . Monthly,
   price sixpence . . .'
2  Hodson, 1989, pp. 160–161; Kingsley, pp. 81–82; Millburn, J.R. Benjamin Martin author,
   instrument-maker, and 'county showman'. (Leyden, 1976); Millburn, J.R. 'Martin's Magazine
   . . .' *The Library*, 5th Series, no. 28 (1973), pp. 221–239.
3  Maxted, p.167 says that he was active until his death in 1798. Millburn, J.R. 1973, op. cit.,
   says he died in 1793.

Hodson no. 230.

**38**

**THOMAS KITCHIN**                                    **1763**

Size: 25.55cm x 19.6cm.

Robert Dodsley, publisher of *England Illustrated* had no serious success with his earlier
topographical work (**26**) but he seems to have planned, nevertheless, a similar work since
a contract survives, in which a writer, John Campbell, was to provide him with a suitable
text.[1] Nothing seems to have come of it but another attempt did have more success. Sheets
for the new work were printed by William Strahan in October, 1763, according to their
ledgers, and its issue was announced in the same month.[2] On 1 February, 1764 it was

announced that the issue of the work in twelve monthly parts would commence; in spite of Dodsley's death in September, 1764 the final part came out in January, 1765.

The title-pages of the set in the two-volume form are dated 1764 and *England Illustrated* appears to be another example of a work published at the end of one year (December 1st) which has the following year's date.[3] Issue in book form before serialisation is complete is something of a rarity.

The maps were engraved by Thomas Kitchin and have some similarities to those for *The Large English Atlas* (**34**); the continuation of the Yorkshire shore in an easterly direction, omitting Spurn Head suggests that Moll-based Read plates (**27**) were near at hand also. The text was an updated amalgamation of texts drawn from Camden, Speed and *Magna Britannia*. The Lincolnshire map and text end volume one.

**Later history**
In June, 1765 an atlas version of the above was advertised by James Dodsley, his younger brother and partner since 1755, who had taken over the stock. The larger work with text was available until at least 1781.[4] James preferred wholesale rather than retail bookselling; in that capacity he sold 18,000 copies of Burke's *Reflections on the French Revolution* in 1790. He died in 1797 a very rich man and his portrait by Reynolds survives.[5]

**EDITION**
(**i**)  *A New/MAP of/LINCOLN/SHIRE,/Drawn from the best/Authorities:/ By Tho$^s$. Kitchin Geog$^r$./Engraver to H.R.H. the/Duke of York./* (Ee). Compass (Ed). *British Statute Miles 69 to a Degree/10* [= 3.25cm] (Ae). *Remarks./7* explanatory lines and symbols (Ab). The inner frame is shaded to show latitude, etc. at 2′ intervals; numbered at 10′ gaps. *Meridian of London* (De, between the inner and outer frames) marks a line drawn down the plate, which avoids the land areas.

*England Illustrated, Or, A Compendium Of The Natural History, Geography, Topography, And Antiquities Ecclesiastical and Civil, Of England and Wales. With Maps of the several Counties, And Engravings of many Remains of Antiquity, remarkable Buildings, and principal Towns. In Two Volumes. Vol. I. London: R. and J. Dodsley, in Pall-Mall. MDCCLXIV.*

**Copies**  BL 190 b 8 and G 3220–1; CUL LL.12.31 and Huntingdon 31.9; Bod Allen 31 etc.; Leeds UL W 71; Nottingham UL s/DA 620.E6; NLS A44 e 13–14. (BL 3355 (15); LAO Scorer Map 16; LRL Map 801; Grantham RL).[6] The county map and the text for Lincolnshire (pp. 402–426) are held in LAO (FL L942 ENG) and LRL (L9).

*Kitchin's English Atlas: Or, A Complete Set of Maps Of All The Counties of England and Wales. Containing all the Cities, Towns, Parishes, Rivers, Roads, Seats, and, in General, every other Particular that is usually sought for, or to be found in, Maps. The Whole engraved . . . By Thomas Kitchin, Geographer . . . London: Printed for J. Dodsley, in Pall-Mall.* [1765].[7]

**Copies**  BL Maps C.25.a.2; CUL Atlas 6.76.2; Bod Allen 36; Leeds RL RQ 912.42 K648.

**References**
1    BL Egerton MS 738, f.6; cited by Hodson, 1989, p. 167.
2    The Strahan ledgers – BL Add.MS 48800 f.95$^v$. It was first announced in *Public Advertiser* for 26 October, 1763.
3    The announcement in October, 1763 referred to the availability of the two volumes from 1 December, 1763 at 'bound 2l.16s and Half-bound 2l.12s.6d.' The 12 monthly parts were 4s. each; Strauss, R., op. cit., p. 380.

4   The atlas was first advertised in *Public Advertiser*, 28 June, 1765 at 10s.6d. bound in red
    leather. Hodson, 1989, p.170 quotes further advertisements in that journal from 1771 to 1780.
5   Maxted, p. 68; Knight, C. op. cit., pp. 212–213.
6   The loose sheets could have come from the atlas form of 1765.
7   See note 4 above.

Hodson nos. 231–232.


## 39

**EMANUEL BOWEN**                                                          **1763**

Size: 51.1cm x 41cm.

Almost before *The Large English Atlas* had been completed and in spite of its protracted
birth, plans for a new atlas were put in hand with the plates reduced in size to two thirds
of those in the larger work. Robert Sayer's short involvement with *The Small British
Atlas* is, no doubt, explained by this scheme for a new work with its potential appeal to
buyers who either could not afford or did not need such a large work as the newly
completed Bowen atlas. Sayer was also involved with work on a new smaller atlas to
rival that of Rocque with plates engraved by Ellis (**40**). The new work was published as
*The Royal English Atlas*, probably in 1763, and, although the plates were engraved very
quickly, there is no sign of haste in the craftsmanship. There is, of course, nothing original
in them. Some of the county plates were not entrusted to Emanuel Bowen; Gibson and
Thomas Bowen engraved three each and Kitchin engraved five. The use of Royal in the
title refers to the size of paper used for the printing and has nothing to do with royal
patronage.

The names in the imprints of the first editions show that the chief names of the time
(in terms of English map-making) worked together on the project. Most of the names
have been met with already; Sayer, Bowles and Overton especially form a related group
whose family mapping products cover more than a century. The Henry Overton here is
the third of the name, being nephew of the second Henry (who had died in 1751) and
responsible for selling the Speed plates to the Diceys by 1754.

### Later history
The plates remained within the ownership of the above group for some years and were
re-issued several times in the late 1770s, after Sayer formed a partnership in 1774 with
John Bennett, who provided a new drive.[1] Bennett died in 1787, a year after the
partnership had probably been broken; Sayer did not re-use the plates again before his
own death in 1794. The plates probably passed into the possession of Bowles and Carver,
who made no real use of them (or the other plates in their ownership[2]) before being sold
at the Wilkinson sale at Sotheby's in 1825. At that sale Martin acquired sufficient sheets
of very mixed ages and states to make some complete atlases, and, with new title-pages,
he issued them in c. 1828.[3]

### EDITIONS

(i)   *An ACCURATE MAP of/LINCOLN SHIRE/Divided into WAPONTAKES;/ Drawn
from the best Authorities, and Illustrated with Historical/ Extracts relative to its Produce/
Trade & Manufactures:/Describing also the Church Livings/Charity Schools &c./By
Eman: Bowen/ Geog$^r$. to his late Majesty./* (Ea). Compass (Ad). S$^t$ *Mary the Cathedral
Church/of Lincoln.* (Ae, an engraving of the Cathedral from the south side is below, and,
below the engraving, a note on its clerical establishment). *British Statute Miles 69 to a*

*Degree/15* [= 8.4cm] (Ce). Plate: *N°. 22.* (Ae, OS). *Printed for T. Kitchin at the Star Holborn Hill, J. Bowles & Son & Mess. Bakewell & Parker in Cornhill, H. Overton without Newgate, T. Bowles S'. Pauls Chur: Yard, R. Sayer & J. Ryal in Fleetstreet.* (Ae–De, OS). The inner frame is shaded to mark latitude, etc. every minute and numbered every 5. *Longitude from London* (De). The area around the map has notes on the history and topography of the county.

*The Royal English Atlas: Being A New and Accurate Set Of Maps Of All The Counties of South Britain, Drawn from Surveys, and the best Authorities; Divided into their respective Hundreds, And Exhibiting All The Cities, Towns, Villages, Churches, Chapels, &c. Particularly . . . Adorned with Views of all the Cathedrals . . . To The Whole is prefix'd, A General Map of England and Wales; Comprehending All The Direct And Principal Cross Roads . . . By Emanuel Bowen . . . Thomas Kitchin . . . and Others. The Whole Comprised in Forty-Four Sheet Maps. London: Printed for Thomas Kitchin, on Holborn-Hill; Robert Sayer, in Fleet-Street; Carington Bowles, in St. Paul's Church-Yard; Henry Overton, without Newgate; Henry Parker, and John Bowles, in Cornhill; and John Ryall, in Fleet-Street.* [1763].[4]

**Copies**   BL Maps C.26.c.9; CUL Atlas 33.76.4 and 3.76.3; Bod Allen 32; Leeds UL W 70; Manchester RL BR f 912.42 B7. (LRL Map 104; Grimsby RL Bowen 1763).

**(ii)**   The plate number has been deleted and the imprint changed to: *London. Printed for Rob'. Sayer & John Bennett, N°. 53, Fleet Street, John Bowles, N°. 13, Cornhill, & Carrington Bowles, N°. 69, S'. Pauls Church Yard, as the Act directs. 1st. of June 1777.*

Issued as separate sheets only.[5]

**Copies**   (LAO Scorer Map 21; LRL Maps 161 and 584; Grimsby RL BOWEN 1777; Boston RL Map 26).

**(iii)**   Bowles' fore-name has been corrected to *Carington* in the imprint.

*The Royal English Atlas: Being A New And Accurate Set Of Maps Of All The Counties In England And Wales, Drawn From The Several Surveys which have been hitherto Published, With a general Map of England, and Wales. From the latest and best Authorities . . . By Emanuel Bowen . . . Thomas Kitchin, and Others. London: Printed for and Sold by Carington Bowles, at his Map and Print Warehouse, at N° 69, in St. Paul's Church Yard.* [1778].[6]

**Copies**   BL Maps C.29.c.2. (BL 3355(1)).

*The Royal English Atlas: Being . . . Maps Of All The Counties of South Britain, Drawn from Surveys, and the best Authorities; Divided into their respective Hundreds, And Exhibiting All The Cities, Towns, Villages, Churches, Chapels, &c. Particularly Distinguishing more Fully and Accurately the Church Livings, Than any other Maps hitherto Published. Adorned with Views of all the Cathedrals . . . London: Printed and Sold by Robert Wilkinson, at No. 58, in Cornhill, Successor to Mr. John Bowles, deceased.* [1779].[7]

**Copy**   CUL Atlas 3.77.3.

*The Royal English Atlas; Being Accurate Maps of all the Counties In England and Wales: From numerous Surveys; shewing their respective Hundreds, &c . . . Also, Several New Maps, &c . . . The Whole comprized in Fifty Sheet maps. By Thomas Kitchin, Geographer, and Others. London: Printed for R. Sayer and J. Bennett, Map, Sea Chart, and Print-Sellers, in Fleet-Street.* [1780].[8]

**Copy**   Bod Toynbee 3723.[9]

*The Royal English Atlas: Being A New and Accurate Set of Maps Of All The Counties in*

*England and Wales Drawn from the several Surveys which have hitherto been published
. . . Carington Bowles . . .* [1781].[10]

*The English Atlas; Or, A Set of Maps Of All The Counties In England And Wales, Drawn
From The Best Authorities; Containing All The Cities, Towns, Villages, and Churches,
Chapels, Many Noblemen's and Gentlemen's Seats &c. Each Map Is Illustrated With A
General Description of the Borough and Market Towns; The Number of Members
Returned to Parliament; Of Land, State Of Inhabitants . . . London, Sold By R. Martin,
Bookseller, No. 47, Great Queen Street, Lincoln's-Inn Fields.* [1826 or just afterwards].[11]

## Reproductions
(i) The Royal English Atlas . . . with an Introduction by J.B. Harley and Donald Hodson.
(Newton Abbot, 1971). (ii) A greatly reduced version of state (ii) illustrates an article by
John E. Holehouse in *Lincolnshire Life,* Vol. 13, no. 4 (June, 1973), pp. 24–25.

## References
1  Hodson, 1984, p. 187.
2  Hodson, 1984, p. 189 suggests very few copies were sold, largely because Bowles was already
   wealthy and had no urge to compete with the new atlases of John Cary.
3  Hodson, 1989, pp. 123–125. On p. 124 a part of the sale catalogue is reproduced, with
   manuscript notes of the prices realised for the large plates and the smaller. (See also **34** above
   and its note 4).
4  Chubb, p. 179 suggested 1762, while later bibliographers have thought that 1764 is more
   likely. In the notes to the facsimile edition (see above) Harley and Hodson make the case for
   1763. The order of names in the imprint on the title-page is varied so that each bookseller sold
   copies with his name printed first. The two CUL copies are examples of these variant title-page
   imprints. The contents do not vary so it has not been thought necessary to record as separate
   works all the combinations of names in the imprints of the atlases.
5  Only recorded in loose sheet form, which must have been put out soon after the date on the
   map and quickly corrected for the atlas form. The atlas issued by Martin is known in three
   examples: BL Maps C.26.c.10; RGS 3 H 6 and one owned by Mr. C.A. Burden. This atlas is
   made up of a variety of loose sheets acquired by Martin in 1825; all the three atlases have this
   state of the Lincolnshire plate.
6  Some plates are dated 1778.
7  Wilkinson took over the premises at 58 Cornhill in 1779. (Maxted, p. 248). The atlas probably
   appeared soon after he was established.
8  Chubb, pp. 183–184 quotes an example of this title provided by a librarian in Cambridge. The
   book was ascribed to 1780; though the reasoning is not clear it is generally accepted.
9  A copy in Birmingham RL (AE 912.42) lacks many plates, of which Lincolnshire is one.
10 Hodson, 1974 p. 54 records a re-issue of the 1778 edition put out by Carington Bowles in
   1781. The only copy is in private hands and has not been examined.
11 Martin acquired the remaindered sheets in October, 1825. The leaf bearing the off-set title-page
   in the BL copy (see note 5) has the watermark date 1826. The miscellaneous nature of the
   sheets used to make up these atlases is indicated by the Sussex plate having a watermark date
   of 1792, all the Lincolnshire sheets being state (ii) and many maps still having plate numbers.

Chubb nos. CCXVIII and CCXIX.

**40**

**JOSEPH ELLIS**                                                        **1765**

Size: 25.4cm x 19.3cm.

John Ellis was an engraver, whose career seems to have spanned the fifty years between 1750 and 1800.[1] Joseph Ellis is now put forward as the engraver of the plates of the *New English Atlas* on the basis of recently discovered evidence.[2] This seems to be the only county map series produced by Ellis and the county plate is almost an exact copy of the Kitchin maps for *England Illustrated* (**38**), with several features identical in wording and placing on the plate.

**Later history**
The work went through several editions over a period of thirty years from the offices of Robert Sayer and Carington Bowles. On Sayer's death in 1794 the plates passed to Laurie and Whittle, who produced one final edition but without changing the title-page, although some county plates were altered to include the new owners' names, e.g. Gloucester in the BL copy.

   In some issues the maps are printed back to back; as a result Lincolnshire appears on the verso of the Leicestershire map. An asterisk * is given after the shelf number of such examples.

**EDITIONS**

(i)   *A Modern MAP of/LINCOLN-/SHIRE,/Drawn from the/latest Surveys; /Corrected & Improved/ from the best Authorities./J. Ellis sculp[t]*. (Ea). *British Statute Miles 69 to a Degree/10* [= 3.2cm] (Ae). *Remarks/7* lines of text and symbols. *Printed for Carington Bowles in S[t]. Pauls Church yard, & Rob[t]. Sayer in Fleet Street* (Be–De, OS). Plate number 28 (Ea, OS). The inner frame is marked (by shading) with latitude, etc. at 2′ intervals and numbered every 20 minutes. *Meridian of London* (De, between inner and outer frames); there is a line down the map marking the meridian and another line across the map at 53° [North].

*The New English Atlas; Being The Completest Sett of Modern Maps Of England and Wales, With their Adjacent Islands, Accurately Drawn from Actual Surveys by the most able Geographers. And Engraved in the best Manner by J. Ellis and Others, On Fifty-Four Copper-Plates. Containing the Particular County Maps . . . Also Five Large Elegant and Useful General Maps . . . London: Printed for Robert Sayer, at the Golden Buck, near Serjeant's Inn, Fleet-Street, and Carington Bowles, near the Chapter House, St. Paul's Church Yard. MDCCLXV.*

**Copies**   BL Maps C.24.aa.2. (LAO Ex 32/2/159 and Scorer Map 17; LRL Map 797).[3]

*Ellis's English Atlas: Or, A Compleat Chorography Of England and Wales: In Fifty-Four Maps. Containing more Particulars than any other Collection of the Same Kind. The Whole Calculated for the Use of Travellers, Academies, and of all those who desire to Improve in the Knowledge of their Country. From the latest Surveys of the Several Counties; With the Addition of Four Maps . . . Engraved by, and under the Direction of, J. Ellis. London: Printed for Carington Bowles, next the Chapter-House, in St. Paul's Church-Yard and Robert Sayer, at the Golden Buck, near Serjeants Inn, in Fleet Street. Price 15s. in the Red Leather for the Pocket. MDCCLXVI.*[4]

**Copies**   BL Maps C.25.a.4 *; Leeds UL W 72 *.

*Ellis's English Atlas . . . Printed for Robert Sayer . . . Carington Bowles . . . M DCC LXVI.*

**Copies** CUL Atlas 7.76.1; Bod C.17.d.2 *.

*Ellis's English Atlas . . . In Fifty Maps . . . Printed for Robert Sayer . . . and Carington Bowles . . . Price 10s.6d. in Red Leather for the Pocket. M DCC LXVI.*

**Copies** BL Maps 11 a 8; Birmingham RL A 912.42; RGS 8 B 7.

*Atlas Britannique, Ou Chorographie Complette De L' Angleterre Et De La Principauté de Galles. En 54 Cartes Executées par les plus Habiles Graveurs, Et dans lesquelles on a réuni tous les Détails qui peuvent intéresser les Amateurs de la Geographie. A Londres: Chez Robert Sayer, Marchand de Cartes & d' Estampes. M DCC LXVI.*

**Copy** BL Maps C.22.c.1.

*Ellis's English Atlas . . . In Fifty-Four Maps . . . London: Printed for Carrington Bowles, Map and Print-Seller, at N°. 69, in St. Paul's Church-Yard. 1768. Price 15s. in Red Leather for the Pocket.*

**Copy** Hull RL (Map Cupboard).

*Ellis's English Atlas . . . Printed for Carington Bowles . . . 1768 . . .*

**Copy** Leeds UL W 73.

*Ellis's English Atlas . . . In Fifty Maps, Containing . . . Printed for Robrt [sic] Sayer, map and Printseller, No. 53. in Fleet Street; Thomas Jeffery's Geographer to his Majesty the Corner of St. Martin's-Lane, Charing-Cross; A. Dury in Duke's-Court, St. Martin's-Lane, and at the Map and Print Shop No. 92, under the Royal-Exchange, Cornhill. 1768.*

**Copies** CUL Atlas 6.76.1; Leeds RL SR912.42 EL59; Norwich RL R912.42 (S.C.3.22).

*Ellis's English Atlas . . . London: Printed for Robert Sayer, Map and Printseller, No.53. in Fleet-Street; and at the Map and Print Shop No. 92, under the Royal-Exchange, Cornhill. 1773.*

**Copies** Bod Allen 38, etc.; Liverpool UL K.4.1 (7); Birmingham RL A912.42; NLS Newman 628.

*Ellis's English Atlas . . . 1773. [1775].[5]*

**Copies** BL Maps C.24.c.10; CUL Atlas 7.77.1.

*Ellis's English Atlas . . . London, Printed for R. Sayer and J. Bennett, Map, Chart, and Print Sellers, No. 53, Fleet-Street. 1777.*

**Copies** BL Maps C.26.d.29; CUL Atlas 6.77.3; Leeds UL W 74; RGS 8 B 2.

*Ellis's English Atlas . . . In Fifty-Four Maps . . . The Maps are placed in Alphabetical Order, viz . . . Printed for Carington Bowles, at his Map and Print Warehouse, No. 69, St. Paul's Church Yard, London. [Price 15s. in Red Leather for the Pocket.] [1785].[6]*

**Copy** BL Maps 198 d 25 *.

*Ellis's English Atlas . . . In Fifty Maps . . . Carington Bowles . . . (Price 10s.6d. in Red Leather for the Pocket). [1786].[7]*

**Copy** RGS 8 B 3.

(ii)   The plate number has been changed to *25* (Ea, OS).

*Ellis's English Atlas: Being Accurate Maps Of All The Counties In England And Wales, According To The Latest Surveys. Calculated for the Use of Travellers, &c. and contains more Particulars than any other Atlas of the Kind, Engraved By, And Under The Direction Of J. Ellis: In Fifty Maps. London: Printed for Robert Sayer, No. 53, Fleet-street. [1796].[8]*

**Copies** BL Maps C.29.d.13; RGS 8 B 4.

**References**

1   Gough, 1768, pp. 46, 267 and 703 describe maps by J. Ellis; only in the index does Gough refer to John Ellis; Darlington, I. and Howgego, J. Printed maps of London, c.1553–1850. (London, 1964), p. 132 record maps up to 1786 and one further map by J. Ellis (with B. Baker) in a map of 1800 (p. 154).
2   Hodson, 1974, p. 54; first suggested on the basis of an advertisement issued by Bowles and Carver in 1795. Joseph Ellis is in the list of subscribers to Bowen's *Atlas Anglicanus* found in the CUL example of the 1767 edition. Maxted, p. 74 refers to John and Joseph Ellis, both engravers and both in Islington in this period, but it seems likely that there was only one engraver there.
3   The loose sheet maps could have come from any edition but the last.
4   A copy owned by Mr. C.A. Burden has the price on the title-page changed by hand to 10s.0d. This implies it was issued some time before the price change noted in later editions.
5   This version includes a chart of distances dated 1775.
6   Not recorded until Hodson, 1974, p. 55. Ascribed by the BL to 1789 but Hodson's date is now generally accepted.
7   Presumed to have been issued in 1786, shortly after the previous edition; the only change relates to the reduced price.
8   Actually issued by Laurie and Whittle who bought the plates in 1794. The general map in the RGS copy is dated 1796.

Chubb nos. CCXXVII–CCXXX. (He does not quote a 1765 edition nor any after 1777.

# 41

## PIETER MEIJER                                                          1766

Size: 18.8cm x 16.85cm.

These maps form a series copied, as the map-title acknowledges, from Bowen's map for *The Natural History Of England* (**37**). The Dutch work containing the maps by Meijer also includes in translation the topographical text that accompanied the English original. It seems that there was about three years between the appearance of the county maps in their first issue in parts in England and the preparation of the equivalent Dutch plates; for instance, the Sussex map appeared here in 1756 and the Dutch version is dated 1759. It is possible, therefore, that there might have been a contractual arrangement with Martin and Owen, the English undertakers in the project, since the project proceded in tandem; the first Dutch plates were being engraved before the English original work was completed. The statement in the title: *Tweede Afdeeling* (second edition) must refer to the *Natural History of England* as being the first edition. Although the map is dated 1766 there is no evidence that the map was sold separately before the issue in volume form.

Part one of the two volume work was published in 1763; as in the Martin work, Lincolnshire's map and text appeared in volume two, which was completed in 1770. There was no other edition.

## EDITION

(i)   *LIN=/COLNSHIRE,/Verdeeld in deszelfs Rechts:/gebieden, berattende alle de/Steden Burg-en Markt=/vlekken, met de Wegen, Af:/standen, enz. Opgessteld door/Eman. Bowen Landbeschryrer/ van zijn Britsche Majesteit./Te Amsteldam by Pieter Meijer/ Uitgegeven 1766./* (Ae). *Verklarung.*[= Explanation]/3 lines with symbols for Towns returning MPs, Market towns and Post stages. *Rechtsgebieden* [List of Wapentakes with key letters A–Z, a–g]. (Ab–Ad). Compass (Ed). *Engelsche Mylen/12* [= 2.5cm]/*Duitsche Mylen/2¼* (Ee). The inner frame is shaded and divided at 1' intervals and numbers appear every 10 minutes. *Lengte van London* (De, between frames).

4. 1766   MAP 41 – PIETER MEIJER

*Algemeene Oefenschoole Van Kosten En Weetenschappen. Tweede Afdeeling. Behelzende De Natuurlyke Historie Des Aardryks. Tweede Deel. Te Amsterdam, By Pieter Meijer op den Dam. MDCCLXX.*

**Copy**   BL Maps C.29.d.16.

Not in Chubb.


## 42

**EMANUEL BOWEN**                                                    **1767**

Size: 31cm x 22.4cm.

The last atlas with which Bowen was involved consists of maps further reduced from those in *Royal English Atlas,* while keeping all the main features of that and *The Large English Atlas.* Once more the finely engraved plates are surrounded with notes on the county. No new features are added, while some have been omitted because of the smaller space available. Bowen died in poor straits in 1767. He was succeeded by his son Thomas, who had engraved plates for others from the 1760s but he also died in poverty (in 1790 in a Clerkenwell workhouse).[1]

The plates were issued in parts; three maps each having the same plate number added to the maps were included in each part, except that in the first (Berkshire, Cheshire and Buckingham) there were no numbers. There are, therefore, no plates with the number one.[2] There were eventually fourteen parts containing 42 maps for the English counties, and including a general map for England and two sheets covering north and south Wales. The other counties issued at the same time as Lincolnshire, and thus having the same plate number, are Norfolk and Nottinghamshire. The atlas version was issued in 1767. The maps have been reduced to a third of the size of those in *The Royal English Atlas* and a fifth of those in the *The Large English Atlas.*

That *The Large English Atlas* was a success is indicated by the number of copies that have survived. Further evidence is indicated by the way Kitchin and associates provided not one but two versions in smaller formats; the survival rate may be an indication that neither the *Royal* nor the atlas under review shared the same public favour.

**Later history**
Kitchin used the plates in an edition in 1777, in association with Andrew Dury, following which the plates passed to Carington Bowles and then to his successors, Bowles and Carver. Each new owner issued an edition. In about 1794 some plates emerged again and there may have been an edition in a final atlas form.

**EDITIONS**

(i)   *LINCOLN SHIRE/Divided into/WAPENTAKES,/Exhibiting the City, Borough and/Market Towns &c./with Historical Extracts, relative to its/Trade & Manufactures./Describing also the Church Livings; with/Improvements not Inserted in any other/Half Sheet County Map, Extant./By Eman: Bowen/Geographer to His late Majesty,/ & Thoˢ. Bowen./* (Ea). *Explanation/* 13 lines of symbols and their meaning. (Ac). *British Statute Miles 69 to a Degree/15* [= 4.7cm] (Ca). Compass (Ad). *Nᵒ. 10* (Ea, OS). The inner frame is marked every minute of latitude, etc., and numbered at 5′ intervals, the latter corresponding to a graticule, which does not cover the land areas. *Long: from London* (De, between the inner and outer frames).

*Atlas Anglicanus, Or a Complete Sett of Maps of the Counties of South Britain; Divided into their respective Hundreds, Wapentakes, Wards, Rapes, Lathes, &c. Exhibiting the Cities, Boroughs . . . The whole Illustrated with Historical Extracts . . . With Various Improvements, not Inserted in any other Sett of Half Sheet Maps extant. To which is added a Correct Map of the Roads of England . . . By the late Emanuel Bowen, Geographer to His Majesty George II^d and Thomas Bowen. Printed for T. Kitchin, N^o. 59, Holborn Hill.* [1767].³

**Copies**  CUL Atlas 5.76.1; Bod C.17.c.1 and Allen 35; Leeds UL W 75; Birmingham RL AF912.42 BOW.⁴

*Atlas Anglicanus . . .* [1770].⁵

**Copies**  BL Maps C.24.f.22; RGS 3 H 14.

**(ii)**  The plate number has been removed. An imprint has been added: *Printed for Tho Kitchin N^o. 59 Holborn Hill London.* (Ce, OS).

*Atlas Anglicanus . . . By . . . Emanuel Bowen . . . and Thomas Bowen. 1777. Printed for T. Kitchin, N^o. 59, Holborn Hill. & Andrew Dury, Duke's Court, S^t. Martins Lane.*⁶

**Copies**  BL Maps C.27.c.2; CUL Atlas 6.77.2; Bod Allen 41 and 42; Brighton RL 912.42 B67; Chesterfield RL 912.42. (LAO Scorer Map 18).

**(iii)**  The original cartouche with the title has been replaced by an oval with a new title: *BOWLES' S/NEW MEDIUM MAP/OF/LINCOLN SHIRE, /DIVIDED into its WAPON-TAKES;/Exhibiting the Roads, Towns and Villages;/with their Distance from London,/Church Livings, Seats of the Nobility, and/Historical Remarks./LONDON:/Printed for the Proprietor/ Carington Bowles, N^o. 69 in S^t.Pauls Church Yard.* (Ea). The imprint now reads: *Published as the Act directs, 3 Jan. 1785.* (Ce, OS). In the Explanation the figure *17* has been replaced by *133* and *from London* added below *Measur' d Distances.* Mileage figures have been added at many places, e.g. *109* at Holbech, *97* at Bourn, *89* at Stamford. Mileages along the main roads, e.g. *7* (above Market Deeping) and *22¼* (near *S^th Witham*) have been deleted. The compass has been altered from a simple cross in a circle to a decorated cross with four extra points in a circle.

*Bowles's New Medium English Atlas; Or, Complete Set of Maps of the Counties of England and Wales: Divided Into Their Respective Hundreds . . . London: Printed for the Proprietor Carington Bowles, at his Map and Print Warehouse, No. 69, in St. Paul's Church Yard. M DCC LXXXV.*

**Copies**  BL Maps C.25.a.5; CUL Atlas 6.78.7; Bod Allen 44; Leeds UL W 76; Birmingham RL A 912.42 BOW. (LRL Map 575).

**(iv)**  The imprint in the title now reads: *LONDON: Printed for the Proprietors Bowles & Carver, N^o. 69 in S^t Paul's Church Yard.* The dated statement (Ce, OS) has been deleted.

*Bowles's New Medium English Atlas . . .* [1794?].⁷

### References
1  Moreland and Bannister, p. 166.
2  Chubb, pp. 192–193 thought that the plate numbers must refer to pigeon holes in the printing office. Sussex, probably by an oversight, did not have a plate number.
3  As the title indicates Emanuel Bowen was already dead and the work was completed by his son, Thomas, who probably engraved the title-page. The work probably appeared later in the year of Emanuel's death. The subscription list that accompanies the CUL copy names many well-known members of the English map trade. A further note records that one subscriber, Dr. Dodd, was hanged in 1777. This refers to William Dodd (b.1729 in Bourn), who was patronised by Lord Chesterfield, in whose name he issued a note for £4,200. In spite of most of the money

being repaid he was tried for forgery, found guilty and hanged; Dr. Johnson and others failed in their appeal to the King.

4    Although the Birmingham copy is dated 1797 on the title-page this was carefully added by hand. The contents are only consistent with this edition.

5    While the Lincolnshire plate remains unchanged, many other plates have had their plate numbers removed and some had imprints added. Hodson, 1974, p. 58 suggested this now generally accepted date.

6    Dury was a mapseller and engraver active from 1771 to 1778. (Maxted, p. 70).

7    Bowles & Carver were in business from 1793 to c.1832. Their catalogue of 1795 lists the atlas and the availability of loose sheets. Hertfordshire is recorded in this state (Hodson, 1974, p. 58); Warwickshire (Harvey, and Thorpe, p. 106), etc. Kingsley, pp. 90–91 notes that Folio Fine Art Ltd. (June, 1969 catalogue) had several counties in this altered state. Lincolnshire may exist in this state also.

Chubb nos. CCXXXII–CCXXXIII (only the 1767 and 1777 issues).


## 43

**THOMAS KITCHIN**                                                                    **1769**

Size: 27.85cm x 23.1cm.

Although the title of the work from which these maps come – *Kitchin's Pocket Atlas* – might suggest that the maps were small the Lincolnshire plate is a slightly larger version of that prepared by Kitchin for *England Illustrated* (**38**); it is an almost exact copy of the earlier map.

There seems to be no obvious gap in the market which Kitchin should feel so strongly needed filling that a new set of plates should be prepared. The *England Illustrated* plates were still available and comparatively new. That the sales were quite low is, perhaps, shown by the note by Chubb[1] that only one copy was known in 1927 (and that was found on a Cambridge bookstall). The plates passed to Carington Bowles who re-issued them in one more edition.

### EDITIONS

(i)    There is no title; *LINCOLN/SHIRE* is engraved across the face of the map (Bb–Db and Bd–Dd). *Scale of British Statute Miles./14* [= 5cm] (Ea). *Remarks./Lincoln is a City/9* lines of symbols and text (Ee). Compass (Ed). The inner frame is shaded to show latitude, etc. at 1' intervals and numbered every ten minutes. *Min$^s$. of Longit W. from London* (Ae); a similar note (Ee) substitutes E. for W.

*Kitchin's Pocket Atlas, of the Counties of South Britain or England and Wales, Drawn to One Scale: By which the true proportion they severally bear to each other may be easily ascertained, with the Measured Distances from London by the nearest Roads annexed, to all the Cities, Borough & Market-Towns in the Kingdom. Being the first Set of Counties, ever Published on this Plan, 1769. London Printed for the Author, T. Kitchin Engraver, Map & Printseller, N$^o$ 59 Holborn Hill; & J. Gapper, Map & Printseller, N$^o$ 56, New Bond Street.*

**Copies**    BL Maps C.7.b.26; CUL Atlas 6.76.3 and 6.76.4; Leeds UL W 77. LRL has a copy of the map bound in *A Description Of England And Wales . . . Vol. V. London . . . Newbery and Carnan . . . M DCC LXIX.* (Shelf mark L.A9). The volume includes text for Lincolnshire. The anonymous ten volume work does not call for maps in its contents list; a full set is in CUL – XXVII.103.1–10.

(ii) *BOWLES'S REDUCED MAP OF LINCOLNSHIRE.* added (Ca, OS). A plate number *31* has been added (Ea, OS).

*Bowles's Pocket Atlas of the Counties . . . London. Printed for & Sold by the Proprietor Carington Bowles, N°. 69 in S*ᵗ. *Pauls Church Yard.* [1780].²

**Copies**   BL Maps C.21.bb.7; CUL Atlas 7.78.26; NLS Newman 833; Manchester UL AMGS A4.

**References**

1   Chubb, p. 195.
2   Professor Newman in a note in his copy (now in NLS) doubted the date 1785 (suggested by Chubb) on the basis of an advertisement for the work in *Bowles Post-chaise Companion* (in volume two of the first edition), which might, he thought, imply publication before 1780. Since Bowles' volume was published in 1781 (see Appendix 11 (ii)) the date generally agreed (1782) is at least a year too late and Professor Newman's suggestion should be accepted. The work is advertised in *Carington Bowles New And Enlarged catalogue, 1782* [cover title], p. 176; 'octavo, half-bound, 11.1s.' (Copy in CUL – s.696.c.78.1). The same catalogue quotes prices for *The Large English Atlas, The Royal English Atlas* and *Ellis's English Atlas* at 3l.13s.6d., 2l.12s.6d., and 15s. respectively.

Chubb nos. CCXXXV and CCLVI.

**44**

**ANDREW ARMSTRONG**                                                             **1779**

Size: 206cm x 151.2cm (approx.).

Captain Andrew Armstrong was responsible for the first survey of the county at a scale of one inch to a mile (1:63000). As the title on the map makes clear, the work took the three years from 1776 to 1778. Shortly after, or perhaps overlapping the Lincolnshire work, he surveyed Rutland at the same scale.

In 1759 the Royal Society of Arts (RSA) offered 'a sum not exceeding one hundred pounds, as a Gratuity to any Person or Persons, who shall make an accurate actual Survey of any County'. The Society was anxious to promote the large-scale surveying of all counties that had not already been surveyed on that basis. At that time few counties had been surveyed at a large scale and using more up-to-date techniques. Additionally, there had not been any national survey since that of Saxton, except for the work undertaken by John Ogilby in association with his road book *Britannia* (1675). Before 1750 only eight English counties (and London) had been surveyed at a scale of one inch to a mile or larger, with Sussex surveyed at ¾″ to a mile.[1] Cornwall was the first to be treated to surveying at the scale that became the basic standard (until recent times) when J. Gascoyne published his map in nine sheets in 1700. Budgen's map of Sussex (1723) was criticized severely by Richard Gough in 1780[2] but, as Kingsley points out, not in his earlier account in 1768.[3] It is probable that Gough's view had changed when a new survey by Yeakell and Gardner began to appear in 1778.[4] Henry Beighton proposed to make a survey of Warwickshire in 1720 but it did not appear until 1728.[5] John Kirby surveyed Suffolk between 1732 and 1734 (issued 1736)[6]; Senex' map of Surrey in 1729, William Gordon's maps of Bedfordshire (1736) and Huntingdon (1730–1) and Seller's map of Middlesex (first edition in 1724) complete a picture of almost furious activity during the first third of the eighteenth century. The first very large-scale map of all in England was not of a county but of the Fens – mapped by Jonas Moore and published on sixteen sheets at two inches to a mile in 1684.[7]

One scheme, proposed but never carried through, was for the mapping of Lincolnshire. In 1730 Edmund Weaver wrote to William Stukeley while sending him a copy of his proposals for 'making and publishing, by subscription, an Actual Survey of the County of Lincoln . . .'[8] The scale suggested was one mile to an inch and it would be accompanied by an Index Villaris, coats of arms and explanatory notes. The intended price was to be five shillings when taking out a subscription and a further five shillings on receipt of the whole map pasted down on canvas, or a final half a crown if supplied in sheet form. Noblemen who wished to have their coats of arms engraved around the edge would make an initial payment of one guinea and a similar final payment. It will be recalled that Blome was an enthusiastic supporter of this method of raising funds for his mapping projects (**17** and especially its note 3). Nothing much seems to have come of his idea, but Weaver was still printing his proposals through the 1730s and 1740s. In spite of making the subscription rate one pound paid in two halves he had only raised £48 at the time of his death in 1748 and the scheme died with him. Weaver was a licensed physician at Frieston, near Caythorpe. From 1723 until his death he published his annual almanac entitled *The British Telescope: Being An Ephemeris . . . with an Almanack . . .* and it was in this publication that he advertised his map proposals, his astronomical and medical services. He also made barometers, undertook surveying work and, from 1737, he was licensed to teach in the school at Caythorpe.[9]

After the activity of the 1720s and 1730s there is quite a long gap largely, one assumes, because of the heavy financial outlay involved. Something of the costs was referred to in discussing the work of Thomas Jefferys, the engraver and surveyor. Jefferys went bankrupt largely because of his over-commitment to county surveying in the period of the RSA's offer of an award. The map of Sussex by Thomas Yeakell and William Gardner only covered half of that county before work stopped through shortage of subscriptions. Kingsley quotes their prospectus in which they estimated that the whole undertaking 'will employ us, from first to last, six years, and cost more than 2,400£ for Surveying, Drawing and engraving . . .'[10] The estimate they made was that 400 subscribers would defray the costs; only 250 were forthcoming in the event. That proposal was made in June, 1778 roughly a year after Armstrong's son, Mostyn John, had issued his ideas for the surveying of Norfolk.[11] In his case Armstrong must have seriously underestimated the costs; he believed the cost would be at least £450, but he would close the subscription list when it reached 300 names, each of whom would have been expected to pay one guinea when registering for a set and another guinea on receipt. The work was never completed although there were 220 subscriptions by April, 1778 and a year later there were an additional one hundred subscribers. The extent of the son's under-estimation of the costs is provided by his father's own submission to the Royal Society of Arts of the costs he incurred in his Northumberland survey. The account is reproduced in Harley's discussion of the RSA's involvement with the large-scale mapping of Great Britain and shows: Expenses of survey (Sept. 1766–March, 1769) for himself, one assistant, two labourers and horses: £350; 9 copper plates, each 24″ by 18″: £14.10.0; For Engineers, by agreement: £90; Paper for 500 maps: £27; Printing copper plates: £20; Printing the Index, 500 copies: £10; Printing and advertising the Proposals, etc: £5. Total: £516.10.0.[12]

Because of the RSA's financial inducement there was an upsurge of work in this field. Jefferys worked on a number of projects and was soon followed by the Armstrongs, father and son. The first county surveyed by Andrew Armstrong was Durham, engraved on four sheets published in 1768, the engraver being Thomas Jefferys. Three versions exist for 1768, issued by three different publishers. John Cary and J. Christopher produced a new edition in 1791. Between 1766 and 1769 Armstrong was joined by his son, Mostyn John, in a survey of Northumberland, which was published on nine sheets in 1769. A single sheet reduction came out in 1770 and went through a further three editions between 1781 and 1796. The pair next turned their attention to a survey of Berwickshire, which appeared on four sheets in 1771, with a reduced single-sheet version the following year. This concentration on three Border counties has led to the suggestion, for lack of other biographical facts, that Andrew Armstrong came from a Borders family.[13] By 1773 their six-sheet mapping of the Lothians had been engraved by Thomas Kitchin and, in 1774, their final work of collaboration appeared – the mapping of Ayrshire. In this case the reduced single-sheet edition appeared first in 1774 and the fuller edition in 6 sheets a year later. The engraver of the large Ayrshire sheets was Stephen Pyle, who later engraved the Lincolnshire sheets.[14] Pyle also engraved in 1779 *A MAP of the Great LEVEL of the/ FENS,/. . .* which included parts of Lincolnshire as far west as Stamford and north to Tattershall. This map was drawn by Kinderley and 'corrected by M.J.A.' – surely the initials of Armstrong's son, Mostyn John.[15] The map appears in the fifth volume of *History And Antiquities Of The County of Norfolk . . .* which Armstrong's son had a large hand in preparing.

After 1774 the Armstrongs do not appear to have worked together again, although they advertised together, and Strawhorn feels forced to conclude that by then the father was convinced that his son was not a very good surveyor and that they should go their separate ways. Mostyn John moved to Peebles and produced a county survey in 1775, which Pyle engraved on two sheets. A charge of indolence might have been brought against the son, since he produced no more county maps, in spite of the issue of proposals for mapping Norfolk and a further scheme for the mapping of Cambridgeshire.[16] Mostyn John Armstrong settled in Norfolk, married there and was made a Lieutenant in the Norfolk Militia by Lord Orford, who was intended to be the dedicatee of the Norfolk map. Later he was described as County Surveyor for Norfolk.[17] Ironically, the only English county that had not been surveyed at one inch to a mile by 1800 was Cambridgeshire.[18] Mostyn died in 1791.[19]

Andrew Armstrong moved south after the completion of the map of Ayrshire, commencing his work on Lincolnshire in 1776. When the eight sheets were published in 1779 they were followed the year after by the single sheet required to cover Rutland. No more map work is recorded after these two projects. Curiously, no estate surveys or other mapping activity by father or son are known before the maps of Durham and Andrew is not known to have undertaken any other type of surveying after 1780.

Biographical details about Andrew Armstrong are so few that he has always seemed a very shadowy figure. In 1934, W. King-Fane asked his readers for details about Armstrong.[20] Two years later J.B. King gave the only reply; namely that Captain Andrew Armstrong was buried at St. Michael's in Stamford on 6th January, 1784.[21] Further detail is added, citing Harrod, that he was buried at the eastern end of the church-yard, but no stone marked the grave.[22] St. Michael's parish register also records that Armstrong was 72 at the time of his death. Of his early life we have few facts. Moir says he is entered under the 32nd Foot as Lieutenant as from 31 August, 1756, that when the regiment was reduced in numbers he went on to the half-pay list in 1763 and is entered thus in the Army Lists up to 1796.[23] A misreading of the Army Lists must be the cause of this erroneous statement. When Armstrong applied to the Royal Society of Arts for a grant in respect of the Northumberland map he described himself as 'a Lieutenant on half pay from the 32nd Regiment'.[24] There is no reason to doubt this; the 32nd Regiment expanded with the raising of a second battalion in Scotland to cope with French incursions being made in Minorca. In 1763 the 2nd Battalion, which never served abroad, was disbanded at exactly the time that Armstrong says he went onto the half-pay list.[25] It seems likely that he stayed in Scotland or the Borders and within three years began the task that occupied him from 1766–1769, namely, the surveying of Durham and Northumberland. After the map of Northumberland came out Armstrong always described himself as Captain, and it is possible that he was promoted captain in one of the many militia regiments of the time, although evidence is lacking. When Armstrong applied to the RSA for their bounty for completing the Northumberland survey he was not aware that the RSA had withdrawn its offer for large-scale mapping. In a further letter, he offered more details of the claim he was making.[26] His Northumberland map was hung in the Society's Great Room for a month and he was finally rewarded with £50, the only sum he ever received from that body.

We know nothing of the form the father's surveying training took. The map of Lincolnshire is, of course, a great achievement but there is a lack of detail and there is a sufficient degree of error to suggest that he had become somewhat casual in his method. Stukeley's map of the Fens of 1723 provides us with far greater detail than Armstrong and sets a standard, based on close local knowledge, that is not found in Armstrong's work. To achieve the detail that Stukeley shows over the whole county area would imply much greater attention to detail and accuracy than Armstrong was able to bring to bear. Of Armstrong's detailed methods we can only make informed guesses. In the Companion to the map of Peebles his son, M.J. Armstrong, gives some idea of his method and says the textbook used as a model for himself and his father was Murdoch Mackenzie's book on maritime surveying. They would have practised a form of triangulation as their map of Durham with its triangulation diagram proves they were familiar with the system. In the case of his Lincolnshire survey, unfortunately, Armstrong *pere* has left no clues as to the location of the triangulation points he used in the county, in contrast to Joseph Lindley, whose Surrey map (1789–90) had an accompanying memoir listing 85 principal stations.

That his methods were at fault can be deduced from some of the errors that appear. He seems to have gone astray in the area south of Horncastle, for instance; Langton is badly misplaced in a position due west of Martin and due south of Horsington, when Horsington is west of Langton and Martin is only a little east of due south. A similar group of errors occur further east, with the positions of East Kirkby, Hagnaby and Stickford, compounded by placing the village of East Kirkby where the church and house are and calling the real site of the village 'South Kirkby'. A mistake that was copied by Cary through many of his atlases and their editions was the inclusion of *Hughington*; this spurious place is near Heighington, which is also marked; the closeness of the two names is clearly the result of some confusion in the surveyors' notes. It is also instructive to lay the Ordnance Survey one inch to a mile map over Armstrong's map. In many cases Armstrong's accuracy is all that one could expect; in others there is a difference of as much as 50% (for instance, between Mareham on the Hill and Hameringham). Error in the siting of Anderby and Hogsthorpe has been noted although, at the same time, the depiction of the coastal detail accords with the documentary evidence.[27]

Shading is used so that high ground might be suggested; the 'Cliff' is thus indicated. In the Wolds its use adds little to one's understanding of the land's profile; its use in a line from Wainfleet almost to Boston and, as a 'flattened' oval, enclosing the land between Spalding to a point east of Gedney leads one to doubt Armstrong's ability. These are hardly hilly parts of the county. The town plans of Louth and Grimsby are the first in the public domain of those two places. Those of Spalding and Boston had been preceded by those of Grundy and Hall respectively. Cary, who surveyed the roads himself on behalf of the Postmaster-General, showed the 'hills' between Spalding and Long Sutton on his maps for the new edition of Camden's *Britannia* of 1789. One suspects the Armstrong map was near Cary's desk.

Such examples should not be exaggerated for the general standard is good for its time; Mack describes Armstrong as 'the learned Andrew Armstrong' and, while he was was sorry that Armstrong had not seen fit to record the names of the old fields and other features in his Berwick map, he still points out in his discussion of the Northumberland map how much that has come down was due to Armstrong's diligence in noting otherwise unrecorded names.[28] The detail included in the Northumberland map is considerably more than the Lincolnshire plate exhibits and this suggests that corners were cut in the latter work and/ or Armstrong's enthusiasm for the project had waned.

In 1778 was published 'An Index To Captain Armstrong's Map of Lincolnshire . . .'

to be given free to subscribers for the final work, due for issue in January, 1779.[29] The index listed all the villages on the map, their respective wapentakes and the grid reference, using the letters engraved around the sides. The map, as the imprint shows, was published as planned in 1779. A further subscription list was published separately, probably much nearer the final time of publication since this version ran to 39 pages (as opposed to the 9-page section in the village index); again the names of the subscribers, their addresses and how many copies they had ordered are listed. There are 294 names and the total number of copies was close to 400; in the earlier example there were 236 names and the number of copies ordered totalled 281. Many prominent Lincolnshire names are included, headed by The Duke of Ancaster, to whom the map is dedicated. Sir Cecil Wray, M.P. and Charles Pelham, M.P. were the largest private subscribers with ten copies each, but the Board of Customs wanted twelve copies. Two sets were wanted by Sir Joseph Banks; Louth Corporation ordered three. Other interesting names among the subscribers are George Tennyson (of Market Rasen), John Grundy and Guy's Hospital, whose ownership of land near the modern Sutton Bridge is marked on the map.[30]

The first proposals issued by Andrew Armstrong may have appeared in the early months of 1776; he had, by then, already waited on Sir Joseph Banks hoping for his approval of the mapping proposals. In a letter of 24 January, 1776 he reminds Banks of this, presumably, abortive attempt at an interview and submits with his letter 'a proposal from my son at Edinburgh setting forth the Conditions and Manner of Survey which if they are thought proper, may be printed . . .'[31] In his reply two days later, Banks upbraided Armstrong for failing to wait for an opinion on the surveys before sending proposals. Sir Joseph went on to regret that he found the surveys deficient and that he needed to know more of the mode of survey and to see the instruments used in the mapping process: 'I should not chuse to encourage any Survey, which is not grounded upon several large Triangles, which can only be accurately measured by instruments of the best Construction.'[32]

One may only presume that Armstrong did provide Sir Joseph with the necessary assurances since, as we have seen, Banks finally ordered two sets of the finished product. Nevertheless, one might have thought that the maps already produced by the Armstrongs would have provided sufficient evidence of the basic surveying abilities being put to use in the new survey. The following eighteen months must have been spent in the surveying work, including the establishment of the triangulation points on which Banks set so much store. At that point two advertisements appeared in the *Lincoln, Rutland and Stamford Mercury* in the issues for 19 and 26 June, 1777. The words used were: 'Proposals for publishing by Subscription, Accurate Maps of the Counties of Lincoln and Norfolk, to be taken from actual Surveys, and laid on a Scale of one Inch to a Mile. By Captain Armstrong, and Son. Each Map will be elegantly engraved and printed on eight sheets of Imperial Paper. Price to Subscribers Two Guineas. Subscriptions are taken by the Surveyors now in Lincolnshire and Norfolk, by the Printer of this Paper. and Mr. Nott, in Stamford, where Proposals at large may be seen . . .' Later evidence suggests that Armstrong was not given to spending too much money on advertising; certainly the subscriptions that came in must have been sufficient for the father to progress with his surveying in Lincolnshire. When Armstrong's son wrote from Edinburgh in January, 1776 he must have been near the end of his own survey of the Great North Road from London to Edinburgh, since that work appeared in the June following. This road book must have been well received since it was revised for an issue in August; in a later state dated 1st January, 1777 all the works of father and son are listed, including Lincolnshire in eight sheets at two guineas.

The map was issued on eight sheets, which are numbered in their outer margins. The date in the map's imprint (20 January, 1779) must be close to the actual date of publication, as only one announcement appeared in the *Lincoln, Rutland and Stamford Mercury,* that for 25 February, 1779 and it said: 'London, Jan 30, 1779. Just published, Price Two Guineas. The Map of Lincolnshire, on eight Sheets:– 6F. 8I. by 4F.8I. With each Map is given a Useful Index, Distances, Tide-Tables, &c Sold, and delivered to Subscribers, (on Payment of the second Part of the Subscription) by Captain Armstrong, at Mr. Wolf's, Copper-Plate Printer, Exeter-Street, Strand, and by no-one else in London. Captain Armstrong has now brought down a Number of the Maps; and Subscribers, or others, may have them at the principal Booksellers in Stamford, Boston, Grantham, and Lincoln where different Ways of fitting them up may be seen.' Undoubtedly Pyle was engraving the sheets as the drawings became available during 1778 and the printing was undertaken by the Mr. Wolf, mentioned in the advertisement. The Lincolnshire map was submitted to the Royal Society of Arts for a grant but this was turned down in December, 1779. Armstrong may have been strapped for cash later since he offered to sell the plates for the Northumberland map to the RSA for £20. The offer was, however, refused.[33] It is perhaps worth recording that by the time of the final disbursement in 1809 the RSA had given £460 for work on thirteen different county maps and that only Donn (Devon) and Burdett (Derbyshire) received the full sum of £100.

One further advertisement of interest appeared in the *Lincoln, Rutland and Stamford Journal* on August 3, 1780, viz. 'Maps. Just Published, A New and Correct map of Rutland, on one Sheet of Royal Paper, agreeable to Proposals given out by Captain Andrew Armstrong. Price to Subscribers 10s. 6d. [who] . . . may have them by applying to W. Harrod, Printer and Bookseller in the High Street, Stamford . . . or any of the Maps of the Counties the Captain has published, either in Sheets or mounted . . . catalogues may be had gratis.' Unfortunately none of the catalogues nor the Rutland proposals have yet been found.

There is no documentary evidence to suggest how well Armstrong's work was received. The plates were never used again, although the plates for other counties at this scale were quite often used more than once. For instance, Faden re-issued in 1804 the four sheets covering Huntingdonshire that Jefferys had prepared in 1768.[34] When Lord Brownlow wrote to Lord Mulgrave, Master General of the Ordnance in 1817 on behalf of the gentry of the county who wanted to make a case to the Ordnance Survey to survey Lincolnshire out of turn he wrote how it was necessary 'to obtain a good Map of this County, inasmuch as the only work of the kind that exists is a Map, executed about 60 years ago by Armstrong very incorrectly done at the time and now from the number of changes occasioned by inclosures, drainages, &c become positively useless'.[35] The map was, in fact, only 38 years old and allowance must be made for Lord Brownlow's desire to see his scheme succeed. The map did undoubtedly fill a gap; perhaps it did quickly become dated as suggested, but, in the event, the Ordnance Survey map was not as great an improvement as the gentry had hoped, a fact that was, at least tacitly, acknowledged by the O.S. in the adoption of new methods of surveying, engraving and presentation by the time of the issue of sheets for Cambridgeshire in 1830.[36]

## EDITION

(i) *Map/OF/LINCOLN-SHIRE,/(Comprehending/LINDSEY, KESTEVEN & HOL-LAND,)/Surveyed in the Years, 1776, 7, & 8,/By CAP$^T$. ANDREW ARMSTRONG./Engraved by Stephen Pyle./MDCCLXXVIII./* (Ea, in an oval wreath, surrounded by an engraving of Dunston Pillar with a hunt in the foreground, sheet II). *Engraved by S. Pyle, Angel Court, Snow Hill.* (Ee, OS, sheet VIII). *Publish'd according to Act of Parliament, Jan$^y$. 20$^{th}$. 1779.* (Ce, OS, sheet VIII). *To the Most Noble/ROBERT BERTIE,/Duke of Ancaster AND Kesteven,/MARQUIS and EARL of Lindsey,/Baron Willoughby of Eresby./Hereditary Lord Great Chamberlain of ENGLAND;/Lord Lieutenant and Custos Rotulorum of the/County of Lincolnshire,/This MAP, is humbly Inscribed,/by his Graces most obedient/and most Humble Servant,/A. Armstrong./Scale of Statute Miles 69½ to a Degree./1 + 8* [= 22.8cm] (Ae, with coat of arms above, sheet VII). Compass (Ea, below title, sheet II). *EXPLANATION./*18 lines of symbols and their meanings (Aa, sheet I).

*Plan/of/SPALDING./Scale of Chains/8* [= 3.4cm] (Aa of plan; the plan occupies the lower right hand corner of sheet VIII). *A/PLAN/ of/GRIMSBY./* (Ae of plan, which occupies the lower right hand corner of sheet II; there is no scale and the plan is orientated with west at the top). *PLAN OF/LOUTH./* (Aa of plan, which occupies the upper right hand corner of sheet IV; below is an an engraving with the caption: *A View of the Ruins of LOUTH ABBY, founded in 1139.*). *PLAN/of/ BOSTON* (Ae of plan, with the town's coat of arms below; *SCALE OF CHAINS./20* [= 6.7cm.] (Ab–Ac of plan; the plan occupies the lower right hand corner of Sheet VI and the upper right hand portion of sheet VIII). Roman numerals appear at the upper outside corner of all eight sheets and outside the lower right hand corner of sheets I–IV. The inner frame is shaded to indicate the minutes of degrees. A graticule covers most of the map; the lines across are at five minute intervals and those down the map at ten minute gaps; between the inner and outer vertical frame lines are the letters A to O, and between the horizontal frame lines are letters A to H. *Meridian of Greenwich* (between the lower frame lines of sheet VIII). The line passes just west of Holbeach and Whaplode and east of Louth.

The sheets are found variously put together, a possibility suggested in the advertisement issued on its publication; no distinction is made here between the various formats of the sets in their present states.

**Copies** BL 3355(16) and K 19.5 TAB; CUL Atlas 2.76.6; Bod Maps C.17.a.6, etc.; LAO FL Maps 5 and 2 Anc 5/22/4–7 (5 sheets only); LRL Map 879; Grimsby RL Armstrong (RM 94).

**References**

1  Harley, J.B. The re-mapping of England. *Imago Mundi*, Vol. XIX (1965), p. 57; Ravenhill, W. The Honourable Robert Edward Clifford, 1767–1817: a Cartographer's Response to Napoleon. *The Geographical Journal*, Vol. 160, No. 2 (July, 1994), pp. 159–172 discusses the background to mapmaking at the end of the eighteenth century, prior to the creation of the Ordnance Survey and (pp. 162- 163) contrasts the French approach to national mapping with the county by county approach adopted in England.

2  Gough, 1780, Vol. II, p. 297.

3  Kingsley, p. 57.

4  ibid., pp. 91–94 describes the four sheets that appeared by 1783. It was never completed.

5  Discussed in Harvey and Thorpe, pp. 20–35 and 93–94.

6  Criticised by Arthur Young 'as a miserable one'; *Royal Society of Arts Transactions*, 1782–3, p. 10.

7  Rodger, E. p. 8; Lynam, E. Early Maps Of The Fen District. *Geographical Journal*, Vol. 84 (1934), pp. 420–423; Lynam, E. Maps of the Fenland. The Victoria County History of . . . Huntingdon, Vol. III. (London, 1936), pp. 291–306.

8  The letter and proposal are printed in: Stukeley, W. Diaries and Letters, Vol. III. *Surtees Society*, Vol. XXXI (1888), pp. 266–267.

9  Capp, B. Astrology and the Popular Press. (London, 1979), pp. 52, 242–243 and 336; LAO Sub VIII/(49). I am grateful to Miss Weaver of East Bridgford, who provided me with certain details of Weaver's life.

10  Kingsley, p. 93 quotes the annotated copy of Gough's British Topography, 1780, p. 475 held in the Bodleian Library.

11  *Norwich Mercury* May, June and July, 1777.

12  Harley, J.B. and Phillips, C.W. The Society of Arts and the surveys of the English counties. *Journal of the Royal Society of Arts*. Vol. CXII (1963–64), pp. 43–46, 119–124, 269–275, 538–543. Armstrong's account is illustrated on p. 271.

13  Strawhorn, J. An introduction to Armstrong's map [of Ayrshire]. *Ayrshire Archaeological and Natural History Society*, Second Series, Vol. 5, (1959), pp. 232–255.

14  Pyle is recorded as active between c.1770 and 1780 (Maxted, p. 181). His last work seems to have been the 4 sheets of Co. Antrim of 1780, noted in Rodger (as items 728–729).

15  CUL Views X.4.38 is a copy of the map; BL 290 h 24 of the book.

16  Rodger, p. 46, items 826 and 836, where 1778 is quoted for both counties; Chambers, B. M.J.

Armstrong in Norfolk. *Geographical Journal*, Vol. 130 (1964), pp. 426–431 shows that Norfolk proposals appeared locally from May, 1777.

17  Norwich City Directory, 1783, and also on a map of Great Yarmouth (1779) copied by M.J. Armstrong from an earlier drawing (not his own).

18  Harley, J.B. 1965, op. cit., p. 57.

19  Chambers, B. 1964, op. cit., p. 431.

20  King-Fane, W. Pre-Ordnance maps of Lincolnshire. *Lincolnshire Notes and Queries,* Vol. XXII (1934), pp. 37–46, 57–61 and 71–77; the request for information on Armstrong's life and parentage is in the same volume (p. 48).

21  *Lincolnshire Notes and Queries,* Vol. XXIII (1936), p. 104, citing Lincolnshire Parish Registers, Vol. V, p. 274 under St. Michael's, Stamford.

22  Harrod, W. History of Stamford. (London, 1785), Vol. II, p. 315; the second edition of 1822 says that no stone had been erected.

23  Moir, D.G. The early maps of Scotland to 1850 . . . Vol. I. (Edinburgh, 1973), p. 136. Moir also states (p. 120) that Capt. Armstrong died in 1794 (he cites Mrs. Chambers' 1964 article, which does not refer to Capt. Armstrong's death anywhere).

24  RSA. GB. A,83.

25  Swiney, G.C. Historical records of the 32nd (Cornwall) Lt. Infantry. (London, 1893), p. 45.

26  RSA. Min. Comm (Polite Arts), 10 Dec., 1779.

27  Owen, A.E.B. Beyond the Sea Bank: Sheep on the Huttoft Out-marsh . . . *Lincolnshire History and Archaeology.* Vol. 28 (1994), p. 39.

28  Mack, J.L. The Border Line from the Solway Firth to the North Sea . . . (Edinburgh, 1926), pp. 65–66 for discussion of the Berwickshire map and pp. 50–63 and *passim* for discussion of the Northumberland map.

29  Grimsby RL X000:912 ARM; Nottingham UL Li 1.P 50 DYK.

30  BL K. Top. XIX. 19.

31  John Rylands UL. Eng. Ms. 700.2.

32  John Rylands UL. Eng. Ms. 700.3.

33  RSA Min. 6th Feb., 1782.

34  Rodger, p. 11.

35  LAO 4 BNL 14, quoted in The Old Series Ordnance Survey Maps Of England And Wales . . . Volume V . . . (Lympne, [1987], p. viii.

36  ibid., p. xxviii.

# 45

## ANDREW ARMSTRONG                                                    1780

Size: 66.4 x 51.2cm.

We have seen that, in a number of cases, Armstrong's large-scale maps had a smaller-scale counterpart and Lincolnshire also fitted into this pattern. Nineteen months after the date on the eight-sheet map a single sheet was put out. Although the map carries no name for a publisher the circumstantial evidence indicates that it was issued by Carington Bowles.

The large lettered title has disappeared as well as the encircling engraving of Dunston Pillar. The town plans have also had to make way for other items, but room has been found for a view of Boston Stump (from the west) and Lincoln from the south-east. The latter is, in fact, a copy of S. and N. Buck's view of Lincoln from the south-west published in 1743! The material gathered under the heading 'Explanation' is also dispensed with but space has been created for a chart of mileages from Lincoln to many towns in the county as well as Hull, Newark and *Linn*. The map bears the eight numbers, which correspond to those on the original sheets. The graticule has been retained, but the letters in the borders have not survived. The dedication has been removed and the space it formerly occupied now has the title and author information.

The plate was engraved by J. Thompson. In 1779 he had engraved a map of the city of Norwich and, apart from that and the present map, he is only recorded as engraving one other map – of the St. Pancras district of London – in 1804. The village of 'South Kirkby' has been removed and East Kirkby indicated in a more correct position. All attempts to indicate the higher land levels have been abandoned and there is no shading or hachur.

## Later history

The plate was re-issued in 1781 by Bowles and Bennett as a separate sheet and as part of Bowen's *Large English Atlas*. In one of the copies of the latter Armstrong's contribution is acknowledged on the title-page, not only for the use of the reduced Lincolnshire plate but for the inclusion also of the single sheet which was all that was necessary for the map of Rutland. Armstrong had completed that survey in 1779 and the map was published in the following year.

In 1787 the map was again re-issued, this time by Sayer alone as Bennett had retired in the year before; he died in December, 1787. Curiously, the single sheet version that Sayer put out in 1787 was not the version used in the issue of *The Royal English Atlas* that Sayer issued in the same year but the 1781 state. Sayer died in 1794 and his plates became the property of Laurie and Whittle, who were to be sold his stocks on favourable terms, as laid down in his will.[1] They issued the map on 12 May, 1794, less than three months after the will was proved. They may have continued to sell copies for some years, but probably not after 1812 when Laurie retired and the firm's name changed, a change that one would expect to have been reflected on the plate. By then, of course, there were several rival large sheet maps available, especially those of John Cary and Charles Smith.

## EDITIONS

**(i)**  *The MAP of/LINCOLN-SHIRE/Reduced from the Actual Survey made/in the Years 1776, 7 & 8,/and Published, Jan$^y$. 20 1779, on Eight Sheets,/By Cap$^t$. Andrew Armstrong,/I. Thompson sculp./ London Published as the Act directs, Aug$^t$. 20, 1780 Price 5$^s$.6$^d$.* (Ae). An engraving of Lincoln, with the title below *SOUTH EAST VIEW of the CITY of LINCOLN* (Ea). Compass (Eb). An engraving of Boston Stump (from the west), with the title below *BOSTON CHURCH* (Ed). A table with the heading *DISTANCES from LINCOLN/To the following Place'*s (sic) (Ac). The inner frame is shaded to mark latitude, etc. at 1′ intervals, the numbering occurring every 5 minutes. *Meridian of Greenwich* (De, between frames). A graticule is formed of lines drawn across the map at very odd and irregular intervals; there are four vertical lines at the 20′ East, 0°, 20′ and 35′ West marks, while across the map they occur at 52° 56′, 53° 13′ and at 53° 33.3′.

Issued as a folded sheet in a cardboard holder on which the label reads: *LINCOLN/SHIRE/Printed for Carington Bowles in S$^t$. Pauls Church Yard./*

**Copy**  Nottingham UL Li 1.B8.D80.

As above but the cover's label reads: *LINCOLN/SHIRE.*

**Copy**  Bod Allen 234.

**(ii)**  The price has been blacked out, implying a later issue.

Issued as a single sheet.

**Copy**  A Private Lincolnshire Library.

*The Large English Atlas . . . London: Printed for Robert Sayer, Map, Chart, and Printseller, No. 53, Fleet-Street. MDCCLXXXVII.*

**Copy**  BL Maps C.10.d.17.[2]

(iii)   The imprint now reads: *London, Publish'd by R Sayer & J Bennett No. 53 Fleet Street, as the Act directs, 24 June, 1781.* (Ae).

Issued as a single sheet.

**Copies**   BL Maps C.24.b.18; CUL Map C.70.78.1; Bod Gough Maps Lincolnshire 14; LAO YARB 4/27/3.

(iv)   The imprint now reads: *London, Publish'd by R. Sayer, Printseller, No. 53 Fleet Street, as the Act directs, 24 June 1787.* (Ae).

Issued as a single sheet.

**Copies**   LAO Scorer Map 20; LRL Map 394, etc.; Lincoln Cathedral (2 copies); Grimsby RL Armstrong RM 38; Louth RL Map A6.

(v)   The imprint now reads: *Published 12th. May, 1794 by LAURIE & WHITTLE, 53, Fleet Street London.* (Ae).

Issued as loose sheets.

**Copies**   BL Maps 3355 (22); CUL Maps aa.17.G.5; Bod Maps C.17.a.6; LRL Maps 60 and 223.

**References**

1   PRO PROB 11/1242/105–6, proved 17 Feb., 1794.
2   Reference has already been made to the curious use of an early (1781) state in a 1787 atlas although Sayer issued the single map with his imprint in the same year. Hodson, 1989, pp. 141 and 147 provides an analysis of another version of the atlas, which also used the 1781 Armstrong map but with a title-page of 1767 vintage. That example was sold at Sotheby's as lot 136 on 2 March, 1971; its present whereabouts are unknown.


# 46

**THOMAS CONDER**                                                                                    **1784**

Size: 19.3cm x 16cm.

Conder was an engraver and bookseller; as a map engraver he was active from 1780 to 1801, but he remained in the book trade up to 1819. The son of John Conder, D.D., he was apprenticed to a stationer in 1765 but turned over to Thomas Kitchin's son as an engraver in 1766.[1] His son, Josiah, was also a bookseller, independently from 1789 and then taking over his father's business in 1813. Thomas died in 1831, aged 84.[2] Conder's name does not appear on the plate until it has reached a late stage in its use.

The map owes a considerable debt, not surprisingly, to those engraved by Kitchin (**38** and **43**) and exhibits no new features. The Lincolnshire map was engraved on a single plate with Lancashire, which occupied the left hand portion.

Alexander Hogg published the work below; his methods and issue of part-works – for which he was notorious – were discussed earlier (**32**). Hogg employed hack writers to produce his texts and, Timperley writes: 'when the sale of a book began to slacken [Hogg] immediately employed some scribe to make him a taking title, and the work, though not a line was altered, was brought out in a new edition.'[3] Walpoole seems to have been used by Hogg as a compiler of material written by others. Hogg was predominant in the issue of works in parts and these works are no exception. The Lincolnshire texts have the number 40 on the first page of text in the lower left corner and this may correspond with the part number.

Although the plates were used in different 'new' works over a period of more than ten years the maps underwent no change. The later editions of the text did not have maps.

## EDITIONS

(i)  *A NEW Map of/LINCOLNSHIRE/Drawn from/the latest & best/ AUTHORITIES./* (Ee). Compass (Ed). *British Statute Miles/15* [= 3.55cm]. (Ae). *Remarks/*five lines of symbols and explanation (Ab). The coat of arms of Lincoln (Ea). The inner frame is marked with degrees of latitude, etc. at 2' intervals but numbers occur only every ten minutes. *Meridian of London* (De, between the inner and outer frames). Two county maps are engraved side by side, with Lancashire to the left of the county plate. Above the top of the whole plate is: *Engraved for WALPOOLE'S New & Complete BRITISH TRAVEL-LER.* The first four words appear above the Lancashire plate. Along the bottom of the whole plate appears: *Published by ALEX^R. HOGG N^o. 16, PATERNOSTER ROW.* When the maps are separated only the address appears below the Lincolnshire map. The remaining area around the edges of the plate has a leafy decoration.

*The New British Traveller; Or, A Complete Modern Universal Display of Great-Britain And Ireland: Being A New, Complete, Accurate, And Extensive Tour Through England, Wales, Scotland, Ireland, the Isles . . . dependent on the Crown of Great Britain. Comprising all that is worthy of Observation in every County . . . Being . . . An actual and late General Survey, accurately made by a Society of Gentlemen . . . And Including A Valuable Collection of Landscapes, Views, County-Maps . . . The Whole Published Under The Immediate Inspection Of George Augustus Walpoole, Esq. Assisted . . . respecting Wales, by David Wynne Evans . . . Scotland, by Alexander Burnet . . . Ireland, &c. by Robert Conway . . . London: Printed For Alex. Hogg, At The King's-Arms, N^o 16, Paternoster-Row. 1784.*

**Copies**   CUL Atlas 4.78.5; Leeds UL W 81; Lincoln Cathedral L Qq.2.14; John Rylands UL NQ 27 A7 (L R48246); Brighton RL F914.2 W16. (LAO Scorer Map 22; LRL Maps 82 and 132; Louth RL Map A26).

*The New British Traveller . . . Paternoster-Row.* [1790].[4]

**Copies**   BL 10348 l.3; CUL Atlas 4.78.7; Bod GA Gen Top b.34.

(ii)  *T. Conder Sculp^t.* added (Ee, OS). The words along the top of the plate and the leafy frame have been removed. The imprint below the joint plate has been re-engraved to read: *Published by ALEX^R. HOGG at the Kings,* (sic) *Arms N^o. 16, PATERNOSTER ROW.* When the maps are separated the letters up to *Kin* remain with the Lancashire plate.

*The New British Traveller . . .* [1790].

**Copies**   Bod Allen 43. (LAO ATH 2/276 L910).[5]

*The New And Complete English Traveller: Or, a New Historical Survey and Modern Description of England and Wales. A work calculated equally to please the Polite – entertain the Curious – instruct the Uninformed – and direct the Traveller . . . Written And Compiled . . . By A Society of Gentlemen . . . Revised, Corrected, And Improved, By William Hugh Dalton . . . London: Printed for Alex Hogg, No. 16, Paternoster-Row . . .* [1794].[6]

**Copy**   Leeds UL Special Collections – Geography J – 1q.

(iii)  The imprint below the plate has been removed – only the name of the engraver remains outside the frame-lines.

*The New And Complete English Traveller . . .* [1794].

Copies   BL 10348 1 8. (LRL Map 81; Grantham RL; Gainsborough RL Map 7).[7]

## Reproduction
(i) A reduced form of the map, without the material outside the frame was used to illustrate: Baxter, K.C. Celia Fiennes in Lincolnshire. *Lincolnshire Life*, Vol. 26, no. 7 (Oct. 1986), pp. 46–47.

## References
1   McKenzie, D.V. Stationers' Company Apprentices 1701–1800. (Oxford, 1978).
2   Maxted, p. 50; DNB.
3   Timperley, p. 838.
4   The date on the title-page has been scratched off, causing a 'thin' at that part of the paper. Several commentators have suggested 1790 for its issue, but the work itself provides no clues.
5   The LAO example consists of the Lincolnshire text with the county map.
6   The date on the engraved second title-page.
7   The Gainsborough map has been trimmed to the outer neat line; it is impossible to say, therefore, the exact state of its original form.

Chubb CCLI and CCLII (the 1784 and 1794 editions only).

## 47

**JOHN CARY**                                                                    **1788**

Size: 26.35cm x 21.25cm.

John Cary, a prolific engraver of maps, set new quality standards in the engraving of maps, coupled with accuracy. The popularity of his work is indicated by the large number of editions the bulk of his atlases passed through, the number of different works he produced (**49–50, 57, 61, 65, 80, 144**) and the quantities of copies that have survived as a consequence. All his plates had long lives, which continued in most cases long after his death and it has been suggested that they were only melted down in the first world war.[1]

Cary was born in c.1754 into a Wiltshire family but, like two of his three brothers, he set up in business in London. He was apprenticed to the engraver William Palmer from 1770–77 and the first map to bear his signature seems to have been a canal map of 1779. Apart from the many atlases he produced and the multi-sheet map of England and Wales (**144**) one other large work deserves mention. In 1815 he issued *A Memoir To The Map And Delineation Of The Strata Of England And Wales, With Part Of Scotland. By William Smith* . . . Smith is regarded as the father of English geology and this is the first work to show the geological strata of the country as a whole. Lincolnshire is shown on a number of sheets; the northern section (as far south as Grasby) is on sheet VI (with a little on sheet V); the greater part of the county is on sheet IX, the western edge being on sheet VIII.[2]

One brother, Francis, was apprenticed as an engraver but to James Taylor, while William, who served his time with Jesse Ramsden the renowned instrument maker, made his reputation as a maker of globes.[3] Another brother, George, later joined John as his partner following his move to new premises in 1820, necessitated by the burning down of his earlier shop and also by the increasing success of his business.[4] The pair published the map of the Bedford Level (1829) prepared by Samuel Wells; the map includes the area of the county south of the 'New South Holland Drain' plus part of Cowbit Fen.[5]

The last two decades of the eighteenth century saw the use of book production by subscription reach its highest point. 319 titles came out using this method between 1780 and 1790 and in the last decade of that century the figure reached was 357; in contrast the figure for the whole of the period from 1901 to 1971 is 327.[6] *Cary's New and Correct English Atlas* conforms to this popular practice. It was issued in twelve parts, starting probably in September 1787, but delays meant that the intention to produce four maps in each of the monthly parts could not be adhered to and it was not completed until early in 1789. Each part was to cost two shillings for uncoloured maps or three shillings and sixpence if coloured. However, in part two prices for coloured maps were reduced by a shilling. Burden deduced, from the copies of the first eight parts acquired by the British Library[7] in 1987, that Lincolnshire was probably in part ten together with maps of Dorset, Rutland and Yorkshire.[8] It is not known exactly when part ten appeared; it is likely to have been late in 1788, since it was advertised in *The Morning Chronicle and London Advertiser* for November 17, 1788. The title-page appeared with part one and this explains why the atlases, which could only have appeared when all the plates had been engraved, bear the date 1787. All the plates are dated September, 1787 in spite of the actual date on which the individual map appeared.

A page of letter-press about the county with a list of its most famous places was issued with the map. Five distinct versions of this text appeared with the map dated 1787 and a further two with the map dated 1793; it is possible to arrange these in a chronological order since the text was corrected and expanded with each appearance. The atlas was in print until 1808; by then the worn state of some plates led Cary to make a complete set of new county plates for what is bibliographically a new work in 1809 (**65**). For all the success of this and the later atlases, which were not confined to British subjects, his sole reward was to receive the gold medal of the Royal Society of Arts in 1804 for the publication of Joseph Singer's large-scale survey of Cambridgeshire in four sheets.

The source of Cary's map of the county is the single sheet version of Armstrong's map (**45**). One glaring difference is that the zero meridian is shown as *1° Longitude West from London*. It is curious that the usually meticulous Cary did not correct this error, while altering much else. Nor was the correction made at the time of re-engraving for the new 1809 edition or during its later life of more than sixty years. The same mistake is perpetuated on other Cary plates, but they were eventually put right (**65**). There is nothing innovative in terms of content, except that the errors that were incorporated in Saxton's plate and copied by everyone else until Armstrong are not repeated here (the siting of Scampton, the appearance of Sutton St. Leonards, etc.). Armstrong's 'invention' of Hughington is included by many nineteenth century copiers of Cary's works. There is a quality to the engraving that led to his being categorised as the founder of the modern English school of cartography.[9] In the later editions of this and his other English atlases there was a continuing attempt to keep the maps up-to-date, particularly as regards the condition of the roads, the result partly of his employment in the 1790s by the Postmaster-General to measure the mailcoach routes. At that time and indeed up to the time of the introduction of the penny post in 1840, it will be recalled, mail was charged according to the computed miles letters had travelled. It was necessary, therefore, to have accurate assessments made of the miles between the main post towns. This task fell to Cary, the results of which were incorporated in his county maps and also in *Cary's New Itinerary*, which then became the cause of the celebrated court case of plagiarism when Cary sued Longmans as publishers of Paterson's *A new and accurate description of . . . roads in England . . .* (often known as Paterson's Roads). Although Cary won the case after two years in the King's Bench he was only awarded one shilling damages.[10] Cary claimed that he had measured 'upwards of Nine Thousand Miles' in the course of his work[11] and a further One Thousand Miles later.[12]

## EDITIONS

(i)  *LINCOLNSHIRE/By JOHN CARY Engraver./* (Ae, the title forms a central bar across the compass with the rest below the compass). *Statute Miles 69½ to a Degree/10* [=

3.35cm] (below the title). *London: Printed as the Act directs September 1ˢᵗ. 1787 by J. Cary, Engraver Map & Printseller the corner of Arundel Street Strand.* (Be–De, OS). The inner frame is marked with latitude, etc., the divisions being at 2′ intervals, numbered every ten minutes. *Longitude West from London* (De, between the inner and outer frames). The latter appears to the right of the 1° mark (an error for the 0° meridian).

*Cary's New and Correct English Atlas: Being A New Set of County Maps From Actual Surveys. Exhibiting All the Direct & principal Cross Roads, Cities, Towns, and most considerable Villages, Parks, Rivers, Navigable Canals &c. Preceded by a General Map of South Britain, Shewing The Connexion of one Map with another. Also A General Description of each County, And Directions for the junction of the Roads from one County to Another. London. Printed for John Cary, Engraver, Map and Print-seller, the corner of Arundel Street, Strand. Published as the Act directs Septʳ. 1ˢᵗ 1787* [i.e. 1789].

**Copies**  BL Maps C.24.f.1 etc.; CUL Atlas 5.78.1 etc.; Bod Allen 45 etc.; Nottingham UL Special Collections G 5512; Leeds UL W 83 etc.; LAO 950.2. (LAO FL 912; LRL Map 792 etc.; Louth RL Map A23; Grantham RL (2 copies); Boston RL Map 24).

*Cary's New and Correct English Atlas . . . 1793.*[13]

**Copies**  BL Maps C.24.f.23; Bod Allen 54; Leicester RL L 912; Leeds RL Q1 912.42 C 258.

**(ii)**  Imprint now reads: *London. Published Jan.1.1793 by J. Cary, Engraver & Map seller Strand.* (Ce, OS). More directions have been added where roads leave the county, viz. *fr. Oakham* (twice), *fr. Uppingham, fr. E. Retford, fr. Bawtry, to Peterboro'*. South of Stamford *to/London* has been changed to *to/Northampton/the London Rᵈ*. The inclusion of *to Wisbeach* has led to *NORFOLK* being re-engraved in smaller letters and moved eastwards, *CAMBRIDGE SH.* becoming *CAMBRIDGE SHIRE*, and *Clowes Cross* being deleted. The roads to the west of Stamford extend further westwards and *Tynwell* has been deleted. *Bridge/Casterton* has been altered to *Gᵗ/Casterton*. Lower case letters have been added to roads where they leave Lincolnshire. *Deer/Park* has been added at Gautby. The Newark-Leasingham and Louth-Saltfleet roads have been upgraded; the [Wisbech] road at Long Sutton has been re-engraved further west and extended southwards. Boundaries between Yorkshire and Nottinghamshire and Leicestershire and between Norfolk and Cambridge are marked by short lines of dashes. Folkingham has been downgraded from market town status (i.e. lower case instead of upper case lettering) while Burton [on Stather] has been upgraded. The words *now Inhabited* have been added after Sunk/Island (in the Humber).

*Cary's New and Correct English Atlas . . . London. Printed for John Cary, Engraver & Map-seller, Nº. 181, near Norfolk Street, Strand. Published as the Act directs Janʸ. 1ˢᵗ. 1793.*

**Copies**  Bod Allen 55 and Maps C.17.c.4; Nottingham UL Dept. of Geography G 5512; Birmingham RL AQ 912.42; NMM 912.44 (42): 094 (E9200). (BL 3355 (21); LAO Foster L 950.2; LRL Maps 754 and 754a; Louth RL Map A19).

*Cary's New and Correct English Atlas . . . 1793.* [1795].[14]

**Copies**  BL Maps C.24.f.15; CUL Atlas 5.79.1; Bod Johnson Maps 9; RGS 7 E 2; Leeds UL W 86; British Geological Survey ZF 912 (420/429) CAR 1793.

*Cary's New and Correct English Atlas . . . 1793.* [1802].[15]

**Copies**  CUL Atlas 5.80.5; Keats House Library 3893.

*Cary's New and Correct English Atlas . . . 1793.* [1804].[16]

**Copy**  Leeds UL Case K 30.

*Cary's New and Correct English Atlas . . . 1793.* [1808].[17]

**Copy**   CUL Atlas 5.80.4.

**References**

1   Harvey and Thorpe, p. 116.
2   A set is in British Geological Survey (Map Case). Another set (not available because of conservation work) is held by BL – B 347 (3). The work as a whole was dedicated to Sir Joseph Banks.
3   Smith, D. The Cary family. *Map Collector* No. 43 (Summer, 1988), pp. 40–47.
4   The fire was reported in *Gentleman's Magazine*, 1820, p. 81.
5   BL Maps 16.a.14, etc.; CUL Maps C.36.82.1, etc.; Grimsby RL F301:912 WEL; Louth RL L 912. The map made a final appearance in 1878; red lines have been drawn to mark the railways, but it still retains the names of G. & J. Cary as the publishers. (Copies in Nottingham UL Li C17.E 29 and Li C17.E 78).
6   Wallis, P.J. Book subscription lists. *Library*, 5th series. Vol. XXXI. (1974), pp. 255–286.
7   BL Maps C.18.c.7.
8   Burden, E. Cary's New and Correct English Atlas. *Map Collector* No. 56 (Winter, 1991), pp. 32–37.
9   Fordham, F.G. 1914, op. cit., p. 86.
10   The case is summarised in Fordham, F.G. John Cary Engraver, Map, Chart and Print-seller . . . 1754–1835. (London, 1925), pp. xvi–xviii. Smith, D. op. cit., 1988, pp. 41–42 refers to the case in discussing the sources of Cary's own plagiarism. Cary himself summarised the case in his *New Itinerary* – 1806 edition, pp. 909–912.
11   The advertisement leaf of the 1798 edition of *Cary's New Itinerary.*
12   The advertisement leaf of the 1802 edition of *Cary's New Itinerary.*
13   Cary began to redate the plates prior to the re-issue of 1793. In some copies Durham only is dated 1793 while the others are still dated 1787.
14   The map of Dorset is dated 1795. Fordham, F. G. op. cit., 1925, p. 25 regards the date as an error (presumably for 1793). In most examples paper with 1794 in the watermark has been used.
15   The CUL atlas has one map (Bucks.) with the watermark dated 1802. In the Keats House copy nearly all the maps have such watermarks, including Lincolnshire.
16   Several leaves bear watermarks dated 1804.
17   Leaf 7/8 is watermarked 1808.

Chubb CCLX–CCLXIII.

# 48

**JOSEPH JOHNSON**                                                          **1788**

Size: 15.7cm x 9.5cm (page size: 19.1cm x 11.2cm).

J.A. signed the preface to this little county geography book and he is identified with John Aikin (1747–1822), who was born at Kibworth Harcourt (Leics.). He was firstly apprenticed to an apothecary in Uppingham but when his father was appointed to be classical tutor at the newly opened Warrington Academy he was then apprenticed to a surgeon and went on to study medicine at Edinburgh and take an M.D. at Leyden. He practised initially in Great Yarmouth and then in London, after being forced to move because of his dissenting opinions. After suffering a stroke he took up writing and seems to have turned his hand to anything that came to hand, editing editions of Goldsmith, Thomson, Pope, Cowley, etc.[1] Through living in the Manchester area he became a friend of Joseph Priestley and Thomas Pennant. One of his more useful geographical works was *A description of the country from thirty to forty miles round Manchester* (1795). He has the distinction of being cited in *Das Kapital*.[2] The undertaker of the work, Joseph Johnson, is also credited as the engraver of the maps; he was the first to publish Cowper's Poems and Beckford's *Vathek*.[3] Thomas Bensley flourished as a printer from 1774 and died in 1789 and his son, also Thomas, specialised in fine well-illustrated volumes and soon invested in the new printing developments; he printed *Antiquities of Westminster* (1804–7), illustrated with the first British examples of the lithographic process; he tried Koenig's first mechanical press (1811) and installed a steam press (in 1814).[4]

Although it was assumed that the first edition of 1788 did not have maps – they are not called for – a copy with a full complement of maps has been found. Maps were not called for on the title-page until the third (1795) edition, in fact. The outline maps have no scales, compasses or frame-lines; the only features are rivers, market towns and the names of the county's three parts. The volume was issued until 1809 with maps; the larger edition of 1818 does not have the county maps.

## EDITIONS

(i)   *LINCOLNSHIRE*. (Ca).

*England Delineated; Or, A Geographical Description Of Every County In England And Wales: With A Concise Account Of Its Most Important Products, Natural And Artificial. For The Use Of Young Persons. London: Printed By T. Bensley, For J. Johnson, St. Paul's Church-Yard. M,DCC,LXXXVIII.*

**Copy**   CUL Atlas 7.78.24.[5]

*England Delineated . . . Persons. Second Edition, With Additions And Corrections. London . . . M,DCC,XC.*

**Copies**   BL 10348 c 12 and Maps C.27.b.50; CUL Atlas 7.79.19 and 7460.d.26. (Gainsborough RL Map 4; Grimsby XOOO: 914 LIN consists of the map and the county text – pp. 196–206).

*England Delineated . . . For The Use Of Young Persons. With Outline Maps of all the Counties. Third Edition, Considerably Improved. London: Printed For J. Johnson, St. Paul's Church-Yard. M.DCC.XCV.*

**Copies**   BL 291 d 13; CUL Atlas 7.79.21; Leeds UL W 103; Nottingham UL LT 109.GB/A 4 (Briggs Collection).

*England Delineated . . . Fourth edition, Considerably Improved. London: Printed For J. Johnson, St. Paul's Church-Yard, By T. Bensley, Bolt Court, Fleet Street. 1800.*

**Copies**   BL 796 e 55; CUL Atlas 7.80.37; Liverpool RL G 9806.

*England Delineated . . . Fifth Edition . . . 1803.*

**Copies**   BL 10348 cc 2; CUL 8474.d.134; Bod Vet A6.e.471; RGS 262 b 9.

*England Delineated . . . Sixth Edition . . . London: Printed for J. Johnson, St. Pauls Church-yard. 1809.*

**Copies**   BL 577 c 2; Bod Vet A6.e.829; RGS 262 b 10.

### References

1   Aikin. L. Memoir of John Aikin. 2 vols. (London, 1823).
2   Frankenberg, R. John Aikin (1747–1822) – doctor and philosopher. *Memoirs and Proceedings of the Manchester Literary and Philosophical Society.* Vol. 106 (1963–1964), pp. 74–93. In Part 1 of *Das Kapital* Marx says: 'What would the good Dr. Aikin say if he could rise from his grave and see the Manchester of today'.
3   Maxted, pp. 18–19.
4   DNB.

5   Other copies examined have no maps – BL 10348 e 13; Nottingham UL LT109 GB/A 4 (Briggs
    Collection).

Chubb CCLXXXVI–CCLXXXIX (2nd, 3rd, 4th and 5th editions).


## 49

**JOHN CARY**                                                          **1789**

Size: 53.35cm x 39.7cm.

After completing his own atlas in 1789 Cary was required to engrave a new set of much
larger county plates using drawings provided by E. Noble. The maps were for a new
edition of Camden's *Britannia*, freshly translated by Richard Gough (1735–1809), the
eminent antiquarian and collector. It is thanks to Gough that we know so much about
early mapmakers; his collections of maps (housed in the Bodleian Library) remain a
prime source for British cartographical research. By 1789 the last edition of Camden's
*Britannia* had been out of print for more than a dozen years (**19 (iii)**). No other work
contained so much material of importance to county historical research and the new
translation indicated its continuing value.

Cary's maps are based on a variety of sources and he seems to have made as much
use as necessary for his purpose of the large number of one inch to a mile surveys that
had appeared in the previous thirty years. Lincolnshire is based on the larger map of
Armstrong and this is most clearly shown in his adoption of hachur to indicate the
'spurious' hills of Holland, north of Boston and between Spalding and Holbeach and the
inclusion of the mythical village of *Hughington*. That the work took Cary some time and
effort is revealed by the proof map of the county corrected in Gough's hand with upwards
of fifty errors of topography or nomenclature. Careless press-work is shown in the
pagination of the Lincolnshire text; two pages (223–224) have numbers already used in
the preceding Rutland text.

John Nichols, the printer (1745–1826), had a varied career. As an antiquarian he edited
or wrote more than fifty works and he was sole manager of *Gentleman's Magazine* from
1792. He was master of the Stationers' Company in 1804. His printing office was twice
burnt down in 1786 and again in 1808. To have made ready the three folio volumes of
*Britannia* within three years of the first fire argues a man of determined character.[1] Payne
(1719–1799) worked with the Robinson family in publishing the new edition of Camden.
George Robinson (1737–1801) had the biggest wholesale business in London by 1780.
His brother John and his son, also George, joined him as partners in 1784 but by the end
of 1811 they were all dead and the firm had foundered.[2] The Robinsons are also listed
among the booksellers who sold copies of Young's *General View of the Agriculture of
... Lincoln* (**55 (i)**).

### Later history

The Robinsons sold their rights in the work to John Stockdale (1750–1814) and Nichols
reports that Gough was so annoyed by the new arrangement that he refused to work on
a new edition.[3] Stockdale is supposed to have completed the editorial work himself and
the work, when re-issued in 1806, had expanded to four folio volumes. This new edition
was issued in parts; the first was issued on 27 February, 1806 and the second on 10 April
and, it seems, monthly thereafter.[4] The four volumes complete were to be four guineas
or six guineas on superior paper and the maps coloured. Stockdale (1749–1814) had come
to London as a blacksmith but secured a job as porter with the bookseller John Almon.
As he worked up in the firm he made contacts with the many important Americans, whose

work Almon published. When Almon retired in 1780 he was unwilling to sell to Stockdale, who (he thought) had cheated him[5] and the business went to Hansard. In retaliation Stockdale set up three doors away in Piccadilly and was soon publishing the works of three Americans, who each became President of the new republic, Benjamin Franklin, Jefferson and John Adams.[6] In 1806 he published a large-scale map of Scotland and, in 1809, a map of England and Wales on 20 sheets. Engraved by Neele it is closely related to Cary's 1794 *New Map Of England and Wales* . . . Lincolnshire appears mainly on sheet XI; the northern part is on the East Riding sheet (VII) and the coastal area is sheet XII. Since each plate normally contains a number of other counties separate consideration here is felt to be beyond the present work's scope. Sheet XI, for instance, covers the area from Stafford to Leake and from Belton (in Axholme) to Stoneleigh in Warwickshire.[7] Stockdale died bankrupt in 1814. Cary and Stockdale together published the 'Camden' plates as an atlas in 1805 with a second issue four years later, priced at seven guineas.[8]

## EDITIONS

(i)   *MAP/OF/LINCOLNSHIRE,/from the best/AUTHORITIES./Engraved by J. Cary./* (Ae). *E. Noble delin & Curavit.* (Ee). *REFERENCES to the WAPONTAKES/* numbered list of 31 names (Ea). The map is as (ii) below but without the changes noted there.

Proof set marked with Gough's own corrections and notes, not all of which have been incorporated.[9]

**Copy**   Bod Gough Gen Top 372–4.

(ii)   Mileages have been added to the *Hibalstow-BARTON* road. The following places have been added: *GRIMSBY, Croxton, Ashby, Hainton, Burgh on Bain, Stainton Vale, Caulwell, Scamblesby, Reepham, S. Carlton, Sandtoft, Aukeley* [west of Haxey], *Spanby, Osbournby, Kirkby Underwood, Scothern, E. Stockworth, Irford/ Abbey. Statute Miles 69½ to a Degree./10* [= 6.45cm] added above imprint (Ee). The inner border has been marked with shading to indicate each minute of latitude, etc., numbered at 5′ intervals. *Longitude West from London* (De, between the inner and outer borders, and to the right of the 1° mark, which should, correctly, be the meridian). Many place-names have been changed: *CAISTOR* (originally *CAISTOK*), *Corringham* (*Corrington*), *Pilham* (*Filham*), *Wrawby* (*Wrawleby*), *Great Cotes* (*Great Coats*), but *Little Coats* remains uncorrected,[10] *Hogthorpe* (*Hogth*), *Barling* (*Batling*), *Brothertoft* (*Brothertof*), *S$^t$ Mary or S. Kelsey* (*S$^t$ Mary*) in spite of *South/Kelsey* being marked further south, *Barnetby le Wold* (*Barnetby*), *S$^t$/James Deeping* (*James Deeping*), and *West Stockworth* (*West Stuckworth*). *Stannigate* has been altered from *Stonnigote* [i.e. Stenigot]. *Frieston* (near Boston) has been altered to *Freston*.

*Britannia: Or, A Chorographical Description Of The Flourishing Kingdoms Of England, Scotland, And Ireland, And The Islands Adjacent; From The Earliest Antiquity. By William Camden. Translated From The Edition Published By The Author In MDCVII. Enlarged By The Latest Discoveries, By Richard Gough . . . In Three Volumes. Illustrated With Maps, And Other Copper-Plates. Volume The Second. London: Printed By John Nichols, For T. Payne And Son, Castle-Street, St. Martin's; And G.G.J. And J. Robinson, Pater-Noster-Row. MDCCLXXXIX.*

**Copies**   BL Maps C.26.c.2; CUL Atlas 4.78.3; Bod Σ 10.33; Leeds UL W 92; Lincoln Cathedral L. Hh 1.1–3; Wisbech Museum Press 1.18. (LRL Map 948; Grimsby RL CARY RM6; Boston RL Map 25).

(iii)   *Published by John Stockdale Piccadilly, 26$^{th}$. March 1805.* has been added (Ae, below title).

*New British Atlas, Being A Complete Set Of County Maps, on which are delineated all the Roads, Cities, Towns, Villages, Rivers & Canals; Together with Correct General Maps of England, Wales, Scotland & Ireland. London Printed for John Stockdale, 1805.*

**Copies**  Bod Maps C.15.a.1. (BL 3355 (25); LAO Exley 32/2/158; LRL Map 791, etc.; Grimsby CARY RM 121).[11]

*Britannia . . . by Richard Gough. The Second Edition. In Four Volumes, Illustrated . . . Volume The Second. London: Printed for John Stockdale, Piccadilly; By J. Nichols and Son, Red Lion Passage, Fleet Street, 1806.*

**Copies**  BL 192 f 4; CUL Atlas 4.80.8–11; Bod Allen 64 and G.A. Gen Top b 22–25; LRL A.9 (S.R.); Loughborough UL Special Collections 942; Sheffield UL **F914.2(C). (LRL L.9 consists of the map and the county text pages (331–392) only).

*Britannia . . . 1806.* [1807].[12]

**Copy**  Sheffield RL 942 STF

*New British Atlas . . . 1805.* [1809].[13]

**Copies**  BL Maps 20.e.19; CUL Atlas 1.80.4; Bod Allen 62; Leeds UL W 109; RGS 1 C 166; Sheffield UL Dept. of Geography. (Louth RL Map A9).

*Britannia . . . 1806.* [1812].[14]

**Copy**  (LAO FL Box 912 is the map with the county text pages).

## Reproduction
(i) The map in state (ii) has been reproduced with *J. Cary 1789* (Ae, OS) and the imprint: *Holderness Publications, Burton Pidsea, East Yorkshire* (Ee, OS). It was probably published in the 1960s and the only copy seen is in a private Lincolnshire Library.

## References
1  Maxted, pp. 161–162; DNB.
2  Maxted, pp. 172 & 191–192.
3  Nichols, op. cit., Vol. 8, p. 482.
4  Bod Gough Gen Top 66–69 consists of the first four parts; the dates of issue are taken from the advertisement leaves in Part I; in the Gough collection (Bodleian Gough Gen Top. 375) are two examples of a proposed title-page, and some plates. The title-pages are different from those finally printed; that only a few plates are included may be evidence to support the theory that Gough became dissatisfied with Stockdale and as a result there is no great quantity of proof material to equal that kept from the 1789 edition.
5  Almon, J. Memoirs of an eminent bookseller. (London, 1790), p. 69; Timperley, Vol. II, pp. 822–823 gives an obituary for Almon, who died in 1805.
6  Maxted, p. 215; Stockdale, E. John Stockdale of Piccadilly: Publisher to John Adams and Thomas Jefferson, pp. 63–87, in: Myers, R. & Harris. M (eds.). Author/Publisher relations during the eighteenth and nineteenth centuries. (Oxford, 1983).
7  Copies in BL (Maps 149.d.19) and CUL (Atlas 1.80.6[1]).
8  Bod. Allen 62 – the original title label has been remounted inside the front cover.
9  For example, Gough wanted *Creton* to be changed to E. Bitham (sic), *Maplethorpe* to be changed to Mablethorpe, *S.* and *N. Cockering* to be increased to Cockerington; *Sutterby* to be corrected to Snitterby, and *Shelland* to be Skelland (it should, of course, be Snelland).
10  This is but one example of Cary copying Armstrong while Gough thought them both wrong.
11  Unless watermark evidence suggests otherwise a loose sheet could come from any edition from 1805 onwards; the examples noted here lack that evidence.
12  An index leaf (signature 7M) is watermarked 1807.
13  Watermark dates of 1808 on many plates would suggest a new issue and in some copies the general map is a new state with the date 1809 in the imprint.
14  The map in the LAO example is watermarked *J. WHATMAN/1812.*

Chubb CCLXXI and CCLXXII (the editions of *Britannia*); CCCXIX (New British Atlas).

50

**JOHN CARY**                                                          **1789**

Size: 14.45cm x 9.3cm.

In what was a prolific period Cary also prepared a set of miniature county maps that were meant for use by travellers on the widening network of turnpike roads. The first set was prepared in 1789, appeared in book form on 1 January, 1790 and was in print more or less continuously for fifteen years. A new set was engraved in 1806 (**61**) and a new re-engraving appeared in 1822 (**80**). Chubb, having seen the editions of 1790 and 1791 only, assumed that all these later versions that Fordham recorded came from a single set of plates.[1] At first the atlas was issued as a separate item but many examples were bound together with Cary's *New Itinerary*. This work first appeared in 1798 and describes a vast number of the country's roads in words, giving mileages, places passed, suitable inns, and the most notable properties with owners' names along each route; the format is identical to that of the atlas. The book's authority depended on the survey of roads that Cary made for the Postmaster General during the middle of the 1790s (see **47**).

The maps were revised considerably during the years 1790–1792 but, thereafter, the plates remained in the same state and the 1791 title-page remained in use until the plates were replaced for a new edition in 1806. There are two versions of nearly all printings; in the notes below (a) means a thicker paper was used and Lincolnshire has Middlesex on the verso; (b) indicates the use of thinner paper with the maps single-sided.

**EDITIONS**

(i)   *LINCOLNSHIRE* (in a panel placed above the map, Ba–Da). *by J. Cary* (Aa, OS, to the immediate left of the title panel, with *Engraver* (Ea, OS, to its right). *London: Published Sep*[r].*1, 1789 by J. Cary Engraver N*[o]. *188 Strand.* (Be–De, OS). *Statute Miles/10* [= 1.4cm] (Ae). Projecting from the title panel is 'half' a compass, i.e. eight points between west and east are indicated. A panel below the map has 3½ lines of places in the county and their distances from London. On the second line appears *Atford*. Latitude and longitude are not marked.

Issued as a set of cards.

**Copy**   The private collection of Mr. Tony Burgess.

(ii)   In the lower panel *Atford* has been corrected to *Alford*.

*Cary's Traveller's Companion; Or. A Delineation of the Turnpike Roads Of England And Wales; shewing the immediate Rout* (sic) *to every Market and Borough Town throughout the Kingdom. Laid down from the best Authorities, On A New Set Of County Maps. To which is added, An Alphabetical List of all the Market Towns, with the Days on which they are held. London. Printed for John Cary, Engraver, Map & Printseller, Strand. 1*[st] *Jan*[y]. *1790.*

**Copy**   (a) NLS Newman 741.

(iii)   *Deer Park* has been added in the place of *Gautby*.

*Cary's Traveller's Companion . . . 1790.*

**Copy**   (a) The private collection of Mr. D. Hodson.

(**iv**)   *Lessington/Pa.* has been deleted with the park symbol (south of *Market/Raisin*). *Gautby* has been restored and *Deer Park* removed.

*Cary's Traveller's Companion . . . 1790.*

**Copies**   (a) CUL Atlas 7.79.5; Bod Map C.17.f 2/1 and Allen 48; Leeds UL W 93; RGS 260 C 1; (b) BL Maps C.24.b.33; CUL Atlas 7.79.1.

(**v**)   A road has been added from *Bridgend Causeway* to the 100 mile mark on the Bourn-Sleaford road; this has necessitated the 'removal' of *Aslackby* to the west side of that road. *Dowsby* and *Billingboro'* have been added as well as *Stoke* [i.e. Stoke Rochford] below the park symbol that was already engraved. *Bolingbroke* has been re-engraved in the 'hamlet' size lettering.

*Cary's Traveller's Companion . . . 1791.*

**Copies**   (a) Bod Allen 47; (b) Bod Allen 52.

(**vi**)   The year in the imprint is now *1792.* Below the name *Glanford Bridge* is added *or Briggs* (sic). The road at *Lit/London* [Long Sutton] has been extended eastwards and *Cross/Keys Ho.* added. The roads at Market Deeping and Crowland have been extended south-wards and joined together (erroneously) at newly engraved *Glinton*, inside the panel below the map. More places have been marked, e.g. *Welton, Saltfleetby, Grimoldby, Broughton.*

*Cary's Traveller's Companion . . . 1791.*

**Copy**   (a) Liverpool RL G 9251.

(**vii**)   The Wragby-Louth road has been downgraded. The road from Sleaford-Tattershall has been upgraded.

*Cary's Traveller's Companion . . . 1791.*

**Copies**   (a) Bod Allen 49; (b) Norwich RL R912.42 (S.B.3.27).

(**viii**)   *Gautby* has been removed and, in its place, *Deer/Park* has been re-engraved.

*Cary's Traveller's Companion . . . 1791.*

**Copies**   (a) BL 118 a 19; CUL Atlas 7.79.2 and 7.79.3; Bod Allen 51, etc.; NLS MS 6589; York RL 912; RGS Fordham 55 etc. (b) Leeds UL W 94.

*Cary's Traveller's Companion . . . 1791.* [1796].[2]

**Copy**   (a) Bod Allen 50.

*Cary's Traveller's Companion . . . 1791.* [1797].[3]

**Copy**   (a) BL Maps C.24.a.20.

*Cary's Traveller's Companion . . . 1791.* [bound with *Cary's New Itinerary . . . 1798.*]

**Copy**   (b) A private library in Lincolnshire.

*Cary's Traveller's Companion . . . 1791.* [1801]. [4]

**Copy**   (a) The private collection of Mr. D. Hodson.

**References**

1   Chubb, p. 220, and quoting Fordham, F.G. Notes on British and Irish itineraries and road books. (Hertford, 1912).

2   Several leaves have the date 1796 in their watermarks.

3  The end-papers are watermarked 1797.
4  One end-paper is watermarked 1801.

Chubb CCLXXIII–CCLXXV.

## 51

**JOHN LODGE**                                                    **1789**

Size: 32.85cm x 26.95cm.

John Lodge was an engraver, whose name first appeared on maps for *Gentleman's Magazine* from 1754, work he continued until 1772. He was apprenticed to Thomas Jefferys. His work for *Political Magazine* appeared between 1782 and 1790 and he died in 1796.[1] That journal undertook to issue a complete set of county maps all engraved by him. The Lincolnshire plate is closely copied from the Ellis plate (**40**) including the eastwards extension of the Yorkshire coast to the right edge of the plate, omitting Spurn Head. Although no atlas with a title- page has been found the circumstantial evidence for an edition c.1795 is strong; an advertising leaf listing the maps and the price (10s. 6d; Coloured 15s) is known in two privately held copies; both have the leaf pasted on the contemporary board covers. Its publisher is unknown.

John Murray was the founder of the famous publishing firm that has survived as a separate imprint until the present time. Born in 1745 in Edinburgh as John Macmurray he had a spell in the Marines until, buying William Sandby's business in 1768, he dropped the Mac from his surname and, helped by an inheritance, founded several literary magazines. In 1785 he took over the *Political Magazine* and oversaw completion of the map series. He died in 1793;[2] his son, also John, succeeded and moved the firm to the Albemarle Street address in 1812 from which the works of *inter alia* Jane Austen, Crabbe, Borrow and Darwin were first to see the light of day and which the firm still occupies. The firm's later fortunes were firmly based on a wide range of *Traveller's Handbooks* covering all areas of the British Isles and much of Europe, which appeared in the second half of the nineteenth century. Lincolnshire appeared in 1890 (**123 N**), with a second edition in 1903.

### EDITIONS

(i)  *A NEW MAP/ OF/ LINCOLNSHIRE,/ FROM THE BEST/ AUTHORITIES./* (Ea). *London, Published as the Act directs, April 30, 1789, by J. Murray Nº. 32 Fleet Street.* (Be–De, OS). *British Statute Miles 69 to a Degree/10* [= 4.25cm] (Ae). *J. Lodge Sc.* (Ea, OS). *Remarks/7* lines with symbols (Ab). Compass (Ec). The inner frame is divided into minutes of latitude, etc. every 2′ and numbered at 10′ intervals. *Meridian of London* (De, between the inner and outer frames). *Political Mag. Ap.89.* (Ea, OS).

*The Political Magazine And Parliamentary, Naval, Military and Literary Journal For April, 1789. Illustrated by a Map of the County of Lincoln . . . Printed for J. Murray, No. 32, Fleet-Street . . .*

**Copies**  Bod Hope Adds 151–169; Birmingham RL AQ 912.42. (Grimsby RL RM 111). BL PP 3357 is an example of the work, but the Lincolnshire map is missing.

(ii)  The imprint, the engraver's note and the reference to the *Political Mag* have been removed.

A collection of maps without title-page. The title on the covers of the privately owned copies referred to above reads: *Atlas Of Great Britain And Ireland; Containing A Whole*

*Sheet Map of England; A Whole Sheet Map of Scotland; A Whole Sheet Map of Ireland, and the following County Maps . . .* [1795].[3]

**Copies**   BL Maps C.24.e.7; CUL Atlas 5.79.3; Bod Allen 56; Leeds UL W 102.

**References**
1   Maxted, p. 140.
2   Maxted, p. 159; Timperley, Vol. II, pp. 780–781 provides an obituary.
3   In the CUL copy a manuscript note refers to an example then owned by Sir George Fordham dated 1795 in the watermark. Chubb gives 1795 as the date of publication, giving no evidence for the suggestion but presumably aware of the Fordham information.

Chubb CCXLIX–CCL.

## 52

**JOHN HARRISON**                                              **1791**

Size: 47.6cm x 33.45cm.

The maps in this series of county maps were issued over a four year period, the bulk of them bearing dates between September, 1787 and the end of 1789.[1] At first the maps came out at monthly intervals. After December, 1787 none appeared until April, 1788 but, with four maps dated May, two June, three July and one each for August and September the series was back on such a schedule. A similarly erratic performance in the following year meant that by the end of December, 1789 29 maps had appeared in 28 months. With five more maps in the first six months of 1790 only three plates out of the final set were not engraved – Yorkshire, Lincolnshire and North Wales (South Wales was never engraved but Monmouth shared the Gloucestershire plate). Who engraved the plates of Yorkshire and Lincolnshire is not noted on the maps. The North Wales plate was engraved by G.S. Allen from drawings by John Haywood, who had prepared the material for nearly all the earlier maps.[2] The reasonable assumption is that Allen did all three plates as they are all dated February, 1791 and the atlas must have appeared soon after. The map is influenced by Emanuel Bowen, whose *Atlas Anglicanus* is closely followed.

Hodson had suggested that the map of Hertfordshire (July, 1788) could have appeared in an untraced edition of Tindal's translation of *The History of England* by Rapin de Thoyras, issued by John Harrison.[3] A new edition of Rapin's *History* was advertised in March 1787 and it was intended to issue the second edition 'with Tindal's continuation' and 'with this work will be given a complete British Atlas . . .'[4] In the event there were no maps of Scottish counties. Two sets of maps were found in 1990 with the maps all dated before July, 1790 (Cornwall); a third set made up at much the same time contains all but the Cornwall map.[5] However, several maps are dated quite differently from those noted by Chubb; the earliest (Monmouth & Gloucestershire) was dated November, 1786 and has been altered to Nov. 1st, 1787 – suggesting that Harrison had made a fresh start with the series. Since all known examples of Rapin's work only contain maps prepared by July, 1790 it follows that Lincolnshire (February, 1791) will not normally occur in that work.

Hodson noted[6] that Harrison advertised in *The Morning Herald* of 2 May, 1787 'Specimens of Maps of the Counties . . . printed on white satin, with crimson ink, and the Hundreds, &c. beautifully coloured, at 10s. each.' On the next day his advert in *The Public Advertiser* clarified the offer; 'a map of Great Britain and Ireland printed on white satin to be used for needlework in ladies Schools which at the same time would help with geography'. No example has been found so far. Three county plates only were available for this purpose at that time.[7]

It seems that the atlas when issued in 1791 did not sell well and a new edition at a reduced price appeared in the following year.[8] The atlas seems to have remained on sale for a number of years. There is some evidence that amended plates appeared on the market over twenty years later – see state (ii) below.

## EDITIONS

(i)   *A MAP OF/LINCOLNSHIRE/divided into the/WAPONTAKES or HUNDREDS/Engraved from an actual Survey/With Improvements./* (Ae). *Drawn and Engraved for I. Harrison, N°. 115 Newgate Street, as the Act Directs, February. 1. 1791.* (Be–Ce, OS). *20/British Statute Miles 69 to a Degree* [= 10.5cm] (Be–Ce). Compass (Eb). The borders carry numbers every 5 minutes and a graticule covers the map, picking up the numbering; the inner frame is shaded to show each minute.

*Maps Of The English Counties; With The Subdivisions Of Hundreds, Wapontakes, Lathes, Wards, Divisions &c. To Which Are Added Two Folio Pages Of Letter-Press, To face Each Map; Descriptive Of The Extent, Boundaries, Rivers, Lakes, Canals, Soil, Mines, Minerals, Curious Plants, Husbandry, and every Curiosity that is Nearly Connected with the Maps. The whole containing the most useful, entertaining and instructive Selection of the Kind, that has yet appeared at the Price of 3l. 9s. London: Printed by and for John Harrison, No. 115, Newgate-Street, M,DCC,XCI.*

**Copies**   BL 143 cc 12; CUL Atlas 2.79.4; Leeds UL Special Collections Geography C-1.2 Fol. (BL 3355 (20)); LRL Map 815; Grimsby RL Harrison 1791; Nottingham UL Li 1.B9.D91; Manchester UL Map C274).

*Maps Of The English Counties . . . Price $2^L.2^S$. London. Engraved and Printed for John Harrison, N° 115, Newgate Street. 1792.*

**Copy**   RGS 7 G 12.

*Maps Of The English Counties . . . 1792. [1794].*[9]

**Copy**   Bod Allen 53.

(ii)   The last two digits of the date in the map's imprint have been deleted.[10]

### References
1   Chubb, pp. 224–225 lists all plates with full details of dates, etc.
2   Only four maps omit the reference to Haywood.
3   Hodson, 1977, p. 68.
4   *The Morning Chronicle, and London Advertiser* for March 7, 1787. I am grateful to Mr. Eugene Burden for telling me of this advertisement.
5   Two sets were found in Sheffield RL. One set of maps occurs in Volume 1 of the 4-volume set of Rapin de Thoyras, 1785–1789. The other set is in a mixed volume of maps by Harrison and maps copied from d'Anville; it has no title-page, but it is bound in uniform style with the Rapin de Thoyras set and is labelled on the spine as forming part of that set. The shelf marks are 942 STF. A third set is in Canterbury Cathedral Library (W/G – 7 – 31). Mr. Clive Burden has acquired a set of Rapin's work; volume one (with the maps) in the John Kelly translation, volume two in the Tindal translation (but with a variant title-page) and the 'third' is a copy of volume 1 of Tindal's translation but now dated 1789 instead of 1785.
6   Hodson, 1977, p. 68.
7   Essex, Leicester & Rutland (both April, 1787) and Nottinghamshire (January, 1787).
8   The BL and CUL copies have 1791 title-pages with the price roughly changed to 3l.3s. The price was further reduced for the 1792 issue (q.v.).
9   The Lincolnshire map has 1794 in the watermark.

10  Hodson, 1977, p. 69 refers to an atlas, then in private hands and subsequently broken up, in which the last two digits of the date on the Hertfordshire's map's imprint had been imperfectly erased. It is not clear if the erasure was made on the paper or the maps came from plates suitably treated; the former seems the more likely since the plates, if they had survived, could have been properly amended with a new date or no date at all.

Chubb CCXCI–CCXCII.


## 53

**BENJAMIN BAKER**                                             **1793**

Size: 22.8cm x 18.3cm.

Baker was active from 1792, when he began to prepare plates for *The Universal Magazine Of Knowledge And Pleasure* . . . For the greater part of this period until his retirement in c.1824 he was engaged in engraving the earliest maps of the Ordnance Survey.[1] Possibly his last work for the Ordnance Survey was the engraving of the first edition of the sheets for Lincolnshire and Rutland, which was considered 'a great advance in topographical science and art'.[2] A series of county maps was produced in the early years of the above journal (**35**).

The Lincolnshire map appeared in the issue for April, 1793 and is a slightly reduced version of Cary's map for his *New and Correct English Atlas* . . . Pages 281–282 of the *Universal Magazine* . . . have the heading 'An Account of Lincolnshire: With a neat and accurate Map of that County'. The publisher of the periodical, William Bent, was also a bookseller in Paternoster Row.

### Later history
The plates appeared in an atlas produced in 1804 at 12 Shillings[3] by William Darton (1755–1819) and Joseph (or Josiah, depending on the work of reference) Harvey (1764–1841). They were print-sellers and publishers from 1787 and particularly involved in the publishing of books for children.[4] Darton's son, also William, set up in 1804 the firm of Darton and Clark, which became Darton & Son – see **75** below. By 1807 the plates had been acquired by Laurie and Whittle, who had set up in business together in 1794 and published a series of works using the plates of earlier engravers, particularly those acquired from the sale of Robert Sayer's goods. The plates passed subsequently to Laurie's son who remained in business firstly with Whittle until the latter's death in 1818 and then by himself until his own death in 1858. The firm still survives after amalgamation with Imray, Norie and Wilson to form Imray, Laurie, Norie & Wilson, Ltd., specialising in sea-charts.[5]

### EDITIONS

(i)   *LINCOLNSHIRE* (in an oval)/*Engraved by B. Baker: Islington./Scale of Miles/10* [= 2.9cm] (Ae). The inner frame is divided every two minutes of latitude, etc., with the numbering occuring at 10′ intervals, between the inner and outer frames. *Longitude East from London* (De, between the inner and outer frames). *Longitude West from London* (Be, between frames). There is no compass.

*The Universal Magazine Of Knowledge And Pleasure . . . And Other Arts and Sciences . . . Vol. XCII. London: Published Under His Majesty's Royal Licence, by W. Bent, at the King's Arms, Pater-Noster Row. M DCC XCIII.* [The issue for April, 1793].

**Copies**   CUL T.900.d.75.92; Bod Per 2705 e 552. (BL 3355 (23)).

*Maps Of The Several Counties And Shires In England, With The Principality Of Wales, &c . . . London: Printed And Sold By Darton And Harvey, Gracechurch-Street. 1804.*

**Copies**   CUL Atlas 7.80.36; British Geological Survey: Conybeare Geological Maps 1804.

(ii)   A 16-point compass has been added; the oval containing the county name forms the east-west axis (Ae). An imprint has been added: *Published. October 13ᵗʰ. 1806 by LAURIE & WHITTLE, Nᵒ. 53 Fleet Street. London.* (Be–De, OS). Many places have been added: *Clea Ness, Gibraltar/Point, Guthram* (west of Spalding), *Lutton, Moulton, Langtoft, Casewick, Easington* (i.e. Essendine), *Carlby, Black/Bull Inn* (s. of Colsterworth), *Kirkby Underwood, Pickworth* (west of Folkingham), *Hanby Grange, Ropsley, Culverthorpe, Barkston, Syston, Belton* (with its park), *Stoke* (i.e. *Stoke Rochford*), *Harrowby, Stubton, Hougham, Dunsby, Quadrang* (sic), *Garrick, Winceby, West Ashby, Gunby Hall, Hangham, Buckland, Anderby, Hagnaby, Great/Carlton, Grimoldby, Munckton, Wyham, Tupholm, Burton/upon Stather, Waddingworth, Kirk/stead, Moor Tower, Repham* (i.e. Reepham), *Branston, Dorrington, Navenby, Temple/Bruer, Somerton/ Castle, Doddington, Bracebridge, Summer/Castle, Haddon/Inn* (on Lincoln-Saxilby road), *Somerby* (near Gainsborough), *Corringham, Keadby, Appleby, Halton, Whitton, Horkstow, Elsham* (and Park), *Thornton/College, Caenby/Hall, Somerby* (near Glandford/Bridge), *Middle/Raisin, Snarford, Maltby, Ludford, Stainton/Vale, Brocklesby/Hall* (and Park) and *Grimsthorpe Castle.* Many rivers are now named, e.g. *Glen, Welland R., Witham* (between Lincoln and Newark), *Louth R.* Canals are marked and named, e.g. *Caistor, Grantham, Stainforth & Keadby.* At Burwell and *Barling* has been added *and Park.* Below *Burton* has been added *and Hou* and below *Scrivelsby* and *Gautby* has been added *and Hall.* Below *Revesby* has been added *and Abbey.* Many roads have been upgraded; marked by deepening one of the double lines that are used to delineate them. The mileage at Spalding has been increased to *101* from 100 and at Holbeach and Wainfleet to *109* and *131* respectively; at Sleaford and Folkingham mileages have been reduced by one to *115* and *106* respectively and at Caistor has been removed altogether. As the road from Grantham crosses the county border *from Nottingham* has been added. Lower case letters have been placed at the county boundary where the important roads leave the county.[6] Several names have been altered, e.g. *Hainton* (previously Hampton); *Maplethorpe (Mablethorpe* – sic!); *Botham (Boatham* = Boultham); *Berkwirth (Barkworth)*, and *Uselby (Usselby).* Binbrook and Kirton [in Lindsey] have been re-engraved in smaller lettering. *Sunk/Island* has now been joined to mainland Yorkshire and the words *now Embank'd* added. *Water* has been added in the Humber near *Barton.*

*Laurie And Whittle's New And Improved English Atlas, Divided Into Counties: Shewing Their Respective Situations, Boundaries, And Extent, Produce, Mines, Minerals, Trade, And Manufactures; Also the Cities, Market And Principal Towns, Hills, Rivers, Navigable Canals, &c.&c. With The Turnpike And Principal Roads . . . Together With . . . Antiquities, Monasteries, Castles, Seats, And Parks . . . London: Printed And Published By Robert Laurie And James Whittle, No. 53, Fleet-Street. 1807.*

**Copies**   CUL Atlas 6.80.9; Bod Allen 65 and J Maps 2; Leeds UL W 111. (LRL Maps 200 and 200a; Grantham RL).

*Laurie And Whittle's New And Improved English Atlas . . . 1807.* [1810].[7]

**Copy**   BL Maps 19.a.7.

*A New And Improved English Atlas, Divided Into Counties: With The Turnpike And Principal Roads, Accurately Laid Down From The Most Recent Surveys And Authorities. London: Printed And Published By James Whittle And Richard Holmes Laurie, No. 53, Fleet Street. 1816.*

**Copy**   Bod Map C.17.d.14.

An atlas without title-page. [1821].[8]

**Copy**   The private collection of Mr. T. Burgess.


(iii)   The imprint on the map now reads: *Published by R.H. Laurie, N°. 53 Fleet Street, London.* Many place-names have been added: *Legsby, Clea, Thedlethorpe, LINDSEY, HOLLAND, KESTEVEN* (with dashes to mark the boundaries), *Sutton/S<sup>t</sup> James, Kings/Hall* (south of Moulton), *Podge/Hole* [i.e. Pode Hole], *Gantlet, Frampton, Honeld/Ho., Tetney R., Thoresby, Utterby, Holton, Nettleham, Holton le Moor, Dunholme, Stainton, Scopwick, Cuxwold, Aylesby, Keelby, Thornton/Curtes, Althorpe, Coddington, West/Butterwick, Scotton, Laughton, Aukeley, Hemswell, Willing-/ham, Bratleby* (sic), *Thorpe* [on the Hill], *Normanton, Claypole, Bathing/Ho* (near present-day Skegness).

*A New And Improved English Atlas . . . 1846.*

**Copy**   The private collection of Mr. T. Burgess.

### References

1   Maxted, p. 9.
2   Portlock, J. E. Memoir of the life of Major General Colby. (London, 1869), p. 95. Cited in: The Old Series Ordnance Survey Maps of England And Wales . . . Vol. V. Lincolnshire, Rutland, And East Anglia . . . (Lympne, [1987]), p. viii.
3   The cover of the BGS copy bears the original label with the price.
4   Maxted, pp. 61 and 105.
5   Robinson, A.H.W. Marine cartography in Britain. (Leicester, 1962), pp. 123–124, 126 and 188–190.
6   This feature is taken directly from Cary and allows maps of adjoining counties to be matched and the common roads linked.
7   The General map is dated 1810 and some watermarks appear to be dated 1810 also. Mr. Donald Hodson's copy has a watermark date 1809 on the Lincolnshire map.
8   The map of Wales has the imprint: *Published 12<sup>th</sup>. October, 1821 by RICH<sup>D</sup>. LAURIE, N°. 53, Fleet Street, London.*

Chubb CCXCIII–CCXCIV. (The Baker edition and the 1807 only).


## 54

**JOHN FAIRBURN**                                              **1798**

Size: 8.95cm x 6.1cm.

Fairburn was a maker of games for children and the maps below are part of a card game to teach basic English geography entertainingly. The maps are closely copied from *Cary's Traveller's Companion* (**50**); they show the market towns, rivers, a few villages and the main county roads. The title quoted for the 'work' below is taken from the booklet, which accompanies the cards and give the rules of play.

## EDITIONS

(i)   *LINCOLNSHIRE* (in a panel, forming part of the outer frame) (Ba–Da). *Scale of Miles/10* [= 0.8cm] (Ce). Compass (Aa).

[*The Junction Of The Counties In England and Wales, With the principal Town in each County, and its Distance from London. Intended To Render Fairburn's Game Of English Geography Clearly Understood . . . Published by John Fairburn, 146, Minories, 1798.*]

**Copy**   The private collection of Mr. C.A. Burden.[1]

**(ii)**   The adjoining counties have been numbered (although they are not named), starting with 1 for Yorkshire, then Norfolk (2) and proceding westwards and northwards ending with (7) Nottinghamshire. Roads have been added: Louth-Saltfleet; Wrangle-Wainfleet, and continuing in a north-easterly direction; from [Long Sutton] eastwards; two roads from *Glandford/Briggs*, one to Burton [on Stather] and another via Kirton to Harpwell; Horncastle-Sleaford. The boundaries between the counties, which adjoin Lincolnshire, have been extended.

*[The Junction Of The Counties . . . 1798.].*[2]

**Copies**   The private collections of Mrs. D. Green and Mr. C.A. Burden.

**References**

1   This may be a proof set since it lacks one of the essential elements of the game – see state (ii).
2   There are two versions of these maps; Mrs. Green's set is printed on paper and plate marks are clearly visible, while Mr. Burden's is on cards, which have been trimmed to the outer frame lines.

# 55

## SAMUEL JOHN NEELE                                    1799

Size: 29.7cm x 21.95cm.

Neele (1758–1824) had a long career as an engraver from 1782 to his death and his work is met again in the maps he and his two sons, James and Josiah, produced (**66, 70,** and **77**). He was a specialist in producing local maps, including those for other works in the series of agricultural reports that Young prepared in the 1790s. One of his earliest pieces is the map of Stamford (based on Speed's 1611 map), used to illustrate Harrod's *The Antiquities Of Stamford . . .* (London, 1785).[1] He also produced at least two maps of the Bedford Level, one in 1789 for Charles Nelson Cole and another in c.1793 in support of a bill for making a new cut from Eau Brink to Kings Lynn.[2] *A Plan of the RIVER NENE AND NORTH LEVEL and Part of SOUTH HOLLAND . . . BY JOHN RENNIE . . . 1813* . . . is another later example of his work on the same area.[3] A further map engraved by Neele: *A MAP of the SOUTH DRAINAGES of LINCOLNSHIRE.* appears in Young's volume and shows an area from Great Steeping south to Wisbech and west to Market Deeping. Perhaps his largest work was the engraving for John Stockdale of *MAP OF ENGLAND & WALES, from the latest Surveys . . . 1809*. This multi-sheet work shows Lincolnshire on plate XI, with the east coast on sheet XII and the north of the county on plate VII.[4]

   The county map in Young's report is the first attempt to suggest the county's physical structure and show the scil types. No other map of the county showing such information appeared until Cary's post-1820 versions of the map in *New English Atlas* (**57**). Cary's sheet only indicated round the edges of the county map area the names of the geological strata. Not until 1860 is there a map of the whole county on one sheet coloured to show the various formations (**114 B**). Neele's map is also one of the first to show the usage of the Greenwich meridian. In one of the other maps in the volume, that showing the drainages of the south of the county, the Greenwich meridian is also supposed to be used; in that case, however, the meridian is marked (in the border) due south of Crowland, many miles to the west of its true position and its place on the map described below.

   Arthur Young (1741–1820) was a famous writer on agricultural matters (although not very successful as a farmer). He became Secretary to the Board of Agriculture in 1793, perhaps as the result of the success of his *Travels in France* published in the year before. For the Board he produced a number of other county reports besides Lincolnshire for which Neele engraved the maps.[5]

### Later history
The plate re-appeared unchanged in the later editions of Young's account. In its final appearance the plate was amended during transfer to a lithographic stone. The revisions on the map are very similar to those carried out on the Baker plate in 1807 (**53 (ii)**).

The author of the article in the Royal Agricultural Society volume for 1843 was Philip Pusey (1799–1855). On the death of his father (1828) he took over the estates at Pusey (Berks.), having had an undistinguished career until then; after Eton and Oxford (no degree) he had married and lived in Rome. On return he became an M.P. and won and lost two seats before becoming one of the county members for Berkshire in 1835, a seat held for 17 years until he changed political horses. An FRS (1830) he took a leading part in 1838 in setting up what became the Royal Agricultural Society in 1840. Also in that year he was a founder member of the London Library.[6]

### EDITIONS

(i)   *Map/of/the Soil of/LINCOLNSHIRE* (in an oval)/*British Miles/ 10* [= 3.9cm] (Ea). *Explanation of the Colours*/4 lines of what the four colours mean. (Ae). *Neele sculp*$^t$. *352 Strand* (Ee, OS). There is no compass. The inner frame is divided to show each minute of latitude, etc. with the numbering every 10 minutes. *Meridian of Greenwich* (De, between the inner and outer frames).

*General View Of The Agriculture Of The County Of Lincoln; Drawn Up For The Consideration Of The Board Of Agriculture And Internal Improvement. By The Secretary To The Board. London: Printed By W. Bulmer And Co. For G. Nicol, Pall-Mall . . . And The Board . . . Sold By G.G. And J. Robinson, Paternoster-Row; J. Sewell, Cornhill; Cadell And Davies, Strand; W. Creech, Edinburgh, And John Archer, Dublin. M,DCC,XCIX.*

**Copies**   BL 988 g 9; CUL XVII.49.15; Bod Gough Lincolnshire 13; LAO L338.1 YOU; Liverpool Athenaeum 630.6; Grimsby RL X000:630 AGR; Reference libraries in Lincoln, Louth, Grantham and Stamford. (BL Maps 3364(2)).

*General View Of The Agriculture Of The County Of Lincolnshire. Drawn Up For The Consideration Of The Board Of Agriculture And Internal Improvement. By The Secretary Of The Board. Second Edition. London: Printed For Richard Phillips, Bridge Street; Sold By . . . Mozeley, Gainsborough; Newcomb, & Drakard, Stamford; Drewry, & Brooke, Lincoln; & Ridge, Newark; By B. M$^c$Millan, Bow Street, Covent Garden. 1808. [Price Twelve Shillings in Boards.]*

**Copy**   Stamford RL L 63.

*General View Of The Agriculture . . . Second Edition. London: Printed For Sherwood, Neely and Jones, Paternoster-Row: Sold By G. and W. Nicol, Pall-Mall; Mozeley, Gainsborough; Newcomb, and Drakard, Stamford; Drewry, And Brooke, Lincoln; And Ridge, Newark. 1813. [Price Twelve Shillings in Boards.]*

**Copy**   LRL L 63.[7]

### LITHOGRAPHIC TRANSFER
(**A**)   The title now reads: *Map/of/the Soil of/LINCOLNSHIRE/by Arthur Young,/1799./* (Ea). The reference to Neele has been replaced by: *Standidge & C$^o$ Litho: London.* (Ee, OS). The following places have been added: *Brocklesby/Park, Swinthorpe, Wyham,*

*Ormsby, Gayton, Elkington, Tathwell, S. Thoresby, Dorrington, Digby, Rushington* (sic), *Temple Bruer.* Several names have been re-drawn: *Butterwick, Middle Rasen, Bennington, Tattershall, Navenby. Glanford/Bridge* has been changed to *Glandford/Bridge* and *Withan on the Hill* has been altered (but remains incorrect) to *Wythan on the Hill. Caiston* has been changed in type size and to *Caistor.* As a result of the lithographic transfer process a gap of 1mm. between the words *NORTHAMPTON SHIRE* and the lower inside frame line now appears.

*The Journal Of The Royal Agricultural Society Of England. Volume The Fourth. 1843. Part II. Practice With Science. London: John Murray, Albemarle Street. MDCCCXLIII.* [Pages 287–315 comprise an article by Ph. Pusey, M.P.: *On the Agricultural Improvements Of Lincolnshire.* The map faces p. 303].

**Copies**   BL Ac 3485; CUL P 440.c.29.4; Bod Radcliffe Soc 19195 e 298/4.

*On The Agricultural Improvements Of Lincolnshire. London. MDCCCXLIII.* [An offprint of the above article, but now with the pages separately numbered [3], 4–32. The original source of its publication is recorded on the title-page verso].

**Copy**   Grimsby RL XOOO: 630 PUS.

### Reproductions
(i) The 1813 edition of Young's report was re-issued by David and Charles (Newton Abbot, 1970), with the map. (ii) An American reprint of the 1813 edition also appeared in 1970 (New York, A.M. Keeley).

### References
1   The map (erroneously) has 1600. Examples in BL 980 e 13, CUL Ll.8.37–38 and Bod Gough Lincoln 2.
2   BL Maps 1308 (8); CUL Views X.4.36 and Maps d.36.78.1; Cambs. CRO R.59/31/40/17 are examples of the 1789 map; examples of the c.1793 map are CUL Views X.4.37 and Cam. a.793.1 and Pembroke College, Cambridge – e.3.1.
3   The map accompanies *Reports As To The Wisbech Outfall . . .* (London, 1814). There is a copy of the report in LRL (L. Fens 627.5). BL Map 3355 (2) is a loose copy of the map.
4   Incomplete set in Hull RL 912.42; plate XI is missing. Complete sets in BL (Maps 149.d.19) and CUL (Atlas 1.80.6¹).
5   DNB; Gazley, J.G. Life of Arthur Young. (Philadelphia, 1973).
6   DNB., Vol. XLVII (1896); Clarke, E. Philip Pusey. (London, 1900).
7   One other copy is known – in a private Lincolnshire collection.

## 56

## CHARLES SMITH                                                    1801

Size: 49.2cm x 43.5cm.

This very fine series of large county plates was engraved by an Islington firm of engravers, Jones and Smith, the form in which their names are given for Lincolnshire and seven other counties; on other maps they have the name Jones, Smith & Co. or Jones, Smith & Bye. Their forenames have not been discovered. The Charles Smith who was the entrepreneur in the publication of this atlas is not connected to Smith, the engraver. Charles Smith was a stationer, who sold maps and globes from an address in the Strand from 1800 to 1852.[1]

    The atlas was the inspiration for the similarly sized atlas that Cary began to prepare almost simultaneously; the quality of the engraving by Smith and Jones challenged even the best that Cary could achieve. Cary's work was not completed until 1809 (**57**), although the Lincolnshire plate was engraved in 1801 and the first issues of the work in parts came out in that year also. There are undoubted similarities between Smith's atlas and the maps contained in Gough's edition (1789) of Camden's *Britannia*. This can be seen particularly in the many little hachured circles representing hills dotted all over the map. Even where one least expects such marks – in the Isle of Axholme – there is such a sign around Burnham (south of Epworth), where, in fact, the land is 40 feet above sea level, while the surrounding area is only a few feet above. Such features are taken directly from Cary, but there is also evidence that other sources have been used; unlike Cary and Armstrong, for instance, the mythical village of Hughington (near Heighington) does not appear. Perhaps Harrison's map was a secondary source for material taken from the Bowen large sheets.

The maps were almost certainly available as single sheets from the time of their first being engraved. The maps were also issued in parts; a full set of the parts in their original covers is held in the BL.[2] Each part consisted of three maps; there were fifteen numbers 'price 6s.6d. neatly coloured, or 8s. handsomely stained'. The frequency of issue was, presumably, intended to be monthly, beginning in 1801 – the date on all the county plates – and ending in 1802 or very early in 1803.[3] However, the final number contains a note to the subscribers, dated Feb 24, 1804 and, of course, 1804 is the date on the title-page of the first atlas appearance. Lincolnshire was in part ten together with the maps of Cornwall and Durham. Almost as soon as the parts came out Smith began to revise the plates but there seems to have been nothing systematic about it. The dates on the plates were gradually changed to 1804, but the watermark evidence indicates haphazard compilation, a trait common to all the surviving later atlases, which have plates with a mixture of dates.

One point of interest is the early use of the longitude of Greenwich, but the effect is somewhat marred in Lincolnshire's case as Smith has numbered the meridian '1' instead of zero; this error was not corrected until the 1808 edition. It is, of course, evidence of Smith's copying of Cary's 'Camden' plates.

### Later history

The atlas went through various correcting processes during its first years, but then settled down to a series of largely unchanged editions until the 1840s. Almost continuously throughout this period the maps were put out as single sheets, available in cardboard holders. After Smith relinquished his business interests the maps were issued by Wyld and, in their last manifestation, by W.H. Smith, the railway bookstore firm (at that time), and no relation of the originator.

### EDITIONS

(**i**) *A/NEW MAP/of the COUNTY of/LINCOLN/Divided into Wapontakes &c./LONDON/ Printed for C. SMITH N°. 172 Strand./January 6ᵗʰ. 1801./* (Ea). Compass and, below it, *Smith & Jones sculp, Pentonville.* (Aa). *EXPLANATION/*followed by 10 lines of descriptions and symbols (Ee). *REFERENCE to the WAPONTAKES &c./LINDSEY DIVISION./ Contains/*followed by 31 numbered names, divided into the three parts (Ad–Ae). The inner frame is marked off at 1′ intervals, with numbering every 5 minutes. *1°. Longitude West from Greenwich* (De, between the inner and outer frames).

Original parts as issued in oblong blue folio covers. *No. 10. Containing Maps of the Counties of Cornwall, Durham and Lincoln.* The rest of the title follows closely that of the title-page of the complete atlas (see below), but with minor changes of punctuation.

**Copy**  BL Maps MT. 6 a 1.

*Smith's New English Atlas Being a Complete Set of County Maps, Divided into Hundreds*

*On which are delineated all the Direct and Cross Roads Part of which are from actual Measurement Cities, Towns, and most considerable Villages, Parks, Gentlemen's Seats, Rivers, and Navigable Canals: Preceded by A General Map of England And Wales. On which the principal Roads are carefully described, For the purpose of facilitating the Connexion of the respective Maps. The Whole Accompanied By An Index Villaris, Containing Upwards of Forty Thousand Names of Places mentioned in the Work, with Reference to their Situation. London. Printed for C. Smith, Mapseller, N°. 172 (Corner of Surrey Street,) Strand 1804. Tomkins fecit. Vincent sculpsit.*

**Copies**   CUL Atlas 2.80.4; Guildhall L. Store 421.5; Middlesbrough RL 912. (LRL Map 578).

**(ii)**   *Great Parsand* has been altered to *Great Porsland* and *Gout* has been altered to *Gowt* in *Frith/Anthony's Gowt* (north-west of Boston)

*Smith's New English Atlas . . . 1804 . . .*

**Copy**   BL Maps 20.e.10.

**(iii)**   *S‹t›. Guthalne/* † (sic), *Snake* (south of Whaplode), and *Grimsthorpe* (entered a second time north of Bourn) have been deleted.

*Smith's New English Atlas . . . 1804.*

**Copy**   BL Maps 24.e.29.

**(iv)**   *Littleworth Drove* has been added along the Spalding-Deeping road. *Great Pointon* has been altered to *Great Ponton* and the by-road from the Great North Road extended eastwards.

*Smith's New English Atlas . . . 1804.*

**Copy**   Bod Allen 59.

**(v)**   The date has been changed to *1804* in the title.

*Smith's New English Atlas . . . 1804.*

**Copy**   The private collection of Mr. D. Hodson.

**(vi)**   *Strangle* (south of Boston) has been altered to *Struggs/Hill.*

*Smith's New English Atlas . . . 1804.*

**Copy**   Liverpool Athenaeum 910 11F.

**(vii)**   The date has been changed back to *1801* in the title. The following places have been added: *Bennington/Br., Leak Common, Black Dike, Wrangle/Common, Gibralter* (sic – on land below Wainfleet; it already appears off the coast), *Keal/Coats, Wildmore/ Fen* and *Great Postland* (*Great Porsland*). The following names have been deleted: *Ramsey* (near Strangle, north-east of Algarkirk), *Market* (south of Wigtoft). Many place names have been changed, e.g. *Tidd* (in both *Tidde/S‹t›. Giles* and *Tidde S‹t›. Marys*), *Swayton* (*Swarton*), *Wilsford* (*Willisford*), *Revesby/Abbey* (*Abbey/in Ruins*), *Enderby/Mavis* (*Enderby/Mauvaise*), *Langrelt/Ferry* (*Langrik/Ferry*), *Leak* (*Leake*), *Titton/Hall* (*Teton/ Hall*).

*Smith's New English Atlas . . .1804 . . .*

**Copies**   Leeds UL W 105. (BL Map 3355 (24)).

**(viii)**   *CAISTOR* has been altered to *Caistor.*

*Smith's New English Atlas . . . 1804.*

**Copy** Liverpool RL Gf 299.

(**ix**) The year in the title reverts to *1804*. The letter *e* has been added on the Bourn to Stamford road; the letter on The Great North Road as it enters the county from the south has been changed from *e* to *f*; and *f* has been changed to *g* on the road near *S. Witham*. The mileage (from London) at Newark is now *124¼* (formerly *116*; the mileages of the intervening places have not been changed to harmonize). Where the *Stainforth & Keadby Can*[al] leaves the county *to the River Don* has been added. Similarly *to Nottingham* has been added where Grantham Canal leaves the county.

*Smith's New English Atlas . . . 1804 . . .*

**Copies** CUL Atlas 2.80.5; Bod Allen 60 etc.; NLS EU 185; Leeds UL W 106; Nottingham UL s/G 5512 oversize X; RGS 1 C 71. (LAO FL Maps 6; LRL Map 129; Grantham RL Map 11; Grimsby RL SMITH 1804; Gainsborough RL Map 10).

(**x**) A further line has been added under the title: *2nd. Edition Corrected to 1808*. The mileage in the *EXPLANATION* now reads *HORNCASTLE/ 136¼* (formerly *139*). Mileages on the map have been changed at Horncastle (as above), Tattershall (now *128* instead of *130*), Bourn (reduced from *98* to *97¾*), Louth (*153* instead of *148*), Barton (*167* from *168*), Saltfleet (reduced to *160*) and Wainfleet (*131½* instead of *132¼*). *1°* (between the lower inner and outer frames) has been altered to *0*. In the corresponding position at the top of the map the *1°* has been deleted. The *2°* mark has been altered to *1° West of Greenwich* between the lower frame lines and to *1°* between the upper frame lines. Where the road to [Wisbech] leaves the county an extra line has been added so that the legend reads: *from Holbeach/to Wisbeach/14 Miles*. Below Dunham has been added: *from Lincoln/to East Retford/24 Miles. Long Sutton* has been changed to *Sutton/S<sup>t</sup>. Mary*. In the list of Wapontakes (Ad–Ae) line 11 now reads: *Louth Eske Hundred* and line 2 *Yarborough D°*.

*Smith's New English Atlas . . . 1804. 2nd. Edition, Corrected to 1808 . . .*

**Copies** Leeds UL W 107; NLS Newman 361. (LRL Maps 138A and 138B; Grimsby RL SMITH 1808).

Issued as a loose map, mounted and folded in a contemporary cardboard case with a label: *LINCOLN*. [1808].

**Copy** LRL Map 224.

*Smith's New English Atlas . . . 1808 . . . [1809].*[4]

**Copy** CUL Atlas 2.80.10.

*Smith's New English Atlas . . . 1808 . . . [1812].*[5]

**Copies** BL Maps 24.e.21. (LAO FL Maps 7).

A loose sheet watermarked *J. WHATMAN/1816*.

**Copy** LRL Map 138.

(**xi**) The last line of the title has been altered to: *3rd. Edition Corrected to 1818*. The roads from Louth-Scartho (via Ludborough) and Bridgend Causeway westwards through Threekingham to Grantham have been regraded as mailroads. *Caistor* has been re-engraved as *CAISTOR*.

*Smith's New English Atlas . . . 1818 . . .*

**Copy** CUL Atlas 2.81.4.

Issued in a cardboard holder, labelled *LINCOLN*, the map mounted and folded.

**Copy** The author's collection.

*Smith's New English Atlas . . . 1818 . . .* [1820].[6]

**Copy** BL Maps 24.e.35.

**(xii)** The imprint now reads: *LONDON/Printed for C. SMITH Nº. 172 Strand./Corrected to/1821./* (Ea).

Issued as folding maps in six boxes, each map being mounted and folded and having a label bearing the county name protruding above the top edge as a contents guide. [1821].[7]

**Copy** The private collection of Mr. D. Hodson.

Issued as a mounted and folded map in a cardboard holder, labelled *LINCOLN*. [1821].

**Copy** LAO Ex 942.50.

*Smith's New English Atlas . . . 1821 . . .* [1822].[8]

**Copy** Admiralty L. Ve 58.

*Smith's New English Atlas . . . 1821 . . .* [1823].[9]

**Copy** CUL Atlas 2.82.12.

*Smith's New English Atlas . . . 1821 . . .* [1824].[10]

**Copy** The private collection of Mr. C.A. Burden.

*Smith's New English Atlas . . . 1821 . . .* [1825].[11]

**Copy** Bod Map C.17.a.25.

**(xiii)** The last lines of the title now read: *Corrected to/1827./* The reference to the engravers has been removed (Aa). The 'lakes' of the East Fen have been deleted. Several new roads across the West/East Fen area have been drawn in, notably Friskney-Stickney, Sibsey-Revesby with a linking road to Boston through Antons Gowt, and Kirton Holme-Coningsby with a spur to Leeds Gate. The road from *Gantlet* (north of Donington) via Sutterton to Algarkirk has been upgraded and continues on a new line around the western end of Fosdike Wash to Holbeach. The road from Spalding-Crowland has been redrawn south of Cowbit, no longer following the line of the *River Wash*. Roads from Burgh-Skegness and Wainfleet-Skegness and to *New Hotel*, Louth-Alford (via N. and S. Reston), Wragby-Louth, Tattershall-Horncastle and Spilsby-Revesby have all been added. *New Hotel, Ferry* (on the north bank of the Humber), *HULL* and a park at Riby are all new additions.

Issued folded and mounted in cardboard holder, with label: *LINCOLN*. [1827].[12]

**Copy** Grimsby RL XOOO:912 SMI.

**(xiv)** The date in the title has been changed to *1829*.

*Smith's New English Atlas . . . 1827 . . .* [1830].[13]

**Copy** The private collection of Mr. T. Burgess.

**(xv)** There is a new title: *A/New MAP/of the COUNTY of/LINCOLN/ Divided into Wapontakes/and the Parliamentary Divisions/Printed for C. SMITH Nº. 172 Strand./Corrected to/1832./* (Ea). The Explanation has been extended with five more lines and symbols referring to County Elections, Polling places and Boroughs returning one or two members, etc. (Ac). As a result there are pink squares at Lincoln and Sleaford (County elections), pink stars at Lincoln, Grantham, Stamford and Boston (two members each), a blue circle at Grimsby (one member), and pink Maltese crosses (polling places) at 19 towns. *NORTH/ DIVISION* and *SOUTH/DIVISION* have been added.

*Smith's New English Atlas . . . 1832 . . .*

**Copy**   CUL Atlas 2.83.10.

(**xvi**)   The date in the title has been changed to *1834*.

*Smith's New English atlas . . . 1834 . . .*

**Copy**   (LRL Map 190a).

Issued in a green card holder. A sticker below the map's title reads: *Sold by J. Wyld, Geographer to His Majesty/CHARING CROSS, EAST/nearly opposite Northumberland House/LONDON./* The holder has a label with *LINCOLN* in the centre, surrounded by details similar to those on the sticker on the map.[14]

**Copy**   BL Maps 28.a.32.

*Smith's New English Atlas . . . 1834 . . . [1835].*[15]

**Copy**   RGS 1 C 72.

(**xvii**)   The date in the title has been changed to *1838*.

*Smith's New English Atlas . . . [1839].*[16]

**Copy**   Bristol RL W.51.

*Smith's New English Atlas . . . [1841].* [17]

**Copy**   The private collection of Mr. E. Burden.

(**xviii**)   The date in the title has been altered to *1843*.

Issued as a mounted and folded map. A label recording: *Sold by J. Wyld . . .* (as above) covers much of the map's title but the wording remains clearly visible. [1843].

**Copy**   Grimsby RL X000:912 SMI pamphlet.

(**xix**)   The date in the title changed to *1844*.[18]

(**xx**)   The date in the title has been altered to *1846*.[19]

## LITHOGRAPHIC TRANSFERS
(**A**)   State (xii) with the phrase *Corrected to* and the date removed from the title. Railways 1, 3–4, 6–17 and the Spalding section only of iv have been marked by barred lines, in often erroneous positions. The ANBEJR extends beyond Grantham to Boston via Folkingham.

Issued in a dark green cover with *LINCOLNSHIRE* in gold on the front. [1853**].

**Copy**   Bod. Johnson Maps 61.

(**B**)   State (A) has been lithographically transferred. Railway 18 has been added and the southern part of iv removed. *Railroads* and their symbol have been added to the Explanation. Several lines have been redrawn nearer to their correct positions, e.g. the ELR now passes through Firsby and the MSLR through *Market Raisin*; there is a spur at Ulceby where the lines from Grimsby and [New Holland] join. A solid black line has been overdrawn on all railway lines in state A.

Issued in cardboard cover with a James Wyld standard label (see xvi) with another superimposed label with *LINCOLN-SH.* [1853**].

**Copy**   LRL Map 872.

**(C)**   The map in state (xii) has been transferred, with a new title: *MAP OF/ LINCOLN/SHEWING ALL THE/RAILWAYS & STATIONS/and the Parliamentary/Divisions/LONDON./SMITH & SON, 63 CHARING CROSS./* (Ea). *EXPLANATION/* (Ee) has been redesigned, with the third line showing the symbols for railways (solid single black lines) and stations (large black circles); broken black lines are used for railways *in construction*. The symbols for parks, canals and navigable rivers have been removed. The two-line reference to the figures on turnpike roads has been replaced by the same words on three shorter lines. The *Explanation of Signs* continues (Be) with new symbols for parliamentary data; the Maltese crosses have been removed from the map. *POLLING PLACES./* followed by two columns listing the places under the three county divisions has been added (Ac). A new imprint reads: *London, Published by Smith & Son, 63 Charing Cross.* (Ce, OS). Railways 2, 19–28 have been added with the names of the railways; railways 29–31 and 33–34 are shown as 'under construction'; line 31 is marked from Sutton S$^t$. Mary (instead of [Sutton Bridge]). The railway from Grantham-Boston has been removed.

Issued in a green and cream cover with the cover title: *SMITH & SON'S NEW SERIES OF COUNTY MAPS. LINCOLNSHIRE. SMITH & SON, MAP & GLOBE PUBLISHERS, 63, CHARING CROSS. S.W. SIXPENCE.* [1866**].

**Copy**   CUL Maps c.34.07.70.

Issued as a loose sheet. [1866**].

**Copies**   BL Map 3355 (34); Bod Maps C 17:39 (4).

### References
1   Maxted, p. 201.
2   BL MT. 6.a.1.
3   The Lincolnshire plate in the set of parts is watermarked 1802.
4   The Lincolnshire map is watermarked 1809.
5   Lincolnshire and several other maps appear on paper watermarked J. WHATMAN/1812.
6   Pages watermarked J. WHATMAN/1820.
7   Only three boxes have survived. Since the fourth box contains counties N-S Lincolnshire must have been in the third box.
8   Most maps are watermarked 1822.
9   Although the paper used for the Lincolnshire map is watermarked 1821, several plates are watermarked 1823.
10   The Yorkshire plate is watermarked 1824. In this volume all the plates are folded so that the book is 'quarto' size.
11   Maps with watermark dates of 1825 are included.
12   In the 1827 *New Pocket Companion* (also issued by Smith) an advert leaf refers to 'single-sheet maps of all the counties 3s.6d. each. In Cases for Travelling 5s. Ditto Smaller size, 1s. each'. The map described here is, therefore, the second of the forms offered. The smaller must refer to Smith's later work (**82**).
13   This example has maps variously dated 1818–1830 and watermarked 1825–1830. The four maps dated and watermarked 1830 are Chester, Dorset, Buckingham and Middlesex.
14   For information on the Wyld family see Smith, D. The Wyld family firm. *Map Collector.* No. 55 (Summer, 1991), pp. 32–38.
15   The general map and twelve county maps are dated 1835. The squares (see 1832 editions) are here coloured purple, the Maltese crosses orange and the stars purple. Chubb recorded only one copy of an atlas dated 1834 – then in the possession of Sir George Fordham. The copy in question is now in RGS and, as Chubb's own description makes clear, could only have been put together in 1835 or shortly after.
16   The general map is dated 1839; others vary between 1832–1838.

17  The maps of the counties of Chester and Wiltshire are dated 1841.
18  Kingsley, p. 119 records a loose sheet from an unrecorded atlas of Sussex with the date altered to 1844. Chambers, p. 94 records an example of Bedfordshire issued as a folding map dated 1844. Lincolnshire may exist in either or both states.
19  Hodson, 1978, p. 81 records an example of Hertfordshire (in BL) with the date altered to 1846, issued as a folding map, sold separately. Michael, D.P.M. The Mapping of Monmouthshire (Bristol, 1985), p. 52 lists 1846 as the year of issue of one of that county's maps. Chubb (CCCXVII) assumed an atlas was issued in 1846 as he knew of several loose sheets dated thus. Lincolnshire may exist thus.

Chubb CCCXI–CCCXVII.


# 57

## JOHN CARY                                                                 1801

Size: 54.5cm x 48.8cm.

Cary's largest map of Lincolnshire was part of a series, which culminated in 1809 in *Cary's New English Atlas*. Only slightly larger than the plate for Camden's *Britannia* it is very closely related to it, perpetuating one or two errors from his earlier maps - the inclusion of *Hughington* (near Heighington) and showing the 0 meridian as 1° west of Greenwich.

The maps were issued in parts originally advertised at 5/6 each. The 1806 edition of *Cary's New Itinerary* refers to the availability of the parts at 7s. a part or 8s. fully coloured. The notice further reports: 'Ten Numbers Are Published; The Whole To Completed (sic) In Fifteen.' The succeeding list of the first ten numbers shows that Lincolnshire formed part of the first issue together with the maps of Bedford and Buckinghamshire. The single maps were then 3/6 each. Lincolnshire's map is dated Sept. 28 1801 but so were twenty-two others. Northamptonshire was dated 21 Dec. 1801, but the next date to appear on any plate is 1 Nov., 1805. It appears likely that Cary originally decided to give all the plates the same date (as he had with the *New And Correct English Atlas* in 1787) and then perhaps changed his mind when half way through and began dating the plates as the engraving was completed. The 1806 notice indicates that the issue of ten parts had taken five years, since the three plates forming part ten are dated July, 1806. A further three are dated November of that year (part 11) with another three dated 1 March, 1807 (part 12). After March, 1807 there is a single plate in the following November, two together in June, 1808, the two plates covering Wales in Feb., 1809 and the concluding general map dated 1 June, 1809. Yorkshire was covered in four plates thus bringing up the complement to the proposed 45 map sheets. Presumably the volume appeared shortly after the later date.

The probable explanation for the delay in preparing the final five plates (1808–9) is that Cary had been very busy, amidst other activities he was re-engraving plates for a new version of *Cary's Traveller's Companion* (1806) (**61**). In 1808–09 he was also preparing for a revised *New and Correct English Atlas* (**65**).

The title of the atlas is used by Cary to advertise his usefulness to the Postmaster-General in surveying the country's roads and, therefore, the quality of his maps. The numerous changes on the plate between 1801 and 1809 suggest careless preparation since some corrections made in the earlier states repeat some of those which were necessary at the preparatory stage of the 1789 edition of *Britannia*. See also Addenda, pp. 411–12.

### Later history
The atlas appeared at various intervals until 1834 the year before Cary died. Although

one further atlas in the Cary form appeared in 1842 there were two issues in 1835 and 1836 in *Lincolnshire in 1835* and *Lincolnshire in 1836*, and the plates, curiously and inexplicably, have the name of Greenwood of Hull as the engraver. Saunders had no access to a county plate at the time of this publication and the availability of Cary's plate must have come opportunely for him. John Greenwood appears to be unrelated to the Greenwood brothers whose work is noted below (**87**). He set up in c. 1830 in Hull, where he was listed as an engraver on copper and wood and a copper-plate printer. By 1840 he had ceased to operate. Curiously although he calls himself engraver he must have had the maps lithographically transferred although he was always described as a copper-plate engraver and letter-press printer. In 1832 he wrote *The Trent And Humber Picturesque Steam-Packet Companion . . . by J. Greenwood.* This had no map, but the second edition (1833) included one engraved by himself, which shows the Humber and lands on either side, as far south as Gainsborough.[1] His most notable work is *Pictures of Hull with 70 illustrations* (1835). Thereafter, Cary's plates were among much else that George Cruchley acquired from Cary's heirs. G.F. Cruchley (1796–1880) had trained with Aaron Arrowsmith, founder of a noted form of engravers and set up by himself in 1823; he produced a number of attractive maps mostly of London and its environs.[2] Apart from two or three atlas editions the plates were largely issued as single folded maps in a variety of covers. A curiosity of all the transfers is that Grantham is shown with two railway stations throughout all issues of the plate; the town did have two stations but once the GNR became established (after 1852) the present site developed as the station for all the converging lines. Conversely, Lincoln, which did have two stations throughout the period of these maps, is only shown as having one. The two stations at Stamford were always properly marked.

Although Cruchley announced on the separately issued maps that they were available 'geologically coloured' no example of Lincolnshire has been found. Use of the plates passed to the Edinburgh firm of Gall and Inglis, following an auction of Cruchley's stock in 1877. They followed Cruchley's practice of issuing single maps in covers for intending railway travellers but at the end of the century they had begun to issue some of the first cycling maps (usually with Cruchley's name on the covers to capitalise on a well-known trade name). Gall and Inglis remained for many years in business as map providers.[3]

## EDITIONS

(i)  *A/NEW MAP/OF LINCOLNSHIRE,/WITH ITS GREAT, AND/SUB DIVISIONS INTO HUNDREDS,/EXHIBITING/Its Roads, Rivers, Parks &c./By JOHN CARY, Engraver,/1801.* (Ae, in an oval). *SCALE/10 Miles* [= 6.45cm] (Ee). *London: Published by J. Cary Engraver & Mapseller, N°. 181 Strand, Sept. 28 1801.* (Ce, OS). Compass (Ed). *REFERENCES/to the/HUNDREDS, or WAPONTAKES.*/followed by 31 names separated into the three county divisions (Ea). The inner frame is shaded to mark each minute of latitude, etc., with numbering at 5′ gaps. *1° Longitude West from Greenwich* (De, between the inner and outer frames).

*Cary's New English Atlas; Being A Complete Set of County Maps, from actual Surveys, Corresponding In Size With His General Atlas; On which are particularly Delineated Those Roads which were measured by Order of the Right Honourable the Postmaster General, By John Cary, As well as all others, both Direct and Cross; Rivers, Navigable Canals, Parks, Heaths, Commons, &c.&c. Exhibiting Also The whole of the Market and Borough Towns, Parishes and Hamlets, as well as Places of inferior note . . . London: Printed for J. Cary, Engraver and Map-seller, N°. 181 near Norfolk Street, Strand 1809.*

**Copies**   CUL Atlas 2.80.3; Bod Allen 67; Birmingham RL E912.42 Skett 1; Cardiff RL G9.8.

**(ii)** *Botsford* has been changed to *Bottesford.*

*Cary's New English Atlas . . . 1809.*

**Copies** BL Maps 13 f 16; Leeds UL W 115; NLS Newman 362. (LRL Map 753; Gainsborough RL Map 6).

*Cary's New English Atlas . . . 1811.*

**Copy** Cardiff RL 940 Print Room.

**(iii)** The Morton-Edenham road and the southern spur to Lound have been reduced from a major to a minor road, while the road from Bourn to Edenham has been upgraded to main road status.

*Cary's New English Atlas . . . 1809.*

**Copies** London UL xf 53 KVC 1 Car. (Grantham RL Map 7).

**(iv)** Place-names have been changed: *Kirton (Kirkton* (in Holland) previously), *Hacconby (Heckingby).* Donington and Folkingham have been re-engraved in upper case letters (to indicate market towns), while Bolingbroke has been altered from upper to lower case. *Overthorp, Parks* and *Nethergate* have been added (all west of Haxey). The roads from Threekingham-Grantham and Irby-Grimsby (now passing through the B of Bradley and having milestones 15 and 17 only) have been upgraded to post roads, while the roads from Threekingham-Folkingham and Wragby-N. Willingham and the spur to Hainton have been reduced to minor roads.

*Cary's New English Atlas . . . 1809.*

**Copies** Brighton RL 912.42 C25 Bas Fol; Middlesbrough RL. (BL Map 3355 (39)).

**(v)** Place-names have been corrected: *Kadby* (= Keadby, previously *Headby*), *Doddington (Dodington), Thorpe Tinley* (sic – previously two separated places *Thorpe* and *Tilney*), *Boultham (Boatham), Aisthorpe (Aistrope), Scothorn (Scottern), Kirmington (Kermington)* and *N^{th} Cockerington* and *S^{th} Cockerington* (the final syllable was lacking before). *Raithby* has been added (south-west of Louth). In the list of wapentakes *Alsacoe* has been corrected to *Aslacoe* and *Grantham* from *Crantham. G^t.* has been added above *GRIMSBY. Wisby* has been deleted. The *Stainforth/ & Keadby Canal* and the unnamed Caistor Canal have been added. The hachur on the road from Boston to Stickford has been removed.

*Cary's New English Atlas . . . 1809.*

**Copies** RGS 1 B 97. (Northamptonshire AO Map No. 1238).

*Cary's New English Atlas . . . 1811.*

**Copy** Bod Allen 74.

**(vi)** The date in the title has been altered to *1811.* The date in the imprint has been altered to *Apr. 28 1811.* The meridian has been altered to *0* in the upper and lower margins, but *East from Greenwich* remains between the lower frames. *II°* has been deleted (Ae) and *I° Longitude West from Greenwich* added (Be, between the inner and outer frames). The remains of a church symbol at Bottesford – largely removed in state (ii) – have been fully taken out.

*Cary's New English Atlas . . . 1811.*[4]

**Copies** BL Maps 13.f.18; CUL Atlas 2.81.7; Leeds UL 115a.

*Cary's New English Atlas . . . 1811.* [1812].[5]

**Copy**   RGS 1 B 98.

(**vii**)   *Billingham* has been corrected to *Billinghay. Thornton Curlis* now appears correctly as *Thornton Curtis. Ancholme Level* has been added alongside the canal (north of Brigg).

*Cary's New English Atlas . . . 1811.* [1812].[6]

**Copy**   Bod Allen 75.

(**viii**)   The date in the title is now *1818.* The date in the imprint now reads: *Jan*[y]. *1*[st]. *1818.*

*Cary's New English Atlas . . . 1818.*

**Copies**   CUL Atlas 2.81.6; RGS 1 B 99; Wisbech Museum. Press 2.72.

(**ix**)   The date in the title now reads: *1820.* The date in the imprint has been removed. Notes on the geological strata have been added around the edges of the map together with oblong boxes.[7] *Note The shaded part represents the Marshes, and/Fenland, formed over the regular Stratification/by deposits from the Sea, and the Vegetables of/ Peat, which is called Black Land/* added (Ee, above scale).

*Cary's New English Atlas . . . 1820.*[8]

**Copy**   (LRL Map 120).

(**x**)   The date in the title has been altered to: *1821.*

*Cary's New English Atlas . . . 1821.*

**Copies**   RGS 1 B 100. (Nottingham UL Li 1.B8.E 21).

(**xi**)   The date in the title now reads: *1824.*

*Cary's New English Atlas . . . 1824.*

**Copies**   West Sussex CRO. PM 35. (Grimsby RL CARY 1824).

*Cary's New English Atlas . . . 1824.* [1825]. [9]

**Copy**   West Sussex CRO PM 34.

Issued as a folded map mounted on linen in a cardboard holder labelled *LINCOLNSHIRE.* [c.1824].

**Copy**   CUL Maps d.18.G.65.

(**xii**)   The date changed in the title.[10]

(**xiii**)   The date in the title has been changed to *1828.*

*Cary's New English Atlas . . . Strand. Corrected with Additions to the Year 1828.*

**Copy**   BL Maps 24.e.34.

*Cary's New English Atlas . . . 1828.* [1831].[11]

**Copy**   York RL 912.

(**xiv**)   The date in the title has been altered to *1831.* The area of Fosdike Wash has been re-drawn, to allow the inclusion of a new (spurious) main road from Kirton to Holbeach. A road from Spalding to Bourn has been added.

Issued as a folded, mounted map in a green mottled cover, with a label: *CARY'S/NEW*

*MAP/OF THE COUNTY OF/ LINCOLN,/DIVIDED INTO HUNDREDS;/*[6 further lines]/ *LONDON:/PUBLISHED BY G. AND J. CARY,/86, ST. JAMES'S STREET./*

**Copies** Bod Allen 238; Nottingham UL Dept. of Geography Map drawer 121 (Map F 133); LAO 3 Anc 5/90 and 3 Anc 5/113.

(**xv**) The area of East and West Fen has been redrawn, with many new roads added, e.g. Coningsby-Stickney, Tumby-Spilsby, Revesby-Westby Bridge, Coningsby-Sibsey with a spur to join the new road from Stickney, etc. Other roads upgraded to main road status include: Bratoft-Skegness, Kyme Tower-Wainfleet, Scartho-Louth, Wragby-Louth, via Hainton, with a connection to N. Willingham. The symbols for fens have been replaced by marsh symbols. *New/Bolingbroke* has been added. *EXPLANATION./*[8 lines of symbols and their meanings] (Aa). Maltese crosses have been added at 15 places; Maltese crosses with an arrow symbol have been added at Lincoln and Sleaford. Stars have been added: Grimsby (1) and Lincoln (2) indicating the number of MPs.

*Cary's New English Atlas . . . 1834.*

**Copy** BL Maps 20.e.26.

(**xvi**) The date in the title has been altered to *1842*. Stars have been added at Grantham.

Issued as a separate folded map with an index tab, in boxes with other counties. Lincolnshire is numbered 21 on the filing tab. [1842].

**Copy** BL Maps C.29.a.6.

## LITHOGRAPHIC TRANSFERS

(**A**) The base map is state (vi), without the name of Ancholme Level. The original title has been removed (Ae) and replaced by *LINCOLNSHIRE/1836* (Aa). The *REFERENCES/to the/HUNDREDS or WAPENTAKES/* has been transferred from (Ea) to (Ae) and appear as two short columns of names instead of one longer one. An imprint has been added: *LONDON & LINCOLN. J. SAUNDERS, JUN$^R$ 1836.* A signature has been added: *Engraved by J. Greenwood, Hull* (Ee, OS). The Scale does not appear now in a shaded panel (Ee). *S.E. VIEW OF LINCOLN CATHEDRAL.* (below an engraving, Ea). The geological notes and rectangles have all been removed. The following places have been added: *Kingston upon Hull, R. Ouse, New Holland, Sutton/Bridge Hotel, Sutton Bridge/Embankment, Whaplode/Fen, Cowbit/Fen, Moulton/Fen, Guthlacs Cross, Boston Deeps, Tetney Haven, Donna Nook, Gibraltar P$^t$.* Other changes include: a bridge across *Cross Keys Wash* (re-engraved), labelled *To Lynn*, with the Nene outlet redrawn; *THE RIVER HUMBER* has been replaced by *RIVER HUMBER* in larger letters, with the addition of two ferries to Hull and the mileages; *Insthorpe/Oakes* and the associated woods have been removed and *Pickworth* put in their place; *Muston* and its church sign have been re-engraved, with a simple cross; *MARKET/RASEN* (formerly *MARKET/ RAISIN*) has been re-engraved. Several roads previously upgraded, e.g. the Louth-Thoresby-Grimsby and Wainfleet-Skegness are in their original state.

*Lincolnshire In 1835: with the Rivers Humber, Trent & Witham Displayed in A Series Of Views Accompanied By An Explanatory And Illustrative Description. London: Published By John Saunders, Jun. and Sold by all Booksellers in the County. 1835.*

**Copies** BL 796 i 12; Sheffield RL 942.53T; LAO SR 942.08 and FL (PM) L 942.075. (LAO Sc 942.52; Stamford RL – bound into their copy of Howlett's *A Selection Of Views Of The County Of Lincoln . . .* 1805).

(**B**) *1836* has been removed from the title (Aa). New roads have been added: Revesby-Westley Bridge; Coningsby-Stickney, Fosdyke-Holbeach, etc. New place-names include: *Eastville/Chapel, Butterwick/ Fen, Hunstans/Charity, Leakefold/Hill, Wrangle/ Lowgate, New Bolingbroke, Carrington, Carrington Ho., Cowbit/Fen, S$^t$ Guthlacs Cross* and *Fenside* has been added at both *Sibsey* and *Norlands* (both north-west of Sibsey village). Other changes include *FOLKINGHAM* (formerly *FALKINGHAM*), *Essendine* (*Easingdon*), *Ryhal* (*Ryal*), *Pode/Hall* (*Podge Hall*), *Washingbrough* (*Washingburg*), *Tydd S$^t$ Mary's* (*S$^t$ Marys Tid*), *Wash/Way* (*Wash/House*) and *Martin* (*Merton*); the removal of *Tucklac's/Cross* (sic) and *Hughington*; the redrawing of Fossdyke Wash to allow the new road to be drawn in and redrawing the Nene estuary, which has led to the replacement of *Cross/Keys* by *Sutton/Bridge Hotel* and the re-engraving of *Cross Keys Wash* (all on the north side of the bridge).

*Lincolnshire In 1836: Displayed In A Series Of Nearly One Hundred Engravings, On Steel And Wood; With Accompanying Descriptions, Statistical And Other Important Information, Maps, &c. Lincoln, Published By John Saunders, Jun. MDCCCXXXVI.*

**Copies**   Bod G.A. Lincs 4$^0$ 52; LAO Sc 942.52; Louth L 907; Gainsborough RL L 9.

(**C**)  The base map is state (xiv). The new title is: *CRUCHLEY'S/MAP OF/LINCOLN,/Showing all the/ RAILWAYS & NAMES OF STATIONS,/ ALSO THE/TELEGRAPH LINES & STATIONS,/Improved from the/ ORDNANCE SURVEYS./ THIS MAP MAY BE HAD GEOLOGICALLY COLOURED./ PRICE 3/6. IN SHEET./ LONDON. PUBLISHED BY G.F. CRUCHLEY, MAP-SELLER & GLOBE MAKER, 81, FLEET STREET.* (Ae). Railways 1–18 and J have been added as black lines with large black dots for stations with railways and stations named in a sans-serif typeface. Railway v consists of two parallel lines, suggesting it was then under construction. The Nottingham-Grantham line is shown terminating at the Wharf station in Grantham. Two lines have been added to the Explanation relating to railway and telegraph symbols. The dots for stations with telegraph facilities have 'sun rays' projecting from them. Several places outside the county have been added: *RETFORD, NOTTINGHAM, DONCASTER, KINGSTON upon HULL, WISBEACH, LYN REGIS* (sic), *SOUTHWELL* and *PEAKIRK. Pier* and *Goxhill/Ferry* have been marked in the Humber, near New Holland.

Issued in a brown-red marbled cover bearing the title *LINCOLNSHIRE* in blue letters. [1854**].

**Copy**   LRL Map 236.

From an undiscovered atlas. [1854?].

**Copy**   (BL 3355 (31)).

(**D**)   Railways 19–24 have been drawn in. The line to Sleaford is now shown properly as a curve from the GNR at Barkston instead of making a junction south of Grantham with the GNR. Railway v has been filled in as a solid line. The line to Holbeach is wrongly aligned, curving north-eastwards from Spalding and crossing the main road from Spalding twice. A second N has been crudely inserted at *LYN REGIS.*

Issued in a card cover with the title: *CRUCHLEY'S RAILWAY AND TELEGRAPHIC County Map of LINCOLN . . . SIXPENCE COLORED . . . G.F. CRUCHLEY . . . 81, FLEET STREET . . .* [1857**].

**Copy**   BL Maps 197.a.64.

Issued with the title *CRUCHLEY'S RAILWAY & TELEGRAPHIC MAP OF LINCOLNSHIRE Showing all the RAILWAYS & NAMES OF STATIONS . . .* and dated (Ae, OS). The card cover has the title: *CRUCHLEY'S RAILWAY MAP OF THE COUNTY OF LINCOLNSHIRE SHOWING ALL THE RAILWAYS AND NAMES OF THE STATIONS, IMPROVED FROM THE ORDNANCE SURVEY . . . SIXPENCE COLORED. PUBLISHED BY G.F. CRUCHLEY . . .*[12]

Issued with the title *CRUCHLEY'S RAILWAY MAP OF LINCOLNSHIRE* and dated (Ae, OS). The cover title is *CRUCHLEY'S RAILWAY MAP OF THE COUNTY OF LINCOLN-SHIRE SHOWING ALL THE RAILWAYS AND NAMES OF THE STATIONS* . . .[13]

(E)   Plate no. *21* has been added (Ee, OS, sidewards on). Railway 24 has been re-drawn so that there is a continuous line at Sleaford, where, previously, a short spur terminated in Sleaford from the 'main' line which passed Sleaford to the south. *S$^t$.JAMES/DEEPING STA.* has been added; Stallingborough station has now been named and Thornton re-named *THORNTON/AB$^Y$ STA.* The Horncastle line has been redrawn; it terminates south of Thornton and passes through the u of Roughton instead of the initial capital. The Edenham line has been redrawn (although still in error); it curves away from the GNR south-eastwards from Careby. A spur has been added at Barkston from the GNR to the Boston line.

*Cruchley's Railway And Telegraphic County Atlas Of England And Wales. London: Published by G.F. Cruchley, Map Publisher, And Globe Manufacturer, 81, Fleet Street.* [1858].[14]

**Copy**   BL Maps 18.d.17.

(F)   Railways 26–27 have been added. Line 23/27 has been redrawn; it curves south from Spalding and remains south of the road to [Sutton Bridge]. The stations on the new lines are marked but not named.

Issued in a card cover which bears the title: *CRUCHLEY'S RAILWAY & TELEGRAPHIC MAP OF LINCOLN, Showing all the RAILWAYS & NAMES OF STATIONS, ALSO THE TELEGRAPHIC LINES & STATIONS, Improved from the ORDNANCE SURVEYS. THIS MAP MAY BE HAD GEOLOGICALLY COLOURED PRICE 3/6 IN SHEET. LONDON. PUBLISHED BY G.F. CRUCHLEY, MAP SELLER & GLOBE MAKER, 81, FLEET STREET.* [1862**].

**Copy**   CUL Maps 34.045.70.

(G)   The map is in state (A) with a new title. *MORRIS & C$^{O'S}$./Railway & telegraphic Map/OF/ LINCOLNSHIRE,/PUBLISHED WITH THEIR/ GAZETTEER & DIREC-TORY,/OF THE COUNTY,/HOUNDSGATE, NOTTINGHAM./1863./* (Ae). Railways 25, 28, C, F–G and K have been added. Lines 29–30, 32–33 are shown as being built. The name *LYNN REGIS* has been redrawn. Line 31 is shown as under construction but from Long Sutton. The SYR at Keadby has been extended westwards to Doncaster. Stations have been named on all the newly opened lines except at Long Sutton. The notes on geology have been removed except for one rectangle and two lines of text near Grimsby. In the list of Hundreds (Ea) *Soke* has been added after Grantham, Winnibriggs has been re-engraved as *Winnibr$^{gs}$. & Threo, Haverstoe* has been added to Bradley, *Loutheske* has been redrawn as *LouthEske, Graffo* added to Boothby, *Lango* has been extended to *Langoe* and *Nefse* reduced to *Nefs*. The new title has meant that railway 5 has been shortened. Railway v and the plate number have been removed. The *O* in *WAPONTAKES* has been replaced with *E* (Ea).

Although the title on the map suggests that the map was issued with *Morris & Co's Gazetteer* . . . it is only known as a single loose copy.[15]

**Copy**   CUL Maps bbb.11.

(H)   The map in state (F) with the changes to the list of Hundreds and most of the railway changes of state (G) added. The line from Stamford towards Melton Mowbray is the elongated form of state (F). The geological data re-appear. Railways 29 and 32 are now solid lines to show they had opened. The plate number has been restored. The panel in which the scale appears (Ee) has been redrawn; it now measures 8.65cm instead of 8.45cm.

Issued in a card cover with the title: *CRUCHLEY'S RAILWAY AND TELEGRAPHIC County Map of LINCOLN. N.B. These excellent County Maps . . . SIXPENCE COLORED . . . G.F. CRUCHLEY, Map Publisher and Globe Manufacturer, 81, FLEET STREET . . .* [1866**].

**Copy**   LRL Map 190.

**(I)**   The title has been changed to: *CRUCHLEY'S/RAILWAY AND STATION/MAP OF/LINCOLN,/Showing all the/RAILWAYS & NAMES OF STATIONS,/ALSO THE TURNPIKE ROADS, GENTLEMENS SEATS, &c&c/Improved from the/ ORDNANCE SURVEYS./THIS MAP MAY BE HAD GEOLOGICALLY COLOURED./PRICE 3/6 IN SHEET./LONDON. PUBLISHED BY G.F. CRUCHLEY, MAP SELLER & GLOBE MAKER, 81, FLEET STREET/* (Ae). A further line has been added to the *EXPLANATION* – a symbol for *RAILWAYS IN PROGRESS*. Railways 30, 33, 35 and 36 are now shown as completed. Railways 34, 39 and xv are shown as 'in progress'; railway 31 continues to be shown as reaching Wisbeach from Long Sutton and 'in progress' while the Wisbech-Peterborough section is marked as operational; railway 37 is shown leaving Firsby in the direction of Spilsby and 'in progress' but after Holton Holgate the symbols almost disappear. *NORTH/ DIVISION, MID/DIVISION* and *SOUTH/DIVISION* have been added.

Issued in a card cover with a label: *CRUCHLEY'S MAP OF THE COUNTY OF LINCOLN, SHOWING THE NEW PARLIAMENTARY DIVISIONS, CHIEF PLACES OF COUNTY ELECTION; ALSO THE POLLING PLACES, WITH THE RAILWAYS & STATIONS. SIXPENCE COLOURED . . . CRUCHLEY, 81, FLEET STREET.* [1868].[16]

**Copy**   Grimsby RL X000:912 CRU pamphlet.

**(J)**   The title has been moved (from Ac to Ea) and the list of wapentakes removed entirely. The scale (without its enclosing panel) has been removed to just below the title and reads: *10 Miles* [= 6.35cm]. *EXPLANATION* has been moved (from Aa to Ae). Railways 31, 34, 37–40 are now shown as in use, with marks from Long Sutton to Wisbeach removed. Railways 41, 43–44, xvi–xvii are shown as in progress, while railway xv is now shown by two parallel lines.

Bound in a card cover with the title: *CRUCHLEY'S RAILWAY AND STATION MAP OF THE COUNTY OF LINCOLN. SIXPENCE COLORED. N.B. These superior County Maps . . . CRUCHLEY, Map & Globe Maker, 81, Fleet Street, LONDON, E.C . . .* [1874**].

**Copy**   Grimsby RL X000:912 CRU.

**(K)**   The title has been revised. The new one follows that of state (I) . . . *ORDNANCE SURVEYS/LONDON: PUBLISHED BY GALL & INGLIS, 25, PATERNOSTER SQUARE./EDINBURGH. 6 GEORGE STREET./* Railways 41 and 43 are shown as open, with stations named. Railways xvi and xvii are marked by two parallel lines. The scale (Ea) is now enclosed in a frame with vertical shading. Railways D and N have been added. *GEDNEY HILL STATION* appears on the Spalding-March line south of Postland.[17]

Issued in a card cover. The plain white label on the cover has *Lincoln* written by hand. [1877].[18]

**Copy**   The author's collection.

Issued in a card cover with the cover title as in state (J) except that the words *IMPROVED EDITION.* have been printed at the top. [1877].

**Copy**   CUL Maps 34.046.70.

*Cruchley's Railway And Telegraphic County Atlas Of England And Wales. London: Published By G.F. Cruchley, Map Publisher, And Globe Manufacturer, 81, Fleet Street.* [1879].[19]

**Copy**   Leeds RL F912.42 C887.

(**L**)   Railway 44 has been added with dots for the (unnamed) stations; railway xvi has been removed.

Issued in a card cover with cover title as in state (J) except that Cruchley's name and address have been replaced by: *Gall & Inglis. Edinburgh: Bernard Ter. London: 25 Paternoster Sq.* [1880**].

**Copy**   Bod Map C17: 39 (17).

(**M**)   Railways 42, 45–46 are marked in. A southern avoiding railway line has been drawn at Lincoln. *STA.* has been added to the four stations on the Louth-Mablethorpe line, two of which have been resited. Railway D has been removed. The map is surrounded by advertisements, those along the bottom being upside down in relation to the map. *GEDNEY HILL/STA* has been removed.

*The 'Finger Post' Bicycle Road Guide, Containing The main Roads Of England, Wales, And Part Of Scotland, Specially Compiled For The Use Of Bicyclists And Tourists. Accompanied With Road Map. Separate Maps of all the counties in England . . . may be had uniform with this Map, each containing the Road Guide, or the Maps can be had separate, price 1s. each. To be obtained . . . E. Harrison & Co., The West End Bicycle And Athletic Outfitters, 259 Oxford Street, London . . .* [1883**].

**Copy**   BL Maps 9.a.21.

Issued in a card cover with the title: *CRUCHLEY'S RAILWAY AND STATION MAP OF THE COUNTY OF LINCOLN. SIXPENCE COLORED. NOTE. – These superior County Maps . . . Gall & Inglis. Edinburgh: Bernard Ter. London: 25 Paternoster Sq.* [1883].[20]

**Copy**   BL Maps 197.a.65.

It is possible that the same map appeared in atlas form. A copy of *Cruchley's Railway and Telegraphic County Atlas of England and Wales* (dated in a contemporary manuscript hand 1883) is recorded.[21]

(**N**)   The map has been overprinted with the names of the new parliamentary divisions following the Representation of the People Act of 1885 in the form *HOLLAND/OR SPALDING*. The stars (showing the number of MPs) have been reduced to one each at Grimsby and Lincoln. In *EXPLANATION* a line *Boroughs returning Two Members None* has been added. Railways 48–49 have been added. *G 3* added (Ae, OS).

*Cruchley's Railway And Telegraphic County Atlas . . .* [Actually issued by Gall & Inglis]. [1888**].

**Copy**   The private collection of Mr. D. Hodson.

Later examples of the Lincolnshire plate issued in card covers may exist. Maps exist with the following or variant cover titles: (a) *Cruchley's County Map Of . . .*; (b) *Price Sixpence. Cruchley's County Map Of . . . for cyclists, tourists, &c Coloured. Gall & Inglis . . .* (c) *Cruchley's County Maps Of England . . . For Cyclists, Tourists, &c.* [22]

Such maps are also known with a variety of code numbers (Ae, OS).

**References**

1   Copies: Grimsby RL X000:942 GRE; Louth RL GL 9. The map and the booklet were issued by J. Drury of Gainsborough, perhaps a relative of J.W. Drury, the Lincoln printer. The 1833 edition was reissued as *Malet Lambert Local History Reprints, No. 13* (Hull, [1981]).
2   Smith, D. George Frederick Cruchley 1796–1880 *Map Collector* No. 49 (Autumn, 1989), pp. 16–22; Darlington, I. and Howgego, J. op. cit., catalogue items between nos. 299 and 382 (not consecutively).
3   Gall & Inglis, publishers 1818–1960 (cover title). (Edinburgh, 1960).
4   The examples in BL and Leeds both lack title-pages, but the Lincolnshire maps are similar to that in CUL Atlas 2.81.7.
5   The title-page and some maps are watermarked 1812.
6   The title-page and maps (including Lincs.) are watermarked 1812.
7   21 counties (in six parts) appeared with various colours showing the geological strata and keys (with numbers) added round the edges of the map area. They are variously dated 1819–1824 and incorporate the studies of William Smith, the 'father' of English geology. The maps were entitled: *Geological Map Of* . . . Lincolnshire did not appear in this form but the geological notes in numbered boxes date from this period.
8   No atlas dated thus has yet been found.
9   Several maps have been redated 1825.
10  In York RL (Y 912) is a copy of Yorkshire dated 1825; Berkshire is known dated 1826 (Burden, p. 83). Lincolnshire may exist dated thus.
11  The Lincolnshire map has a watermark date of 1830 while several plates have been redated 1831.
12  Lancashire, dated October 1$^{st}$. 1857, is held by Manchester RL – 385.094272 Cr 1. Other counties are known in this state, but dated differently, e.g. Berkshire (BL Maps 1420 (7)), with 1$^{st}$ January, 1855.
13  Lincolnshire has not been found thus, but Staffordshire, dated January 1$^{st}$. 1855 is in John Rylands UL (R 108250).
14  Several plates are dated 1858.
15  There are copies of the *Directory* . . . in LRL (L 929), Stamford RL (L 929) and Grimsby RL (X000:058 MOR). Maps are absent and do not seem to be called for. Later editions (1866 and 1869) do not have maps either.
16  The cover title suggests the map was prepared just after the Second Reform Act of 1867. The railway data support 1867. Examples are known of the map with the title as quoted but in covers without the reference to the parliamentary divisions; in the case of the map of Cornwall (John Rylands UL (R108256)) the cover has the heading *IMPROVED EDITION*; Lincolnshire may exist thus. The Cornwall map may be dated to c.1865, since several railways are marked by broken lines to indicate proposed lines; e.g. the Launceston, Bodmin & Wadebridge Railway is marked, part of which was never built and much of the rest only opened 29 years after the Act in 1894.
17  From the opening in 1867 the station was known as French Drove and only became Gedney Hill officially in July, 1938. (Goode, C.J. The Great Northern and Great Eastern Joint Railway . . . (Hull, 1989), p. 27).
18  Gall & Inglis did not obtain the Cary/Cruchley plates till they were sold in 1877. Gall & Inglis occupied the George Street premises only until 1878 when the firm moved to Bernard Terrace. (Gall & Inglis, publishers. op. cit., p. 7).
19  Several commentators suggest 1879, based on railways in their counties.
20  This map was re-issued in a re-folded form in covers provided by Cook, Hammond & Kill, Ltd. 47 & 49 Tothill Street, Westminster, S.W. sometime after 1923.
21  Burden, p. 87, in the possession of Messrs. Tooley and Adams in August, 1992.
22  Of types (a) and (c) Mr. E. Burden notes examples of Berkshire with these or similar titles, the latter dated to c.1895; Kingsley, p. 321 notes a copy of Sussex of type (b), dated to c.1890. A similarly titled map of Hampshire is part of CUL Maps 34.049, which also has four other counties with variant cover titles.

Chubb CCCXXXIII–CCCXXXVII (atlas versions only 1809–1834).

## 58

**ROBERT BUTTERS** 1803

Size: 12.25cm x 8.9cm.

Robert Butters was a printer active in the Fleet Street area from 1785 to 1808.[1] The atlas is his only venture into this particular field. He was involved in the issue of the maps prepared for *The Political Magazine* . . . (**51**); although the earliest maps of that series were issued by John Murray, the thirteen plates issued from October, 1789 until the series was complete (December, 1790) came out under Butters' auspices. The present maps, based on Cary's maps for *Traveller's Companion* . . . (**50**) may have been Butters' attempt to repeat Cary's success.

John Hatchard (1769–1849), the publisher of *The Picture Of England*, set up as a bookseller in 1796, moving to 173 Piccadilly in the following year.[2] He moved to 190 Piccadilly in 1801, was joined by his son in the business in 1808 and they moved to the address at 187 Piccadilly in 1820, which the firm still occupies.

The maps had only a short life in the atlas and William Green's little book. Although Lincolnshire is printed with north at the top many other counties are oddly engraved; e.g. Sussex is engraved with the north to the right while the place-names are read normally from left to right; and, the general map of England and Wales, which acts as a frontispiece, is engraved upside down, i.e. the map has south at the top, but the county names are readable without turning the map.

### EDITION

(**i**)  *LINCOLNSHIRE*. (Be–De, OS). Compass (Ad). *Scale/10 Miles* [=1.4cm] (Ae).

*An Atlas Of England. Containing A Map Of England . . . London. Printed And Sold By R. Butters, No, 22, Fetter-Lane, Fleet-Street. [Price 6s. plain, and 10s. coloured.]* [1803].[3]

**Copy**   CUL Atlas 7.80.4.

*The Picture Of England Illustrated with correct colour'd maps of the Several Counties. By The Author of . . . London: printed for J. Hatchard, Bookseller to his Majesty, Piccadilly. 1803.*[4]

*The Picture Of England Illustrated with correct colour's Maps of the several Counties. In Two Volumes By William Green . . . Vol. II . . . London: Printed for J. Hatchard, Bookseller to his Majesty, Piccadilly. 1804.*

**Copies**   BL 797 f 11–12; Leeds UL W 108.

*An Atlas . . . 1805.*

**Copy**   BL Maps C.27.b.44.

**References**
1   Maxted, p. 36; Todd, p. 31.
2   Brown, P.A.H. London Publishers and Printers, c.1800–1870. (London, 1982), p. 85 suggests he set up directly in Piccadilly in 1797; Maxted, p. 105 records his first shop at 2 Monmouth Court in 1796 and provides other details; DNB.
3   It is generally assumed that the atlas appeared in 1803, perhaps simultaneously with *The Picture Of England* and certainly before the 1804 edition of that work, since several plates have undergone revision before 1804.
4   The only recorded copy was broken up; Mr T. Burgess obtained the Kent map and the title-page and Mr. E. Burden the Berkshire plate.

Chubb CCCX and CCCXVIII.

## 59

**JOHN LUFFMAN**                                                           **1803**

Size: A circle 6.1cm in diameter.

Luffman was a printer, engraver and publisher, who was in business from 1776 until 1820, in spite of a period around 1793 when he was a bankrupt.[1] He was responsible for the engraving of the map of Rutland by Armstrong, issued in 1781. The present curious circular maps had only a short life of about three years. Although used to illustrate a geographical textbook the value of the maps must have been small, since there is only space for a few towns and roads; there are no rivers or other physical features. A curved scale would be difficult to use also!

Lackington, Allen, which took over the publication in 1806 was founded in 1774 by James Lackington. Originally a shoemaker in Somerset he made his way to London after a period in Bristol, where he first started buying books at market stalls, because he was too ashamed to enter bookshops. In London a legacy of £10 enabled him to rent a shop, where he, at first, continued to sell shoes as well as books. Soon he concentrated on books, insisting on ready money only from his customers no matter their nobility of rank. He retired in 1798 and died in 1815. The firm remained in business until George Lackington retired in 1826. They listed over 800,000 books in their 1803 catalogue.[2]

### EDITIONS

(i)   *LINCOLNSHIRE* (top, between the inner and outer circular borders). *Sends 12 members to Parl.^mt* (left, between the frames). *Lincoln Co. Town 133 Miles/from London.* (right between the frames). *Scale of 20 miles* [= 0.95 cm approx.] (bottom, between the frames). *21* (centre above the outer frame). *Sold by J. Luffman 28, Little Bell Alley, Coleman Street, London.* (centre, below the outer frame). Compass (below the words *North/Sea*).

*A New Pocket Atlas And Geography Of England And Wales, Illustrated With Fifty-five Copper plates, Shewing the Great Post Roads with the Towns & Villages situated thereon: Also, A description of the Air, Soil, Productions, and Manufactures as well as the number of Hundreds, Cities, Boroughs, Market-Towns, Parishes, Houses & Inhabitants. By John Luffman, N°. 28, Little Bell Alley, Coleman Street, London. 1803.*

**Copy**   BL Maps 24.aa.17.

*A New Pocket Atlas . . . London Printed for Lackington Allen and C°. Temple of the Muses Finsbury Square. 1806.*

**Copy**   Bod Johnson Maps 257.

(ii)   The plate number has been re-engraved; formerly it was 7mm. above the top border of the map and now it is almost touching the outer ring of the border.

*A New Pocket Atlas . . . Coleman Street, London. 1803.*

**Copy**   RGS 262 a 6.

*A New Pocket Atlas . . . London. Printed for Lackington, Allen and C°. Temple of the Muses Finsbury Square. 1806.*[3]

**Copies**   BL Maps 1.a.4; CUL Atlas 7.80.3; Bod Allen 63; Leeds UL W 110; William Salt L (Stafford) Strong Room Box 1.

**References**

1   Maxted, pp. 142–143.

2   ibid., p. 132 for George and John Lackington; Brown, p. 109; for James Lackington's life:
    Knight, C. op. cit., pp. 282–295.
3   Chubb No. CCCXX refers to the possibility of an 1805 edition since he knew of an example
    of Essex dated thus. This map has not been found nor has such an atlas turned up. Since the
    county maps are not dated on the plate this might suggest that Luffman was preparing a new
    form of the atlas.

Chubb CCCVIII–CCCIX, and CCCXX.

## 60

**BARTHOLOMEW HOWLETT**                                                     **1804**

Size: 30.8cm x 23.55cm.

William Miller (1769–1844), who was the publisher of the map, used here as the frontispiece of a book of plates, was a bookseller and stationer, who set up on his own in 1790 and became one of the most popular traders of his time. He paid £4,500 for the copyright of Charles James Fox's *History of the Reign of James I*. He retired in 1812, selling to John Murray, junior, for £3,822–12–6 the stock and house in Albemarle Street, still occupied by that firm.[1] He was also the publisher of Edmund Turnor's *Collections for the History of the Town and Soke of Grantham* in 1806.

According to the half-title the plates were engraved by Benjamin Howlett and presumably he was responsible for the map also. Howlett (1767–1827) was born in Louth and apprenticed to John Heath (engraver to the King in 1794), who was his senior by only ten years.[2] Howlett had already retired at the time of his comparatively early death.[3]

The plate is a very pleasing map based on Cary's 1801 map; it only includes the market-towns, the main roads and rivers with a few of the more important parklands. In fact, the map acts as a secondary index to the plates, since only the houses, parklands and towns are noted that are the subject of a plate or plates. The map thus avoids the feeling of clutter when so many village names are crammed in; the generous scale improves the clarity.

The work was issued in parts; although an original prospectus has not been found the eleventh number announced in late 1801: 'price, Three Shillings and a few on smaller Paper, Price two shillings and Six-pence . . . the whole engraved by Bartholomew Howlett, late Pupil of Mr. Heath, from drawings by Turner, Girtin, Corbould, Nattes, &c . . . [because of] new arrangements [the publisher] pledges to deliver the future Numbers to Subscribers every two months'.[4] As the Advertisement in the complete volume makes clear the work had had its problems while being prepared, one of which concerned complaints from subscribers that only a single sided leaf of text accompanied each plate. As a sop to the complainants Howlett and Miller extended the text by using the spare space on each leaf's verso. Since the dedication is dated 1800 it is reasonable to presume the first part of the work was issued in that year. If bi-monthly issue had been maintained there were presumably two plates every two months with four pages of letterpress, thus ensuring completion early in 1805 – there were 50 plates and the same number of text pages. That the map was only prepared at the end is made clear in the Advertisement in the book, which suggests such a final addition if there were sufficient subscribers. At the back of the final book form the list of subscribers runs to c.260 names; only two booksellers took six copies and Sir Joseph Banks and Lord Yarborough were the only two who subscribed more than one copy.

## EDITIONS

(i)   *THE/COUNTY/OF/LINCOLN/* (Ed). *Scale of Miles/1+5598O* [5 miles = 3.65 cm.]

(Ae). Compass (Ea). The inner border is marked off and numbered every five minutes of latitude and longitude, except that the meridian is not 'numbered'. An error occurs with the figure for latitude, marked as 35° (Ad and Ed) instead of 53°.

A proof copy, lacking many of the final details of state (ii). [1804].[5]

**Copy**  BL Maps 3355 (26).

(**ii**)  The imprint has been added: *Published January 1st. 1805, by William Miller Albemarle Street.* (Ce, OS). The parklands/stately homes have now been named and the following other places or features have been added/named: *Winteringham, River Trent, Belleau, Burwell Park* (and the park), *Husey Tower* (sic), *Dunston/Pillar, Coleby, Little Paunton, Woolsthorpe, Bowthorpe/Oak, Torksey Castle, Scampton, Stow Church, Gate/Burton, Spital/Chapel, Lynn/Deeps, Boston/Deeps* and *Sudbrooke/Holm. Foss Dike* has been re-engraved, *Castle* has been added at Tattershall, and the eastern-most Old Witham R. has been renamed *New Witham R.* The road at Cross Keys Wash has been extended for a short distance on to the Norfolk side.

*A Selection Of Views In The County Of Lincoln Comprising The Principal Towns And Churches. The Remains Of Castles And Religious Houses And Seats Of The Nobility And Gentry; With Topographical And Historical Accounts Of Each View. London: Published By William Miller, Albemarle-Street. 1805.*

**Copies**  BL 10358 h 3 and G 4176 (2) – 192 d 16 is a large paper copy; CUL Views a.70.80.1; Bod Gough Lincolnshire 2; LAO 942.51 FL 1a Extra Large; LRL (Pye) P.L.92 and A 72 (this copy has Turnor's book on Grantham, 1806 (also published by Miller) bound at the back); Louth RL L 91 and GL 940. (LRL Map 761a).

### References
1   Maxted, p. 154.
2   Bryan's Dictionary of Painters And Engravers. New ed. (London, 1927). Vol. III, p. 25.
3   ibid., Vol. II, p. 79.
4   *Lincoln, Rutland and Stamford Mercury* for 11 December, 1801.
5   Since the plate in its final published form was dated 1 January, 1805, the proof must have been prepared in the preceding year.

**61**

**JOHN CARY**                                                                 **1806**

Size: 13.4cm (to top of title panel) x 9.2cm.

These maps supersede the series first issued in 1789 (**50**). The maps and features of the earlier series are faithfully copied and served Cary until a third series was engraved in 1822 (**80**) to replace the present maps; the only change is to the plate's imprint date.

In the following notes (a) indicates maps printed on thicker paper with the map for Middlesex on the verso; (b) represents maps printed on thinner paper single-sided.

## EDITIONS

(**i**)  *LINCOLNSHIRE* (in a panel (Ca), half a compass with the 7 points between *W* and *E* projects from the upper border of the panel). *By J. Cary* (Aa, OS). *Engraver* (Ea, OS). *London Published July 1. 1806 by J. Cary Engraver No 181 Strand* (Be–De, OS). *Statute Miles/10* [= 1.4cm] (Ae). There are four lines of places with their distances from London in a panel below the map. The modern Colsterworth is spelt *Coltsworth*.

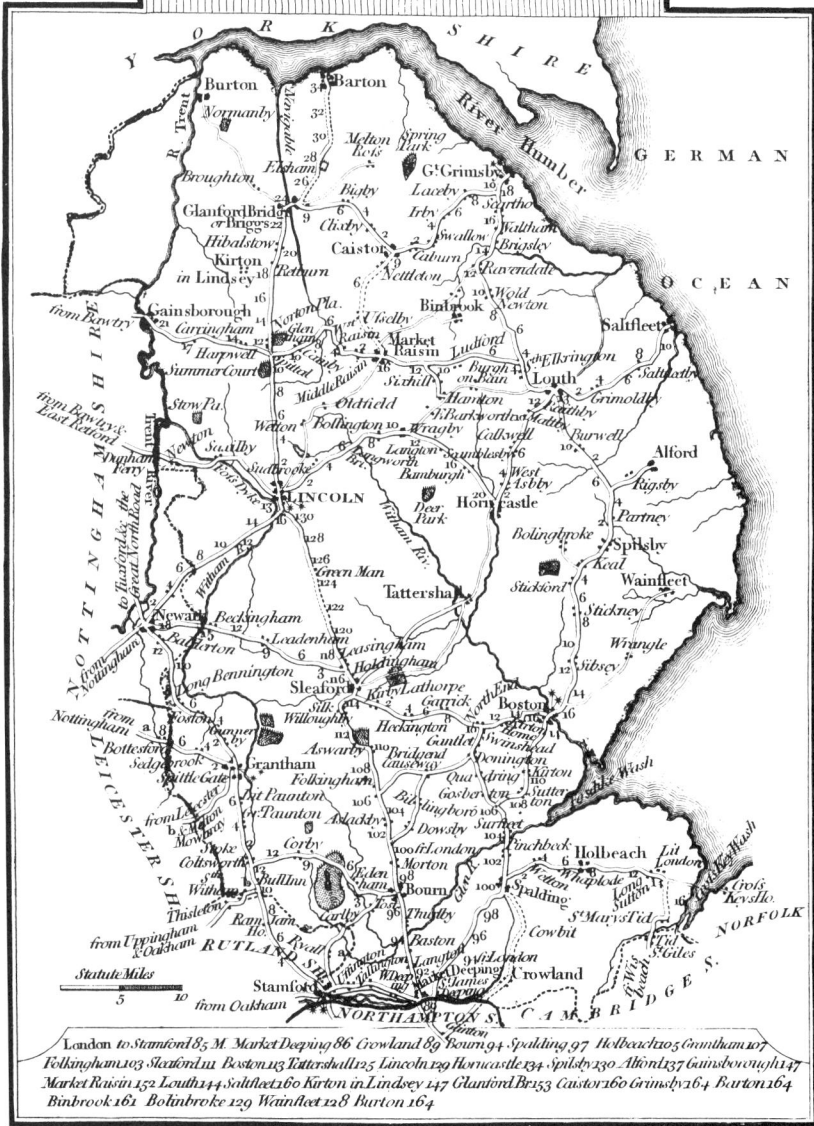

5.  1806   MAP 61 (i) – JOHN CARY.
Reproduced at 117 per cent

*Cary's Traveller's Companion, Or, A Delineation of the Turnpike Roads Of England And Wales; shewing the immediate Route to every Market and Borough Town throughout the Kingdom, Laid down from the best Authorities, On A New Set Of County Maps. To which is added, An Alphabetical List of all the Market Towns, with the Days on which they are held. London. Printed for John Cary, Engraver, and Map-seller, Strand. 1806.*

**Copies**   (a) CUL Atlas 7.80.2; Birmingham UL r DA 652.

(**ii**)   *Coltsworth* has been re-engraved as *Coltersworth.*

*Cary's Traveller's Companion . . . 1806.*

**Copies**   (a) RGS 9 A 61 and 260 c 3; (b) BL Maps 14.dd.33; Bod Maps C.17.f2/2; Leeds UL W 95; Leicester UL H 942.31 CAR; Birmingham RL A 912.42.

*Cary's Traveller's Companion . . . 1806.* [1807].[1]

**Copy**   (b) Bod G A Eng Rds 8º 55 (1).

(**iii**)   The date in the imprint has been changed to *1810.*

*Cary's Traveller's Companion . . . 1810.*

**Copies**   (b) BL Maps 28.c.19; Bod Allen 68 etc.; Leeds UL W 96; RGS 260 C 4 and F 57; Guildhall L S 388/1.

(**iv**)   The date in the imprint has been changed to *1812.*

*Cary's Traveller's Companion . . . 1812.*

**Copies**   (b) CUL Atlas 7.81.4; Bod G.A. Eng Rds 8º 72; Leeds UL W 97; RGS 9 A 6, etc.; Norwich RL 912.42.

(**v**)   The date in the imprint now reads *1814.*

*Cary's Traveller's Companion . . . 1814.*

**Copies**   (a) CUL Atlas 7.81.10; (b) BL Maps 14.dd.34; CUL Atlas 7.81.2 and 7.81.23; RGS 260 C 6 and F59; Edinburgh RL DA 668.4; Maidstone, Centre for Kentish Studies 912.

*Cary's Traveller's Companion . . . 1814.* [1815].[2]

**Copies**   (b) Leeds UL W 98; Leicester UL H 942.31.

(**vi**)   The date in the imprint has been changed to *1817.*

*Cary's Traveller's Companion . . . 1817.*

**Copies**   (b) BL Maps 28.c.8; CUL Atlas 7.81.11; Bod Allen 79; RGS 260 C 7; LAO Ex 942.6, no. 26; Guildhall L S 388/1.

(**vii**)   The date in the imprint has been changed to *1819.*

*Cary's Traveller's Companion . . . 1819.*

**Copies**   (b) CUL Atlas 7.81.18; RGS 260 C 8; Sheffield UL **912.42 (C3); Birmingham RL A 912.42; Bournemouth RL 942T.

(**viii**)   The date in the imprint has been changed to *1821.*

*Cary's Traveller's Companion . . . 1821.*

**Copies**   (b) BL Maps 14.dd.35; CUL Atlas 7.82.43; Bod Allen 84 and 85; RGS 260 C 9.

**References**
1   The Yorkshire map is watermarked 1807.
2   Both copies are both bound with *Cary's New Itinerary . . . 1815.*

Chubb CCLXXVI–CCLXXXII (see **50**, note 1 above).

**62**

**G. COLE and JOHN ROPER**                                              **1807**

Size: 23cm x 17.35cm.

Apart from the statement at the base of the plate that Roper engraved the plates from Cole's drawings there is very little information about either. Roper engraved plates for Wilkinson's *Atlas classica* and he also engraved *PLAN OF LOUTH* that accompanies *A Short Account of Louth Church . . .* put out by J. & J. Jackson of Louth.[1] The publishers Vernor, Hood and Sharpe were only partners for a short time. Thomas Vernor was probably the senior partner in this venture since he had been a bookseller since 1766; Thomas Hood became his partner in 1794[2] and these two began the issue of *The British Atlas* in parts on 1 October, 1804.[3] A number of other partners were also involved, the most notable being Longman and Rees (forerunners of the modern firm of Longman, Green), E.W. Brayley and J. Britton (responsible for the research and text for *The Beauties of Britain . . .* with which the atlas was associated although never forming a part of that work), and Mr. Faden, who was a major mapseller in London. Sharpe joined Vernor and Hood in 1806. Hood died in 1811, the year after the completion of the atlas.

   *The Beauties of England and Wales* originated from suggestions by Vernor and Hood.[4] Britton reports '*the Antiquities of England* by Grose had recently issued in periodical numbers, and several booksellers in Paternoster-Row had produced topographical works of slight pretensions to merit. Amongst them was a folio, published in sixpenny numbers by Alexander Hogg . . . Very illiterate and careless'.[5] Britton and Brayley accordingly undertook to prepare their own work. Although the maps state that they were made to accompany the descriptive work the texts and maps were not issued together. The intention was to publish monthly parts at 2s.6d. per copy (in demy octavo) and 4s. (in royal octavo), each made up of five sheets of letter-press and three prints; six numbers would form a volume and it was expected that there would be six volumes. The first part appeared on 1st April, 1801 and contained Bedfordshire with the other counties to follow alphabetically. The plates were given 'in promiscuous order'.[6] Sales soon developed from the small quantities at the start (1000 small and 250 of the larger format were printed of the first part); by the end of the fifth volume (1804) the sales were 500 copies of the larger size but 3000 of the smaller. By then they had found that they could not produce text quickly enough to maintain a monthly schedule.

   Britton and Brayley divided the country between them for the counties outstanding after volume six and Britton undertook to survey Lincolnshire, which was completed in two volumes, dated 1807 (but published in 1808[7]), and records that among those who provided information were Lords Brownlow and Yarborough, Sir W. Earle Welby, E.J. Willson and Thos. Espin. Britton actually advertised in the *Lincoln, Rutland and Stamford Mercury* on 16 October, 1807 that *A New Topographical History Of Lincolnshire* 'to form the Ninth Volume of the Beauties Of England and Wales' was being prepared and to make the work the more accurate he asked for material to be sent to his London address.

The project was completed (after Brayley felt unable to continue) in 1816 when there were twenty-five volumes in all, which had cost 'upwards of 50,000 *l*'.[8] The first maps were not engraved until five volumes had already appeared and this new feature was announced in part 32 of *The Beauties Of England And Wales* in May, 1804. The maps also appeared at two or three monthly intervals in a separate part work entitled *The British Atlas*, the first part in the October following; parts had either two county maps and one town or city plan or three county maps. Parts I–VII have the maps and the cover bearing identical dates, but thereafter the pattern was more irregular; the Lincolnshire map appeared as part XVI dated Dec. 1807, the same date as that on the map; it appeared with Nottinghamshire (no date) and Norwich (dated November). The last part was dated October 1, 1808. In total there were 58 county maps and 21 maps of cities and towns, which were reduced from 'original surveys . . . published by Mr. Faden. whose permission was *exclusively* granted . . .'[9]

The county map is based loosely on Smith's map of 1801, but the delineation of Sunk Island, now joined to mainland Yorkshire, is unique and suggests models drawn from local surveys. The map lacks any other distinctive feature.

### Later history
The basic map appeared in a number of topographical works until the 1830s; the plates then passed to Tallis & Co. In its various groupings and names this firm issued the map as an accompaniment to Thomas Dugdale's *England & Wales Delineated . . .* from 1835 onwards. There were final appearances in the late 1850s as lithographic transfers issued in railway atlases by two separate firms. The sales must have been low (only a few copies are recorded) and the contemporary rivals (Sidney Hall and Cruchley particularly) prevailed. In all the variety of changes of ownership one feature remained unchanged on the map. At *Fosdyke Wash* and *Cross Keys Wash* the original *Ferry* markings remained unaltered, in spite of the installation of bridges at both places by 1830.

### EDITIONS

(i)  *LINCOLNSHIRE* (Ca, in a panel, projecting slightly above the upper frame line). *Engraved by J. Roper, from a Drawing by G. Cole.* (Ae, OS). *to accompany the Beauties of England & Wales.* (Ee, OS). *London: published for the Proprietors, by Vernor, Hood & Sharpe, Poultry, Dec.ʳ 1. 1807.* (Be–De, OS, but set lower than the signatures). *Scale/10* [=2.75cm]/Compass (Ea). *EXPLANATION/City as___LINCOLN/* [19 further lines of symbols, etc] (Ae). *DIVISIONS./1 Lindsey/2 Kesteven/3 Holland/including 27 Hundreds,/Sokes and Wapontakes./* (Ee). The inner frame is marked off in minutes of latitude, etc. and numbered every 10 minutes. *0 Mer. of Greenwich.* (De, between borders). *1° West of Greenwich* (Be, between borders).

*No. XVI. December 1, 1807.] [Price 2s.6d. Containing Maps of Lincolnshire and Nottinghamshire, With A Plan of Norwich. The British Atlas; Comprising A complete Series of County Maps; And Plans Of Cities and Principal Towns; Intended To Illustrate And Accompany The Beauties Of England And Wales . . . London: Printed for the Proprietors, And Sold By Vernor, Hood, & Sharpe, 31, Poultry; Longman & Co. Paternoster-Row; Cuthell & Martin, Holborn; J. & A. Arch, No. 61, Cornhill . . .*

**Copies**   Bod Map C.17.c.5.[10] (LRL Map 760).

*The Beauties Of England And Wales; Or, Original Delineations, Topographical, Historical, An Descriptive, Of Each County. Embellished With Engravings. By John Britton, F.S.A. Lincolnshire . . . London: Printed by Thomas Maiden, Sherburn-Lane, For Vernor, Hood, And Sharpe; Longman, Hurst, Rees, & Orme; Cuthell & Martin; W.J. & J. Richardson; J. And A. Arch; J. Harris; And B. Crosby. 1807.*

**Copy**   Sheffield RL 914.253 ST.[11]

*The British Atlas; Comprising A Complete Set Of County Maps, Of England and Wales; With A General Map Of Navigable Rivers And Canals; And Plans Of Cities And Principal Towns. London: Printed For Vernor, Hood, And Sharpe; Longman, Hurst, Rees, And Orme; J. Harris; J. Cuthell; J. Cundee; W. Faden; J. And A. Arch; Crosby And Co.; J. Richardson; And J. M. Richardson. 1810.*

**Copies** BL Maps 11.b.3 and Maps 47.d.3; CUL Atlas 6.81.10; Bod Allen 70, 71 and 72 (Allen 71 is a large paper edition); Liverpool RL G 302.

(ii)   The signature has been extended: . . . *G. Cole, under the direction of J. Britton.* (Be, OS). Hachur has been added in the Wolds and around Grimsby. Kirton [in Lindsey], Epworth and Burgh [le Marsh] have been re-engraved in upper case letters. The following features have been added: *Ancholme Riv.*, *Clee thorpe*, *Ludd R.*, *Bollington Abbey* (with symbol), *Drunken/Thoresby* [= North Thoresby], *Abbey Ruins* (north of Louth), *Canal* (by North Cotes), *Castle* (above Newark), Easton House, Well Head (west of Bourn, with symbol), *& Park* (at Burwell, with the area defined), *T.G.* [= toll gate] near Bourn and at Aslackby, *Car Dyke* (in Gothic letters, west of Baston; it already appears north of Bourn, etc. Several place-names have been re-engraved: *CAISTOR* (previously *CAS-TOR*), *Bottesford* (*Botsford*), *Bigby* (*Rigby*), *Stoke Rochford* (*N. Stoke*), *Grt Ponton* (*Grt Penton*), *Evedon* (*Ewerdon*), *Bicker* (*Bickner*), *Haconby* (*Heckingby*), *Donna Nook* (*Donna Hook*), etc. *Hobhole Drain* (with a branch to join Medlam Drain) and *South Forty/Foot* have been added. The Deeping-Boston road has been upgraded to mail coach road. The *Old Sea Bank* and *New Sea Bank* (both north of Holbeach) have additional shading.

*The British Atlas . . . 1810.*

**Copies** BL 7 TAB 47; CUL Atlas 6.81.2; Bod Montagu Illus 41; Leeds UL W 116; Birmingham UL Dept. of Geography; Brighton RL Q 912.42 B77. (LAO Scorer Map 23).

*English Topography; Or. A Series of Historical And Statistical Descriptions Of The Several Counties Of England and Wales. Accompanied By A Map Of Each County. By The Author Of . . .* [Joseph Nightingale]. *London: Printed For Baldwin, Cradock, And Joy, Paternoster Row. 1816.*

**Copies** BL Maps 16.b.2; CUL Atlas 5.81.3; Bod Allen 78 and Maps C.17.c.17; Leeds UL W 121; John Rylands UL Q942 F (17519).

*Lincolnshire; Or, original Delineations, Topographical, Historical, and Descriptive, Of That County. The Result Of personal Survey. By J. Britton. Illustrated With Seven Engravings And A Map. London: Printed For J. Harris, Corner Of St. Paul's Churchyard. 1818.*

**Copy** LRL L 9 (AD 00106177). Grimsby RL has a copy of the book but only the lower half of the map survives.

*A Topographical And Historical Description Of The County Of Lincoln; Containing An Account Of Its Towns, Cathedrals, Castles, Antiquities, Churches, Monuments, Public Edifices, Picturesque Scenery, the Residences Of The Nobility, Gentry, &c. Accompanied with Biographical Notices of Eminent and Learned Men to Whom this Country has given Birth. By John Britton, F.S.A. Illustrated with Seven Engravings and a Map. London: Printed For Sherwood, Neely, And Jones, Paternoster Row; And George Cowie And Co. Successor To Vernor, Hood, And Sharpe, 31, Poultry . . .* [c.1820].[12]

**Copies** LRL L. A9; Grimsby RL X000:914 BRI; Boston RL L 9.

*English Topography; Or, Geographical, Historical, And Statistical Descriptions Of The Several Counties Of England and Wales. Accompanied By An Accurate Map Of Each*

*County, From An Actual Survey. By The Rev. J. Nightingale . . . London: Printed for James Goodwin, Upper Thames Street; And Thomas McLean, Haymarket.* [1827].[13]

**Copy**   Leicester UL F. H942 NIG.

(**iii**)   The imprint and signature have been removed.

*English Topography . . .* [1827].

**Copy**   RGS 7 D 21.

(**iv**)   Plate number *24* in large figures has been added (Ea, OS).

*English Topography . . .* [1827].

**Copies**   BL Maps 16.c.30. (LAO Scorer Map 27).

Issued as an atlas, lacking title-page, but with the title blocked on the front cover: *A County Atlas Through England And Wales. 58 Colored Maps. Price One Guinea.* [1831?].[14]

**Copy**   The private collection of Mr. C.A. Burden.

(**v**)   *Polling Booths* and a Maltese cross have been added (Ce, OS). Maltese crosses have been added to many places. Two stars at Grimsby (for number of MPs) have been reduced to one.

*Curiosities Of Great Britain. England & Wales Delineated, Historical, Entertaining & Commercial. Alphabetically Arranged By Thomas Dugdale, Antiquarian, Assisted by William Burnett, Civil Engineer.* [Vol. II]. *London, Tallis & C⁰. Green Arbour Court, Old Bailey. 1835.*

**Copy**   Birmingham RL T 942.09 DUG.

(**vi**)   A railway is marked by ++++++++ from [Peterborough] to Boston; the line, marked by black dots, continues to [New Holland] through Louth and avoiding Grimsby. At the southern end it is entitled *Boston Railway.*

Issued in a part work. [1842?].

**Copy**   Bod G.A. Gen Top 8⁰ 451–452.

*Curiosities Of Great Britain . . . Vol. II. London Published By John Tallis. 15 Sᵗ. John's Lane, Smithfield.* [1843].[15]

**Copies**   CUL 8460.c.16–18 and Atlas 7.84.9; Bod G.A. Gen Top 8⁰ 1168–1173 (bound as six volumes); Birmingham RL T942.09 DUG; Grimsby RL 914.2 DUG. (LRL Maps 185 and 781).[16]

(**vii**)   Below the note on Polling Booths, *Railway* ++++ has been added (Ce, OS). *Mail/Steam Packet to Hull* above a broken line has been added in the Humber.

*Curiosities Of Great Britain . . .* [1845].[17]

**Copies**   BL 010358 r 7; Bod Allen 107; Birmingham UL r DA625 D9 and r DA 640 D8. (LRL Maps 199, 199a, 199b; Louth RL Maps A 22).

(**viii**)   Below the symbol for railways *Drawn & Engraved for Dugdale's England & Wales Delineated* has been added (Ce, OS).

*Curiosities Of Great Britain . . .* [1846].[18]

**Copy**   (LAO Scorer Map 43).

## LITHOGRAPHIC TRANSFERS

(**A**)   The words and symbols below the lower frame line have been replaced by: *Pub. by Henry George Collins, Paternoster Row.* The plate number has been removed. *Railways* [followed by a thick black line) has been added at the bottom of the Explanation (Ae). Railways 1–4, 6–18, K and N have been added with directions. The dotted line from Boston to the Humber shore remains. In the upper panel to the right of the county name four stars have been added. *N. Holland* [= New Holland] has been added.

Atlas without title-page. [1854**].

**Copy**   CUL Atlas 7.85.9.

(**B**)   Railway 21 has been added (incorrectly; the Sleaford line branched from the GNR at Barkston, not Grantham, as drawn). Plate number *23* has been added (Ea, OS). The publisher's name has been deleted.

*Collins' Railway & Pedestrians Atlas Of England Containing Forty-Three Maps, With All Railways and Roads accurately laid down. Two Shillings & Sixpence. London: Darton And Co., 58, Holborn Hill.* [1858].[19]

**Copies**   BL Maps C.15.b.15; CUL Atlas 7.85.10

### References

1   LRL L 001 Lincolnshire Gleanings, Vol. XV. LRL Map 675 is an example of the map without the booklet. The booklet appeared in two versions; one shilling and, with map, 1/6.
2   Maxted, pp. 114 and 233.
3   A set of the work in its original parts shows Part I dated thus.
4   Britton, J. [Autobiography]. Part 2. A Descriptive Account of the Literary Works . . . (London, 1849), pp. 48–49.
5   ibid., p. 49.
6   ibid., p. 50.
7   The introduction is dated July 25, 1808.
8   Britton, J. op. cit., pp. 57–58.
9   ibid., pp. 63–64.
10   A second set of the parts is in private hands. The Oxford set lacks part XIII.
11   This volume must be an example of a private owner binding the map with the county volume (XI), since a map is not called for.
12   The publishers were in business under the quoted names from 1813 to 1822. There is no internal evidence to suggest a more exact date. Another edition has no reference to a map in the title and such volumes should not contain a map; an example of this edition is LRL L.9 (AD 00106165). Later editions contain maps **70 (i)** and **70 (ii)**.
13   Ascribed to 1827 by Harvey and Thorpe, p. 123 and followed by all later students. McLean was at the Haymarket address from 1822, the date of the watermark of the Yorkshire map in the RGS copy – see state (iii) below.
14   The Lancashire plate shows the Liverpool and Manchester Railway, which was fully opened on 1 December, 1830. The volume has the owner's signature and date (1834), but 1831 seems more likely, since state (v) must post-date the 1832 Reform Act.
15   This work was issued in parts with continuous pagination (p. 1580 refers to an event in 1841) and usually bound in three volumes; only the first has a volume number on its title-page. Occasionally the work is bound in two volumes. Most sets have two title-pages in each volume – one engraved (quoted here) and a longer letter-press version, often having different imprints, which reflect the history of the publisher during the period of issue in parts. The imprint in some third volumes has L. Tallis as publisher; this is Lucinda, widow of John Tallis (died 1842), who ran the business until sons, John and Frederick, took over. The Bodleian set (G.A. Gen. Top. 8° 451–452) is in two volumes only and contains the covers of the original parts, which are all identical and include a space left to be filled in by hand with the part's number. It is not known in which part Lincolnshire occurred. Mr. T. Burgess also has a two volume set but with the second title-page: *An Alphabetical Chronology Of Remarkable Events . . . By*

*Leonard Townsend. London: Tallis And Co., 1835.* Similarly titled sections are more often found at the end of the third volume in three volume sets. Later editions in the 1840s are also found with the Archer maps (**112**).

16 One general map is dated 1843. The CUL copy has a letterpress title-page still dated 1835. The Grimsby copy has a manuscript note of ownership dated 29th July, 1843.
17 When sold as a full set there are eleven volumes (the sets in BL and Bod). In other cases the work is bound in three or five (the two Birmingham sets are both in five). Chubb records a four volume set formerly owned by Sir George Fordham.
18 Published after state (vii) presumably and before 1846 when the first county railway line opened, from Newark to Lincoln. A copy of the work with all the maps in this state used to be in Watford RL (Hodson, 1977, p. 94).
19 The BL copy has the receipt date on all plates – 18 AU [18]58.

Chubb CCCXXXIX, CCCLIII–CCCLIV, CCCLXIX (the [1827] edition ascribed by Chubb to 1820), CCCCLXV, CCCCLXVa and DXXXIX.

## 63

**CHARLES COOKE**                                                                   **1808**

Size: 12.3cm x 9.85cm.

Charles Cooke was primarily a bookseller, who succeeded to his father's business in Paternoster Row in c.1789; he remained there until 1817.[1] In the edition attributed to 1822 he was described as 'the late C. Cooke'. *The Modern British Traveller . . .* was a popular series of little books giving historical and travel information in an easily palatable form. The general series title was *Cooke's Topography Of Great Britain . . .* and there were two versions, the cheaper being priced 1s.6d and a superior 'on large wove vellum paper' and 'with a coloured map 2s.6d.' The work was issued in parts, county by county, initially following the 'Camden' plan by starting with Cornwall, Devon and Dorset. Each county was complete at first in one part and only the one shilling and sixpence version was advertised. However, parts 8 and 9 dealt with Lancashire, thus putting up the price to three shillings. Several other larger counties were also given the two volume treatment. Cooke was a regular advertiser in *Lincoln, Rutland and Stamford Mercury*, mainly of novels and classics in pocket formats and at cheap prices. In the issue for 14 February, 1806 the first number – Cornwall – was announced; the wording of the details of the contents of each volume closely mirrors the title-pages of the volumes. The second announcement on 21 November, 1806 lists 9 counties, 7 single volumes on southern counties plus the two on Lancashire. By 15 April, 1808 22 volumes were available, Yorkshire, Kent and Middlesex each being in two parts. On 26 August, 1808 the final counties were promised on a monthly basis on the first of each month. Finally on 18 November, 1808 a further advertisement and list showed that Lincolnshire and Derbyshire had been issued since the previous announcement, but which came first is not made clear. Oddly, Lincolnshire is sometimes noted as volume XXIX – at the time of the November advertisement only 28 volumes were listed including Lincolnshire. The little pocket volumes are variously combined; Lincolnshire is found bound with each of its neighbouring counties, including Yorkshire, and also with Lancashire and Derbyshire. The engraver of the map is unknown, but the map is based on Cary's map of 1791 (**50 (vii)**).

### Later history

The maps appeared in two series as above. Some examples have one or two half title-pages giving one or both of the series titles. The copyrights were assigned by Cooke's executors to Sherwood, Neely and Jones; this publishing firm had premises close by in Paternoster Row from 1809 and, with a variety of other partners, Sherwood continued in business until 1849 when the firm was taken over by Piper and Thomas. Among the many works issued by this firm was *Historical And Descriptive Sketches Of . . . Horncastle* by George Weir 1820[2] and 1822.[3]

## EDITIONS

(i)   *LINCOLNSHIRE* (in a panel, projecting above the upper frame line, Ba–Da). *Statute Miles/10* [= 1.45cm] (Ae). Compass (Ea). *DIVISIONS. /1 Lindsey./2 Kesteven/3 Holland/ including 27 Hundreds/Sokes and Wapontakes/* (Ed). *The City is denoted by red, and the respective Divisions of the County by different Colours/which distinctions are peculiar to the superior Edition./* (Be–De, OS). The inner border is marked with minutes of latitude, etc. every 5′ and numbered every 10′. *0 Mer. of Greenwʰ*. (De, between borders).

*Topographical And Statistical Description Of The County Of Lincoln. Containing an Account of its Situation, Minerals, Agriculture . . . To which is prefixed, A Copious Travelling Guide, Exhibiting, The Direct and principal Cross Roads, Inns and Distance of Stages, Noblemen's and Gentlemen's Seats. Forming a Complete County Itinerary . . . By George Alexander Cooke . . . Illustrated with A Map Of The County. London: Printed for C. Cooke, No. 17, Paternoster Row, by Brimmer and Co. Water Lane, Fleet Street, And sold . . .* [1808].[4]

**Copies**   Bod G.A. Gen Top 16° 138 (bound with Cambs.); LAO Sc 52/1; LRL L91; Grimsby RL X000:914 COO; Gainsborough RL L9 (bound with Notts.); Birmingham RL A 942.09 (bound with Notts.). (LRL Map 88).

*Topographical . . . Description Of . . . Lincoln . . .* [1810].[5]

**Copies**   BL 291 a 11 (bound with Leics.); Bod G.A. Gen Top 16° 257 and Allen 58 (bound with 3 other counties).

*Topographical . . . Description Of . . . Lincoln;* [the title continues as above but with minor changes of wording, use of upper case and punctuation] *By G.A. Cooke, Esq. Illustrated with a Map Of The County. London: Printed, by Assignment from the Executors of the late C. Cooke For Sherwood, Neely, And Jones, Paternoster-Row; And Sold . . .* [1822].[6]

**Copies**   LAO FL 950.2; Nottingham UL Li 1.D28 COO; LRL L 91.

*Topography Of Great Britain, Or, British Traveller's Pocket Directory; Being An Accurate And Comprehensive Topographical And Statistical Description Of All The Counties . . . Illustrated With Maps Which Form A Complete British Atlas. By G. A. Cooke, Esq. Vol. XVII. Containing Leicestershire And Lincolnshire. London . . .* [1822].[7]

**Copy**   LAO Sc 942.52.

(ii)   Plate number *13* has been added (Ea, OS).

*Topographical . . . Description Of . . . Lincoln . . .* [1822].

**Copies**   LRL L 91; Grimsby RL X000:914 COO; Grantham RL L 9.

*Gray's New Book Of Roads. The Tourist And Traveller's Guide To The Roads Of England And Wales, And Part Of Scotland, On An Entirely New Plan. Whereby The Different Lines Of Route Leading To Any Required Point . . . Are . . . Shewn . . . The Great Roads . . . And The Cross Roads . . . Forming Altogether, A Comprehensive, Useful, And Perspicuous Itinerary. By George Carrington Gray. London: Printed For Sherwood, Jones And Co. Paternoster Row. 1824.*

**Copies**   BL Maps 24.a.34; CUL Atlas 7.82.7; Bod Allen 88 and G.A. Eng. Rds 16º 138; Leeds UL W 136; Leeds RL 912.42; Edinburgh RL DA 668.4.

*Topography Of Great Britain . . . Vol. XVI . . .* [title of the whole volume]; the Lincolnshire section has the title-page: *Topographical . . . Description Of . . . Lincoln . . . London . . . Sherwood, Neely, And Jones . . .* [1830].[8]

**Copy**   Leeds UL W 133.

### References

1   Maxted, p. 50; Nichols, Vol. 8, p. 488.
2   Bod G.A. Fol. A 72; Nottingham UL Li 1.D14 WEI; LRL L Gran; Gainsborough RL L Horn 9; Grimsby RL H 360:942 WEI; Wisbech Museum BB5.25.
3   Louth RL L Horn 9.
4   Chubb assigned this series to the period 1802–1810; as the above notes make clear publication began in 1806 (the watermark date on the LRL copy's title-page).
5   The county text contains a reference (p. 52) to the removal of the cathedral's spires in 1808. Chubb (CCCV) suggests the whole work was completed in 1810 and this is usually accepted.
6   C. Cooke ceased in business in 1817 and died before 1822 when Sherwood, Neely and Jones became Sherwood, Jones & Co. The latest date found in the text (p. 70) is 1817.
7   LAO Sc 942.52 has the general title-page quoted; the county section is preceded by its own title-page of usual type. The reference in the title suggests the maps could form an atlas; an atlas without title-page was held by Maynard & Bradley in 1988. I am grateful to Mr. E. Burden for this information.
8   The frontispiece is an engraving of Boston Stump by Storer of Cambridge and *Pub^d. by Sherwood, Gilbert & Piper 1830.*

Chubb CCCV, CCCLXXXVI, CCCXCII and CCCCXVII.


## 64

**H. COOPER**                                                              **1808**

Size: 17.8cm x 10.3cm.

Cooper was an engraver in Chancery Lane, who produced other maps at this period. Richard Phillips (1767–1840), born in London, came from a Leicestershire family and founded the *Leicester Herald* in 1792 but the following year he was sentenced to 18 months in Leicester Gaol for selling Paine's *Rights of Man*. After his sentence his house and offices burnt down (1795) and, aided by the insurance money, he moved to London. He began publication of *The Antiquarians' Magazine and Monthly Magazine*.[1] He was later Sheriff of London and Middlesex (1807) and received a knighthood in the following year.[2] Although bankrupted in 1811 he continued in a smaller way until he retired in 1823.

The map is loosely based on the small maps of Cary and has no special distinction. The work in which the maps appeared, had a life of more than thirty years and passed through several hands.

### EDITIONS

(i)   *LINCOLNSHIRE* (in a rectangle)/*in which every Parish & Place is laid down/containing upwards of 40 Houses./Cities . . . . 1/Boroughs . . . 4*/[8 further lines of statistics] (Ae). *British Miles/10* [= 1.8cm] (Ad). Compass (Ea). *Published January 1, 1808, by R. Phillips, Bridge Street Blackfriars London.* (Be–De, OS). *Cooper del^t et sculp^t.* (Ee, OS). The names of the Wapentakes are listed under the names of the three parts in three columns (Be–Ee). [Plate] *XXI.* (Ea, OS).

*Atlas Of The British Islands; Comprising Forty-Six Maps Newly And Originally Drawn From The Population Returns, And Other Modern Authorities. London: Printed For Richard Phillips, Booksellers. Price Ten Shillings plain, And Twelve Shillings coloured. 1808. [J.G. Barnard, Printer, Snow-hill.]*

**Copy**   CUL Atlas 7.80.30.

*A Topographical Dictionary of the United Kingdom; Compiled From Parliamentary, And Other Authentic Documents And Authorities; Containing Geographical, Topographical, & Statistical Accounts Of Every District, Object, And Place In England, Wales, Scotland, Ireland, And The Various Small Islands Dependent On The British Empire. Accompanied By Forty-Six maps, Drawn Purposely For This Work, On An Original Plan. By Benjamin Pitts Capper . . . London: Printed For Richard Phillips, Bridge-Street, Blackfriars. [Price 25s. in boards, or 30s. with the plates coloured; the Maps coloured and done up separately, price 12s. half-bound.] 1808. J.G. Barnard, Printer, Snow-Hill.*

**Copies**   BL 290 c 11; Bod G A Gen Top 8° 356 and Allen 113; Leeds UL W 112; NLS Newman 831. (LRL Map 196).

(ii)   The note on the engraver has been deleted (Ee, OS).

*A Topographical Dictionary . . . 1808 . . .*

**Copy**   Leeds UL W 113.

*A Topographical Dictionary . . . Accompanied By Forty-Seven Maps . . . With Additions And Corrections, And the Population Tables Published In 1812. By . . . Capper . . . London; Printed For Longman, Hurst, Rees, Orme, And Brown, Paternoster Row. 1813.*

**Copies**   BL 10348 e 7; CUL S 696.c.81.20; Bod G A Gen Top 8° 1565; Leeds UL W 114.

(iii)   The imprint reads: *Published by G. & W.B. Whittaker. 13 Ave Maria Lane 1824.* Be–De, OS). In the lines of information below the title figures have been changed, thus: *Houses . . . 53, 813/Inhabitants . . . 283, 058/Acres of Land.1, 758, 720/ . . .* (Ae). About 38 place-names have been added, e.g. *Holywell, Bigby, Snarford*, etc. *Moulton* has been added, near Pinchbeck (Moulton is already in its correct place, east of Spalding); *Sutton St./Edmunds* has been added in error for Sutton St. Marys. *Conings/by* has been added, south of Tattershall, while it is also included, north of Tattershall.

*A Topographical Dictionary Of The United Kingdom; Containing Every City, Town, Village, Hamlet, Parish, District, Object, And Place, In England, Wales, Scotland, Ireland, And The Small Islands Dependent. With Forty-Seven Maps. By . . . Capper . . . London: Printed For Geo. B. Whittaker, Ave-Maria-Lane. 1825. Price 30s. in boards, or 31s. 6d. bound.*

**Copies**   BL 10347 k 4; CUL S 696.c.82.9; Leicester UL H 942 CAP; Sheffield RL 914.2 ST; Manchester Portico L. Y.c.26.

Issued in atlas form without a title-page. [1825].[3]

**Copy**   Leeds UL W 137.

*A Topographical Dictionary . . . 1826.*

**Copy**   BL 10347 k 3.[4]

*A Topographical Dictionary . . . Capper . . . A New Edition. London: Printed For Sir Richard Phillips And Co. 1829. Price 1l.11s.6d. in boards.*

**Copies**   CUL S 696.c.82.8; Manchester Chethams L. U.6.51; Leeds RL 942 C173.

*A Topographical Dictionary . . . 1829 . . . [1834].[5]*

**Copy**   BL 10347 k 2.

*A Topographical Dictionary . . . Capper . . . Fifth Edition. London . . . 1839. Price 1l.11s.6d. in boards.*

**Copy**   Leicester UL H 942 CAP.[6]

**References**

1   Timperley, Vol. II, p. 771; Maxted, p.176; DNB.
2   Timperley, Vol. II, p. 831; *Gentlemen's Magazine*, Vol. 14, new series (1840), p. 212.
3   The maps were available separately as an atlas. All the plates are dated 1824 and have watermarks of 1823 or 1824. This example is bound as a travelling pocket atlas.
4   Two other examples are known in private hands.
5   Bound before the title-page is a catalogue of 16pp. of the firm of Sherwood, Gilbert And Piper, dated May, 1834.
6   Three copies are also known in private hands.

Chubb CCCXXVII–CCCXXXI (1808–1829 editions of the Dictionary).


## 65

**JOHN CARY**                                                              **1809**

Size: 26.1cm x 21.15cm.

The plates for *Cary's New And Correct English Atlas* (**47**) continued in use until 1808. In that year appeared the map for Hertfordshire newly engraved although issued in an atlas with a 1793 title-page and the plate also being dated 1793.[1] However, Lincolnshire and the great majority of other county plates are dated 1809 and closely resemble the plates they were intended to replace. Interleaved with and facing the maps are single sided leaves of letter-press relating to the county similar to those, which accompanied the earlier editions. During the life of the work the text was revised occasionally.

**Later history**
The plates remained in use by Cary until his death in 1835. His son issued two rare editions in the 1840s and the plates then passed, like so many other Cary plates, to G.F. Cruchley, who used them for lithographic transfers from 1863 until 1877; although his stock all passed to Gall & Inglis little use was made of the plates, however.

**EDITIONS**

(i)   *LINCOLNSHIRE* (in a panel, which forms the west-east axis of a compass)/*By JOHN CARY, Engraver./Statute Miles 69½ to a Degree/ 10* [= 3.2cm] (Ae). *London. Published by J. Cary. Engraver & Map seller N°. 181 Strand July 1ˢᵗ. 1809.* (Ce, OS). The inner border is marked with minutes of latitude, etc. every 2′ and numbered every 10 minutes. *1° Longitude West from London* (De, between borders); an error for the meridian.

*Cary's New and Correct English Atlas . . .* [the title follows the form of that in **47** above]. *London. Printed for John Cary, Engraver & Mapseller, N°. 181, near Norfolk Street, Strand. Published as the Act directs July 1ˢᵗ. 1809.*

**Copies**   CUL Atlas 5.80.1; Bod Map C.17.c.6 and Allen 66; RGS 7 E 3 and 260 g 3; Leeds UL W 87; York RL 912; Glasgow Mitchell L. F 912.42 CAR (B92 595).

(ii)   The year in the imprint has been changed to: *1812.*

*Cary's New and Correct English Atlas . . . 1812.*

**Copies**   CUL Atlas 5.81.4; Leeds UL W 88.

**(iii)**   The whole date in the imprint has been removed.

*Cary's New and Correct English Atlas . . . Strand. corrected to 1818.*

**Copies**   CUL Atlas 5.81.5 and 5.81.16; Bod Map C.17.c.15; RGS 260 g 4; Leeds UL W 89; Leicester UL H 942.07 CAR. (LAO Scorer Map 26 (2 copies) and Misc Don 142/1/4/3).

*Cary's New and Correct English Atlas . . . corrected to 1821.*

**Copies**   Cardiff RL; Brighton RL 912.42 C25.

*Cary's New and Correct English Atlas . . . London. Printed for John Cary, Engraver & Map-seller, Nº. 86, Sᵗ. James's Street. corrected to 1823.*

**Copies**   BL Maps C.27.f.20; CUL Atlas 5.82.2 and 5.82.5; RGS 260 g 5.

*Cary's New and Correct English Atlas . . . corrected to 1825.*

**Copy**   Leeds UL W 90.

*Cary's New and Correct English Atlas . . . corrected to 1826.*

**Copy**   Leeds RL Q 912.42 C258.

*Cary's New and Correct English Atlas . . . 1827.*[2]

*Cary's New and Correct English Atlas . . . 1829.*

**Copies**   CUL Atlas 5.82.15; Leeds UL W 91.

**(iv)**   The following new roads have been added: Horncastle-Tattershall, East Keal-Tattershall, Friskney-Tattershall, Revesby-Westley Bridge, Spilsby-Sibsey, Sibsey to a point near Coningsby, Billingboro'-Folkingham and Bourn-Spalding. These roads have been upgraded: Louth-Hainton-Wragby, Hainton-North Willingham, Louth-Markby-Alford, Louth-Utterby-Grimsby, Brothertoft-Tattershall, Wainfleet-Boston and Horncastle-Martin-Coningsby. The delineation of, and references, to East and West Fen have been removed. The road from Sutterton through Algarkirk crosses *Fosdyke Wash* to join the Spalding-Holbeach road. *NEW/BOLINGBROKE* has been added.

*Cary's New and Correct English Atlas . . . 1831.*

**Copies**   CUL Atlas 5.83.5; RGS F 259.

**(v)**   One of the two stars at Grimsby has been removed.

*Cary's New and Correct English Atlas . . . 1831.*

**Copy**   BL Maps 47.e.7.

*Cary's New and Correct English Atlas . . . 1840.*[3]

*Cary's New and Correct English Atlas . . . 1843.*[4]

## LITHOGRAPHIC TRANSFERS
**(A)**   The title, compass and the engraver's name have been removed. The new title is: *RAILWAY & STATION/MAP OF/LINCOLNSHIRE/ WITH/THE NAMES OF THE STATIONS. Statute Miles 69½ to a Degree/10* [= 3.25cm] (Ae). *LONDON. PUBLISHED BY G.F. CRUCHLEY, MAP-SELLER & GLOBE MAKER'* (sic) *81, FLEET STREET.* (Be–De, OS). *21* has been added (Ee, OS but sidewards on). Railways 1–4, 6–17, 19–27 and K have been marked as opened, while railways 29 and 32 are shown as under construction; railway 31 is also shown as under construction but from Long Sutton

instead of *Sutton/Br.* Large dots for stations have been added and the names of places with stations are underlined. This has meant the addition of new names, e.g. *Eastville, Old/Leake.* The names of some railways have been added. There are two lines at Grimsby; the ELR passes east of the MSLR which has a separate terminus. The road south of Market Deeping has been shortened and the direction *to Peterboro'* removed. *KING-STON/upon Hull* has been replaced by *HULL.*

*Cruchley's County Atlas of England & Wales Shewing All The Railways & Stations With Their Names, Also The Turnpike Roads And Principal Cross Roads To All The Cities, Market And Borough Towns With The Distance From Town To Town Delineated On A Series Of 46 County Maps Published By G.F. Cruchley, Map Seller & Globe Maker, 81, Fleet Street, London. 1863*

**Copy**   BL Maps 4.b.3.[5]

(**B**)   Railway 28 has been added.

*Cruchley's County Atlas . . . London.* [1864**].

**Copies**   Bod. Map C.17.d.19; NLS EME b.1.20.

(**C**)   The alignment of railways at Grimsby has been altered. Previously the two lines were kept apart; in the new alignment the line from Ulceby (MSLR) crosses the ELR and terminates at Clee Ness.

*Cruchley's County Atlas . . . London.* [1864].[6]

**Copy**   The private collection of Mr. C.A. Burden.

(**D**)   Railway 30 is shown as under construction to Wrawby.

*Cruchley's County Atlas . . . London. 1864.*

**Copy**   The private collection of Mr. D. Hodson.

(**E**)   The title has been changed to: *LINCOLNSHIRE* (Ae), with the scale moved back to a position below the title. Railways 29–30, 32–36 are shown as opened. Railway 36 is named where it leaves the county. The line south from Long Sutton has been removed. The railways at Grimsby have been 'returned' to states (A) and (B) above. Railway 39 is shown as under construction. The configuration of the land to the east of *Sutton/Br.* has been altered and *Cross Keys/Wash* has been re-engraved (formerly shown as a single line) and *Cross Keys* has been removed to a more easterly position. The Norfolk coast is extended to the edge of the map. There is no imprint. *CROWLAND* has been underlined to show there is a station so named.

*Cruchley's County Atlas . . . London.* [1868**].

**Copies**   BL Maps 4.b.13; CUL Atlas 6.86.5; NLS Map m.10.2; RGS 3 B 1.

(**F**)   Railway 37 has been added. Railway 31 has also been added but from Long Sutton (instead of Sutton Bridge).

Cover title: *. . . Cruchley's New Pocket Companion, Or, Handmaid To Bradshaw And All Other Railway Time-Tables For England & Wales, In Four Divisions, (North-West, North-East, South West, South-East.) Twelve County Maps Each. One Shilling, Eighteen Pence, Coloured. North-East Division: Rutlandshire, Lincolnshire . . . Cruchley, Map & Globe Manufacturer, 81, Fleet Street, London, E.C. . . . [1872].*[7]

**Copies**   BL Maps 3.b.6; CUL Atlas R.6.87.14; Bod Map C.17.d.20.

(**G**)   Railway 38 has been added and 39 is now marked as open.

*Cruchley's County Atlas . . . London.* [1873**].

**Copy**   The private collection of Mr. D. Hodson.

**(H)**   Railway 40 has been added. Railway 31 has been correctly placed. Railways 41 and 43 and xvi are shown as under construction. The imprint has been added in the form of state (A), but with a comma after *MAKER*. A double rule has been added under the county name (Ae).

*Cruchley's County Atlas . . . London.* [1875].[8]

**Copies**   BL Maps 11.b.5; CUL Atlas 6.87.7.

**(I)**   *St. Mary/Tid* has been underlined to show the village has a station.

*Cruchley's County Atlas . . . London. 1875.*

**Copies**   BL Maps 7.b.35; Bod Map C.17.d.17; NLS Map m.10.1.

### References

1   Hodson, 1977, p. 73.
2   Chubb, No. CCLXIX refers to a note concerning this edition in: Fordham, F.G. Hertfordshire Maps and supplement. (London, 1914). No copy has been found.
3   The only recorded copy at Bournemouth RL has been lost. Hodson, 1977, p. 74 in his description of the Hertfordshire map in that atlas shows that the borough boundaries have been marked by pecked lines. On some plates railways had been added.
4   The only recorded copy at Colchester RL has been stolen. The Hertfordshire map was identical to that of the 1840 edition.
5   Two other copies are known in private hands; one lacks the title-page but has the cover title: *Cruchley's Travelling County Atlas of England & Wales With All The Railways & Stations.*
6   State (D) is dated 1864; the progressive changes to the plate between states (A) and (D) suggest that there were three issues within twelve months.
7   The date of receipt in the BL copy is 26 SEP 1872.
8   Presumably the minor change in state (I) was made soon after this state, and probably in the same year. Chubb (DLIV–DLV) notes that the copies in the BL were received in August and October, 1875.

Chubb CCCXXXIII–CCCXXXVII (no edition after 1834); DLII–DLVI (Cruchley editions).

## 66

**SAMUEL JOHN NEELE**                                           **1811**

Size: 23.09cm x 19.5cm.

*Encyclopædia Londinensis* was published between 1810 and 1829 and when complete there were 24 volumes. The dates on the maps vary between 1801 and 1828. The engraver Neele (**55**) used Cary's map from *New and Correct English Atlas* as a model, repeating his error of confusing 1° West with the meridian. John Wilkes (1750–1810) began as a printer in Winchester in 1784 but was something of a shady character, who was fined for piracy in connection with the text of this work. In 1790 with Peter Barfoot he issued *The Universal British Directory* . . . whose title is close to the present work's sub-title.[1] James Adlard, the printer, began in 1792 and the firm finally closed in 1900.[2]

### EDITION

(i)   *LINCOLNSHIRE/Published, November 1ˢᵗ. 1811, by Adlard & Cᵒ. Ave Maria Lane.*

(Ce, OS). *Neele sculp.* (Ee, OS). *British Miles/10* [= 3.2cm] (Ae). Compass (Ea). The inner frame is marked with each minute of latitude, etc. and numbered every 10. *Longitude West from London* (De, between borders); there is no number in the lower border, but the upper border has the erroneous 1° for the meridian.

*Encyclopædia-Londinensis; Or, Universal Dictionary Of Arts, Sciences, And Literature . . . Embellished By A Most Magnificent Set Of Copper-Plate Engravings . . . Arranged, By John Wilkes . . . Volume XII. London: Printed For The Proprietor, By J. Adlard, Duke Street, West Smithfield. Sold At The Encyclopædia Office, 17, Ave-Maria-Lane; By White And Co. Fleet-Street; And By Champante And Whitrow, Jewry-Street, Aldgate. 1814.*

**Copies**   BL 12220 p; CUL XXII.4.12; Bod 399 d 159; Leeds UL Spec. Coll. Encyclopedias. (LAO Scorer Map 24; LRL Maps 757 and 767)

**References**

1   Norton, J.E. Guide to national and provincial directories. (London, 1950), pp. 32–35; Nichols, op. cit., Vol. VIII, p. 475.
2   Maxted, p. 2.

Chubb CCCXLV.

<h1 style="text-align:center">67</h1>

**JOHN ALLEN**                                                      **1811**

Size: 8.75cm x 6cm.

Allen (1796–c.1851) was an engraver based in Surrey, whose major work seems to have been the engraving of Rowe's *London*, which appeared in 1811 also.[1] He based the present geographical cards on the Cary plates for the 1806 edition of *Traveller's Companion . . .* with fewer place names and an apparently greater emphasis on the county's roads. The card maps were unknown until they were described by their purchaser in 1983.[2] Since then two sets have come to light, which are different although both bear the same date of publication. Until the discovery of the card issues of 1811 the maps were only known in the form of two little atlases put out by J. Thomson in 1823 and by Orlando Hodgson later in the 1820s.

## EDITIONS

(i)   *LINCOLNSHIRE* (in a panel forming part of the border, Ba–Da). *Pub by J ALLEN, No. 3 Hampden Str. Summers Town, 1811.* (Ba–Da, below the border). *Also by R ROWE, No. 19 Bedford Str. Bedford Row, London.* (in a panel forming part of the border, Ae–Ee). *Scale of Miles/10* [= 0.8cm] (Ae). Compass (Ad).

*A Geographical Game – Allen's English Atlas – being a set of County Maps on Cards showing the whole of the Turnpike Roads, Great Rivers, Navigable Canals &c. adapted for the Instruction Of Youth In English Geography. Pub. as the Act directs June 4th. 1811.*[3]

**Copy**   The private collection of Mr. C.A. Burden.

(ii)   The date in the imprint (Da) has been removed. The second imprint has been replaced by: *This County contains 1,783,--- (the final digits are illegible because of faulty engraving) Acres, of 630 Parishes, 31 Market T. 1 City,/222,551 Inhabitants, and sends 12 Members to Parliament./* (Ae–Ee). Figures for the distances from London have been added to the market towns.

*A Geographical Game – Allen's English Atlas . . . 1811.*

**Copy**   The private collection of Mr. C.A. Burden.

**(iii)**   The imprint (Ba–Da) has been removed. The engraving of the acreage figures has been tidied up and now reads: *1,783,680*. The names of the adjoining counties have been added and the boundaries between them are marked by dotted lines. A number of parks have been drawn, six of which have lower case letters a–f placed near them.

*The New English Atlas, Being A Complete Set Of County Maps, Neatly Coloured; Exhibiting The Whole Of The Turnpike Roads, Cities, Market Towns, Great Rivers, With The Distances From London; Also, The Number of Acres & Inhabitants, Members of Parliament, &c.&c. London: Printed For The Proprietor, J. Thomson . . . 1823.*

**Copy**   Bod Map C.17.f.10.

*The Pocket Tourist & English Atlas, being A New and Complete Set Of County Maps . . . Great Rivers, and Navigable Canals, with . . . & Inhabitants, &c.&c; Including a Copious Topographical Account Of each County. London: Printed For O. Hodgson, Maiden Lane Wood S^T. [1827].*[4]

**Copies**   BL Maps 2.aa.4; Bod Map C.17.f.9.[5]

#### References

1   Tooley, R.V. *Dictionary of Mapmakers* (Tring, 1979), p. 11.
2   Beresiner, Y. A New Source For Orlando Hodgson's County Maps. *Map Collector*, No. 30 (March, 1983), pp. 40–41.
3   This title is taken from the booklet containing the game's rules.
4   Hodgson only occupied the premises in Maiden Lane from 1825–1828 (Brown, p. 91).
5   The Bod copy lacks the title-page and the preliminaries; it is bound in green leather with the cover title *POCKET/TOURIST.*

Chubb CCCLXXI (the [1827] edition ascribed to 1820).


### 68

**JAMES WALLIS**                                                         **1812**

Size: 23.5cm x 18.9cm.

James Wallis seems to have been an engraver from 1789 until c.1812; he was bankrupted in 1801 and, having paid a dividend, set up again but was bankrupted a second time in 1805.[1] He somehow survived to produce the plates in the present atlas but also a second series on a larger scale, which was issued simultaneously (**69**).

The rights in Wallis' maps passed briefly through the hands of P. Martin and then Simpkin and Marshall before becoming the property of William Lewis. Lewis began business in 1796 as a master printer, continuing to trade by himself until 1838, when he was joined by his son.[2] *The New Traveller's Guide* passed through several editions until 1836. There is some evidence that the plates may have passed to John and Frederick Harwood by 1840. John was in business alone from 1822–1829; from 1830 until 1844 he worked with Frederick Harwood and, thereafter, he continued by himself until 1856.[3]

#### EDITIONS

**(i)**   *LINCOLNSHIRE* (within a panel outside the inner frame of the map (Ba–Da). *London Published by J. Wallis, Engraver, 77, Berwick Str. Soho.* (Be–De, OS). *Scale of Miles/8* [= 1.05cm] (Ae). Compass (Ea). *Explanation----  City thus LINCOLN . . .*

[occupying three lines in a panel between the bottom of the map and the lower neat line). The inner border is marked and numbered every 5 minutes. *0° Mer. of Green^h*. (De, between borders).

*Wallis's New Pocket Edition of the English Counties or Travellers Companion in which are carefully laid Down all the Direct & Cross Roads, Cities, Townes, Villages, Parks, Seats, and Rivers. with a General Map of England i* (sic) *Wales. London Published by J. Wallis Engraver. Berwick St. Soho. and Sold by Davies & Eldridge Exeter.* [1812].[4]

**Copies**   CUL Atlas 7.81.17; Bod Map C17.f.5 and Johnson f 1401; RGS 9 A 6; Leeds UL W 118.

(**ii**)   Plate number *20* added (Ea, OS).

*Wallis's New Pocket Edition . . .* [1814].[5]

**Copies**   BL Maps 24.a.14; CUL Atlas 7.81.3 and 7.81.22; Bod Allen 73 and Map C.17.f.6; Nottingham UL s/G 5514; Birmingham 9 8 22RL A 912.42; Leeds RL 912.42 W 158. (LRL Map 171).

(**iii**)   The imprint has been replaced by: *London Published by P. Martin 198 Oxford Street.* (Be–De, OS).

*Martin's New Traveller's Guide, or a Pocket Edition of the English Counties, Containing all the Direct & Cross Roads in England & Wales. With the Distance of each Principal Place from London. London Published by P. Martin 198 Oxford Street.* [1816].[6]

**Copy**   The private collection of Mr. C.A. Burden.

(**iv**)   The plate number, the hachur, the shading around the coast, *Caborn, Billingboro* and a river, north of Grimsby have been removed. Re-engraving has led to many alterations affecting names or mileage marks, e.g. at Burton [on Stather], Wainfleet. Caistor, Glanford/ Briggs. There are two stars each (for MPs) at Grimsby and Lincoln. Irby has been re-engraved as a result of the removal of hachur signs. Virtually all roads have been re-engraved and seem wider than in earlier states; *fr. East Retford* now reads *fr E^st./Retford* and placed by the county boundary; similar changes affect the naming of other roads as they enter the county. All the place-names at [Caenby Corner] have been re-engraved with varied spelling; other changes of a like nature occur elsewhere.

*Martin's Sportsman's Almanack, Kalendar, And Traveller's Guide, For 1818; Containing Ample Directions, In Monthly Order, For Shooting, Coursing, Hunting, And Fishing, In all its varieties after the most approved Systems . . . matured by Experience: The Almanack For The Year; Correct Tables . . . A Series of Maps For Every County In England . . . London: Printed by W. Smith, King Street, Seven Dials, For Simpkin & Marshall, Stationers' Court, Ludgate Hill . . . 1818.*

**Copies**   The private collections of Mr. D. Hodson and Mr. C.A. Burden (2 copies).

*Martin's Sportsman's Almanack . . . For 1819, Containing Ample Directions . . . By . . . Major P. Hawker . . . London: Printed For W. Simpkin And R. Marshall, Stationer's Court, Ludgate Hill . . . 1819.*

**Copy**   CUL Atlas 7.81.33.

*Lewis's New Traveller's Guide, or a Pocket Edition of the English Counties, Containing all the Direct & Cross Roads in England & Wales. With the distances of each Principal Place from London. London. Publish'd by W. Lewis, N°. 21 Finch Lane, Cornhill.* [1819].[7]

**Copies**   Leeds UL W 123; LAO 950.2.

(**v**)   The imprint is: *London Published by W. Lewis, 21 Finch Lane.*

*Lewis's New Traveller's Guide* . . . [1821].[8]

**Copy**   BL Maps 24.a.36.

*Lewis's New Traveller's Guide* . . . [1824].[9]

**Copy**   CUL Atlas 7.81.5.

*Lewis's New Traveller's Guide* . . . [1825].[10]

**Copy**   Bod Allen 80.

*Lewis's New Traveller's Guide* . . . [1827].[11]

**Copy**   Bod Map C.17.f.14.

*Lewis's New Traveller's Guide, And Panorama of England & Wales. Containing Forty-Four Superior Maps; With all the direct and cross-roads, and objects of interest, and the distance of each principal place from London, including much information useful to all Travellers. Price Ten Shillings and Sixpence. London. Published by William Lewis. 1835.*

**Copies**   BL Maps 32.aa.4; CUL Atlas 7.83.37; Bod Allen 102.

*Lewis's New Traveller's Guide* . . . *1836.*

**Copies**   BL Maps 32.aa.34; CUL Atlas 7.83.38; Leeds UL W 124; RGS F 155; Guildhall L S 388/1.

**(vi)**   A new imprint of J. and F. Harwood, 26 Fenchurch St. has been added. [1840].[12]

**References**

1   Brown, p. 212; Maxted, p. 237. Brown records a J.W. Wallis at the Berwick Street address in 1825, possibly a relative.
2   Todd, p. 119.
3   Brown, p. 85.
4   Ascribed by Chubb to [1810]; [1812] is now generally accepted and Brown (p. 212) records Wallis at the Soho address only in that year. Watermarks are not visible on any plate.
5   There is no strong evidence for 1814; it is widely assumed, however. CUL's four examples of this atlas show the stages of change from state (i) to (ii). Atlas 7.81.17 has 17 maps without plate numbers, including Lincolnshire; in atlas 7.81.6 only 10 maps lack a plate number and Lincolnshire now has one; atlases 7.81.3 and 7.81.22 have a full complement of maps all with numbers. Various other mixtures are known. It may seem plausible to assume all volumes without plate numbers are [1812] and those with plate numbers throughout are [1814] and the intermediate states are [1813]. Life in print shops and for atlas compilers is not usually so tidy and this remains surmise only.
6   Todd, p. 127 records Patrick Martin residing at 196 Oxford Street from 1813–1818 with his business next door at 198. This edition predates that in the greatly altered 1818 edition and 1816 is suggested accordingly.
7   The preface is dated Oct. 16, 1819. The copies must have appeared as soon as Lewis moved to this address in 1819, since the plates retain the earlier imprint. Within a short period (probably) the new imprints were added to the plates.
8   Four maps have watermarks dated 1821.
9   The Cambridgeshire plate has a watermark date of 1824. In a privately owned copy the Lincolnshire plate has such a watermark.
10   Three maps have watermark dates of 1825.
11   The Staffordshire plate is watermarked BALSTON 1827.
12   Hodson, 1974, p. 107 recorded this variety; only Cheshire has been found in this state, however, but Lincolnshire may exist thus.

Chubb CCCXLIV, CCCLXIV–CCCLXVI.

## 69

**JAMES WALLIS**                                                      **1812**

Size: 26.7cm x 18.3cm.

Apparently simultaneously Wallis prepared this larger set of plates with the smaller maps
(**68**). The plate is based on Cary, even to the extent of placing the county name in a tablet
across the central west-east axis of a compass. This set was published by Samuel
Augustus Oddy, who seems only to have had a short career in this field, firstly with Henry
Oddy from 1809–1810 and then by himself for a further five years.[1] By c.1819 the plates
had passed to G. Ellis, who used them for a few years in one atlas. In 1842 eight plates
resurfaced and were used by the map-seller James Wyld, who intended to issue his own
atlas. The scheme proved abortive.

**EDITIONS**

(**i**)  *LINCOLNSHIRE* (in a tablet forming the west-east axis of a compass)/*Engraved by
J. Wallis/* (Ea). *London Published by S.A. Oddy 1812.* (Ce, OS). *Scale of Miles/12*
[=3.4cm] (De). *DIVISIONS/1 Lindsey./2 Kesteven/3 Holland./including 27 Hundreds/
Stokes* (sic) *and Wapontakes* (Ee). *EXPLANATION/*10 lines of text and symbols (Ae).
The inner border is marked off and numbered every 5 minutes of latitude, etc. *0 Mer. of
Green[h]*. (De, between borders).

*Wallis's New British Atlas containing a Complete set of County Maps. Divided into
Hundreds in which are Carefully Delineated all the Direct & Cross Roads . . . & a
General Map of England & Wales London Published by S.A. Oddy. 1812. Engraved by
J. Wallis.* [1813].[2]

**Copies**   CUL Atlas 4.81.3; Leeds UL W 120; RGS 7 G 17. (LRL Maps 207 and 168;
Grimsby RL WALLIS 1812; Grantham RL; Gainsborough RL Map 5).

*Wallis's Second and Superior British atlas Containing a Complete Set of County Maps.
Divided into Hundreds. London. Published by Ja[s]. Wallis, Berwick Street, Soho.* [1816].[3]

**Copy**   Leeds UL W 120a.

(**ii**)   A line has been drawn up the map, with, at the southern end, *Meridian/of London.*
(Da–De, reading upwards) and *Longitude* (De). Between borders the reference to
Greenwich has been replaced by *West 0 East* (De). Stars have been added to indicate the
number of MPs each borough returns.

*Wallis's New British Atlas . . . 1812.* [1813].

**Copies**   CUL Atlas 4.81.9 and 4.81.2.

*Wallis's Second & Superior British Atlas . . .* [1814].[4]

**Copies**   BL Maps 33.b.41; Bod Allen 76.

*Wallis's Second and Superior British atlas . . .* [1816].

**Copy**   BL Maps 59.c.59.

(**iii**) The imprint has been removed (Ce, OS).

*Ellis's new and Correct Atlas Of England And Wales being An Entire New Set Of County
Maps, Exhibiting All the direct & principal Cross Roads, Cities, Towns, & most considerable Villages, Parks, Rivers, Navigable Canals, &c. accompanied by Letter Press*

*Descriptions of Each County . . . London Publish'd by G. Ellis, Nᵒ. 5. Smith's Square, Westminster.* [1819].[5]

**Copies**   Leeds UL W 125; RGS 7 D 14.

(**iv**)   The engraver's name has been removed (Ea).

*Ellis's new and Correct Atlas . . .* [1819].

**Copies**   CUL Atlas 5.81.13. (Nottingham UL Li. C.17.E.19).

*Ellis's new and Correct Atlas . . .* [1822].[6]

**Copy**   BL Maps 19.b.37.

(**v**)   The imprint has been added: *London. Published by Jaˢ. Wyld. Charing Cross East.* East Fen has been deleted and a number of new roads have been drawn from Tattershall (to Horncastle, to W. Keal, and to Swineshead), from near Revesby to Sibsey with a branch to Friskney. The roads from Boston-Burgh-Skegness, from Wrangle-Burgh, and Sutterton-Holbeach have been upgraded. The boundaries between the three parts have been changed from dots to alternating dots and dashes. Bolingbroke has been re-engraved in a new place as *N./Bolingbroke* and many villages have been added, e.g. *Lound, Carrington, Elsham,* and *Heighington. EXPLANATION* has been re-engraved and the last five lines have been replaced by four lines referring to marks on the maps to indicate polling places; *Population 317,224. Sqʳᵉ. Acres 1,893,100* forms a new fifth line. Diamonds and circles with a dot in the centre, representing places of county elections and polling places, have been added to the map. Grimsby has lost one of its stars. The *o* in *Wapontakes* has been altered to *e* (Ee).

[Wyld's Atlas of English Counties. London. 1842].[7]

**Copy**   BL Map 3355 (30).

**References**
1   Brown, p. 141.
2   Several plates are dated 1813.
3   This volume contains Cambridgeshire and N. Wales dated 1816, maps dated 1814 and 7 dated 1813.
4   Several plates are dated 1814, but not the same ones in each example.
5   The preface is dated July 1st, 1819.
6   The title-page has a watermark date of 1822.
7   Chubb quotes *London Catalogue of Books published in Great Britain from 1814–1846* for an entry: 'Wyld's Atlas of English Counties 4ᵗᵒ.hf.-bd . . . 3.3.0'. It seems certain the announcement refers to a planned work that never came to fruition.

Chubb CCCXLIX–CCCL; CCCLXVII; DV.

# 70

## SAMUEL JOHN NEELE                                          **1812**

Size: 24.8cm x 19.65 cm.

The author of the text that the maps illustrate, John Britton, had produced a topographical and historical description of Lincolnshire *inter alia* for *The Beauties of England . . .* (**62**). The original work did not call for the maps prepared by Roper and Cole and which finally emerged as *The British Atlas.* It is an oddity that the county volume should appear with a map that was not uniform with the volume and needed to be folded into the book while the Roper/Cole plates were still available. The text of the work below is identical to that of the earlier work, including the pagination taken from the continuous numbering of that series. The map by Neele is closely related to his own map for *Encyclopaedia Londinensis* (**66**) but is based on Cary (**65**) and like most deriving from that source contains the error of the Greenwich meridian; that is also an oddity in that Neele's first county map of 1799 (**55**) was an early Lincolnshire map to show the Greenwich meridian. James Cundee, who appears in the map's imprint as publisher, first came to notice in 1799 and ran the Albion Press (named after the first steam press) from 1805–1812.[1] John Cundee is recorded at the same address from 1820–1833 with James Robins. It seems certain that the work by Britton must have appeared after 1813, since George Cowie was reported at various addresses from 1801 but was not at 31, Poultry until 1813, when he began trading as Cowie & Co.[2]

**Later history**
When Robins acquired the plates he used them to illustrate *The New British Traveller* . . . which he issued in parts at Eight Shillings each. The work is similar to the *The Beauties of England* and was intended to be complete in fifteen parts, although advertisements for the plates in atlas form also exist. As the bulk of the plates are dated 1818 it seems likely that the work began to appear in that year and was completed in the following year (as the title-page records). Whether James and Thomas Dugdale, the authors of the texts for this work and that which appeared with the Cole/Roper plates from c.1835 (**63 (ii)**) were related is not clear.

**EDITIONS**

(i) *LINCOLNSHIRE* (in a panel)/*British Statute Miles/10* [=3.25cm] (Ae). *Published by James Cundee Albion Press London January 1ˢᵗ. 1812.* (Ce, OS). *Neele sculpᵗ. Strand.* (Ee, OS).Compass (Ea). The inner border is marked with minutes of latitude, etc. every two minutes and numbered every 10. *1° Longitude West from London* (De, between borders).

Sold singly perhaps from 1812 onwards.[3]

*A Topographical And Historical Description Of The County Of Lincoln; Containing An Account Of Its Towns, Antiquities, Public Edifices . . . The Residences Of The Nobility, Gentry, &c. Accompanied with Biographical Notices of Eminent and Learned Men to whom this County has given Birth. By John Britton, F.S.A. Illustrated with Seven Engravings and a Map. London: Printed For Sherwood, Neely, And Jones, Paternoster Row; And George Cowie And Co. Successors To Vernor, Hood, And Sharpe, 31, Poultry* . . . [1818].[4]

**Copy**    Louth RL L9.

*The New British Traveller, Or, Modern Panorama of England And Wales. Vol. III.* [or 3, in some cases]. *London. Published by J. Robins & Cº. Albion Press, Ivy Lane, Paternoster Row. 1819.*[5]

**Copies**    Leeds UL W 126; RGS 260 g 9–12; Birmingham UL r q DA 30; Wisbech Museum. Press GG.10.20; Manchester RL Q942 D1. (LAO Cragg 2/38 consists of the map and the county text, which runs from pp. 585–620, 421–456; there is clearly an error in pagination (common to all examples seen) since the catchword of p. 620 carries over to p. 421).

(ii)    The imprint has been changed to: *Published by J. Robins & Cº. Albion Press London January 1ˢᵗ. 1818.*

*A Topographical And Historical Description Of The County Of Lincoln . . .* [1818].[6]

**Copies**   LRL L 9; Boston RL L9.

*Robins's Atlas of England And Wales, Accurately Engraved by Neele, from the Latest Surveys. London, Published by J. Robins & Cº. Albion Press, Ivy Lane, Paternoster Row. 1819.*

**Copies**   BL Maps 13.d.8; CUL Atlas 5.81.7. (LRL Maps 172 and 172a).

*The New British Traveller, or, Modern Panorama Of England And Wales. Vol. III. London. Published by J. Robins & Cº. Albion Press, Ivy Lane, Paternoster Row. 1819.* [1820].[7]

**Copies**   BL 10348 f 9; CUL 8500.b.83; Bod Allen 81 and G.A. Gen Top 4º 267–270.

*Robins's Atlas of England . . . 1819.* [1820].[8]

**Copy**   The private collection of Mr. C.A. Burden.

**References**

1   Maxted, p. 87.
2   Todd, p. 48.
3   It seems unlikely that a map prepared in 1812 was not sold before its appearance in book form six years later, but all copies found so far relate to volumes containing the county text. The same applies to the findings of students of the maps of other counties.
4   A similar text by Britton but with the map of Cole and Roper appeared in 1818 and it is assumed that this volume also appeared about that time. The latest date seen in the volume's text is 1803.
5   Most sets have two title-pages; one (quoted here) shorter, engraved and dated; the second longer letterpress and undated.
6   LRL L9 is interleaved with paper watermarked *J. WHATMAN/1818*; it is assumed that volumes were issued after the map had been re-dated.
7   There is a reference to the weather in January, 1820 (Vol. 1, p. viii); in some examples the general map is dated 1820 and the chart of noblemen contains entries for 1820. Mr. C.A. Burden has five of the projected fifteen parts. Assuming the first part was issued in late 1818 or early 1819 and parts appeared monthly the work could not be completed before 1820.
8   The maps of Northamptonshire and East Riding are watermarked 1820.

Chubb CCCLXII (Robins's Atlas); CCCLXIII (Dugdale).

## 71

**ROBERT ROWE**                                                                **1813**

Size: 41.1cm x 34cm.

Rowe (c.1775–1843) is recorded as an engraver and it is possible that he drew as well as engraved the maps here. His involvement with maps has already been recorded (**67 (i)**). The present maps, based on Smith and Jones (**56**), were issued by Rowe for only a few years. They made new appearances in the hands of Henry Teesdale (**92**) from 1829, H.G. Collins from 1852 and, finally, George Philip (**122** and **130**). Altogether through lithographic transfers the maps had a life of nearly fifty years in Lincolnshire's case, although the plates of other counties were used to illustrate local directories and had even longer lives.

## EDITIONS

(i)   *A NEW MAP/of the COUNTY of/LINCOLN/Divided into Wapontakes &c./By R. Rowe.* (in an oval, Ea). *London: Printed for R. Rowe, Nº. 19, Bedford Street, Bedford Row, Janᵧ. 1. 1813.* (Between the inner and outer frames of the oval above). Compass

(Ae). *EXPLANATION./*8 lines of symbols and their meaning (Ee). *REFERENCE to the WAPONTAKES* &c./31 names in two columns under the headings of the three parts (Ae). *Scale/½ + 12 Miles/* [12 = 5.85cm] (Eb). The inner border is marked with every minute of latitude, etc. and numbered every 5. *1°. Longitude West from Greenwich* (De, between frames).

Issued in a cardboard holder; *LINCOLNSHIRE.* appears on a label stuck on another label with a decorative border. Below the border the same imprint appears as on the map, but without the date. [1813].[1]

**Copy**   Grimsby RL X000:912 ROW.

*The English Atlas; Being A New And Complete Set Of County Maps, Divided Into Hundreds. Exhibiting The Direct And Cross Roads, Cities, Towns, Villages, Parks, Rivers And Navigable Canals. With The Distances Marked Between The Towns And From London. Preceded By A General Map Of England And Wales, Shewing The Connections of the respective Maps. By Robert Rowe, Geographer. London: Printed For R. Rowe, 19, Bedford Street, Bedford Row. 1816. Printed by Joyce Gold, 103, Shoe-lane, Fleet-street.*

**Copies**   Bod Allen 77. (LAO Scorer Map 25).

*The English Atlas . . . 1816 . . .* [1821].[2]

**Copy**   The private collection of Mr. C.A. Burden.

**(ii)** *LINCOLNSHIRE.* (Ea, in place of the former oval). *London, Published by Henry Teesdale & Cᵒ. 302, Holborn.* (Ce, OS). *The Stars prefixed to the Towns, Denote the/number of Members returned to Parliament./* (Be). Two stars have been added at Lincoln. The Yorkshire east coast has been completed (previously the oval bearing the title obscured the coastline).

[New British Atlas . . . 1828].

**Copy**   See *The Romance of the Road*, by Cecil Aldin (London, 1928), p. 115; he preferred to have this volume with him to the then modern road maps for its greater accuracy. No copy has since been found. It is assumed that its make-up matches the 1829 edition.

*New British Atlas, Containing a Complete Set of County Maps, On Which Are Delineated All The Principal Cross Roads, Cities Towns & most considerable Villages, Parks, Rivers, Navigable Canals & Railways, preceded by General Maps Of England, Ireland, Scotland, North & South Wales. The Whole Carefully Revised & Corrected To The Year, 1829.* [List of county names in three columns]. *London Published by Henry Teesdale & Cᵒ. 302, Holborn. T. Barnett Sc.*

**Copies**   BL Maps 2.d.1; CUL Atlas 3.82.3; Leeds UL W 140; Manchester RL BR FF912.42 T15; Nottingham UL s/G 5512. (LRL Maps 774 and 774a; Louth RL Map A 11; Grimsby RL Teesdale 1829; Boston RL Map 29).

*New British Atlas . . . To The Year, 1830 . . .*

**Copies**   CUL Atlas 5.83.10; RGS 3 H 1; NLS EMGB b.1.21.

*New British Atlas . . . To The Year, 1831 . . .*

**Copy**   Norwich RL 912.42 (S.F.3.20).

**(iii)**   Two stars each have been added at Grimsby, Stamford, Boston and Grantham. The Grantham Canal has been re-aligned and runs parallel to the Grantham-[Nottingham] road. Parks have been drawn in at least 17 places, including Blankney, Norton Place, Belton (in Axholme) and Belvoir. The *Welland Canal* (i.e. Boston-Spalding) has been added and another from Alford to the sea. The Louth Navigation has been added only as far as $N^{th.}$ *Cockerington.* Roads have been added: Holbeach-Sutterton, but not passing through Fosdike; Kirton Holme-Coningsby and a fork to Tattershall; Hagnaby-Stickford; a network of minor roads around Alkborough and Burton upon Stather, etc. *Stocking Ho.* (near Castle Bytham) and *Newport* (at Lincoln) have been included.

*New British Atlas . . . 1831 . . .*

**Copies**  BL Maps 3.d.39; CUL Atlas 5.83.4; Leeds UL W 140a; Signet Library, Edinburgh 11 912 (42) T 22.

**(iv)**  The note on stars denoting MPs has been replaced by: *Population $N^{th}$. $Div^n$. 173,209. $Asses^d$. Taxes £.34,185./. . . $D^o$. $S^{th}$. $Div^n$. 144,035 . . .$D^o$ . . . $D^o$ . . . £.31,200./* (Be). Maltese crosses have been added to many places. Squares have been placed at Lincoln and Sleaford, large double circles at Stamford, Grantham and Boston, and a triangle (with its left hand half shaded) at Grimsby. Below the title (Ea) has been added: *2 For each Division of County . . . 4/2 $D^o$. Four Boroughs . . . 8/1 for Borough . . . 1/Total Number of members 13/*. The road at [Sutton Bridge] has been extended to the east across Cross Keys Wash; several roads have been added across East and West Fen notably Boston-Revesby and Stickney-Coningsby; Sutterton-Swineshead. Roads have been upgraded: Louth-Thoresby-Grimsby and Boston-Wainfleet. More minor roads have been added: see, for instance, around Goxhill, *Botsford* and Skegness. Among places or houses added are: *New Hotel* (Skegness), *Walcot/Hall* (Alkborough), *Paupers Drain* (Crowle), *Skitter Sands, Goxhill/Ferry* and *Clayhole/ House* (s.e. of Fishtoft). *The Scalp* (formerly *Scap* in the entrance to the Witham at Boston), *SOUTHERN/DIVISION* and *NORTHERN/ DIVISION* have been added. The stars at Grimsby, Boston and Lincoln have been removed. The boundaries of the five boroughs have been marked to show the newly enlarged status; this is especially notable at Boston, and at Stamford, where the area around *St./Martins* now has the county boundary enclosing it. *NEW/BOLINGBROKE* has been added. *Gibraltar/Point* is named but the original finger of land has been re-drawn so that the Point is on an island.

*Improved Edition Of The New British Atlas . . . The Whole Carefully Revised & Corrected To The Year, 1832.* [The list of counties is now flanked by, on the left, *NOTE.* and an explanation of the marking of new boundaries on the maps, and, on the right, *EXPLA-NATION.* with 12 lines showing the symbols applied to the maps. This shows that the shading of the triangle at Grimsby is in error on the map]. *London, Published By Henry Teesdale & $C^o$. 302, Holborn.*

**Copies**  BL Maps 27.d.21; CUL Atlas 5.83.14; Bod Allen 95; Leeds UL W 141.

*Improved Edition Of The New British Atlas . . . 1832 . . .* [1833].[3]

**Copy**  NLS EMGB b.1.13.

**(v)**  Plate number *26* has been added (Ee, OS, but sidewards on).

*Improved Edition Of The New British Atlas . . . 1835 . . .*

**Copies**  CUL Atlas 5.83.5; Bod Allen 100; Leeds UL W 141a; Manchester RL BR f912.42 Te1.

*Improved Edition Of The New British Atlas . . .* [1836] . . .[4]

*Improved Edition Of The New British Atlas . . . 1840 . . .*[5]

*Improved Edition Of The New British Atlas . . . 1842 . . .*

**Copy**  CUL Atlas 4.84.18.

## LITHOGRAPHIC TRANSFERS
**(A)**  Floral decoration has been added to the borders at the corners and the centres of

each side, thus increasing the width of the map by 1.2cm. The title is now in a plain frame. The plate number has been removed. *Railways* with solid black line and dot marked *Stations* has been added (Eb). The imprint now reads: *London, Published for the Proprietors, by H.G. Collins, 22 Paternoster Row.* (Ce, OS). Railways 1–4, 6–14 have been added. The note on population and taxes has been removed (Be). Nearly all the directions where roads leave the county have been removed.

*New British Atlas, Containing a Complete Set of County Maps . . . The Whole Carefully Revised & Corrected. London, Published By Henry George Collins (For The Proprietors) 22, Paternoster Row.* [1849**].

**Copies**   CUL Atlas 4.84.36; Leeds UL W 142. (LAO Scorer Map 44).

(**B**) *THE/BRITISH GAZETTEER./* surrounded by a foliate decoration has been added above the county name (Ea). Railways 15, 17, vii–ix have been added in the form of double lines (to show lines under construction). Railway 17 curves into the Lincoln line, north of Newark. Railway 1 has been extended south-westwards.

*The British Gazetteer, Political, Commercial, Ecclesiastical, And Historical . . . Illustrated By A Full Set Of County Maps, With All the Railways accurately laid down; Forming At Once An Iron-Road Book And County Atlas. By B. Clarke . . . Volume II. London: Published (For The Proprietor) By H.G. Collins, Paternoster Row, 1852.*

**Copy**   CUL S696.b.85.11–13.[6]

An atlas, without title-page. [1852].

**Copies**   BL Maps 8.d.39; CUL Atlas 4.85.2; Bod Gough Adds. Gen. Top. 4° 8; Edinburgh UL F* 24.36.11–4. (Bod C17: 39(3); Grimsby RL Clarke & Collins 1852).[7]

(**C**)   The map is coloured; Lindsey is pale yellow, Kesteven green and Holland pale pink; all divisional boundaries are marked by deeper shades of the colours.

Issued in a cardboard cover, with the title: *Lincolnshire With Its Railways. Sixpence. Published By H.G. Collins, Paternoster Row, London, And Sold Ay* (sic) *W.H. Smith And Son, 136, Strand, London, And At All The Railway Stations. J. Wertheimer And Co. Printers, London.* [1852].

**Copies**   NLS Maps m.51.1. BL Map 3355 (2) is a similar sheet mounted on card; the receipt stamp is dated 13 AP 52 (i.e. 1852).

(**D**)   The decorative foliage around the title has been removed but a triangular label with similar foliage has been stuck on the map above the title to obscure the words *THE BRITISH GAZETTEER*. Railways 15 and 17 have been marked in black to show they have opened. The curve of railway 17 into railway 1 has been deleted and the line extended northwards.

Issued in a card cover with the title: *COLLINS' LINCOLNSHIRE. WITH ITS RAILWAYS. SIXPENCE. PUBLISHED BY H.G. COLLINS, PATERNOSTER ROW, AND SOLD BY . . .* [1853**].

**Copies**   CUL Maps 34.052.70; LAO Scorer Map 44 and Ex 942.50.[8]

Issued in a card cover with the title: *No 21. Price 1s.6d. England Depicted, In a Series of Splendidly Full-coloured Maps. LINCOLNSHIRE. Where . . . all the Railroads . . . are carefully drawn . . . PUBLISHED FOR THE PROPRIETORS, BY H.G. Collins, 22 Paternoster Row.* [1854].

**Copy**   LAO Ex 942.50.

(**E**)   The foliate edges and label (Ea) have been removed. The title is now: *LINCOLN*

(Ea). The imprint reads: *GEORGE PHILIP & SON, LONDON & LIVERPOOL*. (Ce, OS). Several towns outside the county have been named, with the roads that connect them to the county: *Thorne, East/Retford, Tuxford, Bingham* and *Wisbeach*. Railways outside the county have names. Although Oakham is not shown two roads leading to it have been added, with directions. The road at *Cross Keys* has been extended to Lynn. Railway 2 has been redrawn; it now turns towards [Peterborough] and runs parallel to the GNR without crossing it to join the ELR, as before; it is now labelled *From Melton & Leicester*. The double lines from Long Sutton now extend to Wisbech. Railway 20 has been added.

Issued in a card cover with the title: *Philips' Popular Series Of County Maps. Full Coloured And Folded In Case Price Sixpence Each*. [List counties and their numbers]. *21 Lincoln . . . Any Of the Above May Be Had Mounted On Linen & In Cloth Case, One Shilling each. London. George Philip & Son, 32, Fleet Street.* [1856].[9]

**Copy**   LRL Map 235.

(**F**)   Plate number *21* has been added (Ee, OS, but sideways on).

*Philips' County Atlas of England and Wales, Containing Forty-Seven maps. London: Published By George Philip And Son, 32, Fleet Street, And 51, South Castle Street, Liverpool. Edinburgh: Oliver And Boyd. Dublin: William Robertson.* [1857].[10]

**Copy**   The private collection of Mr. T. Burgess.

### References

1   It is assumed that the map was issued soon after the plate was ready; it seems unlikely that the plate would remain unused until its atlas appearance in 1816.
2   The Berkshire map has been redated 1821.
3   The title-page is watermarked 1833. Chubb (No. CCCCXIII) notes a copy dated 1833 in the 1882 catalogue of the Signet Library, Edinburgh. In fact, that atlas lacks a title-page; the watermark date of the Middlesex map (1831) is the latest date found in the work, which is noted under state (iii).
4   Mr. P.J. Cassidy, bookseller of Holbeach, sold a copy of the county map in 1990, which had the watermark date 1836. The map may have come from an atlas not yet found.
5   Chubb (No. CCCCXV) notes a copy so dated belonging to Sir F. G. Fordham.
6   The Lincolnshire map is in Vol. II although Vol. III covers L–Z.
7   In all these atlases the maps are collected to form a [Fourth] volume, volumes I–III containing the text only. The text is separately shelved in the BL at 19348 h 6; similarly for CUL at K.19.28–30.
8   The three examples all differ slightly in the way the lines leaving the county stretch towards the map's margins.
9   It is believed that this map came out in 1856, i.e. before the atlas in state F. This type of map appeared into the 1860s in the case of other counties.
10   The date is recorded in Philip, G. The Story of the last Hundred Years – A Geographical Record. (London, 1934).

Chubb CCCLIVa; CCCCIX–CCCCXV; DXXXIV.

# 72

## AARON ARROWSMITH                                                    1815

Size: 62.7cm x 91.3cm (approx.).

This map is included here for the convenience of map librarians and collectors. It forms part of a larger work, which covered the whole of England and Wales, but Lincolnshire and slightly larger parts of adjoining counties are included on sheet 9. The area covered stretches from Barton on Humber to Paston (near Peterborough) and from Saxby (Leicestershire) in the west almost to Yarmouth in the east. On the outside of the folded sheet is a small version of what the large map encompasses; this thumb-nail outline only notes six towns in the county, plus Oakham and three places in Norfolk; the number 9 appears in the top right corner, although the plate proper is not numbered.

Aaron Arrowsmith (1750–1823) is one of several generations of eighteenth and nineteenth century mapmakers. He was born in Durham and came to London, where (under Cary's aegis) he undertook much of the work on *Great Post-Roads between London and Falmouth* (1782), which Cary engraved. He set up for himself in 1790 and by 1810 he was Hydrographer to the Prince of Wales and in 1820 he received a similar title from the King, specialising in large maps and covering all parts of the globe. Among his works is *A Sketch of/WILDMORE FEN,/WEST FEN & EAST FEN;/with the Marshes/* ... dated by him 1810 and prepared as an accompaniment to Rennie's report on the drainage of Wildmore Fen.[1] The style of engraving in this map is also to be seen in the map, prepared by A. Bower but under Rennie's direction, of the River Witham between Boston and Lincoln (1803).[2] Tooley has described him as 'one of the greatest English cartographers but he has hardly received the attention that is his due'. He remained active until his death and among his pupils was Sidney Hall (**76** and **94**).[3]

Aaron was succeeded by his sons, Aaron and Samuel, who ran the business until the latter died in 1839. George Cruchley served his apprenticeship with the firm before setting up for himself in the 1820s. A nephew of the first Aaron, similarly named, who had been active since 1810 by himself, took over the business, when he acquired the plates, manuscripts and copyrights of his cousins in 1839. He managed the business successfully until his death in 1873; the assets were auctioned in 1874, when the plates were acquired by Edward Stanford. This last Aaron was a a leading figure in the founding of the Royal Geographical Society in 1830, enjoyed the title of Hydrographer to the King in 1832 and was awarded the RGS medal in 1863.[4] Arrowsmith prepared maps for other areas of the country based on Ordnance Survey sheets and was involved with that body over infringement of their copyright.[5] In the case of the present map, Arrowsmith adapted the large Armstrong sheets to his purpose. This is particularly seen in the shading used for the Adventurers Lands south-west of Spalding, which is not found on any other map. He simplified much and this is notable in only naming the East Fen without attempting to show the waters and meres of that area. The inclusion of Hughington, the 'village' near Heighington, is further evidence of his source.

There was a later edition in 1816 and a further revised issue in 1818, but the map sheet for the county appears to be unchanged. Stanford issued the maps as *RAILWAY & STATION MAP OF ENGLAND & WALES IN 24 SHEETS. One Shilling per sheet, plain; Eighteen Pence colored* ... There were editions in 1876 and examples are known dated 1877, 1878, 1882 and 1884.[6] In these versions Lincolnshire occupies plate 10 and on the outside a small map of the county is provided to make clear the contents of the large folded sheet. The plates were also issued by Stanford overprinted in several colours as hunting maps, in which the railways and station names have been added in a prominent red.[7] See also Addenda, pp. 411–12.

## EDITION

(i) *LINCOLNSHIRE* (in a curved line across the centre of the county). *London, Published 10th March. 1815 by A. Arrowsmith 10 Soho Square Hydrographer to H.R.H. the Prince of Wales*. (De, OS). The county's boundary is coloured yellow; the parks green and the boundaries of the adjoining counties green, brown, blue or pink.

Issued as a single sheet, but part of a larger map. The whole work has the title (on plate 3): *MAP OF/ENGLAND AND WALES,/THE RESULT OF FIFTEEN YEARS LABOUR./ Dedicated by Permission to His Royal Highness the Prince Regent,/by H.R. Highness's Dutiful Servant and Hydrographer/A. Arrowsmith./1815./*. The compass appears on plate 6 and takes up all the right hand half of the double folio, being labelled *SCALES SHEWING THE UNEQUAL SHRINKING OF PAPER* above the circle/compass. The map of Lincolnshire is the left half of a double folio plate, [Sheet no. 9] of 15 whole sheets and 3 half sheets. 1815.

**Copies**   BL K 5–74 ENGLAND II TAB. (LAO Misc Don. 471).

Issued as above, but the title has the following addition: *1815./ ADDITIONS TO 1816./*

**Copies**   BL Maps S.T.E. (3); RGS E & W. G 241.[8]

Issued as above, with the addition to the title: *. . . 1816. 1818/* and the imprint (on sheet 9) altered to: *London Published 10 March 1815, by A. Arrowsmith No 10 Soho Square. – Hydrographer to His Majesty*. (Ee, OS).

**Copies**   BL Maps 148.d.25; CUL Maps A.34.81.1.

### References

1   Copies of the book are: BL B 267 (3); Bod Gough Lincoln 3; Cambridge AO R 59/31; LAO 2 BNL 12/3 and Monson 7/16/58; loose copies of the map are BL Maps 3365 (4); RGS 1 C 88 (53); LAO 2 BNL 12/12.
2   LRL Map 650; a second version appeared in 1804 – BL Maps 25.d.4; LAO 3 Cragg 1/37, etc.
3   Moreland and Bannister, p. 176; Tooley, R.V. Aaron Arrowsmith. *Map Collector*, no. 9 (Winter, 1979), pp. 19–22; Maxted, p. 6; DNB.
4   Herbert, F. The Royal Geographical Society's Membership, The Map Trade, and Geographical Publishing in Britain 1830 to c.1930 . . . *Imago Mundi*. Vol. 35 (1983), pp. 67–95.
5   The Old Series Ordnance Survey Maps, with an introduction by J.B. Harley and Y. O'Donoghue. Vol. II. (Lympne, 1977), pp. xxvii–xxxii discuss some of the problems faced by the authorities from unauthorised use of the early O.S. maps.
6   1876 – BL Maps 1220 (17); a full set of 1882 is CUL Maps b.18.G.28.
7   BL Maps 12.b.67 contains two examples: sheets 4 and 14; it is not known which sheet number contains Lincolnshire, since the map covers provide no information on others in the series.
8   CUL Maps 18.G.50 consists of sheets 2,3 and 5 only; the BL example consists of all the plates joined to form a single massive roll.

## 73

## EDWARD LANGLEY and WILLIAM BELCH                          1817

Size: 25.5cm x 17.3cm.

Langley and Belch traded together from 1805 until 1820, at which time they began operating separately for several further years. Langley had his own printing shop in Dorking from 1799 before moving into London.[1] The map is closely related to Cole and Roper's map of 1807 (**63**). The atlas sold for £2–12–6 bound.[2] By 1820 the plates were with Joseph Phelps. He issued one atlas, but there is evidence that the atlas (and perhaps loose sheets) were on sale for quite a few years later.

## EDITIONS

(i) *LANGLEY'S new MAP of LINCOLNSHIRE.* (Aa–Ea, OS). *Printed and Published by Langley & Belch, N°. 173, High Street, Borough, London Nov*. 1*st*. 1817.* (Be–De, OS). *EXPLANATION/*11 lines of symbols, etc. (Ee). *DIVISIONS./1 Lindsey/2 Kesteven/3 Holland/ including 27 Hundreds/Sokes and Wapontakes/* (De). *Scale of Miles/10* [= 2.6cm] (Ce). An engraving of Boston Stump from the south-west, and below (left): *Height 282 feet/ Length 290 feet.* (and below, right): *Boston Church founded A.D. 1309.* (Ae). Compass (Ea). The inner border is marked and numbered every 10 minutes of latitude, etc. *Meridian of 0 Greenwich* (De, between frames).

*Langley's New County Atlas Of England and Wales, Embellished With A Beautiful Vignette To Each Map Exhibiting All The Mail Coach, Turnpike & Principal Cross Roads. With The Cities, Towns, Villages, Parks, Rivers, & Navigable Canals, Peculiar To Each County. Also an Index Map of England and Wales, shewing the Connection of one County with another. London. Published by Langley & Belch, 173, High Street, Borough.* [1818].[3]

**Copies**    Bod J Maps 8; Leeds UL W 122. (LAO Exley 32/2/162).

(ii)    *Old Witham R.* has been added, north-west of Boston; it already appears north-west of Tattershall. Previously the river was drawn with three straight lines below *Chapel Hill* but is now clearly drawn with many short bends.

*Langley's New County Atlas . . .* [1818].

**Copies**    BL Maps C.8.a.19; CUL Atlas 7.82.8. (BL Map 3355 (27); LRL Map 770).

(iii)    The imprint now reads: *Printed & Published by J. Phelps, N°. 27, Paternoster Row, London. 1820.* (Be–De, OS).

*Langley's New County Atlas . . . Joseph Phelps . . .* [1820].

**Copies**    Bod Allen 82. (LRL Map 770a).

*Langley's New County Atlas . . . Phelps . . . Row. London.* [1830].[4]

**Copy**    The private collection of Mr. C.A. Burden.

Issued dissected in a slip case, labelled *LINCOLNSHIRE, Published by J. Phelps, No. 27, Paternoster Row.* [After 1820].[5]

(iv)    The title and imprint have been removed. A label has been added with the county name and a new imprint: *Published by J. Phelps N° 44 Paternoster Row.* [not after 1839].[6]

### References
1   Todd, p. 116; Brown, p. 110.
2   The BL copy has the original label pasted to the cover.
3   Plates are variously dated 1816–1818.
4   Watermarks vary between 1817 and 1830 (Suffolk and Derbyshire).
5   Mr. E. Burden has a copy of Berkshire in this form and Lincolnshire may exist thus also.
6   Kingsley p. 154 records a copy of Sussex in this state. Phelps operated from this new address from 1834–1839 only. Kingsley also records an atlas in private hands with the map of Northamptonshire treated similarly.

Chubb CCCLIX–CCCLX.

**74**

**THOMAS CRABB**                                                      **1819**

Size: 10.45cm x 6.8cm.

Until the discovery in the early 1990s of a part set of cards with these small maps by Crabb the plates were known only by their first user in atlas form, Robert Miller. Chubb assigned the first issue by Miller to 1810 but without giving any evidence for the choice.[1] Whitaker first drew attention to the falsity of the dating, since his copy (now Leeds UL W 117) includes advertisements for the 'Memoirs of his late Majesty . . . George III . . . with . . . likeness of . . . George IV'. Thus, Whitaker's copy must have been prepared in 1820 or later. The maps are copies of the 1806 Cary map.

**Later history**
The plates seem to have passed rapidly from Cragg to Robert Miller, then to William Darton. On several occasions they also appear on cards. In that form the Lincolnshire plate came out with the imprint of Hodgson & Co. This presents a dating problem, since Hodgson & Co. were at the address on the card c.1825. However, the plates had, by 1822, come into the possession of William Darton and the later lithographic transfers were made by that firm's successor. There is, therefore, an element of guesswork in trying to determine the correct order of the appearance of the maps in their manifestations in atlases and on cards. The final appearance as litho. transfers began in the mid-1840s, with the last in c.1850, as illustrations for simple children's books on the geography of England, written by Reuben Ramble, a pseudonym of the Rev. Samuel Clark. Clark (1810–1875) was made to join his father's prosperous basket and brush-making business when 13 but found time to teach himself Latin, Greek, Hebrew, French and German and to be well read in all of them. In 1836 he took himself off to London, where he quickly became Darton's partner. Through F.D. Maurice's influence he matriculated at Magdalen College, Oxford but took 7 years to obtain his degree because of his business activities. He became a curate in Northamptonshire. However, he then turned to education, specialising in geography, and from 1851 until ill-health forced his retirement in 1862 he was Principal of the Battersea Training College. He then returned to pastoral work at Bredwardine and, finally, at Eaton Bishop. Although his partnership with Darton ended in 1843 he wrote little books of geography for children, published by his former associate.[2]

**EDITIONS**

(i)  *LINCOLNSHIRE* (in a panel, with background shading of vertical lines, Be–De). *London. Published by T. Crabb. 15 John Street Blackfriars Road* (Be–De, OS). *Scale of Miles/10* [=1.05cm] (Ea). Compass (Aa). The inner frame is marked every 2′ of latitude, etc. and numbered every 10 minutes. *Long W. from London* (De, in the border).

Issued as a set of geographical cards. [1819].[3]

**Copy**  The private collection of Mr. C.A. Burden.

(ii)  The imprint now reads: *London. Published by R. Miller, 24, Old Fish Street.* (Be–De, OS). Plate number *23* has been added (Ea, OS).

*Miller's new Miniature Atlas, containing a complete set of County Maps, in which are Carefully Delineated All the Principal Direct & Cross Roads, Cities, Towns, Villages, Parks, Seats, Rivers, & Navigable Canals. with a General Map of England & Wales. London. Published By R. Miller, 24, Old Fish Street S<sup>t</sup>. Pauls.* [1821].[4]

**Copies** BL Maps 3.a.71; CUL Atlas 7.81.6; Leeds UL W 117; Leeds RL 912.42 M617.
Issued as cards. [1821].[5]

**(iii)** The imprint has been altered to: *London Published by C. Hinton, 1 Ivy Lane Paternoster Row; & J. Wallis, Berwick Str. Soho.* (Ce, OS).
Issued as cards. [1822].[6]

**(iv)** The imprint has been altered to: *London: William Darton, 58, Holborn Hill.*
Issued as cards. [1822].[7]

*Darton's new Miniature Atlas . . . Published by William Darton, 58, Holborn Hill.* [1822].[8]

**Copies** BL Maps 32.aa.3; CUL Atlas 7.82.27; Leeds UL W 127; NLS Newman 753; John Rylands UL L 912 D256 (R 31818).

**(v)** The plate number has been removed. The imprint now reads: *Published by Hodgson & Cº. 10, Newgate Street.* (Be–De, OS). The scale has been moved (Ae). Compass now (Ee). *Mail Coach R$^{ds}$ thus/* double lines, one thick and one thin (Ea). *50.* [= 50 minutes] has been added in the upper margins at the left and right. The coastal shading has been removed.

Issued as a set of cards. [1825].[9]

**Copy** BL Maps 32.aa.14.

### LITHOGRAPHIC TRANSFERS
**(A)** The map in state (iv), but without imprint, has been transferred to the centre of a larger sheet surrounded by drawings of rural scenes and, upper left, Boston Stump with Town Bridge. The title is no longer in a panel.

*Reuben Ramble's Travels In The Eastern Counties Of England. London. Darton & Clark. Holborn Hill. Nosworthy & Wells, Litho. 24, New Union Str$^t$. Moor Lane, City.* [1845].[10]

**Copy** RGS F 215.

*Reuben Ramble's Travels through the Counties of England. With Maps and Historical Vignettes. Darton & Clark, 58 Holborn Hill, Print And Map Publishers.* [1845].

**Copy** BL Maps 1302 g 2.[11]

*The Child's Treasury Of Amusement And Knowledge; Or, Reuben Ramble's Picture Lessons: London. Darton & Clark, Holborn Hill.* [1845].

**Copy** CUL CCD.7.11.27.

**(B)** The lettering below the Boston illustration, which forms part of the design surrounding the map, has been altered. Instead of italic lettering a roman alphabet has been introduced.

*Reuben Ramble's Travels Through The Counties Of England. With Maps And Vignettes. London: Darton & Clark & Co., Holborn Hill.* [1850].[12]

**Copy** CUL Atlas 7.85.6.

**References**
1 Chubb, p. 253.
2 DNB. Vol. X. (London, 1887), pp. 405–406.
3 One of the cards (Scotland) is dated Jan$^y$. 1819. Crabb is noted in Robson's London Directory (1820) at the address on the card.

4  Whitaker showed that 1820 was the earliest date for his copy. The advertisements in his copy (on p. iv) differ from those in the CUL example, where, on p. ix, occurs an offer of portraits of the royal family, including George IV. The BL copy is similar to CUL's. There were thus two editions in 1820–1821. Which came first is still conjectural.
5  BL Maps 3.a.71 has advertisements for the maps 'on extra-super Large Cards, plain 5s. the set, or 2d. each. The same, coloured, 7s.6d. the set, or 3d. each. A certain number of copies are done up with a topographical account of each county, printed at the back of each map, at an additional charge of 1s.6d. the copy'. Lincolnshire has not been found in any of these versions.
6  Bod J Maps 279 (64) is a coloured example of Warwickshire. No other card is recorded and, possibly, not all counties were issued. Dating is very conjectural; Miller had the plates in 1820–1821 and Darton may have had them in the following year.
7  Bod has five examples (J Maps 65–69), the Yorkshire card differing in not having a plate number. They were probably issued simultaneously with the book, as with Miller (note 5 above).
8  On p. 27 under Yorkshire is a reference to the 'population now (1821) . . .' It is assumed that the volume came out soon after the availability of the census figures. Chubb (p. 255) refers to a copy, which the owner dated to 1825 on the grounds of the inclusion of these figures. It is doubtful if there were two issues. William Darton moved in 1823 to set up in Great Eastcheap, while Darton senior only ever operated from Gracechurch Street (1794–1838; Todd, p.53).
9  There are 45 cards in the BL's incomplete set, with several variations of names in the imprints, suggesting that not all owners of the plates changed the imprint while in their possession. Several cards retain plate numbers. Hodgson & Co. were at this address in c.1825 (see Hodson, 1974, p. 114).
10 All the examples of state (A) were probably issued simultaneously since the maps and the accompanying county texts do not vary. Darton & Clark remained at the Holborn Hill address until 1845. On the cover of the first example a map shows railways in East Anglia, among them the Yarmouth & Norwich Railway (open in May, 1844).
11 Mr. C.A. Burden's copy has the title-page differently arranged.
12 Generally ascribed to 1850 but the evidence is not clear. An example in Bod (Maps C.17.c.21) only has 6 maps. Another in NLS (Newman 592) lacks the title-page. Neither includes Lincolnshire.

Chubb CCCXL–CCCXLII; DXVII.


# 75

**THOMAS DIX**                                                                      **1820**

Size: 42.9cm x 34.8cm.

Darton (son of the Darton of Darton and Harvey, etc.) has already been noted. Thomas Dix (c.1769–1813) worked as a surveyor and is recorded working in Norfolk.[1] From the title-page it seems that Dix died also in that county. In 1805 he published *A treatise on the construction and copying of all kinds of maps*. The maps in the first atlas form date from 1816 to 1821 and Kingsley has inferred that there might have been earlier issues than those in sheet form with Darton's imprint.[2] The maps follow Rowe (**71**), copying Smith (**56**).

The maps appeared twice in atlas form and at least that many times in slip-case forms spread over a period of fifteen years. In c.1853 the plate re-appeared for a Post Office map and in 1877 a lithographic transfer marks its final use.

## EDITIONS

(i)  *A NEW MAP/of the County of/LINCOLN,/Divided into Wapentakes, &c./By/*

*THOMAS DIX./* (in a circle, Ea). *LONDON; WILLIAM DARTON, 58, HOLBORN HILL; JUNE 16, 1820.* (Ce, OS). *EXPLANATION/*11 lines of symbols, etc. (Ee). *REFER-ENCE/to the/WAPENTAKES, &c/*31 numbered names in three groups (Ec & Dd). *SCALE OF MILES/12* [= 6.25cm] (De). Compass (Ac). An engraving and below: *LOUTH STEEPLE, &c from the S.E.* (Ae). *NOTE./* followed by four lines of text (Aa). *NOTE, continued* (Ca) is followed by 10 lines of text. The inner border is marked and numbered with every minute of latitude, etc. *Longitude West from Greenwich* (De, between borders, which shows the erroneous 1° instead of the meridian).

Issued as a loose sheet in a slip-case; the cover title is: *LINCOLNSHIRE LONDON; WILLIAM DARTON, HOLBORN HILL BY THO^S. DIX; Surveyor North Walsham*; there is another (different) label on the back. Stuck on the back of the map there is an advertisement leaf, which says the map 'may be had in sheets, Cases for the Pocket, or, on Canvass with rollers'. [1820].[3]

**Copy**  LRL Map 835.

*A Complete Atlas Of The English Counties, Divided Into Their Respective Hundreds, &c . . . Commenced By The Late Thomas Dix, Of North Walsham; Carried On And Completed By William Darton, London; William Darton, 58, Holborn Hill. 1822.*

**Copies**  BL Maps 1.e.19; CUL Atlas 2.82.9; Leeds UL W 133a; Nottingham UL G 5512 oversize xx; Birmingham RL A F912.42. (LRL Map 775).

**(ii)** The title now reads: *LINCOLNSHIRE,/Divided into Wapentakes &c./AND THE/Parliamentary Divisions./* (Ea). *MARKET TOWNS and MARKET DAYS,/*2 columns each of 17 places and dates, ending with Wragby below the columns (Eb). *REFER-ENCE/to the/Wapentakes &c/*31 names in three sections (Ec-Ed). The imprint now reads: *LONDON: WILLIAM DARTON & SON, 58, HOLBORN HILL.* The *EXPLANATION* has been re-engraved and enlarged, so that there is no area delineated east of the [Nene] (and the place-names have also been deleted); there are now 16 lines of text and symbols (Ee). The meridian is now correctly marked *0* and *2* has been changed to *1* in the upper and lower borders. The compass and the text (Ae and Ca) have been deleted. Crosses have been added for polling places; double rings at Lincoln and Sleaford; diamonds for places returning 2 MPs (4 boroughs) but a triangle at Grimsby (one MP). *SOUTH/ DIVISION* and *NORTH/DIVISION* have been added. The delineation of East and West Fen has been removed to allow the inclusion of new roads in the triangle Coningsby-Boston-Wainfleet. The road from *Gantlet* (near Swineshead) to Holbeach is now continuous and upgraded. Other upgraded roads include: Grimsby-Utterby-Louth, Boston-Skegness and Wragby-S^th Elkington. Many changes have been made in the Holland area, e.g. *Old Guide Ho.* added (near Fosdyke), *Drove/End* added, but *Gedney Drove* and *to Wisbeach* have been removed. *NEW/BOLINGBROKE, Adlingfleet* and *Tetney Haven* are three of the places added. *Kirton* has been re-engraved as *KIRTON/in Lindsey.* Many places have been added in Yorkshire and its county name has been re-engraved.

Issued in slip-case, with labels to the front and back covers, identical to the [1820] edition. [1835].[4]

**Copy**  Grimsby RL X000:912 DIX.

*The Counties Of England: With General Maps Of North And South Wales. London: William Darton And Son.* [1835].

**Copies**  BL Maps C.18.c.9; Bod Allen 101. (Northamptonshire CRO Map no. 3585).[5]

**(iii)** The imprint now reads: *LONDON: JNO. DARTON & Co. 58 HOLBORN HILL.* Railways 1–4, 6–11, 13 have been added; railway 15, named *AMBERGATE/RAIL,* extends beyond Grantham in a straight line passing south of Spalding across the fens, labelled *TO WISBEACH* as it leaves the county. The MSLR at Gainsborough is labelled *FROM WORKSOP*. Railway 4 reaches the coast nearer Barton than [New Holland]. More place-names have been added, e.g. *Halton/Chap.*, *Coleby* (north of *Burton/upon Stather*) and *Gibralter* (sic).

Issued as a folded map inside a cover with the title: *The Post Office Map Of* [Royal insignia] *Lincolnshire. Dedicated To Rowland Hill. Esq^r.* [1850].[6]

**Copy**   The private collection of Mr. M. Stafford.

**(iv)**   The imprint now reads: *LONDON. DARTON & HODGE, 58 HOLBORN HILL.* New railways have been added.

Issued in a cover with the title: *DARTON & HODGE'S Railway, Commercial AND TOURISTS' MAP OF LINCOLNSHIRE. PUBLISHED BY DARTON & HODGE, 58, HOLBORN HILL.* [c.1864].[7]

## LITHOGRAPHIC TRANSFER

**(A)**   The Darton imprint has been replaced by: *W.J. FREEMAN, STEAM-LITH^O. 2, OLD SWAN LANE, UPPER THAMES S^T. E.C.* (Ae, OS). The title now reads: *OFFICIAL MAP OF/LINCOLN* (Ea). Railways 12, 14, 16–43 have been added; the Mablethorpe line (railway 44) has been added but not in the correct alignment. *TO/LYNN* has been added where railway 29 leaves the county. Railway 31 is drawn from Long Sutton towards [Wisbech]. The extension of the Ambergate-Grantham line towards Spalding, etc. has been deleted. The list of Market towns has been reduced to two columns each of thirteen names and the market days altered. Below are five lines on the parliamentary representation for the county (Eb). *New/Holland*, *THE/WASH* and *French Drove* have been added. *KINGSTON/upon HULL* has been replaced by *HULL*. In the *EXPLANATION* two lines of symbols and their significance have been replaced by one line: *Railways & Stations* with symbol. A compass has been added (Ed). *NORTH DIVISION* is coloured green, with the boundary emphasised by a deeper shade of green; *MID* is violet with a deeper shade for the boundary and *SOUTH DIVISION* is lemon with deep yellow edging.

*Lincolnshire. The 'Official' County Map And Guide Shewing Its Railways, Telegraphs, Roads, Rivers, Canals, Market Towns And Days; With Geological, Archaeological, Historical, Statistical, Descriptive, Parliamentary, And Postal County Information For Home Reference, And For Travellers, Tourists, And Men Of Business. London: Offices: 19 Salisbury Street, Strand. Simpkin, Marshall, & Co., Stationers' Hall Court . . .* [1877].[8]

**Copies**   CUL Maps C.34.010.70:1; Bod 10352 bb 2 and G.A. Lincs. 8° 23 (11); NLS Map s.53.1.

### References

1   Eden, P. Land Surveyors in Norfolk, 1550–1850. *Norfolk Archaeology*, Vol. 36, pt. 2 (1975).
2   Kingsley, p. 156 noted that two impressions of Sussex had traces of an erased signature with a possible address in the Strand.
3   It is assumed Darton made use of his plates as soon as they became available. The atlas form was still two years off. The map in its other advertised forms has not been found.
4   Probably issued simultaneously with the atlas (usually put at 1835). The inclusion of election data indicates a post-Reform Act date of issue.
5   Brown, p. 53 notes that W. Darton & Son traded at Holborn Hill from 1830 to 1837. The BL atlas has manuscript lines drawn for railways opened by 1838. The three Lincolnshire examples all show variations of colouring affecting the election and polling symbols.
6   A manuscript note on the inside cover reads: John Carter, Holbeach, Lincolnshire June 24 1850.
7   Burden p. 122 reports maps of Surrey and Cornwall with covers, which list Lincolnshire as issued in this form.

8    The book contains a note of further guides due to appear on 1 Jan. 1878. The guide sold for
     one shilling (cover).

Chubb CCCLXXXVII (1822 Edition) only.


**76**

**SIDNEY HALL**                                                                    **1820**

Size: 12cm x 6.9cm.

Hall (*fl*.1818–1860), the engraver of these plates, was a prolific producer of maps in
numerous editions. The present pocket atlas appeared regularly until 1843 coupled with
a popular road-book. Hall also engraved plates for another long-running series of works
(**94**). The present plates are derived from Cary's 1806 plates, but show carelessness in
copying; far more detail is compressed into the small format than Cary attempted, clarity
being sacrificed. Samuel Leigh had various addresses in the Strand from 1812 until 1832;
he was succeeded by Mary Ann Leigh for two years, when the firm became Leigh & Son
(1834–9) and Leigh & Co. (1840); George Biggs took over the business but soon ceased
to publish the atlas and road-book.[1] William Clowes, master printer, began his career in
1801 and was the founder of the present firm of high quality printers.[2]

**EDITIONS**

(i)    *LINCOLNSHIRE* (in a panel, above the inner frame line, Ba–Da). *Pub. by S. Leigh
18. Strand.* (Ae, OS). *Sid^y. Hall sculp^t.* (Ee, OS). Plate number *22* (Ea, OS). Compass
(Ae). *English Miles/30* [= 3.55cm] (Ce–Ee).

*Leigh's New Pocket Atlas Of England And Wales, Consisting Of Fifty- Five Maps Of The
Counties, And A General Map; With A Complete Index Of The Towns, Villages, Country-
Seats, &c. London: Printed For Samuel Leigh, 18, Strand, By W. Clowes, Northumber-
land-Court. 1820.*

**Copies**    Bod J Maps 200; Leeds UL W 128; Birmingham RL 912.42 Skett 1.

*New Picture Of England And Wales . . . Comprehending A Description Of The Principal
Towns . . . A Synopsis of the Counties. &c., Embellished With Numerous Views, And A
General Map . . . 1820.*[3]

*Leigh's New Atlas of England & Wales. Published by Samuel Leigh, 18, Strand.*
[MDCCCXXV].[4]

**Copies**    BL Maps 24.a.30; RGS F 116.

*Leigh's New Atlas . . .* [MDCCCXXVI].

**Copies**    CUL Atlas 7.82.12; Manchester RL LC912.42 Le1.

*Leigh's New Atlas . . .* [MDCCCXXXI].

**Copies**    Bod Allen 92; Leeds UL W 129; RGS F 117.


(ii)    The imprint now reads: *Pub. by M.A. Leigh 421 Strand.* Canals have been added at
Caistor, Keadby and Grantham. One star has been removed at Grimsby.

*Leigh's New Atlas . . .* [MDCCCXXXI].

**Copy**    Guildhall L. S 388/1.

*Leigh's New Atlas . . .* [MDCCCXXXIII].

LINCOLNSHIRE 22

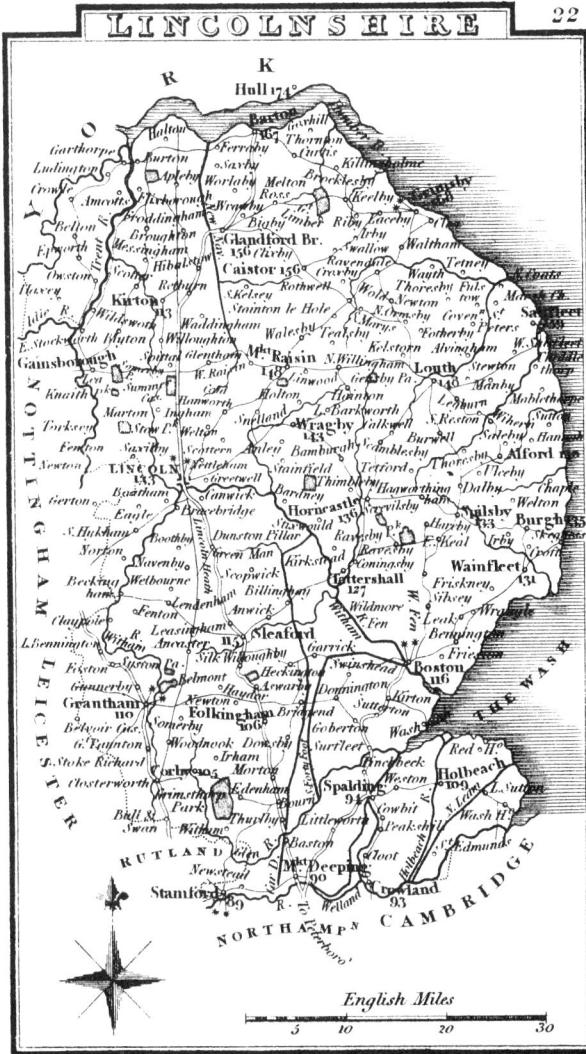

Pub. by S. Leigh 18 Strand.

Sid? Hall sculp?

English Miles

6. 1820 MAP 76 (i) – SIDNEY HALL.
Reproduced at 116 per cent

**Copies**  BL Maps 197.a.8; Bod Gen Top 16° 363; RGS F 118 and F 119; Guildhall L. S 388/1.

*Leigh's New Pocket Atlas Of England And Wales; Consisting Of Fifty-Five Maps Of The Counties, And A General Map: With The Number Of Members Returned To Parliament, And Divisions Of Counties. Corrected since the passing of the Reform Bill. New Edition. London: Printed For M.A. Leigh, 421, Strand. (Removed from No. 18.) 1834.*

**Copy**  Bod Allen 99.

*Leigh's New Atlas of England And Wales. London Published by M.A. Leigh, 421, Strand.* [MDCCCXXXV].

**Copy**  CUL Atlas 7.83.27.

**(iii)**  The note on the engraver (Ee, OS) has been removed.

*Leigh's New Atlas . . .* [MDCCCXXXVII].

**Copies**  BL Maps 24.a.31; CUL Atlas 7.83.15; Leeds UL W 129a.

*Leigh's New Atlas . . .* [1839].

**Copies**  CUL Atlas 7.83.8; Leeds UL W 130.

*Leigh's New Atlas . . . Published by Leigh & Son 421 Strand.* [1840].

**Copies**  Bod G.A. Gen. Top. 16° 364; RGS F 122.

*Leigh's New Atlas . . .* [1842].

**Copy**  Guildhall L. S 388/1.

*Leigh's New Atlas . . .* [1843].

**Copy**  Leeds UL W 131.

### References
1  Brown, p. 112.
2  Brown, p. 40; Todd, p. 40.
3  Chubb, p. 272 refers to a note in the introduction of this title that he (Leigh) projects with Hall 'a series of correct county maps, the same size as this book . . .' It has been taken to imply (by Whitaker, H., 1949, no. 367a) that the county maps appeared in this title. Chubb has been misread. The full text (p. xi of the book's introduction), ends 'County Maps . . . which may be had in a *separate* (my emphasis) volume . . .' There is a copy of the quoted work in LAO (Ex 942.6/34) and it has the general map only (all that is called for). However, Mr. T. Burgess has a copy that contains a set of county maps; apart from a new spine, his copy seems to be unchanged in make-up from its date of issue. It must have been specially prepared, perhaps as a presentation copy, since the contents do not match the title-page. Cruchley was advertising this work (on the back of folding maps) in the mid-1830s at 13s. or 20s. with a full set of county maps. My thanks to Mr. E. Burden, who has the folding map.
4  Virtually all copies of this and subsequent editions have two title-pages; the shorter engraved title-page is used here; the second is longer (letter-press), carries the date and a note on the edition. Most copies are bound with *Leigh's New Pocket Road-Book Of England And Wales* . . . which was sold separately with or without the 55 county maps.

Chubb CCCLXXIII–CCCLXXXII.

# 77

**SAMUEL JOHN NEELE**                                    **1820**

Size: 16cm x 12.8cm.

Neele's work has already been noted. George Byrom Whittaker (1793–1847) was originally a schoolmaster before moving to London and working as a bookseller/ publisher. He obtained the plates that Cooper prepared for Capper's *A Topographical Dictionary* . . . in 1825 (**64 (iii)**), the year after he was sheriff of the City of London. The career of William Pinnock (1782–1843) was similar to G.B. Whittaker. He too was born in Hampshire, became a schoolmaster and then turned to bookselling, firstly at Alton, then in Newbury, before coming to London in 1817. He is described as a master printer.[1] However, from bookselling and publishing he is said to have earned £4–5,000 p.a. until an unwise speculation led to his selling his copyrights to Whittaker. He was well-known for the publication of a series of over eighty catechisms, which covered a wide range of subjects, including geography; it is tempting to think that it was Pinnock's Catechism of Geography that Maggy Tulliver in George Eliot's *Mill on the Floss* seemed to have by heart.[2] These books were prepared in partnership with Samuel Maunder, whose sister he married. His publishing business seems to have foundered since he is also described as selling musical instruments, making globes and running a warehouse for pianos by 1822. The plates and Pinnock disappeared by 1826.[3] The map is an enlarged version of Cary's small maps of 1806 (**61**).

## EDITIONS

(**i**)   *LINCOLNSHIRE* (in a panel, Eb). *Neele & Son Sculp$^t$. 352 Strand.* (Ee, OS). Compass above *Scale of Statute Miles/10* [= 1.9cm] (Ed).

*Pinnock's County Histories, Historical and Topographical; With Biographical Sketches, &c.&c. And A Map Of Each County. In Six Volumes. Vol. III. London: Printed For G. Whittaker And Co., 13, Ave Maria Lane.* [1823].[4]

**Copies**   The private collections of Mr. D. Hodson and Mr. C.A. Burden.

*Pinnock's County Catechisms. The History and Topography Of Lincolnshire . . . Pinnock and Maunder.* [1820].[5]

(**ii**)   *EXPLANATION./7* lines of symbols, etc. have been added (Ea).

*The Traveller's Pocket Atlas Consisting of A Complete Set of County Maps, For England & Wales, On an Original & Improved Plan. The Roads leading to the nearest Towns in the adjoining Counties being delineated On Each Map To which is added an Alphabetical List of the Cities, Boroughs & Market Towns Of England & Wales, their distance from London, Population, &c. London. Published by Pinnock & Maunder, Mentorian Library, 267, Strand. Neele & Son, sc. 352, Strand.* [1820].[6]

**Copy**   CUL Atlas 7.82.26.

(**iii**)   An imprint has been added: *Published by G. & W.B. Whittaker, Ave-Maria-Lane, 1821.* (Ce, OS).

*The Traveller's Pocket Atlas . . . London, Published by G & W.B. Whittaker.* [1821].[7]

**Copies**   RGS 265 a 7. (LRL Map 197).

*The Traveller's Pocket Atlas . . . G & W.B. Whittaker, 1823.*

**Copies**   BL Maps 24.a.39; CUL Atlas 7.82.3 and 7.82.13; Leeds UL W 135.

*The Traveller's Pocket Atlas . . . 1823.* [1824].[8]

**Copy**   Bod Allen 87.

*Pinnock's County Histories. The History And Topography Of Lincolnshire, With Bio-graphical Sketches, &c.&c., And A Map Of The County. London: Printed For G. And W.B. Whittaker, Ave-Maria-Lane. 1824.*

**Copies**   CUL Syn 8.81.59; LAO Sc P.2/10; LRL A 9; Grimsby RL X000:914 PIN.

*Pinnock's History And Topography Of England And Wales . . . London: Printed For Geo. B. Whittaker, Ave-Maria Lane. 1825.*

**Copy**   BL 10348 aa 17.

### References

1   Todd, p. 151.
2   *Mill on the Floss.* (London, Zodiac Press edition, 1951), p. 113.
3   Brown, p. 150; Kingsley, p. 160.
4   Although these volumes are dateable on internal evidence to 1823 they incorporate the map in its earliest form. Each county text has separate pagination; with the addition of individual title-pages they were issued as separate works – see 1824 below.
5   Other counties are recorded with this form of title and Lincolnshire may exist thus. Burden, p. 119; Hodson, 1974, p. 113, and Kingsley, p. 161 note single copies only of their counties.
6   The London map is dated Feb. 1. 1820.
7   All maps except Oxfordshire are dated 1821.
8   Several maps, including Lincolnshire, are watermarked 1824. A similar example is in private ownership.

Chubb CCCXC (1823 atlas only; he refers also to an 1821 example).

## 78

**W.H. REID**                                                          **1820**

Size: 10.35cm x 6.8cm.

The map is based on Cary's maps of 1806 (**61**). Reid is not recorded as a publisher in the standard lists. The plates were only used until 1825.

### EDITIONS

(**i**)   *LINCOLNSHIRE* (in a panel, Be–De). *London; Publish'd by C. Hinton, 1, Ivy La. Paternoster Row; & J. Wallis, Berwick Street Soho.* (Ae–Ee, OS). *Scale of Miles/10* [= 1.05cm] (Ae). *Mail Coach R$^{ds}$. thus*/double line symbol (Ea). Compass (Ee). The inner border is marked with every minute of latitude, etc. and numbered each 5 minutes. *Long. W. fr. London* (De, between borders).

*The Panorama: Or, Traveller's Instructive Guide; Through England And Wales; Exhib-iting All The Direct And Principal Cross Roads, Cities . . . Accompanied By A Description Of Each County . . . London And County Bankers, Members Of Parliament, The Route Of The Mail Coaches, and the Portage of Letters . . . London: Printed by J. Wallis, Berwick Street, Soho; And Published By W.H. Reid, Charing Cross.* [1820].[1]

**Copies**   BL Maps 32.aa.2; CUL Atlas 7.82.46 and 7.82.48; Leeds UL W 132.

(**ii**)   The imprint has been removed.

*The Panorama . . .* [1820].[2]

**Copy**    Bod Allen 83.

**(iii)**    A new imprint now reads: *Published by Hodgson & Cº. 10, Newgate Street.* (Be–De, OS).

*The Panorama Of England And Wales. Published by Hodgson & Cº. 10. Newgate Street.* [1825].[3]

**Copy**    Bod Allen 89.

Issued as part of a set of cards. [1825].[4]

**Copy**    BL Maps 32.aa.14.

**References**

1    The county text occupies pp. 60–64 with the map facing p. 60. On p. 62 is: '... the House of Commons, July the 12th, 1820 strongly reprobating the electioneering conduct of this town' (Grantham).
2    It is generally accepted, for lack of other evidence, that states (i) and (ii) appeared in the same year.
3    Hodgson & Co., were at this address for only a few years in the mid-1820s. 1825 is generally accepted.
4    Several county cards have plate numbers. Lincolnshire may also exist in that state.

Chubb CCCLXXII (1820 edition only).

**79**

**AARON ARROWSMITH**                                                                    **1821**

Size: 47.1cm x 39.2cm.

The work of Arrowsmith has been noted earlier (**72**). This map covers the whole county with Leicestershire, but carefully excludes Rutland. In *Valor Ecclesiasticus* it is accompanied by another similarly sized map, which includes Rutland, Northamptonshire and parts of Hertfordshire, Bedfordshire and Buckinghamshire.[1] Between the two, therefore, the whole of the former Lincoln diocese is mapped. It is difficult to ascribe with any certainty the model for Arrowsmith's engraving; perhaps the nearest was his own map of 1815, but there are several differences, apart from the general removal of much geographical detail from the earlier map to permit the inclusion of the names of all the parishes (in their Latin form). In 1836 the Church Commissioners produced a report that contained another map of part of the old diocesan area, engraved by Samuel Arrowsmith (**106**).

**EDITION**

**(i)** *DIOC'/LINCOLN'/PARS BOREAL'/* (Aa). *Tabula – Juxta Valorem Ecclesiasticum XXVIº. Henrici VIII, Institutum – Geographica* (Ca, OS). *A. Arrowsmith delin.* (Ae, OS). *A.D. MDCCCXXI.* (Ee, OS). *NOTÆ EXPLAN./*16 lines of abbreviations and their meanings (Ab). *DEC' XRIANITAT[S]./Civitas Lincoln'/*21 parish names.(Ac). *DEC' XRIANITAT[S]./Leicestr'/*10 parish names. (Ac, below the Lincoln list). *GRYMMESBY/*4 parish names (Ec). *STAMFORDE/*16 parish names (Ec, below the Grimsby list). All these lists of parish names are within their own separate boxes.

*Valor Ecclesiasticus Temp. Henr. VIII. Auctoritate Regia Institutus. Vol. IV. Printed By Command Of His Majesty King George III. In Pursuance Of An Address Of The House Of Commons Of Great Britain. 1821.*

**Copies**  BL 21 g 15 and 10072;[2] CUL OP.1.16; Bod G.A. Eccl. top. 21–26 = R, etc; Liverpool RL If 30; Sheffield RL 283.42 STF; Canterbury Cathedral L. F14.12. (Bod Gough Misc. Antiq Fol. 3 (No. 305); LRL Map 810).

*Valor Ecclesiasticus . . . 1821.* [1834].[3]

**Copy**  LAO SR.

#### References

1   The second sheet is entitled: *DIOC'/DE BURGO/SANCTI PETRI/antea/LINCOLN'/* and is very similar to the county map, with boxed lists of parishes, imprint, date and heading.
2   There is a third set in BL (CIRC 81.b.60), but it only has the second map of the Lincoln diocese.
3   On the verso of the title is: 'Record Commission 12th March 1831'; it is further dated 1834. The CUL set has on its verso a printed note: 'This Book is To Be Perpetually Preserved In . . . The Public Library, Cambridge, 1822'. Similar notes appear in the Canterbury and BL (21 g 15) copies.

Not in Chubb.

**80**

**JOHN CARY**                                                             **1822**

Size: 13.25cm x 9.2cm.

This is the third and final version of the plate originally produced by Cary in 1789 (**50**) and then re-engraved in 1806 (**61**). In this set publication was shared with his brother George. After Cary's death in 1835 the family probably produced one more edition; the plates were acquired by Cruchley and he used them as the basis for several lithographic transfers until about 1868. Compared with the 1806 version only one more place-name has been added: *Sibsey*. All maps are single-sided, except the Cruchley transfers.

#### EDITIONS

(i) *LINCOLNSHIRE* (in a panel, which protrudes above the upper frame line, being surmounted by a 'half' compass (Ba–Da, OS). *London. Published by G. & J. Cary, Nᵒ. 86 Sᵗ. James's Str.* (Be–De, OS). *10* (plate number, some 2cm. below the lower neat line). *Statute Miles/10* [= 1.4cm] (Ae). Below the map area a framed area includes 3½ lines of county place names and their distances from London (Ae–Ee).

*Cary's Traveller's Companion . . .* [the rest of the title is exactly as engraved for the 1806 series]. *London: Printed for G & J Cary. Engravers. Nᵒ. 86 Sᵗ. James's Street.* 1822.

**Copies**  CUL Atlas 7.82.4; Bod Allen 86; Leeds UL Case K 34; London UL [G.L.] 1822; Norwich RL R912.42; Guildhall L S 388/1.

*Cary's Traveller's Companion . . . 1822.* [1823].[1]

**Copies**  BL Maps 14.dd.36; RGS F 62 and 260 C 10.

*Cary's Traveller's Companion . . . 1824.*

**Copies**  BL Maps 14.dd.37 and 28.c.9; CUL Atlas 7.82.10; Leeds UL W 99.

*Cary's Traveller's Companion . . . 1826.*

**Copies**  RGS 260 C 12; Guildhall L. S 388/1.

*Cary's Traveller's Companion . . . 1828.*

**Copies**  CUL Atlas 7.82.50; Leeds UL W 100; NLS Newman 826; Maidstone. Centre for Kentish Studies. Kg.

**(ii)**  The plate number has been changed to *12*.
*Cary's Traveller's Companion . . . 1828.* [1829?].[2]
**Copies**  BL Maps 24.a.24; Bod Map C.17.e.25; RGS 9 A 62.

**(iii)**  Roads have been added: Louth-Alford; Louth-Scartho; Sibsey-Tattershall; Wrangle-Tattershall; near Sibsey-[Revesby]; Spilsby-Wainfleet; Sutterton-Holbeach-Cross Keys/Ho.; Spalding-Bourn; Heckington-Bridgend/Causeway. One of Grimsby's stars has been removed. The area around Fosdike has been re-engraved to permit the inclusion of a new road. A park has been added at Revesby. The Horncastle-Tattershall road has been upgraded.
*Cary's Traveller's Companion . . . 1828.* [1835].[3]
**Copies**  CUL Atlas 7.82.22; RGS S/G 91.

An edition ascribed to the early 1840s has been suggested on the basis of the inclusion of railways on a loose Surrey map. In such an edition Lincolnshire's plate is likely to be as in state (iii).

**LITHOGRAPHIC TRANSFER**
**(A)**  Size: 12cm x 9.4cm. *LINCOLNSHIRE* (Ba–Da, OS). *21* (Aa, OS). *London. Published by G.F. Cruchley, Map Seller & Globe Maker 81, Fleet Street.* (Be–De, OS). Railways 1–4, 6–26 are drawn in; this has meant the addition of many names as all places with stations are underlined. The four railways entering the county from the west are named. *Hull* has been added. The list of villages in a panel below the map has been removed. The map is on the verso of Leicestershire.

*Cruchley's Railroad Companion To England & Wales Shewing All The Railways & Stations With Their Names, Also The Turnpike Roads To All The Market And Borough Towns With The Distance From Town To Town Delineated In A Series Of 42 County Maps. G.F. Cruchley, Map-Seller & Globe Maker 81, Fleet Street, London.* [1862].[4]
**Copies**  BL Maps 1.aa.29; Leeds UL W 101; Leicester UL AJ 180 (L).

**INTAGLIO PRINTING**
**(iv)**  Size: 13.25cm x 9.2cm. The map in state (iii) has the following alterations. The plate number appears twice now (Aa and Ea, both OS). Railways 27–30, 32–36 have been added; railway 31 has been drawn in from *L. Sutton*. The addition of the new railways has led to the inclusion of more places, underlined to show that they have stations, including Crowland. The map is single-sided.
*Cruchley's Railroad Companion . . .* [1868].[5]
**Copy**  NLS EME.b.5.8.

**References**
1  The Yorkshire map in two examples is watermarked 1823; in one of the RGS copies several other plates are similarly watermarked.
2  The numbers at the bottom of the page only apply to 20 maps and are not applied in any special sequence; they occur only on pairs of facing plates. None of the paper used for this edition is watermarked later than 1827; it is assumed that it appeared very shortly after the first state, also dated 1828. The label of the map-seller James Wyld obscures the date on the title-page of the Bod copy. A different, later, version of the Wyld label covers the title-page imprint on the RGS copy and another in private hands.

3    The suggestion of 1835 for this work comes from a note in the CUL copy, which draws attention to the inclusion of the London & Birmingham Railway (LBR) on the maps of Warwickshire, Northamptonshire, Hertfordshire and Middlesex. Although the act for the line was passed in 1835 it was not opened fully until September, 1838. Oddly, no part of the Great Western Railway is shown on the appropriate maps although it was being built at the same time as the LBR.

4    The county railway data suggest a slightly earlier date. However, the BL copy has the receipt date 24 JU 62 (24 June, 1862).

5    The railways marked were opened in 1867, but other county plates do contain 1868 lines. Mr. E. Burden draws attention to the fact that some plates in the NLS copy (presented by Mr. R.M. Gall Inglis in 1974) have manuscript additions; he suggests that it was marked thus as preparation for an intended edition in the late 1880s. Further support comes from the maps being printed directly from the original plates; i.e. they are not lithographic transfers.

Chubb CCLXXXIII, etc.

# 81

## J. WALKER                                                                          1822

Size: 12.6cm x 8.2cm. 9

Although the county map bears no information on the engraver or the person responsible for the initial drawing the majority of plates in this atlas were engraved by J. Cox from drawings made by J. Walker. On the map of Durham, however, Cox is recorded as the draughtsman. The map is a version of Cary's 1806 maps. One innovation is that the mileage mark against Lincoln is the distance from London, but those against the other towns are their distances from Lincoln. Earlier editions of *Crosby's Complete Pocket Gazetteer*, which first appeared in 1808, do not have county maps. Indeed, the first state below occurs in a single copy of the 1818 edition, whose title-page only calls for two general maps. There is no evidence in that copy to suggest later interpolation of maps; the most likely explanation is that a 'wrong' title-page was bound in at the time of preparation. J. Walker has not been firmly identified. Kingsley equates him with the John Walker, who was Admiralty Hydrographer from 1808 until he died in 1831.[1] Of his four sons, two (Thomas and Michael, who succeeded as Hydrographer) were Admiralty draughtsmen. His other two sons, John and Charles, set up a very successful business together as engravers and their works are recorded later (**102** and **105**). John, who was the eldest, was an engraver to the East India Company under James Horsburgh from 1825 and followed him as Hydrographer on his death in 1836. Whether the father or the son engraved the present series of maps remains unproven. Kingsley also notes a J. and a John Walker, who are potential candidates.[2] Among other later work by J. and C. Walker may be mentioned two railway maps of local interest. They engraved *Cambridge and Lincoln Railway. Engineer, George Watson Buck* . . . in c.1844 and two versions are known.[3] They engraved *Grand Union Railway from Nottingham via Grantham, Folkingham, Spalding, Holbeach & Long Sutton to Kings Lynn . . . Engineer. Chas. Vignoles . . .* probably in 1845.[4]

The printers have been encountered earlier in connection with the production of several works, notably *English topography* (1816) – see **63 (ii)**. John Bumpus, publisher of the map, is believed to be related to the well-known firm of booksellers, J. & E. Bumpus, founded by Thomas Bumpus in 1816 and still in operation. John was a bookseller, who published in 1827 a map of London[5] but drowned himself in 1832.[6] Benjamin Crosby (1768–1815), the compiler, came from Leeds and, in London, worked for other booksellers before setting up his own business, specialising in travelling the

his wares. He has been described as the founder of wholesale bookselling.⁷ After a severe
paralysis the other firms named on the title-page took over his business; William Simpkin
and Richard Marshall had been Crosby's assistants. Crosby died at Louth, a year after
his illness.⁸

## EDITIONS

(i)  *LINCOLNSHIRE./contains 2787 sq. Miles, 237,891 Inhabitants & sends 12 Members to Parliament.* (Ae–Ee, in a panel at the foot of the map). *London Publish'd May 1,
1822. by J. Bumpus, 6 Holborn Bars.* (Be–De, OS). No compass or scale.

*Crosby's Complete Pocket Gazetteer Of England and Wales, Or Traveller's Companion.
Arranged under the various Descriptions of Local Situation . . . Whatever is worthy the
attention of Gentlemen . . . A New Edition, Illustrated By Two Maps. London: Printed
For Baldwin, Cradock, And Joy, Pater-Noster Row; Simpkin And Marshall, Stationers'
Court; And J. Bumpus, Holborn Bars. 1818. [1822].⁹*

**Copy**  The private collection of Mr. T. Burgess.

(ii)  The note on the county and the frame-lines of the panel have been removed, leaving
the title alone (Be–De). The imprint is: *Published 1835, by J. Nichols & Son, 25
Parliament Street.* (Be–De, OS). *The figures show the distances from Lincoln.* (Ba–Da,
OS).

*The Family Topographer: Being A Compendious Account Of The Antient And Present
State Of The Counties Of England. By Samuel Tymms. Vol. V.- Midland Circuit. London:
J.B. Nichols And Son, 25, Parliament Street. 1835.*

**Copies**  BL 577 b 29–35; CUL VIII.19.5–11; Bod GA Gen. Top. 8° 9 V 1149–55;
Leicester UL H 942 TYM; Leeds RL 942 T976.

*A Compendium Of The Ancient And Present State Of Derbyshire, Leicestershire, Lin-
colnshire, Northamptonshire, Nottinghamshire, Rutlandshire, Warwickshire, Cheshire.
By Samuel Tymms. London: J.B. Nichols And Son, 25, Parliament Street; And Simpkin
And Marshall. Price 5s. [1835].*

**Copies**  LRL L9; Louth RL BL 9.¹⁰

*Camden's Britannia Epitomized And Continued; Being A Compendious Account Of . . .
England . . . Vol. V . . . London: Henry G. Bohn, York Street, Covent Garden. [1843].¹¹*

**Copies**  Bod GA Gen. Top. 16° 179–185; Liverpool RL F 1105.

(iii)  *LINCOLNSHIRE.* (in a panel, Ba–Da). The former title, note on the county and
imprint have been removed. The coastal areas have been outlined with further lines of
shading. One name has been corrected: *Weston* (near Spalding, previously *Welton*).

*A Sketch Illustrative Of The Minster And Antiquities Of The City Of Lincoln. Lincoln:
Printed For E.B. Drury, Simpkin And Marshall, Paternoster Row. London. 1835.
[1844?].¹²*

**Copy**  LAO L Linc 910 (Sc.P 5/2.11).

**References**
1  Kingsley, p. 202.
2  ibid., p. 203.
3  LAO Anc VI/C/3d and Anc VI/C/3i, the latter having an extra line for a branch joining
   Stamford to Spalding, via Market Deeping.
4  LAO Anc VI/C/3e.
5  Darlington and Howgego, op. cit., item 306.

6  Timperley, p. 927.
7  Smith, D. Previously Unknown Pocket Gazetteer Found. *Map Collector*, no. 29 (December, 1984), pp. 34–35.
8  Timperley, appendix, p. 11.
9  Only the three general maps and Yorkshire are not dated 1822. Mr. Burgess, the owner of this volume, and Mr. E. Burden first drew attention to the relationship between the maps in Crosby's work and those in the editions produced by Samuel Tymms. *Map Collector*, No. 38 (March, 1987), p. 52.
10  The copy in Louth lacks a title-page and consists of the county section only.
11  The *Family Topographer* consists of 7 volumes issued between 1831 and 1842. The Liverpool set has two title-pages in most volumes and the date in the final volume (seven) is 1843. The Lincolnshire map is in the work whose title is quoted here; a second title-page reads: *A Compendious Account Of The Antient And Present State Of The Midland Circuit. By Samuel Tymms . . . 1835.* The CUL set (VIII.19.5–11) also has this second title-page.
12  The date is uncertain. There is no internal evidence to suggest the text was prepared after 1835 (the restoration of Great Tom in 1835 is mentioned on page 47). As the plate was changed for this volume that must have occurred later than 1842/3; the work was probably prepared as a rival to the handbooks being issued in Lincoln by R.E. Leary.

Chubb CCCCXLI (The Tymms work).

## 82

### CHARLES SMITH                                                            1822

Size: 23.5cm x 18.6cm.

Smith was the map-seller responsible for the large atlas of 1801 (**56**). Pickett whose name appears on the title-page not only engraved that plate but also seventeen of the 43 other plates, the remainder being the work of W.R. Gardner. Pickett has not been identified completely; there are two candidates, A.J. or J.[1] The map is a reduced version of the larger sheet of 1801.

### EDITIONS

(**i**)   *LINCOLNSHIRE* (in a panel with a compass on the top, showing the main points between west and east, Ca, OS). *Printed for C. SMITH, Nº. 172 Strand, 1822.* (Ce, OS). *Scale of Miles/15* [= 4.6cm] (set into the lower neat line, Ce). *Pickett Sculpᵗ.* (Ee, OS). The inner frame is marked with every minute of latitude, etc. and numbered every 5.

*Smith's New English Atlas, Being A Reduction Of His Large Folio Atlas Containing a Complete Set of County Maps, On Which Are Delineated All the Direct & principal Cross Roads, Cities . . . The whole carefully Arranged according to the Stations & Intersections Of The Trigonometrical Survey Of England. London: Printed for C. Smith, Mapseller extraordinary To His Majesty Nº. 172 Strand. 1822. Pickett scrip. et sculp.*

**Copies**   CUL Atlas 6.82.2; Leeds UL W 134.

*Smith's New English Atlas . . . 1825 . . .*

**Copies**   BL Maps 49.a.35; CUL Atlas 6.82.12.

*Smith's New English Atlas . . . 1825 . . . [1827].*[2]

(**ii**)   The date in the imprint has been removed.

*Smith's New English Atlas . . . 1828 . . .*

**Copy**   The private collection of Mr. C.A. Burden.

*Smith's New English Atlas . . . 1828 . . .* [1829].³

**Copy**   Bod Allen 90.

*Smith's New English Atlas . . . 1828 . . .* [1830].⁴

**Copy**   CUL Atlas 7.82.29.

*Smith's New English Atlas . . . 1828 . . .* [1832].⁵

*Smith's New English Atlas . . . 1833 . . .* [1834].⁶

**Copy**   The private collection of Mr. T. Burgess.

*Smith's New English Atlas . . . 1844.*

**Copy**   The private collection of Mr. C.A. Burden.

Issued in a cardboard holder labelled *Smith's Lincolnshire.* [1845?].⁷

**References**

1   Tooley, op. cit., p. 505.
2   Warwick Eardsley, map dealer, broke up an atlas, which contained Yorkshire re-dated 1827. The date on the title-page may not have been altered.
3   The Yorkshire map is dated 1829.
4   The Yorkshire map is dated 1830.
5   Whitaker, H. 1948 op. cit., no. 416 refers to a copy of the Yorkshire map dated 1832. Kingsley, p. 165 suggests that Whitaker's view that there was an atlas dated 1832 may be correct, since the only atlas, dated 1833 on its title-page, contained Yorkshire dated 1834.
6   The Yorkshire map is dated 1834.
7   Burden, p. 129 records a copy for Berkshire; it has additional data compared with the 1844 atlas. Lincolnshire may exist also in this format. Burden also records the issue of Cheshire, dissected and in plain cover, issued c.1834.

Chubb CCCLXXXVIII (1822 edition only).

## 83

**A. M. PERROT**                                                            **1823**

Size: 11.5cm x 6.6cm (whole drawing); 5.4cm x 4cm (map area).

These curious little maps were drawn, presumably, by Aristide Michel Perrot for a topographical survey of Great Britain by George Bernard Depping, a German (b. Munster in Westphalia, 1784) who became a naturalised Frenchman in 1827.¹ He was a professor of German and a member of the French Royal Society of Antiquaries and died in 1853.² Perrot was a geographer, who was a member of a number of learned societies and produced many atlases and geographical books.³ The map, engraved by the Parisian Adrien Migneret (1786–1840), is accompanied by fourteen pages of letterpress in the sixth (and final) volume. The text is a curious mixture of odd topographical and historical notes with nearly five pages devoted to the hunting activities of the fenland people. The map has so little detail it is not possible to guess its source. The whole engraving consists of the lower end of a pyramid, with a spade handle, the map on a scroll in the centre; out of the top appear the upper part of a church tower and spire, a shepherd's crook and a globe, set with stars and *NEWTON* on an equatorial band.

**EDITION**

(i)   *LINCOLN* (in area below the map). *A M Perrot 1823* (below the lower left corner of the engraving). *Mᵉ Migneret Sc* (below the lower right corner of the engraving). *P.80* (top right corner of the page). There is no frame, compass or scale.

*L'Angleterre, Ou Description Historique Et Topographique. Du Royaume Uni De La Grande-Bretagne; Contenant: Les comtés de la principaute de Galles, des royaumes d'Ecosse, d'Irlande et d'Angleterre, les Iles Orcades, Shetland, etc. Par M. G.-B. Depping. Ornée De 75 Cartes Et Vues. Tome VI. A Paris, Chez Etienne Ledoux, Rue Guénégaud, N⁰ 9. 1824.*

**Copies**   BL 010368 q 19; NLS NE 807 e 1.

*L'Angleterre, Ou Description Historique Et Topographique. Du Royaume-Uni De La Grande Bretagne, Par G.-B. Depping; Ornée De 80 Cartes Et Vues. Seconde Edition. Tome VI. A Paris . . . 1828.*

**Copy**   Leeds UL W 138.

*L'Angleterre . . . 80 Cartes Et Vues. Troisième Edition. Tome VI. Bruxelles. L. T. Brohez, Librairie-Editeur, Longue Rue Des Dominicains, N. 744. 1828.*

**Copy**   CUL Lib.8.82.15.

*L'Angleterre . . . Seconde Edition . . . A Paris . . . Ledoux . . . 1835.*[4]

**References**

1  Querand, J.M. La France Littéraire. (Paris, 1828), Vol. II, pp. 383–385.
2  Catalogue General De La Librairie Francaise . . . 1840–1865; ed. by O. Lorenz. (Paris, 1868), Vol. II. p. 82.
3  Querand, op. cit., Vol. VII (1835), pp. 71–73.
4  Whitaker, H. 1948, op. cit., item 456 (p. 135) draws attention to this edition, to the anomalous use of *seconde edition* and notes that a copy was sold by Sotheby's in November, 1940.

Not in Chubb.


## 84

**ORDNANCE SURVEY**                                                      **1825**

Size: 15.9cm x 11.5cm.

The Ordnance Survey authorities were persuaded by the gentlemen of Lincolnshire to survey the county out of turn largely because the gentry felt the need of an up-to-date and detailed map for hunting purposes. At the time of the Napoleonic threat of invasion the decision was made to survey the counties along the southern coast starting with Kent. Later, the need was felt to continue the surveying process so that eventually the whole country would be covered and the plan put into effect was to work from south to north. The processes undertaken by the Lincolnshire hunting gentry and their pporters are fully documented.[1] At the time of the approach to the Ordnance Survey's Master General (Col. Mudge), led by Sir Joseph Banks in 1817, only Essex of the counties north of the Thames had been surveyed and published. Largely because of a promise to raise enough subscriptions in advance from the county's landowners the scheme was agreed. Progress was slow and changes were made in the surveying standards by Major Colby, following the death of Col. Mudge in 1820. Finally, although the date on the maps is 1824, the sheets were made available in April, 1825. It is sufficient here to state that the county was covered in eight sheets, numbered LXIV–LXV, LXIX–LXX and LXXXIII–LXXXVI, working from south to north in pairs. The eight maps were initially only sold as a set.

The early sets were sold in parts, the first eight being the southern coastal counties from Essex to Cornwall, except Hampshire, but with Wiltshire. 'Part the IX<sup>TH</sup>' was Pembrokeshire and 'Part the X<sup>TH</sup>' Lincolnshire.[2] Thereafter, this method of selling was discontinued. In the BL set of original parts Parts I–IX have a cover and an index map, features missing from their set in Lincolnshire's case. However, in what seems to have been a special presentation set an example of the index map has survived (see below). The county was surveyed by the O.S. at scales of 6 inches and 25 inches to a mile in 1883–1888; a revision of these plates was undertaken, in the county's case, from 1898 to 1902. The 'Second Edition, or New Series' of the one inch scale maps was issued between 1884 and 1888; the sheets, which contain parts of the county, are numbered 79–81, 88–91, 101–104, 114–116, 126–129, 143–145 and 157–158.

The index map below is exactly what its name implies. Important towns and main roads are shown; a grid divides the map into eight sections, each bearing the number of the O.S. sheet. Physical features are few; the area covered includes all of Rutland with Norfolk eastwards to Swaffham, Cambridgeshire as far south as Chatteris and Leicestershire westwards to Melton Mowbray.

There is good evidence that the agents of the Ordnance Survey issued the whole of the county on single large sheets prepared from the original constituent parts. Grattan and Gilbert, who had the agency for the City of London in the early 1840s, issued advertisements in which the whole county was offered in three different formats, viz., mounted and in case at 4 guineas; coloured, mounted on a roller and varnished for a further guinea; and, at seven and a half guineas, coloured and mounted on the best mahogany spring roller.[3] In January, 1842 this firm went bankrupt and Letts was given the City franchise.[4] Arrowsmith also had an agency to cover other areas outside the City but it is not clear whether they also offered the county in whole, mounted sheets. In c.1855 Longman advertised such sheets again in various versions and, from the early 1860s Stanford provided the county map in a range of alternative issues. In their catalogue of 1867[5] Lincolnshire was offered as a single large sheet, measuring 82 inches by 63 inches in a case at two guineas, on a roller and varnished for £2.7.6 and on a spring roller at £5.2.6. Colouring cost an extra 10s. Stanford offered the map right through until the 1890s in the same formats and at marginally increased prices; the version on spring roller and coloured cost £5.15s.[6] The rarity of examples, not only of Lincolnshire but nearly all other counties, from any of these various agents suggests that sales were very low and any copies sold were bought for special purposes, whose use eventually resulted in the destruction of the map.

## EDITION

(i) *LINCOLN/SHIRE* (across the map Bb–Db and Bc–Dc). *INDEX* (in a box, protruding from the top border (Ca). There is no scale or compass.

*Part of the General Survey of Great Britain, carrying on by Order of His Grace The Master General of the Ordnance, containing Lincolnshire and Rutlandshire, With Portions of the Adjoining Counties Performed under the Direction of Major Colby of the Royal Engineers, F.R.S. &C. 1824. Jones & Barriff, scrip: et sculp:*

As noted above the BL set lacks the county index map.

In a brown leather folder, with a black leather label on the front, on which is blocked in gold: *ORDNANCE MAP/OF/LINCOLNSHIRE/ AND/RUTLAND/*. The index map is pasted on the inside front cover of a set of the first edition of the county sheets. [1825].

**Copy**   Boston RL Map drawer.

**References**

1   The Old Series Ordnance Survey Maps Of England And Wales Scale: 1 inch to 1 mile

Introductory Essay by J.B. Harley . . . Volume V Lincolnshire, Rutland, and East Anglia . . . (Lympne, [1987]), pp. vii–xxvi. Pages xxxvii–lvi are a very detailed bibliography of the known states of all eight sheets, with locations of copies.

2   The BL shelf number for parts I–X is Maps 148.e.27.
3   Grattan and Gilbert catalogue; kindly supplied by Mr. E. Burden.
4   The Old Series Ordnance Survey maps . . . Volume V, p. xxxiv.
5   Bod Map Library G.2/C.16:10/16. An example of the 1869 catalogue is in NLS and is virtually identical to the 1867.
6   A Résumé Of The Publications Of The Ordnance Survey For England & Wales . . . By Francis P. Washington . . . (Stanford, 1890).

Not in Chubb.


# 85

## WILLIAM EBDEN                                                1828

Size: 43.2cm x 34.7cm.

Ebden produced one set of county maps but was active for a period of over forty years in this field. His name is associated with *Laurie and Whittle's Map of Norfolk and Suffolk* (1811) and he is also noted as late as 1856.[1] The map of Lincolnshire is based on Cary's maps of 1809 (**65**), with the later influence of Rowe (**71**); it is without the error of longitude, but includes *Hughington*, the mythical village. Other errors suggest that there were verbal errors in communicating names to the engraver, e.g. *Tucklass Cross* (for *Guthlac's Cross*) and *Thetford* (for Greatford, pronounced Gretford). However, Ebden's plates were never issued as a set by him. The history of their production and issue is complicated; some plates were put out by Orlando Hodgson, some of which with further new ones were then issued by William Cole; these and a final batch of new plates were issued by S. Maunder. Cole and Maunder both used the address of Hodgson, though by 1828 Hodgson had moved elsewhere.[2] Lincolnshire was one of the final tranche in the plate-making work and only Kent is known to have been issued by all three owners of the plates. The first issue as a full-blown atlas was not until 1833, by which time James Duncan had acquired access to all the plates and Ebden's connection had been removed. Tooley refers to an issue of *Atlas of English counties* by S. Maunder but no atlas has been found.[3] It has been suggested that Duncan had the plates in 1832, thus giving him time to update all the plates with the post-Reform Bill election details.[4]

### Later history
Duncan issued his atlas until c.1845. Thereafter, the plates were used for litho. transfers by H.G. Collins, who also used Rowe's plates similarly (**71 (vi)**). Although issued with Collins' name in the title a later issue in the 1870s was prepared by the well-known firm of Edward Stanford. They had a variety of addresses in Charing Cross from 1853 onwards. Edward Stanford (1827–1904), although born in London, served his apprenticeship as a printer in Malmesbury. On returning to London he first worked for, and then took over the map business of Trelawny William Saunders (1853).[5] In 1854 he became a Fellow of the Royal Geographical Society and began his own series of maps, acquiring Collins' plates in 1858. He retired in 1882 and was succeeded by his son, also Edward, who increased the firm's trade through a valuable ten year contract with the Ordnance Survey from 1885. George Philip and Son took over Stanford's in 1946.

### EDITIONS
(i)   *EBDEN'S/Map of the County of/LINCOLNSHIRE;/Divided into Hundreds/laid*

*down from Trigonometrical Observations/By/W. EBDEN/Hoare & Reeves, Sc. Warwᵏ.* *Cᵗ. Holbⁿ.* (Ae). *London. Pubᵈ. Septʳ. 20. 1828, by S. Maunder, 10, Newgate Street.* (Ce, OS). *SCALE/15 Miles* [= 7.45cm] (Be). *EXPLANATION*/15 lines of text and symbols (Ee). Compass (Ea). *REFERENCES/to the/HUNDREDS OR WAPONTAKES/*31 names (Aa). The inner frame is marked every 2 minutes of latitude, etc. and numbered every 10. *Longitude East from Greenwich* (De, between borders). *1° Longitude West from Greenwich* (Ae, between borders).

Issued as loose sheets. [1828].

**Copy**  Leeds UL W 139.⁶

(**ii**)  The title now reads: *Map of the County of/LINCOLNSHIRE;/ Divided into Hundreds/Containing the District Divisions and other/ LOCAL ARRANGEMENTS/effected by the/REFORM BILL/* (Ae). The imprint now reads: *London. Pubᵈ. by J. Duncan, Paternoster Row.* (Ce, OS). The final two lines of the explanation have been replaced by four lines on the signs used to indicate the new parliamentary representations compared with the previous arrangements. Plain crosses have been added to places of election; a small cross with two stars occurs at those places retaining the same number of MPs; a star and a circle at Grimsby (indicating one MP instead of the previous two); and oblongs at Lincoln and Sleaford (to show places of county elections; the symbols at those two places differ from that specified). *County Members 4. Elections at Polling Places* + (Ce–De).

*A New Atlas Of England And Wales; Consisting Of A Set Of Large County Travelling Maps, Divided Into Hundreds; With The Cities . . . The New District Divisions, Polling Places, Disfranchised and Enfranchised Boroughs, &c. Agreeably Exhibiting On The Map Of Each County Both Its Present And Former State Of Parliamentary Representation. London: James Duncan, 37, Paternoster Row. 1833.*

**Copies**  Birmingham RL A F912.42 NEW; Leeds RL F912.42 D 912. (Grimsby RL DUNCAN 1833 – 2 copies).

*A Complete County Atlas Of England & Wales, Containing Forty Four Superior Maps. With All The Improvements – Projected Or Completed. Divided Into Hundreds, With The District Divisions, and other Local Arrangements effected by the Reform Bill. London: Published By James Duncan, Paternoster Row. Price Two Guineas, Plain, – Four Guineas, Coloured.* [1835].⁷

**Copy**  BL Maps 7.d.22.

*A Complete County Atlas . . . Coloured. 1837.* ⁸

**Copies**  Leeds UL W 167; Glasgow Mitchell L. Sf 912.42 (B 8089).

*A Complete County Atlas . . . 1838.*

**Copies**  The private collections of Mr. D. Hodson and Mr. C.A. Burden.

*A Complete County Atlas . . . Four Guineas, Coloured.* [1840].⁹

**Copies**  BL Maps 7.d.13; CUL Atlas 4.84.23.

*A Complete County Atlas of England & Wales, Containing Forty-Four Superior Maps, With All The Railroads . . .* [1845].¹⁰

**Copy**  RGS 9 F 24.

*A Complete County Atlas Containing Forty Four Superior Maps . . .* [1845].

**Copies**  BL Maps 8.e.7; CUL Atlas 4.84.19; Gloucester RL f 912.42.

## LITHOGRAPHIC TRANSFERS
(**A**)  The title is now: *LINCOLNSHIRE*; (i.e. all the wording before and after the county

name has been removed (Ae). The imprint now reads: *London, Pub. by Henry George Collins, Paternoster Row* (Ce, OS). The last eight lines in Explanation have been replaced by five lines, which refer to the symbols for railways and telegraphs, etc. Railways A–B, F, v, 1–4 and 6–18 have been added, with black dots marking stations and directions for those leaving the county or in adjoining counties. Many roads leading into the adjoining counties have been added, with their directions; towns added outside the county include *Wisbeach, Bingham, East Retford, Tuxford, Hull* and *Pattrington* (sic). *New/Holland* has been added.

*The New British Atlas, Containing A Complete Set Of Maps Of The Counties Of England And Wales, With All Railroads, Telegraph Lines, And Their Stations. The Whole Carefully Revised. London: Published By Henry George Collins, 22, Paternoster Row.* [1855].[11]

**Copy**   The private collection of Mr. C.A. Burden.

**(B)**   The title now reads: *COLLINS'/Railway and Telegraph/MAP OF/ LINCOLN-SHIRE;/Divided into Hundreds/Containing the District Divisions and other/LOCAL ARRANGEMENTS/effected by the/ REFORM BILL/* (Ae). The *O* in *WAPONTAKES* has been replaced by *E* (Ac).

Issued as a loose sheet, possibly in slip cases. [1855].[12]

**Copies**   LAO Ex 942.50; Nottingham UL Dept. of Geography Drawer 121 (F 133).

**(C)**   The title now reads: *Map of the County of/LINCOLNSHIRE;/ Divided into Hundreds/Containing the District Divisions and other/ LOCAL ARRANGEMENTS/* (Ae). The imprint is: *London. Edward Stanford 6 Charing Cross.* (Ce, OS). *Boundaries of Divisions of County* [symbol]/ *Dº. Wapentakes . . ./* (Ae, below the title). The figure for county members has been changed to *6* (Ce). Railways 19–28, 30–38 and I, L, R have been added. The explanation is as state (ii) (Ee). *Withernsea* (Yorks.), *THE/WASH,* and *CROSS KEYS/WASH* have been added.

Issued as a folded map in a cover with the title: *COLLINS' RAILWAY MAP OF LINCOLN SIXPENCE. LONDON: EDWARD STANFORD. 6 & 7, CHARING CROSS; AND SOLD BY . . .* [1871].[13]

**Copy**   Grimsby RL X000:912 COL.

### References

1  Tooley, 1979, p. 185.
2  Smith, D. The early issues of William Ebden's English County Maps. *Imago Mundi*, vol. 43 (1991), pp. 48–58 clarifies the likely early history of the plates.
3  Tooley, 1979, p. 426.
4  Smith, D. op. cit.
5  Hyde, R. Printed maps of Victorian London, 1851–1900. (Folkestone, 1975), pp. 11–13 includes other references.
6  Whitaker's volume consists of a complete set of the maps by Ebden, Hodgson and Cole and variously dated 1824–1828. It is doubtful if it was intended as an atlas; it lacks a title-page. In Smith's article (note 2) he refers to a second set in the hands of a dealer, but subsequently sold as individual maps.
7  The atlas must have been issued between 1832 and that dated 1837. The Durham map shows a line (Stanhope and Tyne Railway) opened in Sept. 1834. 1835 is suggested, therefore.
8  Mr. D. Hodson has a copy dated 1827 on the title-page; since the contents mirror the 1837 issue an error is likely, especially as Duncan did not own the plates in the earlier year.
9  The cover has the title: *RAIL ROAD/COUNTY ATLAS/44 MAPS/1840/.*
10  Although the title-page has gone through slight changes of punctuation all commentators believe the date of issue to be 1845. The order given here is believed to be correct chronologically, based on the addition of railways on other plates.

11   The county railway data suggest a date after 1853 but before 1855. The volume appears to
     have the original binding and includes a map of the siege of Sevastopol (1855).
12   Probably issued after state A but the map has not been updated.
13   The railways are complete to Sept. 1871. Stanford's address of 6 and 7 Charing Cross predates
     their acquisition of no. 8 in 1874.

Chubb CCCCLV–CCCCLVIa (1833 to 1845 editions).


# 86

## ANDREW BRYANT                                                                             1828

Size: the eight sheets vary between 48.85cm and 53.4 cm. in height and 79.2 and 79.65
in width (approx.).

In 1820 Bryant began a survey of Hertfordshire at a scale of one and a half inches to a
mile, the resulting map being published in April, 1822 on four sheets. Hodson has
suggested that the survey was probably the result of a general dissatisfaction with the
products of the Ordnance Survey. Hodson also shows that Bryant even resurveyed that
part of the county that the O.S. had covered so poorly.[1] Bryant then proceeded to survey
other counties and when *The British Atlas, Or A Series Of Maps Of The Counties Of
England And Wales, made From New Surveys . . .* was announced four counties had
already been surveyed; a manuscript note indicates that two other counties were in
preparation, which dates the announcement to c.1826.[2] Lincolnshire was offered on six
sheets at four guineas complete; in fact, the county eventually occupied eight sheets, just
as Armstrong and the O.S. had. A list of subscribers shows the Bishop of Lincoln at the
head. At least one further advertising pamphlet appeared (presumably in 1828, as the
county is the last on the list – 8 sheets at 4 gns).[3] The county map appeared in April, 1828
as the tenth in the series. In the following seven years only three more counties appeared
and the whole plan foundered with the map of Herefordshire in 1835. It is safe to believe
that the subscription income failed to cover costs and the rate of progress was too slow
to allow of the successful completion of the original scheme.

   That Bryant made his own surveys is quite clear when his county sheets are set beside
the Ordnance Survey sheets issued in 1825 and those prepared by the Greenwood brothers
who were making their own survey in the years 1827 and 1828 (**87**). Bryant took three
years over his work on Lincolnshire (1825–1827), beginning soon after the publication
of the O.S. sheets and finishing at almost the same time as the Greenwoods began. It is
remarkable that three such surveys at an inch to a mile (more or less) were under way
within the same decade. The palm for quality and the amount of detail must go to Bryant.
A detailed comparison between Bryant and the original edition of the Greenwoods' larger
map shows that Bryant includes details and many names that do not appear on the rivals'
map. When the Greenwoods issued a revised version of their map they took much from
Bryant; for example, all but one of the names added by them in Elloe already appear on
Bryant's map. The only thing Bryant took from the O.S. was the piano keyboard edging
to his map.

   Very little is known of Bryant as a person or the background to the grandiose project
that, eventually, he failed to complete even after fifteen years of high quality activity.

## EDITIONS

(i)   *MAP/OF THE COUNTY/OF/LINCOLN,/FROM ACTUAL SURVEY/ made in the
Years 1825–26 & 27,/BY A. BRYANT,/Respectfully Dedicated/To The Nobility,/CLERGY
& GENTRY/of the County./ Published/BY/A Bryant, 27, Great Ormond Street, London./*

*APRIL 19ᵀᴴ. 1828/* (Ea). *Published April 19ᵗʰ. 1828, by A. Bryant. 27 Great Ormond Street, London.* (Ce, OS, and also at the foot of all of the single sheets, except the second (which contains the title). *SCALE/ONE INCH TO A MILE/1* + 7 [7= 17.5cm] (Ce). *NORTH- WEST VIEW OF LINCOLN CATHEDRAL* (An engraving of Lincoln Cathedral from the South-west (Ee).*ECCLESIASTICAL DIVISIONS./ The County of Lincoln is in the Diocese of Lincoln,/and Comprises/The Archdeaconries of Lincoln and Stow./*4 columns of the names of the Deaneries and the names of the Wapentakes they enclose (Ad–Ae). *Explanation/*25 lines of symbols and their meanings. (Ac). The location symbols given above assume a complete map of the eight sheets joined together. The border is the piano keyboard type, with an inner frame marked with every minute of latitude, etc. and numbered every 5. *MERIDIAN OF GREENWICH* reads upwards (De) and downwards (Ad).

Issued as loose sheets, usually found joined to form one, two or four larger sheets. 1828.

**Copy**   LAO Scorer 32–38 (sheet one is not included).[4]


**(ii)**   *PART OF/NORTHAMPTON* has been added along with many other extra names, far too numerous to list. Other notable additions include: the River Nene has been extended into the Wash, *Postland* has been altered to *Great Postland*.

Issued as loose sheets . . . 1828.

**Copies**   BL Maps C.23.a.7; CUL Atlas 2.82.10 and Maps B.70.82; Bod C.17.39.d2; LAO SR. (Canon Foster's copy); Guildhall L. 411.6; Scunthorpe Museum.

**References**
1   Hodson, 1974, p. 118.
2   BL Maps 60.e.15 (4).
3   Guildhall L. 411.6.
4   LAO has 2 other sets (Misc. Don. 42 and WG 10/9); neither has been examined since their condition makes them unavailable to students; it can not, therefore, be stated whether they are states (i) or (ii) or, indeed, an undiscovered further state.


# 87

## CHRISTOPHER and JOHN GREENWOOD                                      1830

Size: 202cm x 152.7cm (approx.)

Christopher Greenwood (1786–1855) was born in Yorkshire and was already involved in surveying work by 1815 and, by 1818, had opened an office in London. In 1821 his brother John (1791–1867) joined him. By then eight counties had already been surveyed and published, all at one inch to a mile, except Yorkshire (¾" to a mile).[1] How far the surveys were entirely independent is open to doubt; the surveying of so many counties in such a short period of time suggests Christopher had not been without help from other surveyors. With his brother the work continued and a set of all the counties was offered at 125 guineas, although, in fact, six counties never were surveyed.[2] Individual counties were sold for three guineas each. During the mid-1820s Christopher was also involved with Thomas Sharp and William Fowler in surveying four Scottish counties. Many of the later financial worries were already apparent in this period, since there were very few takers of the full subscription, largely because 125 guineas was considered too much for a set of maps.[3]

Eventually the finances of the Greenwoods were stretched too far and their business seems to have collapsed right at the point of the appearance of the atlas version. When the reduced versions of their maps appeared in atlas form (**102**) they were issued from the premises of the Walkers, John and Charles, who had engraved the plates. It can only be assumed that the Walkers undertook the final production of the full work because the Greenwoods were unable to meet the costs of engraving and they had been forced to give up their own premises. Certainly after 1835 the Greenwoods retired from the production of county sheets and surveying. John, in fact, returned to Yorkshire and worked as a land surveyor from 1838. Christopher was able to re-establish himself as a publisher off Bloomsbury Square in London but the project was short-lived. Some of the plates were sold apparently; the Lancashire plates were issued again in 1836 by George Cruchley.[4] The maps of Durham, Cumberland and Westmorland were re-issued by Christopher, the last appearing in 1850.[5]

As the large sheets indicate Lincolnshire was surveyed by the Greenwoods during 1827–1828; at the same time Christopher was surveying Fife and Mid Lothian and the brothers were also engaged in Northumberland; indeed in 1827, they were also completing the surveying of Shropshire, Glamorgan, Brecon and Radnor. Some discussion and comparisons have already been made with the contemporary work of Andrew Bryant (**86**), with the palm for the degree of detail going to Bryant. The Greenwoods produced a revised edition, which included information, taken from Bryant as he was the only source of the many added names. The map, however, is a very fine piece of engraving and, in their usual hand-coloured state, are understandably highly valued by collectors. The six sheets are usually found mounted as three and labelled *NORTH, CENTRAL* and *SOUTH*.

## EDITIONS

(i)  *Map/of the County/OF/LINCOLN/FROM/An actual Survey/Made in the Years 1827 & 1828,/By/C. & J. GREENWOOD./Most Respectfully Dedicated/TO THE/Nobility, Clergy & Gentry/OF THE COUNTY/By the Proprietors/GREENWOOD & C$^O$./Regent Street, Pall Mall,/London/PUBLISHED/FEB$^Y$. 24$^{th}$. 1830./* (Ea). *London Published by Greenwood & C$^o$. Feb$^y$. 24$^{th}$. 1830.* (Be, and also at the foot of other sheets but not always consistently).*SCALE OF STATUTE MILES/1+7* [7= 17.25cm] (Ee). *NORTH WEST VIEW OF LINCOLN CATHEDRAL* (Ae). Compass (Ac). All locations are based on the six sheets joined together. The border is of the piano keyboard type with inset numbers increasing by 5 each time to mark off the minutes of latitude, etc. Dots in the inner border show the single minutes. *Longitude East of Greenwich* (Ad).

Issued as loose sheets, but most often mounted on 2 or 3 larger sheets. 1830.

**Copies**  Bod Allen 236; NLS EME b.3.60.

(ii)  A large number of new places have been added, e.g. *Irby/le Marsh* (previously *Irby in the Marsh*), *Hareby House, Guy's Hospital/Land* (north of South Holland Drain), *Fleet Haven, Austin Dyke* (east of Spalding).

Issued as loose sheets. 1830.

**Copy**  Birmingham RL Q 912.4253.

(iii)  More place-names, etc. have been added, e.g. Marsh symbols and *East Marsh; or/Long Sutton* has been added to *SUTTON/S$^T$. MARYS.*

Issued as loose sheets. 1830.

**Copies**  BL Maps C.23.c.13; CUL Atlas 1.82.4; Bod Allen 237 and Map C.17:39.d.1; Gainsborough RL; Grimsby RL X000:912 GRE; Nottingham UL Dept. of Geography Drawer 121 (F 133).

**References**

1   BL Maps 60.e.15 (1) is a prospectus for Middlesex, with a note on the availability of Lancashire and Yorkshire. It must have been issued early in 1819 as it omits Cheshire published in that year.
2   BL Maps 60.e.15 (2) is a prospectus for the full set, with Lincolnshire expected to be 4 guineas and on 6 sheets. The conjoint leaf advertises Cheshire and Staffordshire (pub. 1820), but not Wiltshire, which was issued soon after Staffordshire.
3   Harley, J.B. Christopher Greenwood: County Map-Maker And His Worcestershire Map Of 1822. (Worcestershire Historical Society, 1962).
4   ibid; Rodger, p. 14.
5   Rodger, pp. 4, 6 and 26.

**88**

**JAMES PIGOT**                                                               **1830**

Size: 35.6cm x 22.3cm.

Pigot is first recorded in 1794 as an engraver and printer in Manchester, engraving the map for Dean's Manchester Directory of 1804. From 1811 he began his own business and, while a rival of Dean, the two firms jointly produced a directory of Manchester every two years. After Thomas Underhill's failure Pigot was the first to try to produce directories of a national character, setting up for the purpose in London in Basing Lane and using agents to collect data.[1] The firm and its successor, I. Slater, provided directories that covered all English counties and the major cities for well over thirty years. Once the basic data relating to each area had been set in print it was the Pigot policy to issue counties in groups, with or without the major cities within the area, the groups often of an apparently assorted kind. Frequently such directories have been split up and sold as separates and, in such forms, individual counties are found in many local collections. The rarity of some directories suggests that some of the groupings, particularly the alphabetical groups, failed to find a very great market amongst potential commercial users. The role of directory provision for the whole of the British Isles was finally taken over by Kelly's Directories (**116**), who took over the London area after Pigot decided to concentrate on the northern and midland counties. From that base Kelly eventually covered the whole country.

Pigot and Co.'s first directory was *London and Provincial New Directory* for 1822–3, which included Lincolnshire.[2] The third edition, for 1826–7, is, according to its preface 'embellished with a new series of Elegant Maps . . .' but this only covered six of the Home Counties. In 1829 a new title appeared; the *National Commercial Directory* calls for maps for the sixteen counties and North Wales that the work covers. On the title-page it is recorded that the maps were available bound up separately as an atlas. The maps continued to appear in atlas form (the 1835 edition being issued in three parts) and similarly entitled directories. In 1839 Pigot took Slater into partnership. After 1843, the year of Pigot's death, Slater took control until 1857–8. The directories of northern areas still appeared until 1882. The 1921 Directory of Scotland was entitled *Kelly's (Slater's) Royal National Directory.*

The map is based on Cary's map of 1801 (**57**), with additions and, as far as is known, is the first to have been engraved on steel plates. Although the maps always bore a note on the use of steel engraving one suspects that the final issues were lithographic transfers, as the changes to railway lines (states (viii) onwards) are not consistent with alterations on the plate. There are more variations within the full title-pages than are quoted in the following notes.

## EDITIONS

(i)  *LINCOLNSHIRE*. (Ca, in a panel, between the inner and outer frames), *Published by Pigot & Cº. 24 Basing lane, London, & 18 Fountain Sᵗ. Manchester*. (Ce, OS). *Engraved on Steel by Pigot & Son, Manchester*. (Ee, OS). *Scale/10 Miles* [= 3.7cm]. (Ce, in a panel, between the inner and outer borders). *Explanation./14* lines (Ea). Compass (Ea, below the Explanation). *LINCOLN CATHEDRAL* (Ae, below an engraving from the south-west). *REFERENCE to the HUNDREDS./3* columns of names (Ee). The inner border is marked off with every minute of latitude, etc. and numbered at 5′ intervals. *Longitude East from Greenwich* (De, between borders). *Longitude West from Greenwich* (Ae, between borders).

*Pigot And Co.'s National Commercial Directory For 1828–9; Comprising A Directory And Classification Of The Merchants, Bankers, Professional Gentlemen, Manufacturers And Traders, In All The Cities, Towns . . . In . . . Cheshire, Cumberland, Derbyshire, Durham, Lancashire, Leicestershire, Lincolnshire* [and 9 other counties] *And the whole of North Wales . . . Embellished With A Large Map Of England And Sixteen Elegant County Maps . . . Published by J. Pigot & Co. 24, Basing Lane, London, 18, Fountain Street, Manchester . . . And Sold by . . . Simpkin And Marshall . . . Sherwood And Co . . . Winchester And Varnham . . . Effingham Wilson And John Richardson . . . Price . . . bound in Cloth, with Plain Maps, £1.10s. Handsomely bound in Calf, with the Maps Coloured and bound up separately as an atlas, £2.* [1830].

**Copies**  BL PP 2506.u.5; Guildhall L. (LAO FL Box L 910 and Scorer Map 39; Grimsby RL PIGOT RM 28; Louth RL Map A15; Manchester UL Map C 275. LRL L 929 and Boston RL L 929 are examples of the county text (pp. 505–564) with the map).[3]

*Pigot & Coˢ. British Atlas Part Second contains the following Maps viz England.* [Two columns of county names exactly corresponding to the 16 in the directory above]. [1830].[4]

**Copy**  Leeds UL W 143.[5]

(ii)  The East Fen 'lakes' have been removed. *BOLINGBROKE* has been added. Several new roads have been added: Lusby-Brinkhill; Baumber-Fulesby [i.e. Fulletby]; Gringley (Yorks.)-Crowle; Gainsborough-Newark; Bourn-Spalding, etc. Over 30 roads have been upgraded, e.g. Boston-Wainfleet-Skegness, Alford-Louth, Brigg-Kirmington, Horncastle-Spilsby.

*Pigot And Coˢ. British Atlas, of The Counties Of England, With A Map of England and Wales, And A Circular One Of The Counties Round London, To The Distance Of Fourteen Miles; The Whole Engraved On Steel Plates . . . Published By J. Pigot & Co. Basing Lane, London, And Fountain Street, Manchester: And sold by . . . with Coloured Maps, neatly done up in boards £2 10s 0d Extra Coloured, elegantly bound, and the Maps mounted on Cloth 4 0 0 October, 1830.*

**Copies**  Bod Maps C.17.b.5; Leeds UL W 144; Guildhall L. Gr.I.1.1; Manchester UL AMGS B7.

(iii)  NEW has been added in front of *BOLINGBROKE*. Grantham Canal has been marked. *Potter/Hanway* has been changed to *Potter/ Hanworth*.

*Pigot And Co.'s British Atlas . . . Extra Coloured, and the Maps mounted on Cloth 4 0 0 . . . elegantly bound, and gilt 4 10 0 1831.*

**Copies**  CUL Atlas 4.83.5; Leeds UL W 145.

(iv)  One of the two stars at Grimsby has been removed. An extra line: *Polling Places*

. . . Δ has been added to the Explanation. These symbols have been added to 18 places, but not Stamford. A road across the Wash [at Sutton Bridge] has been added and extends into Norfolk. The spelling of *Fulletby* (previously *Fulesby*) has been corrected. *New Holland* has been added above *Skitter/Ferry*.

*Pigot & Co.'s Maps Of The Counties Of Derby, Hereford, Lincoln, Leicester And Rutland, Monmouth, Nottingham, Salop, Stafford, Warwick And Worcester; With A General Map Of Wales: Being An Appendage To The Midland County Volume Of Their National And Commercial Directories.* [1835].[6]

**Copy**   BL Maps 3.d.40.

*Pigot And Co.'s National Commercial Directory . . . Of* [the list of counties is exactly as in the above Map volume] *. . . Price . . . in Cloth, with Plain Maps . . . £1.5s.– Handsomely bound in Calf, with the Maps coloured, and made up as an Atlas, £1.10s. 1835.*

**Copies**   (Boston RL Maps 8 and 23; LRL L929 is the county section (pp. 163–246) with the map in damaged condition; LRL Pye P.L 929 has the work's title-page, etc. but no map; Leicester UL H 942 PIG has two copies of the directory; neither has the maps).

(v)   The imprint has been changed to *Published by Pigot & Cº. 59 Fleet Street & 18 Fountain Sᵗ. Manchester.* (CE, OS). The Witham Navigation from Saxilby to Brothertoft and then to the sea, east of Wrangle is marked by alternating vertical and horizontal dashes.

*Pigot & Co.'s British Atlas . . . Two Guineas and a Half. Pigot And Co. Printers.* [1840].

**Copies**   BL Maps 3.d.38; Bod Allen 104*.

*Pigot And Co.'s Royal National And Commercial Directory And Topography Of The Counties Of Leicester & Rutland, Lincoln, Northampton, Norfolk And Suffolk . . . October, 1840. Published By J. Pigot . . . Price . . . bound in Cloth, with Plain Maps . . . £1 5 0 . . . bound in Calf, with Coloured Maps 1 10 0 . . . Calf Copy (without Maps) and the British Atlas complete 3 0 0.*

**Copies**   Leicester UL H 942 PIG; Hull RL (Local History – Directory).

*Pigot And Co.'s Royal . . . Directory And Topography Of . . . Leicester And Rutland, Lincoln, Northampton, Kent, Middlesex, Norfolk, Suffolk, Surrey, And Sussex . . . A Classified Directory Of London & its Suburbs . . . Manchester And Salford . . . Embellished With Beautiful County Maps; A Map Of . . . London . . . Manchester And Salford. September, 1840. Published By J. Pigot & Co . . . Price . . . bound in Cloth, with Plain Maps £1 5 0 . . . in Calf, with Coloured Maps 1 10 0 Price of the Calf Copy (without Maps) and the British Atlas . . . 3 0 0.*

**Copy**   NLS Ref. 2 C.17.

(vi)   Railway A has been added as a black continuous line.

*Pigot & Co.'s British Atlas, comprising The Counties Of England, (Upon Which Are Laid Down All Railways Completed And In Progress), With Superior Large Sheet Maps Of England And Wales, Ireland And Scotland, And A Circular One Of The County Round London: The Whole Engraved On Steel Plates. Accompanied By Topographical And Statistical Accounts Of Each County . . . Published by Pigot & Co. 59, Fleet Street, London, And Fountain Street, Manchester: And sold by . . . Price, Bound, with Coloured Maps, Two Guineas. Pigot And Slater, Printers.* [1839].[7]

**Copies**   Leeds UL W 145a; Birmingham RL F912.42. (LRL Maps 189 and 771; Grimsby RL PIGOT 1839).

*Pigot & Co.'s British Atlas . . . Pigot And Co. Printers.* [1840].[8]

*Pigot & Co$^S$. British Atlas . . . Price, Bound with Coloured Maps, Two Guineas and a Half. Pigot And Co. Printers.* [1840].

**Copy**  CUL Atlas 4.84.15.

*Pigot And Co.'s Royal . . . Directory And Topography . . . Of York, Leicester & Rutland, Lincoln, Northampton, And Nottingham . . . Comprising Classified Lists Of All Persons In Trade, And Of The Nobility, Gentry And Clergy . . . Including Perspicuous Railway Tables . . . To Which Is Added . . . Manchester And Salford . . . Embellished With Beautiful County Maps; A Large Map Of England . . . And A Plan . . . of Manchester and Salford . . . August, 1841. Published By J. Pigot . . . And Sold by . . . G.G. Bennis, 55 Rue Neuve St. Augustin Agent For Paris And The Continent.*

**Copies**  Leicester UL H 942 PIG; Manchester RL 914.2 P 14.

*Pigot And Co.'s Royal . . . Directory . . . Of York* [and 5 counties as above] *. . . August, 1841 . . . And the Continent. Price . . . Cloth, with Plain Maps £1 5 0 . . . Handsomely bound in Calf, with Coloured Maps £1 10 0 To Non-Subscribers 5s. extra.*

**Copies**  BL PP 2506.u.5; Newark RL R/L 97.

*Pigot And Co.'s Royal And Commercial Directory And Topography Of The Counties Of Warwick, Leicester, Rutland, Lincoln, Northampton, Nottingham, Stafford, Worcester, York. Comprising Classified Lists . . .* [Manchester & Salford are not included, nor the map of England nor the plans of Manchester & Salford]. *December, 1841. Published By J. Pigot & Co . . . in Calf, with the Maps Coloured 1 10 0 To Non-Subscribers 5s. extra.*

**Copies**  LAO SR. 942.50; LRL L 929; Gainsborough RL L 929).[9]

*Pigot & Co$^S$. British Atlas . . . Accompanied By Distance Tables Of The Principal Towns In England, Scotland . . . Smaller County Tables, Containing All the Chief Towns In Each Of The English Shires. Published By Pigot & Slater, Fleet-Street, London, And Fountain-Street, Manchester: And sold by them and . . . Hatchard And Son, Piccadilly. I. Slater, Printer, Fountain-Street, Manchester.* [1843].[10]

**Copy**  CUL Atlas 4.84.16.

*Pigot And Co.'s National And Commercial Directory Of . . . Derbyshire, Leicestershire, Lincolnshire, Nottinghamshire, Rutlandshire, Shropshire, Staffordshire, Warwickshire, Worcestershire, And Yorkshire . . . Embellished With Beautiful County maps, Upon Which Is . . . Every Line Of Railway. Published By I. Slater, Fountain-Street, Manchester, and Fleet-Street, London: And Sold by him . . . Agent For Paris And The Continent. Price . . . Cloth, with Plain Maps 1 5 0 . . . Handsomely bound in Calf, with the Maps Coloured 1 10 0 To Non-Subscribers 5s. extra.* [1844].[11]

**Copy**  Bod G.A. Gen. Top. 4° 77.

**(vii)**  The imprint reads: *Printed by I. Slater, Fleet Street, & 55, Fountain S$^t$. Manchester.*

*I. Slater's New British Atlas, Comprising The Counties Of England . . . Published By Isaac Slater (Late Pigot & Slater), Fleet-Street, London, And Fountain Street, Manchester . . . Hatchard And Son, Piccadilly. I. Slater, Printer, Fountain-Street, Manchester.* [1846].[12]

**Copy**  The private collection of Mr. C.A. Burden.

**(viii)**  Railways 1, ii, 4, 9, and 13 have been added. Railway 4 goes direct to Barton on Humber from Croxton, avoiding New Holland. The signature is now: *Engraved on Steel by I. Slater, Manchester.*

*I. Slater's New British Atlas . . .* [1847].[13]

*I. Slater's New British Atlas . . .* [1847].[13]
**Copy**   BL Maps 3.d.41.

**(ix)**   Railway ii has been removed.
*I. Slater's New British Atlas . . .* [1848].[14]
**Copies**   Bod Maps C.17.b.12; Leeds UL Case K 27.

**(x)**   Railways 10–11 and K have been added.
*I. Slater's New British Atlas . . .* [1849].[15]
**Copy**   CUL Atlas 4.84.30.

**(xi)**   Railways 3, 6–8 and 14 have been added. Railway 4 has been redrawn; it now leaves the MSLR at Habrough and, at a point east of Barrow, turns west to reach Barton. Railway 2 has been added but from Market Deeping to a point between Stamford and Uffington.

*Slater's (Late Pigot & Co.) Royal National Directory And Topography Of Yorkshire And Lincolnshire, Comprising Classified Lists Of Merchants, Bankers, Professional Gentlemen, Manufacturers And Traders . . . The Nobility, Gentry and Clergy; With Historical Sketches of each County, Town & Village . . . With County Maps . . . Published By Isaac Slater . . . And Sold . . . to Subscribers 15 0 Price, to Non-Subscribers 17 6 1849.*

**Copies**   LAO SR. 942.50; LRL L 929; Grimsby RL X000:058 SLA. (Leeds RL M4253 P629). Guildhall L has a volume, lacking title-page, with the map and text of Lincolnshire only.

**(xii)**   The Explanation has been extended by three lines, one with the symbol for railways and two for *Division of the Counties,/according to the Reform Bill.* The line to Barton has been re-drawn again and ends above the town's name.

*Slater's (Late Pigot & Co.) Royal National Directory . . . Of Yorkshire And Lincolnshire . . . 1849.*

**Copy**   CUL L475.c.22.1. [16]

*Slater's . . . Directory . . . Of Derbyshire, Herefordshire, Leicestershire, Lincolnshire, Monmouthshire, Northamptonshire, Nottinghamshire, Rutlandshire, Shropshire, Staffordshire, Warwickshire, Worcestershire . . . Printed And Published By Isaac Slater . . .Sold by . . . Halksworth, 58, Fleet-Street . . . Calf, with Coloured Maps 1 10 0 . . . in Cloth, with Plain Maps 1 5 0 Price to Non-Subscribers, of the Cloth-bound copy Thirty Shillings; of the Calf-bound Two Pounds. 1850.*

**Copy**   BL PP 2506.u.5.

**(xiii)**   Railway 16 has been added and extends to East Retford.
*Slater's . . . Directory Of Derbyshire* [and 11 counties as above] . . . [1850].
**Copies**   Birmingham UL r DA 650; Manchester RL 914.2 S8.[17]

**(xiv)**   Railways 15, 17, B and M have been added. Railways vii–ix have been added but stop short at Long Sutton. Railway 21 is shown as a branch from Grantham to Sleaford.
*I. Slater's New British Atlas . . .* [1850].[18]
**Copy**   The private collection of Mr. C.A. Burden.

*Slater's . . . Directory . . . Of Bedfordshire, Berkshire, Buckinghamshire, Cambridge-*

*shire, Huntingdonshire, Leicestershire, Lincolnshire, Norfolk, Northamptonshire, Nottinghamshire, Oxfordshire, Rutlandshire, And Suffolk . . . Printed . . . By Isaac Slater . . . 1850.*

**Copies** CUL Atlas 6.85.41; Leicester UL H 942 SLA; Guildhall L. LRL L 929 is the map and text of the county only. (LRL Map 105).[19]

*Slater's . . . Directory Of Bedfordshire, Buckinghamshire* [8 counties as above] *. . . Rutlandshire, Suffolk . . . 1851.*

**Copy** BL PP 2506.u.5.

*Slater's . . . Directory . . . Of Bedfordshire, Berkshire, Buckinghamshire, Cambridge-shire, Hampshire, Huntingdonshire, Lincolnshire, Norfolk, Oxfordshire, And Suffolk . . . By Isaac Slater, 37, Fountain-Street, And 36, Portland-Street, Manchester, And 58, Fleet-Street, London . . . 1 5 0 Price, to Non-Subscribers, bound in Calf 2 0 0 . . . bound in Cloth 1 10 0 1852.*

**Copy** Bod G.A. Gen. Top. 4° 1048.

*I. Slater's New British Atlas . . . Fleet-Street, London; Fountain-Street And Portland Street, Manchester: And sold by . . . Hatchard And Son, Piccadilly. I. Slater, Printer, Fountain Street, Manchester.* [1857].[20]

**Copies** BL Maps 4.d.34. (Grimsby RL SLATER 1857).

(**xv**) The imprint has been altered to: *Published by I. Slater Fleet Street London. & 36, Portland St. Manchester.* Railways vii–ix and B have been altered to broken lines. Railway ix now includes an extension from Long Sutton to Cross Keys Wash.

*Slater's (Late Pigot & Co.'s) Royal National Commercial Directory, Containing Every City, Town, And Principal Village In The County Of Lincolnshire. Embellished With A Map Of The County, Faithfully Depicting The Lines Of Railways In Operation Or In Progress Engraved On Steel. Printed . . . By Isaac Slater . . . sold by him and . . . Halksworth . . . London. Price, to Subscribers, 9s.; Non-Subscribers, 11s. 1857.*

**Copy** LRL L 929.[21]

*Slater's . . . Directory And Topography . . . Of Lincolnshire, Nottinghamshire, Rutland-shire & Yorkshire, Containing Classified Lists Of . . . Traders. The Towns Of Hull, Leeds, Nottingham, and Sheffield . . . With Beautiful Maps Of The Counties Of Lincolnshire, Nottinghamshire & Yorkshire . . . Price, bound in Calf, with coloured Maps, to Subscrib-ers 1 10 0 . . . Non-Subscribers 2 0 0 . . . in Cloth, with Plain Maps, to Subscribers 1 5 0 . . . Non-Subscribers 1 10 0 1857.*

**Copy** Grantham RL L 929.

A reproduction of a so-called Pigot's . . . National Commercial Directory . . . Bedford-shire, Huntingdonshire, Cambridgeshire, Lincolnshire, Northamptonshire . . . 1830 was issued in 1992 (Kings Lynn, M. Wilson). No such directory was issued originally; the work consists of the texts of the named counties from various Pigot directories from around 1829 and 1830, with the maps of those counties, all in 1835 states and the imprints removed.

### References

1 Brown, p. 150; Norton, J.E. 1950 op. cit., pp. 43–45; Shaw, G. and Tipper, A. British directories . . . (1850–1950) and Scotland (1773–1950). (Leicester, 1988), pp. 8–9.
2 A photocopy of the Lincolnshire section is in LRL (L 929), provided by Westminster RL. The example in BL (PP 2560.u/s) covers 18 counties and Wales.
3 All loose copies are listed here; they could have come from the atlas version or pre-1831

directories. Leicester UL H 942 PIG is another example of the directory but the maps are missing.

4  The first atlas version had no title-page; the maps were issued in three parts; in the middle of part 3 Pigot's imprint changed to 17 Basing Lane, the address to which he moved during 1830.
5  Mr. C.A. Burden has a second example. The title is on a label stuck to the cover of the section.
6  Although the general map is dated 1830 the newly added data clearly relate to the results of the Reform Act of 1832. These maps must be the work referred to as 'An Appendage' to the directory dated 1835 (see next entry). The BL example consists of three parts bound together; the first two parts have the title as transcribed while the third part's cover has a label: *Pigot & Co. British Atlas Part Three.*
7  One of the general maps is dated 1837. This volume predates those dated 1840 and Whitaker has suggested 1839 for its issue.
8  The dates on the general maps are 1840. Sheffield RL 912.42 STF may be another example but it lacks the title-page.
9  Birmingham UL r DA 650 lacks the maps, although the offsets are visible in each case. All examples are of the county text and map only, the Gainsborough copy having the title-page of the whole work.
10  Pigot died in 1843 and this is the latest likely date for the title-page imprint.
11  Generally, after Pigot's death, Slater's imprint refers to Pigot. This edition must predate the 1846 edition (q.v.) and 1844 is therefore suggested. Bodleian ascribes the work to 1842, but Slater had not assumed full control then.
12  The map of England and Wales is dated 1846.
13  The railways are largely anticipatory. Only the Gainsborough-Grimsby and Lincoln-Newark lines are close to their correct routes, the latter opening in 1846. The general map is still dated 1846.
14  The general map is dated 1847. In the text is: 'A Railway connects Newark with Lincoln; and another is constructing between Gainsborough and Grimsby, by Brigg'. The latter opened throughout its course in April, 1849. 1848 is the probable date of issue.
15  Later than state (ix) but before state (xi), which is dated 1849.
16  A second example is held in a private Lincolnshire library.
17  The Manchester copy lacks the title-page but the make-up and maps conform; the spine title is *Slater's Directory Midland Counties 1850.*
18  The map of Scotland is dated 1850.
19  LRL Map 105 has been closely trimmed, losing the imprint.
20  The acceptance date at BL is 25 MY 58 (i.e. 1858).
21  All the map north of Aslackby is missing.

Chubb CCCXCIX–CCCCI; CCCCXXV–CCCCXXIX.

# 89

## T.L. MURRAY                                    1830

Size: 46.1cm x 35.98cm.

Nothing seems to be known of Murray. The map is a close copy of Ebden (**85**) and contains the same spelling errors of that map, e.g. *Norton* (for Nocton), *Imingham* (Immingham), *Woolstrope* (Woolsthorpe). The mythical *Hughington* remains. Mail roads are marked by triple strands, while in the past the normal practice has been to use one thick and one thin strand for their delineation. The maps were engraved by Hoare and Reeves, who were responsible for the Ebden plates. The atlas only appeared for three years.

## EDITIONS

(i)  *LINCOLNSHIRE.* (Ea). *London. Published May 1ˢᵗ. 1830, by T.L. Murray, 19, Adam*

S$t^l$. *Adelphi*. (Ce, OS). *Drawn under the Superintendence of T.L. Murray* (Ae, OS). *Hoare & Reeves Sc* (Ea, OS). *EXPLANATION./13* lines of symbols, etc. (Ee). *REFERENCE to the HUNDREDS/or WAPONTAKES./31* names divided between the three parts (Ae). *SCALE/10 Miles* [= 5.15cm] (Be). Compass (Eb). The inner frame is marked (without the usual shading) for every minute and numbered at 10' intervals. *Meridian of Greenwich* (De, between borders).

*An Atlas Of The English Counties Divided Into Hundreds, &c, containing the Rivers, Roads, Parks, Parishes, &c. in each, Exhibiting the whole of the Inland Navigation, Rail Roads, &c, And accompanied with Maps of England, Ireland, Scotland And Wales, Projected on the Basis of the Trigonometrical Survey By Order of the Hob$^{ble}$. The Board of Ordnance . . . Hoare & Reeves sc 15, Warwick C$^t$. Holborn.* [1830].[1]

**Copies** BL Maps 20.e.11; CUL Atlas 3.83.8; Bod Allen 91; Leeds UL W 147 and 147a; Birmingham RL AE 912.42. (LRL Maps 819 and 820; Birmingham UL Dept. of Geog. B 45; Grimsby RL Murray 1830).

(ii) The year in the imprint has been altered to *1831*. A canal has been added from Alford to the sea and another from Spalding to Fosdike Wash. From milestone 9 on the Boston-Spilsby road a line labelled. *Hob Hole/Sluice* at its southern end has been added. *Kyme /Tower* (at Bennington) has been removed and *Kyme/Tower* (east of Boston) added.

*An Atlas Of The English Counties . . .* [1831].

**Copies** BL Maps 24.e.32; CUL Atlas 3.83.10; Leeds UL W 148; RGS 3 H 7; Birmingham RL AV 912.42. (Grimsby RL MURRAY 1831).

(iii) The year in the imprint has been altered to *1832*. Below the title has been added: *Explanation/Northern Division . . . Red/Southern Division . . . Yellow/Boroughs returning two Members* [double circle symbol]/*D$^o$. D$^o$. one* [member] ●/*Principal place of Election* □/*Polling Places* ■/ (Ea).

*An Atlas Of The English Counties . . .* [1832].

**Copies** CUL Atlas 3.83.2; NLS EU 16.M.1. (LRL uncatalogued).

**Reference**

1 The maps are dated 1830; the dates on the maps were altered for the 1831 and 1832 editions. Hull RL has a catalogue record of an example of the 1830 atlas but it could not be found (2/94).

Chubb CCCCXIX–CCCCXX (1830 & 1831 editions only).

## 90

**JOHN SAUNDERS** 1830

Size: 42.1cm x 26.25cm.

Thomas Allen's *The History Of The County Of Lincoln* was published in parts, the first of which, as well as the title-page of the first edition, is dated 1830. The original map has no name for the engraver or cartographer, but J. Noble of Hull added his name to signify his ownership in 1833. Joseph Noble was in business in Hull by 1826 as a bookseller and stationer. Later he was noted as a printer (White's Directory of Hull, 1831) and, occasionally, as a newsagent. He probably died in the early 1840s since his wife Mary Frances seems to have carried on the business from 1846 until she retired c.1862. On the cover of the first eleven parts the names of the following engravers are given: Rogers,

Shury, Rolls, Radclyff, and Wallis. it is quite clear from the context, however, that they were engraving the views 'from drawings taken on the spot, by the author'. Whether the Wallis named here is related to the Wallis responsible for the maps of 1812 (**68** and **69**) is not clear, since no further information is given beyond the surname. Although James Wallis is not recorded in business after c.1820 the map for Allen's *History* . . . is quite closely related in its general outline to the larger Wallis plate of 1812. The map was used in a further work relating to the county and issued by Noble (of Hull) in 1833, a year before the completion of Allen's original work.

The author was the son of John Allen, the map engraver (**67**) and was born in 1803. He had begun in 1829 to promote a new part work *An Historical and Topographical Atlas of England and Wales* . . . It was intended to cover all the places from roman times and place their names on the maps in all their forms, whether included in Domesday, etc.[1] By then Allen claimed the authorship of histories of Yorkshire, London and Lambeth. In the event Allen did not live to complete the proposed atlas nor the History of Lincolnshire, since he died of cholera in 1833.[2]

Allen's work was published by John Saunders of Leeds and Joseph Noble (see above). On the back cover of the first part the plan of publication is set out. Each part would have 32 pages of Demy Quarto letter-press, with two elegant views, on fine paper, two shillings. A limited number of examples were to be printed (with India proofs) on superfine Royal paper at four shillings per part. In the only full set of the parts the map is included in part one.[3] It was presumed there would not be more than twenty-five parts in total. One part per month was the intended rate of issue. If the intended rate were maintained from the start it can be assumed that the first issue was made in September, 1830 as the fifth part is the first to be dated, i.e. 1831. The work was also intended by Saunders to form part of a larger work; at the top of the first parts is: 'Saunders' Grand Historical Survey of the Northern Counties, commencing with Lincolnshire and Notting-hamshire. Each county complete in itself and available separately'. The work was dedicated to the Duke of St. Albans, by permission.

The work ran into difficulties even before the death of Allen. If the plan had gone as expected, i.e. a work of twenty-five or less parts, it should have been completed by September, 1832. In fact, only twenty-two parts were finally issued, and the completion date would have, therefore, been June of that year, that is, exactly one year before Allen's death. Apologetic notes appeared in some of the parts and it appears that not all the problems could be laid at Allen's feet. With part XII already dated 1832 the scheme was well behind schedule and G. Virtue took over sole responsibility for sales to non-subscribers. From part 14 John Saunders, junior, seems to have been solely responsible for all aspects of publication and selling. With part 17 Allen's name disappeared from the covers of the parts and the authorship is referred to in the terms 'By The Author of . . .'. The work was finally completed in June, 1834 with the issue of parts 18–22. Also starting with part 14 the work is described as 'Entirely New Edition, Revised And Corrected' – a statement only accurate in that Allen's involvement had ceased and Saunders was now responsible for the final parts and producing a list of errata for the parts already set up. Why examples were issued with title-pages, closely related to the cover of the first part (and, therefore, dated 1830) is not clear. The preface is dated July, 1834, the text includes a reference (Vol. II, p. 360) to the *Lincoln Mercury* of 27th June, 1834, suggesting that the work came out in the summer of that year. The notes below will show the variety of dating on other sets. The whole work sold, in its cheaper form, for £2.12.6.[4]

Saunders was born in Barnstaple (1810) to a long established Devon family. His father was a bookseller in Exeter, with shops in London and Leeds. Saunders, junior, lived in Lincoln with his sister, Mary, during the 1830s before moving to London to undertake literary work, including writing plays.[5] Tennyson said he was a 'man of true dramatical genius'.[6] He died in Richmond in 1885.[7]

Towards the end of the project a second map (**101**) was prepared, which has very few physical features and is intended to show the effect of the Reform Act on election procedures in the county, information usually added to already existing plates. That a new plate was prepared for Allen's History is very suggestive that the larger plate had already been sold, prematurely as it seems. As noted above it seems to have been with Noble by 1833 and was no longer available. As loose copies of the map are very rare it seems certain that the map was not sold separately and only sufficient copies were pulled for the subscribers to the History. Saunders used Cary's larger plates (**57 A–B**) for *Lincolnshire in 1835* and *Lincolnshire in 1836*. When Saunders put out an edition of the *History* in 1838 he used a new map, a reduced version of the Ordnance Survey map of the county, engraved by Stevenson of London (**107**).

## EDITIONS

(**i**)  *LINCOLNSHIRE*. (Aa). *LINCOLN CATHEDRAL* (above an engraving of the cathedral from the south-west, Ea). *20/English Miles 69 to a Degree* [= 8.1cm] (Ee). *REFERENCE/to the/HUNDREDS &c./ NORTHERN DIVISION/1–17/SOUTHERN DIVISION/1–13* (Ae). The inner frame is marked with minutes of latitude at 2′ intervals and numbered every 10′. There is no imprint or compass.

*The History Of The County Of Lincoln, From The Earliest Period To The Present Time. By Thomas Allen . . . Illustrated By A Series Of Views, From Drawings Taken By The Author. Engraved By Rogers, Shury, Wallis, Radclyff, Rolls, &c.* [Vol. 1]. *Leeds: John Saunders, Bond-Street; Joseph Noble, 23, Market-Place, Hull. Sold By Simpkin And Marshall, Stationers'-Court, And Sherwood, Gilbert And Piper, Paternoster-Row, London. MDCCCXXX.*

**Copies**  LAO FL L 942 ALL (Canon Foster's copy); Grimsby RL X000: 942 ALL; Scunthorpe RL L 638 942 Q; Nottingham UL Li 1.D 14 ALL oversize; Nottingham RL R942.53 All; Stamford RL L 9. (Vol. II is dated 1834 in all cases). (LAO Ex 30/2).[8]

*The History Of The County Of Lincoln . . . By Thomas Allen . . . Embellished By Numerous Views . . . London: John Saunders, Junior. MDCCCXXXIII.*

**Copies**  LRL A9 (SR.); Grimsby RL X000:942 ALL.

*The History Of The County . . . To The Present Time; By The Author of . . . Assisted by several Gentlemen in the County, Eminent, either for their well known Literary Abilities, or their extensive Local Knowledge. Embellished By Numerous Views. Vol. 1. London & Lincoln: John Saunders, Junior. MDCCCXXXIII.*

**Copies**  CUL VIII.2.51; Bod G.A. Lincs 4° 62–63; LAO AS 942.51; LRL L 9; Louth RL GL 9.[9]

*The History Of The County . . . Vol. 1 . . . MDCCCXXXIV.*

**Copies**  CUL Syn.3.83.1; Bod G.A. Lincs C 3–4; Birmingham UL r q DA 670.L69; LRL L 9 (SR.) and A 907; Boston RL L 9; Grimsby RL X000:942 ALL.

(**ii**)  An imprint has been added: *Published by J. Noble, N°. 23, Market Place, Hull, for his Gazetteer of Lincolnshire.* (Ce, OS).

*The Gazetteer Of Lincolnshire, Historical, Topographical And Antiquarian; Containing A Complete List Of Every Town, Village, And Hamlet In The County . . . Divided Into Wapentakes; With Population . . . The Distances . . . From London, And The Exact Bearing . . . To Which Is Added A Perfect List Of The Reformed Parliament, With The Numbers Polled By Each, Whether Elected Or Unsuccessful . . . By Joseph Noble. Hull: Printed By And For Joseph Noble, 23, Market-Place; Sold By Noble, Boston; Brooke And Son, And Drury, Lincoln; Rooe, And Wilson, Stamford* [and 19 other booksellers in Lincolnshire and] *London. Simpkin And Marshall. 1833.*[10]

**Copies**   BL Maps 34.a.26; CUL S696.c.83.16; Bod G.A. Lincs 8° 165; LAO Sc 942.50; LRL A 913; Grimsby RL X000:914 NOB.

*The History Of The County Of Lincoln . . . Vol. I. By Thomas Allen . . . MDCCCXXXIII.*

**Copy**   Nottingham UL Li 1.D14 ALL oversize.

*The History Of The County Of Lincoln . . . By The Author Of . . . MDCCCXXXIII.*

**Copy**   Bod G.A. Lincs 4° 62–63.

*The History Of The County Of Lincoln . . . MDCCCXXXIV.*

**Copy**   A private library in Lincolnshire.

### References

1   *Gentleman's Magazine*. Vol. XCIX, part 2 (1829), p. 356.
2   DNB, Vol. I, p. 313; obituary in *Gentleman's Magazine,* Vol. CIII, part 2 (1833), p. 86.
3   Mr. P. Cassidy, Holbeach bookseller, sold a full set in the original covers and I am grateful to him and the present owner for this information. The covers of all the parts were bound in at the end of the second volume of a set in Grimsby RL (X000:942 ALL), but missing in May, 1993. LAO (FL Box 942) has parts 1, 3 and 4.
4   The original paper label survives on the Birmingham UL copy.
5   DNB. Vol. XVII, p. 814.
6   Letters of Lord Alfred Tennyson; ed. by Cecil Y. Lang and Edgar F. Shannon II. Vol. II (Oxford, 1987), p. 96.
7   *Times* obituary, 4 April, 1885.
8   BL has a volume (800 cc 24), but without the map. It does have the covers of the first ten parts as issued in the 4s, version, with the heading *PROOF IMPRESSIONS*; the tenth part is dated 1831.
9   LAO Sc 942.51 and Grantham RL L9 are other examples, but lack this map.
10   Several copies have the original covers printed with a short title and showing that the book sold for 4s.

**91**

**JAMES SANDBY PADLEY**                                                    **1830**

Size: 26.8cm x 23.7cm.

Padley (1792–1881) first came to the county, probably from Nottinghamshire in 1819 to assist in the first Ordnance Survey of the county. Later, he said, he surveyed nearly all the main roads 'from Wragby to Southwell (Notts.) and from Dunsby Lane to Spital'.[1] In 1819 he produced his first map of Lincoln (at 10 chains to an inch); only two years earlier Marrat had produced his first map of the city and, compared with that map, Padley's is less detailed, showing none of the field divisions drawn on Marrat's and, because of the smaller scale, fewer place-names.[2] His much larger map of Lincoln in 1841 is a finer work,[3] which was re-issued in 1842[4] and came out in revised editions in 1851,[5] 1868[6], 1877[7] and a so-called 'Third Edition' in 1883.[8] Padley settled in Lincoln and became Surveyor of County Bridges for Lindsey, after being recommended by Charles Chaplin, M.P. following surveying work he had carried out for him in the Blankney area.[9] In 1826 he produced *PLAN of the FOSSDYKE NAVIGATION from Lincoln to the Trent.*[10] This map was reissued in c.1852 as part of *Handbook Guide to Lincoln and Business Intelligencer . . .*[11] Two years later he produced *PLAN Shewing the LINE of the Proposed CANAL from the RIVER WITHAM to BISHOP'S BRIDGE on the RIVER ANCHOLME . . .* which accompanied *Report On The Practicability Of A Junction Canal Between The Rivers Witham And Ancholme . . .*[12] In the same year he also prepared

*PLAN & SECTION OF THE NAVIGABLE CANAL from the town of LOUTH, TO THE RIVER HUMBER . . . Lithographed by J & J. Jackson, Booksellers, Louth.*[13] In the year after his death (1881, aged 89) his survey *The Fens And Floods Of Mid-Lincolnshire* appeared. It is of special interest to cartographers since he included lithographic transfers of several older maps, the earliest being Hollar's 1661 map of the East and West Fen.[14]

The map below is an odd affair, because the emphasis is on places near the sea, with many villages near the coastal areas but comparatively few names west of a line from Grimsby through Louth, Horncastle, Boston and Spalding. The county's roads are treated similarly. It shows the 'Line of Padley's Canal' from the Witham (south of Fiskerton) and forming a link to the River Ancholme (see above); all the streams running into the canal are a prominent feature, as are the streams in the southern part of Kesteven. In the delineation of the Wash the coast from Fosdyke to Kings Lynn is virtually a straight line. The deep V that usually appears on contemporary maps to indicate the Cross Keys Wash, (which is not mentioned), is hidden behind a line that Padley calls 'Proposed Embankment'. There is nothing in the map to show what other sources Padley drew on.

The Association for the Preservation of Life from Shipwreck was formed following a meeting in Spilsby on 23 January, 1826 on the formal proposal of Rev. T.H. Rawnsley. C.B. Massingberd took the chair and Lord Gwydyr led the subscription list with 20 guineas. A meeting on 28 March, 1826 agreed the rules of the association; Earl Brownlow agreed to be the patron, as Vice-Admiral of the county, and Lord Yarborough became the president. A further subscription list of July, 1827 was headed by Earl Brownlow with £50.[15] Annual reports were issued[16] and these show the Society in a generally flourishing condition. However, financial problems occurred when the need to replace four boats at the same time would have taken all of the £1700 the association had in its reserves. The responsibility for the four stations (at Donna Nook, Theddlethorpe, Skegness and Sutton) was transferred to the Royal National Lifeboat Institution following a meeting at Spilsby on 11 January, 1864, although the Association continued to exist as a purely fund-raising organisation until 1911.[17]

## EDITION

(i)   *Map/Shewing the Stations/of the/SHIPWRECK ASSOCIATION/on the/LINCOLN-SHIRE COAST/1830.* (Ea). *J.S. Padley del.* (Ee, OS). *T. Brettell Lithog: Regent St. Haymarket.* (Ae, OS). There are neither compass nor scale.

*Rules Of The Lincolnshire Coast Shipwreck Association. Instituted July 31st, MDCCCXXVII. Lincoln: Printed By J.W. Drury, High-Street, Cornhill. 1830.*

**Copies**   LAO Ex 942.51; LRL L 614.86.

**References**
1   Padley, J.S. *Fens and Floods of Mid-Lincolnshire.* (Lincoln, 1882), preface (p. v).
2   Copies of the map are: LRL Map 509 and NLS EME.b.1.1.
3   LRL Map 509a.
4   CUL Maps a.70.84.1; Bod Map C17: 70 Lincoln d.1; Nottingham UL Li 3.B8 E42; Grimsby RL L400:912 PAD.
5   LAO SR. Maps Box (942.51); Nottingham UL Li 3.B8.E 51.
6   CUL Maps C.70.86.1; Bod Map C17: 70 Lincoln e.1; Nottingham UL Li3.B8 E68; LAO BM 16 and Exley 30/6; LRL Map 301.
7   In *City of Lincoln Directory, 1877.* (LRL Lin 929).
8   BL Maps 8.b.17; Nottingham UL Li 3.B8 E83; LRL Rolled map; Grimsby RL L400:912 PAD.
9   Binnall, P.B.G. J.S. Padley. *Lincolnshire Life,* Vol. 20, no. 6 (September, 1980), pp. 28–29.
10  LRL Map 674; Louth RL GL Lind 386.
11  Bod Johnson f 1498.
12  LRL L 386.

234

92  TEESDALE

13  LRL Maps 85a and 699.
14  BL 10388 k 15; CUL Cam a.882.4; Bod G.A. Lincs c.15; LAO SR. 942.51. Copies are widely available in the county's reference libraries at Lincoln, Boston, Grimsby, Grantham, etc.
15  LAO 4 BNL 14.
16  BL 8085.c.76 has the annual reports for 1849, 1850 and 1860.
17  Farr, G. The Lincolnshire Coast Shipwreck Association, 1827–1864. (Portishead, Bristol, 1981), p. 14.


92

## HENRY TEESDALE                                                      1830

Size: 19.1cm x 14.9cm.

Teesdale set up on his own in 1828 and acquired the plates of Robert Rowe in 1829 (**71**); in the following year he issued his own set of small county maps. While he issued editions of the Rowe plates at regular intervals, until 1842, he only issued two editions of his own plates, the second in 1843 being the last. Both the Rowe and these smaller plates became the property for a few years of H.G. Collins, who used them for lithographic transfers, as did the later owners; firstly William Somerville Orr and finally John Heywood. Orr worked originally with W.L. Graves until he set up by himself in 1838; he was joined by his son in 1859, the year after he had acquired at auction the Rowe plates and these Teesdale plates from Collins. John Heywood (1804–1864) was a weaver, who set up in Manchester as a stationer in 1846 and made a fortune from making copy-books. His son, also John (1832–1888), carried on the business, adding atlas production to his activities from the early 1860s and using the Teesdale plates to help satisfy the growing need for guide-books. His Excelsior printing works employed 750 people at one time.[1]

The maps have their basis in Cary's smaller plates and a distinctive feature is the use of the Ordnance Survey's piano keyboard decoration for the borders.

## EDITIONS

(i)  *LINCOLNSHIRE* (in a panel, with vertical shading as background, Ea). *London, Published Sept$^r$. 1830, by Henry Teesdale, & C$^o$. 302 Holborn.* (Be–De, OS). *Scale of Miles/10* [= 1.9cm] (Ae). Compass (Aa). The frame, marked in keyboard style, lacks minutes or degrees.

*A New Travelling Atlas, Containing a complete Set of County Maps, on which are Delineated all the Main & Turnpike Roads, Cities, Towns, Parks & Gentlemens Seats Preceded By General Maps of England & North & South Wales. The whole carefully Revised and Corrected to the year 1830. London Published By Henry Teesdale & C$^o$. 302, High Holborn.*

**Copy**  CUL Atlas 7.83.1.

(ii)  The imprint has been removed. Plate number *24* has been added (Ea, OS).

*This Edition Contains All The Railways. A Travelling Atlas . . . to the Year 1843. London Published By Henry Teesdale & C$^o$. 2 Brunswick Row, Queen Sq$^r$. and Sold by D.W. Martin 16 Westmoreland Place, City Road.*

**Copy**  CUL Atlas 6.84.23.[2]

*A Travelling Atlas . . . 1843 . . .*

**Copy**  CUL Atlas 6.84.25.

## LITHOGRAPHIC TRANSFERS

(**A**)   Size: 17.6cm x 14.6cm. A new imprint: *London. Published for the Proprietors by H.G. Collins. 22 Paternoster Row.* (Be–De, OS) has been added. Railways 1–4, 6–7, 9–14 have been marked by single black lines over thinner lines, which were originally cross-hatched. Railway 2 is extended through Market Deeping to join the ELR at Glinton. The plate number and the shading in the title panel have been removed. The piano keyboard borders have been replaced by three lines.

*The Travelling Atlas, of England & Wales, with all the Railways & Coach Roads, The Cities, Towns, Parks & Gentlemen's Seats Preceded by General Maps of England & North & South Wales. The whole carefully Revised and Corrected to the Present Time. London, Published (For The Proprietors) By Henry George Collins, 22, Paternoster Row.* [1849**].

**Copies**   BL Maps 15.b.13; CUL Atlas 7.84.23; Bod G.A. Gen. Top. 8° 79; NLS EME b.1.10.

(**B**)   Railways 8, 15, 17, vii–ix, xi and B have been added. At its southern end the GNR passes through Market Deeping from where a short branch links up with the Lincoln loop line (railway 8).

*The Travelling Atlas . . .* [1850].

**Copies**   Leeds UL W 193. (LRL Map 777).

(**C**)   A new imprint reads: *London: Published (for the Proprietors) by W.S. Orr & Co., 2 Amen Corner, Paternoster Row.* (Be–De, OS). All the railways have been redrawn with single fine lines closely crosshatched. Railways vii–ix and xi have been removed and railway 2 has been extended and 'converted' into railway 5 with the addition of *fr. Leicester*. The railways now shown are: 1, 3–17. The line from Lincoln to the MSLR (east of Brigg) forms a T (instead of a V). The Lincoln-Newark line stops a little west of Newark and *fr. Nottingham* (previously *from/Nottingham*) has been added. Railway 15 has been extended westwards. *New/Holland* has been added.

*The Travelling Atlas . . . Present Time. London. W.S. Orr & Cº. 2, Amen Corner, Paternoster Row.* [1852].[3]

**Copies**   BL Maps 14.dd.14; Leeds UL W 194. (Boston RL Map 14).

(**D**)   Size: 19.1cm x 14.9cm. The imprint reads: *Printed & Published by John Heywood 170 Deansgate Manchester.* (Be–De, OS). The piano keyboard border and the boxed and shaded frame to the title have been restored. Three lines of shading mark the county's coastline. *Donna Hook* (sic), *Wainfleet Harb.* and *LYNN AND/BOSTON DEEPS* have been added and *Ryall* (formerly *Ryal*) corrected.

*The Travelling Atlas . . . Gentlemens Seats. Revised & Corrected by John Heywood, 170, Deansgate, Manchester.* [1858].[4]

**Copy**   Birmingham RL A 912.42.

(**E**)   The imprint is: *PRINTED & PUBLISHED By JOHN HEYWOOD 143, DEANSGATE./& 3, BRAZENOSE Sᵀ. MANCHESTER./* (Ce, OS). The plate number *24* has been restored (Ea, OS). Railways 19 24, 26–27, 29–30 and 33 have been added. Railways 34–35, xvii and the east end of 25 are marked by broken lines. *S.Y. Ry from/ Doncaster, to Doncaster* and *to Lynn* have been added as well as the following place-names: *Cleethorpes, Stabelthorpe* (i.e. Mablethorpe), *Ulceby, Winterton, Winteringham, Scawby Park, Brocklesby/Park, Hackthorn* (the last two with the delineation of the parks). *Retborn* has been redrawn as *Redbourn*. The line at [Scunthorpe] is named *T.A. & G R*[Y]. A dotted line joins Alford to *Stabelthorpe*.

*The Tourist's Atlas Of England And Wales, With Railways . . . Gentlemen's Seats. Revised And Corrected To The Present Time. Manchester: Printed By John Heywood, 143, Deansgate. London: Simpkin, Marshall, And Co.; Hamilton, Adams, And Co.; And W. Tegg. MDCCCLXIV.*

**Copy**   (The author's collection.)[5]

(**F**)   The imprint has been removed. Railways 25, 32 and 34–36 have been drawn in as solid, cross-hatched lines, together with a line from Long Sutton to [Wisbech], labelled *to Ely*. The line from Stamford to [Peterborough] has been redrawn and now runs parallel to the GNR at the southern end. Railway 22 has been deleted. New place-names are: *Epworth, Grainthorpe, Thornton le Fen, Saxby, Ingoldmells P*[t.] *Cleethorpe* has been redrawn, losing the final letter in the process, *Mablethorpe* has been corrected and moved northwards (and, with it, railway xvii) to its correct site and *Donna Nook* has been redrawn and corrected.

*The Travelling Atlas . . . Gentlemens Seats preceded by General Maps of England & North & South Wales. The whole  . . . to the Present Time. Manchester: Printed and Published by John Heywood, 141 & 143, Deansgate. [1868].*

The cover title reads: *John Heywood's County Atlas Of England And Wales, With All The Railways & Coach Roads . . . Corrected To The Present Time. Price One Shilling. Contents:* [list of the maps in three columns] *Manchester: John Heywood, 141 And 143, Deansgate. Educational Department: 141, Deansgate.* [1868].[6]

**Copies**   Gloucester RL 912.42; Birmingham RL A 912.42; NLS 3.e.27.

(**G**)   Railway 22 has been restored. The railway from Stamford joins the GNR above the G of Glinton.

*The Travelling Atlas . . .* [1868].[7]

**Copies**   CUL Atlas r.6.86.17; Bod C.17.e.4; Leeds UL W 195.

(**H**)   The imprint is: *PRINTED & PUBLISED* (sic) *BY JOHN HEYWOOD. 141 & 145 DEANSGATE./& EXCELSIOR WORKS, MANCHESTER./* (Ce, OS). Railways 28 and 37 have been added. *Cross Keys Wash* has been re-drawn because the inlet has been separated from the sea by a 'sea wall' and the 'reclaimed land' marked with marsh symbols. As a further result the railway from Spalding curves south at Long Sutton towards Wisbech, with the line to Kings Lynn redrawn from *Tid S*[t]*. Marys* eastwards. Railways 5 and 17 run parallel at Stamford. *Firsby* and *Croft* have been added. Railway 22 has been redrawn and now passes south of *Red Deer/Park*.

*The Travelling Atlas . . .* [1875].[8]

**Copy**   Nottingham UL s/G 5514.

(**I**)   The imprint is: *Published by John Heywood. Sc. Excelsior Works Manchester.* (Ce, OS). Below the title has been added: *Contains 2,676 Sq. Miles (1, 776, 640 Acres)/ Population (1871) 436,163* (Ea). *The Figures near the principal Towns denote/the distance in miles from London, by road./* has been added (Ee). *Lon 0°33' W.* has been added (Ba and Be). *Lat. 53°14' N.* has been added (Ac and Ec). Railways 38–40 have been added. Railway 42 has been added but from the Grantham-Sleaford line across the Lincoln-Grantham line (both wrongly sited) to the GNR. *Wroot, Sutton/S*[t]*. Edm*[ds]*. Welland R. Timberland, Nocton, Fiskerton, Skellingthorpe, Roughton, Beesby, Cumberworth, Cockerington, N. Cotes, Holton, Haxey, Friskney, Stow, Bennington, Skegness,*

*Winthorpe* and *The Wash* have been added. Hachur has been drawn to indicate the Wolds and the Cliff. Roads have been added between Sibsey and Tattershall, with a branch to Stickford and a road linking the two.

*The Travelling Atlas . . . Published by John Heywood, Sc. Excelsior Works, Manchester.* [1876].[9]

**Copy**   The private collection of Mr. E. Burden.

(**J**)   The imprint is: *PUBLISHED BY JOHN HEYWOOD, 141 & 143, DEANSGATE MANCHESTER.* (Ce, OS).

*The Travelling Atlas . . . Corrected to the Present Time. Price Plain 2/6 Coloured 5/-. Printed & Published By John Heywood, 141 & 143, Deansgate Manchester.* [1880].[10]

**Copy**   The private collection of Mr. D. Kingsley.

(**K**)   The imprint is: *JOHN HEYWOOD, PUBLISHER & EDUCATIONAL BOOK-SELLER, EXCELSIOR BUILDINGS, RIDGEFIELD, MANCHESTER,/AND 18, PATER-NOSTER SQUARE, LONDON, E.C.* (Be–De, OS). Railway 41 has been added.

*The Travelling Atlas . . . to the Present Time. John Heywood,* [the imprint is exactly as that of the map]. [1881].[11]

**Copy**   CUL Atlas 6.87.27.

(**L**)   The imprint now reads: *John Heywood, Publishing. Deansgate & Ridgefield, Manchester.* (Ce, OS). Railways 43–44 has been added. Railway 22 has been removed. *Tetney Haven* has been added. The second line below the title reads: *Population (1881) 469,994/* (Ea).

*John Heywood's County Atlas . . . With . . . Parks and Gentlemen's Seats. John Heywood, Deansgate And Ridgefield, Manchester; And 11, Paternoster Buildings, London.* [1882**].

**Copies**   CUL Atlas 6.88.31; Manchester RL BR 912.42 H14.

(**M**)   The imprint is: *JOHN HEYWOOD, LITHO, MANCHESTER & LONDON.*[12]

**References**

1  Kingsley, p. 192.
2  Mr. C.A. Burden owns another copy, but in the title-page imprint Road is spelt R$^d$. Chubb (p. 288) draws attention to the map of Cambridgeshire in the CUL atlas, which has the London-Cambridge Railway (opened July, 1845).
3  Although no railway is shown here open after 1852 lines on other county plates suggest this edition appeared in 1858. From this edition onwards the cover title is often different from the title-page and imprints may differ between the two. See state (F) below.
4  There is a copy in Gloucester RL but it lacks the county plate; it, like a copy owned by Mr. C.A. Burden, has no title-page but the cover has the title as quoted, but with the imprint still of W.S. Orr. Mr. Burden's copy has the Middlesex map on the reverse of the Lincolnshire map.
5  The map and title-page came from a volume being broken up.
6  1868 is the generally accepted date, but within that year there were several variant editions as indicated. The Gloucester copy has no title-page.
7  The Bod copy has the cover title: *John Heywood's County Atlas . . .* but *London: Simpkin, Marshall & Co.* has been added to the imprint.
8  Other counties show railway data updated to 1875; the Lincolnshire plate shows no railway after May, 1868.
9  1876 has been suggested, based on railway data on other plates; the county lines are only accurately depicted to 1873.

10  This date is based on the railway information on other maps.
11  The atlas is later than state J and 1881 is suggested, as the line from Lincoln-Spalding (opened 1882) is not shown.
12  Mr. T. Burgess owns a map of Kent, which may have come from an atlas in which all plates are altered thus. No date is suggested.

Chubb CCCCXVIII; DXXIX–DXXXI (Teesdale atlas of 1830 and the appearances of the 1848, 1852 and 1868 editions only).

## 93

**R. CREIGHTON**                                                           **1831**

Size: 30.4cm x 23.65cm.

R. Creighton was active as an engraver from 1818, when he drew the map of Lancashire for the Greenwoods,[1] until 1855 (A Map of England & Wales . . .).[2] The engravers, John and Charles Walker, have been mentioned in connection with the plates published in 1822 (**81**) and the Greenwood maps (**87**); they will be dealt with more fully concerning their own atlases of 1837 (**105**). Little seems to be known about the publisher Samuel Lewis; Kingsley records that he died in 1865.[3] Uniform with the very successful *Topographical Dictionary of England* he also produced separate series on Ireland, Wales and Scotland.

The map is based on Ebden (**85**). In many of its appearances the map occurs bound in with the text of the Topographical Dictionary (in Lincolnshire's case, Volume III); however, a fifth volume often formed an atlas supplement, into which the maps have been transferred (with or without a title-page). From 1835 to 1838 the fifth volume often has the second Creighton set of county and town maps (**104**).

### EDITIONS

(**i**) *LINCOLNSHIRE*/Compass (Ea). *DRAWN & ENGRAVED FOR LEWIS' TOPOGRAPHICAL DICTIONARY.* (Ce, OS). *Scale of Miles/ 15* [= 5.6cm] (Ae). *Engraved by J. & C. Walker.* (Ee, OS). *Drawn by R. Creighton.* (Ae, OS). The inner frame is marked every five minutes and numbered every ten minutes of latitude, etc. *Meridian of Greenwich* (De, between borders).

*A Topographical Dictionary Of England, Comprising The Several Counties . . . And The Islands Of Guernsey, Jersey, And Man, With Historical And Statistical Descriptions; Illustrated By Maps Of The Different Counties And Islands . . . By Samuel Lewis. In Four Volumes. Vol. III. London: Published By S. Lewis And Co., 87, Aldersgate-Street. M.DCCC.XXXI.*

**Copies**  BL G 492 and 010493 p. 3; CUL VIII.1.1–4; Bod G11/C 17.2; LAO FL 1a 950.2; Nottingham UL s/DA 640.14 oversize; Norwich RL RT 942. (LRL Map 106; Grimsby RL LEWIS 1831).

*A Topographical Dictionary . . . By Samuel Lewis. The Atlas. London . . . M.DCCC.XXXI.*

**Copies**  CUL Atlas 5.83.23; Bod Allen 93; Leeds UL W 149.

(**ii**)  The following has been added: *PLACE of ELECTION for the NORTHERN DIVISION. Lincoln___POLLING PLACES {* 10 places in three columns (Brigg to Spilsby) *: PLACE of ELECTION for the SOUTHERN DIVISION Sleaford__ POLLING PLACES {* 8 places in two columns (Boston to Spalding) (Aa–Ea, OS).

*A Topographical Dictionary . . . Second Edition. With An Appendix, Describing The*

*Electoral Boundaries Of The Several Boroughs, As Defined By The Late Act. By Samuel Lewis . . . M.DCCC.XXXIII.* [1834].[4]

**Copies**  BL 576 m 8–11; CUL Atlas 5.83.12.[5]

*A Topographical Dictionary . . . Third Edition, With A Supplementary Volume Comprising A Representative History Of England, With Plans Describing The Electoral Divisions Of The Several Counties, And The Former And Present Boundaries Of The Cities And Boroughs. By Samuel Lewis. In Five Volumes. Vol. III . . . M.DCCC.XXXV.*[6]

**Copies**  BL 10353 e 1; Leeds UL W 150; Nottingham UL s/DA 640.L4 oversize; NLS Ref. 2.C.17; London UL f.MU 1 LEW. (Grantham RL).

**(iii)**  The notes on election places have been removed (Aa–Ea, OS).

*A Topographical Dictionary . . . M.DCCC.XXXVII.*

**Copies**  CUL 8474.b.16; Leeds UL W 151; Liverpool Athenaeum 914.2Q; Leicester UL H 942.F LEW.[7]

**(iv)**  *Reference to the Unions./*18 names in two columns, has been added (Ee). The boundaries of the unions (numbered 1–18) have been indicated by dotted lines.

*A Topographical Dictionary . . . Fourth Edition. In Four Volumes. Vol. III. London . . . MDCCCXL.*

**Copies**  BL 10350 h 5; Bod G.A. Gen. Top. 4o 534 = R.Top. 516; Leeds UL W 152; LAO SR. 950.2; Canterbury Cathedral L. B12.22. (LRL Map 106b).[8]

*An Atlas, Comprising Maps Of The Several Counties, Divided Into Unions, And Of The Islands Of Guernsey, Jersey, And Man; With A Map Of England And Wales, And A Plan of London and its Environs. London: Published By S. Lewis And Co., 87, Hatton Garden. M.DCCC.XLII.*

**Copies**  BL Maps 45.a.11; CUL S 696.b.84.8; Leeds UL W 153; LAO FL 1a 950.2 (Extra large); Bournemouth RL R 912.42; London UL f.MU 1 LEW.

*A Topographical Dictionary . . . Fifth Edition. In Four Volumes. Vol. III. London . . . 87, Hatton Garden. M.DCCC.XLII.*[9]

**Copy**  BL Maps Ref. N.3 (420).

*Atlas To The Topographical Dictionary, Comprising A General Map Of England, A Plan of London, And Maps Of The Counties Of* [names of the counties in four columns]. *London: Published By S. Lewis And Co., 13, Finsbury Place, South. M.DCCC.XLIV.*

**Copies**  CUL S 696.a.84.7; Leeds UL W 154; Liverpool UL Y84.5.98; Edinburgh RL DA 670; Birmingham RL AQ 912.42; Sheffield RL 914.2 STQ.

*Atlas To The Topographical Dictionary . . . M.DCCC.XLV.*

**Copies**  CUL Atlas 6.84.11; Sheffield RL 914.2 STQ.

*Atlas To The Topographical Dictionaries Of England And Wales . . . M.DCCC.XLV.*

**Copies**  BL Maps GAZ 420/429; Leeds UL W 155; Hull RL 914.2; Glasgow Mitchell L. 914.203 LEW (B16697); Nottingham UL s/DA 640.L4 (originally owned by Sir Francis Hill).

**(v)** Railways 1, 4, 9, 13 and the western end of B (from Stamford to Market Deeping) have been added.

*A Topographical Dictionary Of England . . . Seventh Edition. In Four Volumes. Vol. III. From Laceby To Ryton-Woodside . . . M.DCCC.XLVIII.*[10]

**Copy**   John Rylands UL Q 942.066 L 588 (20835).

*An Atlas To The Topographical Dictionaries Of England And Wales . . . M.DCCC.XLVIII.*

**Copies**   BL 2061 i; CUL Atlas 6.84.13; Bod Allen 108; Leicester RL 910.342.

*Atlas To The Topographical Dictionary Of England . . . M.DCCC.XLVIII.*

**Copies**   Leeds UL W 156; Leeds RL Q 942 L 588.

*Atlas To the Topographical Dictionaries . . . M.DCCC.XLIX.*[11]

**Copies**   Sheffield UL **Q 914.2 (L); Liverpool RL Q914.2.

**References**
1  Rodger, 1972, item no. 256, p. 14.
2  Smith, D., p. 157 notes that this map was used by Stanford's for transfers of individual counties (see **135** below).
3  Kingsley, p. 203.
4  The map of London is dated 1834.
5  There was probably a separately issued atlas, of which the CUL example may be a case in point, as it has maps only, no text and no title-page. Another copy is held in a private library in Lincolnshire.
6  Volume 5 has the Creighton county and town maps (**104**).
7  The Leeds copy lacks a title-page. The other sets have a fifth volume with Creighton's second set of maps.
8  Only in the case of the BL example is the map in volume 3. In all the others cited plus sets owned by Mr. D. Hodson and Mr. J. Spedding the maps appear in an untitled fifth volume, with the map of London dated 1840. Mr. Hodson has a second set in which the atlas volume (5th) has a separate title-page of the style found in later issues (q.v.). The Leeds copy lacks the text volumes.
9  Leicester UL (Hatton F 942 LEW) is an 1844 edition with a revised letterpress title-page, also described as Fifth Edition . . . but lacking all reference to maps. None is included.
10  A Sixth Edition has not been found; Chubb p. 297 cites Fordham as an authority for its issue in 1845.
11  The BL has a set (2061 i) of the Dictionary with all the text volumes dated 1849 but the atlas dated 1848. Mr. D. Hodson and Liverpool UL have sets of the 1849 Dictionary without maps and there is no atlas volume.

Chubb CCCCXXX–CCCCXXXVI.

**94**

**SIDNEY HALL**                                                                 **1831**

Size: 21.6cm x 19.2cm.

Hall's work has already been recorded in 1820 (**76**). The new maps were originally prepared for the Gorton version of *A Topographical Dictionary* . . . perhaps as a rival to that of Lewis (q.v.) and as a successor to his own *A General Biographical Dictionary*, published in 1828. The work was issued in parts over a three year period as the dates on the maps (1830–1832) testify. The dictionary was not issued again after 1833. Thereafter Hall used the plates to provide a handy atlas for use by the increasingly sizeable public using the railways. Its popularity is evidenced by its appearance at regular intervals, the last being in c.1888. Although Kingsley notes that Hall, after 1850, was meticulous in recording railways only when they were built this is not quite the case with the Lincolnshire maps.[1] The maps from 1846 and for the next forty years continue to include several lines promoted during the railway mania but never built. The reason is that Hall

used for his lithographic transfers plates that had contained the 'spurious' lines; he
reduced the problem to some extent by colouring only those lines that were finally built.

The publishers of the Hall atlases, Chapman and Hall, were well-known for their work
among nineteenth century novelists; with Dickens they made youthful fortunes with the
publication of *Pickwick Papers* (1836). Later they were the publishers of Ainsworth, Mrs.
Gaskell, Bulwer Lytton, Kingsley and Trollope; when the firm became a limited company
in 1880 Trollope became a director. Sidney Hall is not related to the Hall of the publishing
partnership.

## EDITIONS

(i)   *LINCOLN/SHIRE*. (In a plain rectangle, Ea). *ENGRAVED BY SID^Y. HALL*. (Ea,
below the title). *English Miles/10* [= 3.2cm] (Ea, below the signature). *London Published
by Chapman and Hall N^o. 186 Strand Aug^t. 1831*. (Be–De, OS). *REFERENCE TO THE/
HUNDREDS &c./* 31 numbered names divided into the three parts (Ee). Compass (Ae).
The inner frame is marked every two minutes and numbered every ten. *Long^t. East from
Greenwich* (De, between borders). *Long^t. West from Greenwich* (Ae, between borders).

*A Topographical Dictionary Of Great Britain And Ireland, Compiled From Local
Information, And The Most Recent And Official Authorities. By John Gorton . . . The Irish
And Welsh Articles By G.N. Wright . . . With Fifty-Two Quarto Maps, Drawn And
Engraved By Sidney Hall. In Three Volumes. Vol. I. London: Chapman And Hall, 186,
Strand. 1832.*

**Copies**   CUL VIII.7.38–40; Leeds UL W 157. (LRL Map 769; Boston RL Map 22;
Grimsby RL HALL 1831).

*A Topographical Dictionary . . . Vol. II . . . 1832.*

**Copies**   Bod Allen 94; Edinburgh UL ref. 91 (42) 03 GOR.

Issued as an atlas without title-page. [1833].[2]

**Copy**   BL Maps 9.b.54.

*A Topographical Dictionary . . . With Fifty-Four Quarto Maps . . . Vol. II . . . 1833.*

**Copies**   BL 10353 b 46; Bod G.A. Gen. Top. 8° 559–561; Guildhall L S 914/2; Leicester
UL H942 GOR; Manchester Chetham's L. 914.2 (GO 693).

(ii)   The date in the imprint has been changed to *1833*.

*A New British Atlas; Comprising A Series Of 54 Maps, Constructed from the most Recent
Surveys and Engraved By Sidney Hall. London: Chapman & Hall 186, Strand 1833.*

**Copies**   CUL Atlas 5.83.30; Leeds UL W 160; Birmingham RL AQ 912.42; Guildhall
L. Gresham 4.

*A New British Atlas . . . 1834.*

**Copies**   BL Maps 20.a.54; CUL Atlas 6.83.3 and 6.83.4; Leeds UL W 161; Chesterfield
RL L 912.

(iii)   The date in the imprint has been removed.

Issued as a folding map in a green cover with *LINCOLNSHIRE* blocked in gold. [1833].

**Copies**   CUL Maps c.34.033.70 and Maps c.34.029.70; Grimsby RL X000:912 HAL;
Nottingham UL Dept. of Geography F 131 (drawer 121).

Issued as an atlas, without title-page. The cover title is: *SIDNEY HALL'S BRITISH
ATLAS.* [1834].[3]

**Copies**   Leeds UL W 162; PRO F.O. 925/4188.

*A New British Atlas . . . 1836.*

**Copies** BL Maps 20.a.24; Bod Allen 103 and Maps C.15.d.39; Leeds UL W 163; Manchester RL BR q 912.42 H10.

(**iv**)   Railway i is marked by double lines keeping (wrongly) just to the west of the Market-Deeping-Lincoln road.

*A Travelling County Atlas: With The Coach And Rail Roads Accurately Laid Down And Coloured, And Carefully Corrected To The Present Time. Engraved By Sidney Hall. London: Chapman and Hall, Strand MDCCCXLII.*

**Copies** BL Maps 199.f.26 and 199.d.49; CUL Atlas 7.84.13; RGS 9 B 1; Norwich RL R 912.42 (SC. 3.21).[4]

*A Travelling County Atlas . . . Strand. MDCCCXLIII.*

**Copies** CUL Atlas 7.84.5; NLS Newman 509.

Issued, folded in a card cover, about 10.5cm x 9cm, with the cover title *Pocket County Maps. Lincolnshire. Price 1s. Chapman & Hall, 186, Strand.* [c.1845.].[5]

(**v**)   Railways 1, 4, 9–10, 12–13 and the portion of B from Market Deeping to Stamford have been added by hand in blue. Roads have been coloured brown, parks green; the boundary of North Lincolnshire is marked in yellow and of South Lincolnshire in blue; the  latter colours are varied in some copies of some later editions.

*A Travelling County Atlas . . . Coloured And Carefully Corrected . . . 1845.*

**Copies** BL Maps 28.bb.16; Bod Allen 105; Leeds UL W 173; Birmingham RL A 912.42.

(**vi**)   Railway i has been removed. The railways of state (v) have now been engraved, before the application of colour. The names of railways 1 and 4/9/13 have been added.

*A Travelling County Atlas . . . 1845.*

**Copy** CUL Atlas 7.84.7.

(**vii**)   Railways 3, 6–8, 11, 14–17 have been engraved on the plate; railway 12 consists of plain double lines, while all others are double lines, cross hatched. Railways v–viii have been added, but vii only extends to Spalding. Railway 21 has been added as a branch from vii. Railway B (portion) has been deleted. Holland/Ferry has been named. *Notting^m. Newark/& Lincoln Railway* has been added at Newark. *Great Grimsby & Sheffield Junction R^y.* has been added along railway 4/9/13/. *Lincoln Loop/Rail^y.* has been added at Gainsborough and at its southern end. The *Ambergate, Notting^m/& Boston R^y.* has been named at Grantham. *London &/York R^y.* has been added at Newark along the line of the GNR.

*A Travelling County Atlas: With All The . . . Rail Roads . . . Coloured, And Carefully Corrected To The End Of The Last Session. London: Chapman and Hall, Strand 1846.*

**Copy** CUL Atlas 7.84.8.

(**viii**)   All the railway names have been removed except at Newark (N, N & L.R.) and the G.G. & S. J. R.

*A New County Atlas: With All The Coach And Rail Roads . . . To The End Of The Session Of M.DCCC.XLVI. Engraved By Sidney Hall. London: Chapman And Hall, Strand. M.DCCC.XLVII.*

**Copies** Bod Maps C.15.d.26; Leeds UL W 192. (LRL Map 769b).[6]

*A New County Atlas . . . To The End Of The Session Of M.DCCC.XLVI . . . London . . . 1847.*

**Copy** CUL Atlas 7.84.4.

(**ix**) The railway line from Brigg has been joined to the New Holland line with a short spur. The names of railways that were deleted in state viii have been restored. All railway lines are engraved.

*A Travelling County Atlas . . . 1847.*

**Copies** CUL Atlas 7.84.14; Leeds UL W 174.

*A Travelling County Atlas: With All The Railroads Accurately Laid Down And Coloured. Engraved By Sidney Hall. London: Chapman And Hall, Strand. 1848.*[7]

**Copies** BL Maps 34.a.12; CUL Atlas 7.84.6.

*A Travelling County Atlas . . . Strand.* [1850].

**Copies** Bod Maps C.15.e.8; Leeds UL W 175.

(**x**) Railway 2 has been added; the GNR and ELR have been extended southwards and nearly touch the frame-line. The address in the imprint has been changed to: *193 Piccadilly.* Plate number *23* has been added (Ea, between the inner and outer frames).

*A Travelling County Atlas . . . London: Chapman And Hall, 193 Piccadilly. (Late 186 Strand.)* [1852].[8]

**Copy** CUL Atlas 7.85.4.

(**xi**) The GNR has been extended northwards at Newark (crossing the Lincoln- Nottingham line); *G<sup>t</sup>. Northern R<sup>y</sup>.* has been added to replace the former name. *Sta.* has been added along all lines to mark the stations. Railways are now coloured red.

*A Travelling County Atlas . . . Chapman And Hall, 193, Piccadilly.* [1853**].

**Copies** Leeds UL 176. (Grantham RL Map 6).

*A Travelling County Atlas . . . Railroads Accurately Laid Down, And The Boundaries Coloured. Engraved By Sidney Hall. London . . .* [1854].[9]

**Copy** Leeds UL 177.

(**xii**) Railways 19–20 have been added.

*Sidney Hall's Travelling Atlas Of The English Counties With All The Railroads Accurately Laid Down, And The Boundaries Coloured. London: Chapman And Hall, 193, Piccadilly. Price Ten Shillings And Sixpence.* [1856].[10]

**Copies** Bod Allen 109 and G.A. Gen. Top. 8° 1511; PRO MPS 3/76.

*A Travelling Atlas Of The English Counties. By Sidney Hall. With All The Railroads . . .* [1856].

**Copy** PRO MPS (Y) 19.

Issued in a cover with the title: *SIDNEY HALL'S MAPS OF THE ENGLISH COUNTIES, WITH ALL THEIR RAILWAYS AND COUNTRY SEATS. LINCOLNSHIRE. PRICE SIXPENCE. LONDON; CHAPMAN & HALL, 193, PICCADILLY.*[11]

(**xiii**) Railways 21 and 24 have been added from Barkston and railway vii (to Boothby Pagnell only) removed. *GRANTHAM* has been re-engraved; it was previously obscured by the railway extension.

*A Travelling Atlas . . .* [1857].[12]

**Copy**   Leeds UL W 178.

(**xiv**)   Railway 23 has been added. Railway 21 is drawn in by hand in a straight line across Grimsthorpe Park. The GNR has been extended southwards and runs parallel to the Stamford line.

*A Travelling Atlas . . .* [1858**].

**Copy**   The private collection of Mr. D. Hodson.

## LITHOGRAPHIC TRANSFERS
(**A**)   Railway 26 has been added along with railway vii to Spalding again.

*A Travelling Atlas . . .* [1859].[13]

**Copy**   The private collection of Mr. D. Hodson.

(**B**)   Railway 18 has been added.

*Sidney Hall's Travelling Atlas . . .* [1859].

**Copy**   The private collection of Mr. E. Burden.

(**C**)   Railway 22 has been redrawn and keeps to the east of Grimsthorpe Park and stops on the south side of the road beyond. The lettering of *GRANTHAM* has been strengthened.

*A Travelling Atlas . . .* [1860].

**Copies**   BL Maps 25.a.28; CUL Atlas 7.86.5; Leeds UL W 179.

(**D**)   Railway 27 has been added.

*A Travelling Atlas . . .* [1860].

**Copy**   RGS 9 B 2.

(**E**)   Size: 40.5cm x 30.9cm. State (xiv) has been used for the transfer with railways 18 and 23 included. The GNR at its southern end is in its earlier, unextended state, and merges with the line to Stamford. There is now a double line curving from the MSLR into Grimsby.

*The English Counties. By Sidney Hall. With All The Railroads . . . Also General Maps Of Scotland, Ireland, And Wales. London: Chapman And Hall, 193, Piccadilly.* [1860].[14]

**Copies**   BL Maps 9.e.1; CUL Atlas 3.86.1; NLS Map 8.a.15.

*The English Counties . . .* [1862].

**Copy**   CUL Atlas 4.86.15.

Issued in a cover with the title: *LINCOLN/A RAILWAY MAP/BY/ SIDNEY HALL./ CHAPMAN & HALL, 193, PICCADILLY./Sixpence./* [1862].

**Copy**   The private collection of Mr. D. Hodson.

(**F**)   Size: 21.6cm x 19.2cm. Railways 25 and 28 are added to those of state D.

*A Travelling Atlas . . . Boundaries Coloured. London . . . Price Ten Shillings and Sixpence.* [1863].

**Copy**   Leeds UL 179a.

*A Travelling Atlas . . .* [1864].[15]

**Copies**   CUL Atlas 7.86.6; Leeds UL W 180; NLS Newman 507.

*Eastern England, From The Thames To The Humber. By Walter White. In Two Volumes . . . London: Chapman And Hall, 193, Piccadilly. 1865.* [16]

**Copies**  BL 10358 d 23; CUL Ll.50.5–6; Bod G.A. Gen. Top. 8° 122–123; LRL A9; Boston RL L. 91; Grimsby RL X000:942 WHI.

(**G**)  Railways 30 and 32 have been added. Plate number *23* has been removed (Ea). *Nº. 25* has been drawn in (Ee, OS and sidewards on).

*A Travelling Atlas . . .* [1869**].

**Copy**  (LRL Map 759).

(**H**)  Railways 29, 31 and 35 have been added. The Spilsby branch (railway 37) has been added also, but by an apparently heavy hand.[17]

(**I**)  Railways 36–37 and 39 have been added. At Newark a spur links the GNR to the Lincoln-Nottingham line; a spur also occurs at Grantham between the GNR and the Nottingham line. Railway 35 has two spurs on to the Sleaford line at its southern end.

*A Travelling Atlas . . . 193, Piccadilly. Price Ten Shillings and Sixpence.* [1872**].

**Copy**  Bod Map C.15.e.19.

(**J**)  Railway 38 has been added.

*A Travelling Atlas . . .* [1872**].

**Copies**  BL Maps 2.b.18; CUL Atlas 7.87.9.

(**K**)  Plate number *23* has returned in its former position, while plate number 25 has also been retained. Railway 40 has been added. The MSLR at Gainsborough has been extended south-westwards. The alignment of the junction at [Sutton Bridge] has been altered to a curve instead of a T, with the extension to Kings Lynn being turned into a sharper curve to the south-east. The extra spur at Grantham in state I has been removed.

*A Travelling Atlas . . .* [1874**].

**Copy**  The private collection of Mr. C.A. Burden.

(**L**)  Railway R has been added.

*A Travelling Atlas . . . Ten Shillings And Sixpence. 1875.*

**Copy**  Leeds UL W 181.

(**M**)  The plate number (Ea) has been removed.

*A Travelling Atlas . . . 1875.*

**Copy**  Leeds RL 912.42 H 147.

(**N**)  Plate number *23* has been added (Ea, between the frame-lines). Railways 41 and 43 have been added.

*A Travelling Atlas . . .* [1876].[18]

**Copy**  Nottingham UL s/G 5514.

(**O**)  The imprint has been removed (Be–De, OS). Railways 44–46 and (most of) T have been added. The alignment of lines at Grantham has reverted to state I.

*A Travelling Atlas . . .* [1882**].

**Copy**  Lincoln RL Map (Volume) 852.

**(P)**  A spur north-eastwards at Sleaford links railway 21 to railway 45/46.

*A Travelling Atlas . . .* [1887].[19]

**Copy**  CUL Atlas 7.83.12.

**(Q)**  Railways 48–49 have been added.

*A Travelling Atlas . . .* [1888**].

**Copy**  Birmingham RL A 912.42 HAL.

### References

1  Kingsley, p. 391.
2  Several maps are dated 1833.
3  Two maps are dated 1833 in the PRO copy and several have had their dates removed, thus implying production in 1834 or later; a privately owned copy has maps dated 1830–1832 or without dates.
4  The two BL atlases show differences on other county plates. Lincolnshire maps have a red hand-drawn line to separate the northern division from the southern.
5  Lincolnshire has not been found but there are 17 examples of other counties in CUL (Maps c.34.02.63).
6  On the LRL map the 'new' railway lines have not been engraved on the plate but have been hand-drawn.
7  The BL and CUL atlases are not identical. Additions to the plates of other counties suggest the BL volume post-dates the CUL copy and may have appeared in 1850.
8  Chapman & Hall moved to Piccadilly in 1851.
9  Additions on other plates suggest later issue, perhaps [1854].
10  Comparisons between the two Bod. copies show that Allen 109 contains railways on the Leicestershire plate not present on the other, while the position is reversed in the case of Bedfordshire.
11  Lincolnshire has not been found yet. Mr. E. Burden has a copy of Huntingdonshire, the plate being identical to the copy in the PRO. The list inside the cover includes Lincolnshire.
12  Railway 24 has been anticipated. It opened in 1859, i.e. after several other lines not shown on the map.
13  Railway 26 opened in 1860. However, 1858 has also been suggested (Burden, p. 139).
14  This is the generally accepted date based on railway data. There were two editions and the variations on the Bedfordshire plate suggest this is the earlier.
15  1864 is suggested on the basis of the railways on other plates; CUL and Leeds UL both ascribe their atlases to 1866, however.
16  This work was issued in two editions, with and without maps. The Lincolnshire map is found in volume one in most cases, but in volume two in the CUL example.
17  The lower half of such a map is illustrated in Pearson, R.E. and Ruddock, J.G. Lord Willoughby's Railway The Edenham Branch. (Grimsthorpe, [1986]), p. [77]. The atlas from which it came has not been seen; on p. 116 the authors ascribe the atlas to 1865 but it may be later as two railways are shown that were opened in 1867 and the Spilsby line opened in 1868.
18  Although railways 41/43 opened in 1876 (the suggested date), the line as drawn is wrongly sited. It remained uncorrected.
19  CUL's ascription to 1887 is justified by other county maps; the Lincolnshire map shows no line opened after 1882.

Chubb CCCCXLVI (Gorton); CCCCLI–CCCCLIV; CCCCXCVI–DIII.

## 95

JAMES BINGLEY                                                        **1831**

Size: 25.7cm x 19.9cm.

The maps in this series are commonly called 'Moule's' since they were first used in a series of parts written by Thomas Moule. There is some confusion concerning the engravers of the plates; some have the name of J. Bingley (as the first issue of the Lincolnshire plate) and he seems to have engraved the first fourteen counties issued; most of the next group of county plates have the name of W. Schmollinger; while most of the final county plates were the work of John Dower. The Lincolnshire plate, at a later stage of its usage, re-appeared with Bingley's name replaced by Schmollinger's.[1] This change also occurred on the plates for Nottinghamshire and The Environs of London, while the reverse process is noted for Hertfordshire[2] and Warwickshire.[3]

Thomas Moule (1784–1851), a writer of books on heraldry and antiquities, was, for a period of seven years (1816–1823), a bookseller.[4] He then found employment in the General Post Office and simultaneously held the office of Chamber-Keeper in the Lord Chamberlain's Department. With the latter post came an official residence within St. James Palace, where he died in 1851. On his title-page Moule claims the authorship of *Bibliotheca Heraldica* as well as other 'popular topographical works'.

*The English Counties Delineated* was originally issued in parts, a few of which had been found dotted around various libraries; a set of the first 56 parts (of the final total of 67) acquired by the British Library enables the publication history to be more fully revealed. One of the most important values of the BL set is that the original purchasers (the Fifth and Sixth Viscounts Galway) noted the month of issue of each part as received. The first part appeared in May, 1830 when it was the authors's expressed intention to produce six parts 'in about Forty-eight monthly Numbers'. The sequence of publication followed the six judicial circuits, starting with Oxford, followed by the Midland, Home, Norfolk, Northern, Western in that order as far as the county plates were concerned (Westmoreland was the only exception). Interspersed throughout the series of county plates appeared some seven plans of towns and cities and several area maps; they do not follow the judicial sequence of the counties. Lincolnshire's map appeared in part XIII in May, 1831. The back cover of the issue has an Index To The Principal Places In Lincolnshire and a notice: 'In consequence of the abundance of interesting matter in the History of Lincolnshire, it has been found absolutely necessary to form Two Parts of this County. The conclusion will be given in the next Number, to be published on the First of June'. Part XIII consisted of sixteen pages of text and the following part was of the same size and included a plan of Boston, also engraved by Bingley.[5] Even if the set of monthly parts had not been discovered the Boston map is evidence of the preparation of the plates well before the appearance in book form in 1836/7, since one of the three illustrations, which appear on that map, is of Nocton House, which was destroyed by fire in 1834.[6] Monthly issue seems to have been adhered to (not quite strictly); the original plan of 48 numbers was expanded because several counties after Lincolnshire were also spread over more than one monthly number; the final part probably came out in the Spring of 1836. Publication in book form followed shortly after.

The map is based on Ebden (**85**) and its derivative Hall (**94**), which series was being prepared at almost the same time. Moule's knowledge of and enthusiasm for heraldry is reflected in the numerous shields and coats of arms that decorate the maps. His interest in antiquities is mirrored in the use of the neo-gothic elements that frame his maps. While the maps are in themselves not very exciting their decorative qualities give them an appealing unity that has made the series one of the most favoured by collectors (and interior decorators). Of the engravers, Bingley and Schmollinger, very little is known. George Virtue, the publisher, was active from 1823 until 1857, when he was succeeded by his son, James Sprent Virtue (1829–1892), who had opened a United States branch.[7] The latter published *The National Gazetteer* from 1863, which contained maps prepared by William Hughes (**124**).

### Later history

There were editions of *English Counties Delineated* until 1839. The life of the plates was extended to over twenty years by their use in a series of English dictionaries edited by Rev. James Barclay. Barclay's Dictionary was first published in 1774 and there were, at least 23 reprints. The edition by Dewhurst was issued in 1841, while that of Bernard Bolingbroke Woodward appeared in 1848. Apart from serving many years as a clergyman in Edmonton very little seems to be known of him.[8] In some copies of Hume and Smollett's *History of England* as published by Virtue and Co., some of the Moule county maps appear (the numbers vary between different copies); it has been suggested that the maps were not an integral part of the issue (in parts) of the History but were bound in by subscribers because of their uniformity of size.[9] What is clear is that Virtue's use of whatever came to hand at the binding stage produces bibliographical problems. The plates disappeared by 1860, but copies of the maps must have survived in the Virtues' print-shop since some map sheets occur in the 1870s in later issues of Hume and Smollett.

### EDITIONS

(**i**)  *LINCOLNSHIRE* (Ca, between the inner border and the gothic outer frame). *Engraved by JAMES BINGLEY, 57, Charles St. Goswell Road, for MOULE'S ENGLISH COUNTIES*. (Be–De, OS). A view of Lincoln Cathedral from the south-east with the spires still in place, and, below, the bishop's coat of arms (Ea). *REFERENCE/TO THE/ HUNDREDS, WAPENTAKES &c./*31 names arranged under the three parts, viz. Holland, Kesteven, Lindsey (Ee). The Vanbrugh north front of Grimsthorpe Castle, with the Willoughby family arms above (Ae). The Grimsby coat of arms (Aa).

*The English Counties Delineated; Or, A Topographical Dictionary Of England. Illustrated By A Map Of London, And A Complete Series Of County Maps. By Thomas Moule . . . [Vol. II]. London: George Virtue, 26, Ivy Lane Paternoster Row. MDCCCXXXVII.*

**Copy**  Leeds UL W 168.

(**ii**)  Vertical shading has been added to the lower part of the shield, below the view of the Cathedral (Ea).

*No. XIII Price 1s. plain, or 1s.6d. coloured. Moule's English Counties. Lincolnshire. The English Counties Delineated; Or Descriptive View Of The Present State Of England and Wales; Illustrated By A New Map Of London, And A Series Of Forty County Maps, With Vignette Views Of Remarkable Places, And Armorial Decorations, Chiefly From The Seals Of County Towns. Forming Two Volumes handsomely printed in Quarto. By Thomas Moule . . . London: G. Virtue, 26 Ivy Lane; Simpkin And Marshall, Stationers' Court; Jennings And Chaplin, 62 Cheapside . . . 1831. Richard Taylor, Printers, Red Lion Court, Fleet Street.* [Cover title of the original part issue].

**Copy**  BL Maps C.29.b.2.

*The English Counties Delineated . . . Vol. II. MDCCCXXXVII.*

**Copies**  CUL Atlas 6.83.7–8; Liverpool Athenaeum 914.2Q. (LRL Map 173).

*The English Counties Delineated . . . MDCCCXXXIX.*

**Copy**  Cardiff RL.

(iii)  The spires on the Cathedral have been removed.

*The English Counties Delineated . . . MDCCCXXXVII.*

**Copies**  BL 10353 e 5; LAO 950.2; Guildhall L S914/2/MOU.[10]

*The English Counties Delineated . . . MDCCCXXXVIII.*

**Copy**  The private collection of Mr. C.A. Burden.

(iv)  Shading has been added to the remaining shields (Ae and (two) Ad). The panel, which contains the county name, has been shaded and six Tudor symbols have been drawn in. The imprint has been changed to: *Engraved for MOULE'S ENGLISH COUNTIES, by W. Schmollinger.* (Be–De, OS).

*The English Counties Delineated . . . MDCCCXXXVII.*

**Copies**  Leeds RL Q942 M861; NLS Newman 135; Manchester Chetham's L. U.6.7. (BL 3355 (29)).

*The English Counties Delineated . . . MDCCCXXXVII.* [1839].[11]

**Copy**  Norwich RL R T942.

*The English Counties Delineated . . . By Thomas Moule. Vol. II. London . . . MDCCCXXXVIII.*

**Copy**  CUL Atlas 6.83.12–13.

*The English Counties Delineated . . . Vol. II. London . . . MDCCCXXXIX.*

**Copies**  Leeds UL W 169; Birmingham RL HGD AQ 942.09. (LRL (Pye) PL 9 is a copy of the whole of the county text with the county map and that of Boston).

(v)  The imprint has been removed.

*The English Counties Delineated . . . MDCCCXXXVII.*

**Copies**  BL Maps 5.b.4. (LRL L A9 is the county text only, with map).

*The English Counties Delineated . . . Vol. II. London . . . MDCCCXXXVIII.*

**Copy**  Leicester UL H 942 MOU.

(vi)  Plate number *42* has been added (Ce, OS).

*A Complete And Universal English Dictionary, By The Rev. James Barclay, Illustrated By Numerous Engravings & Maps. Revised By Henry W. Dewhurst . . . London, Published by George Virtue, Ivy Lane.* [1841].[12]

**Copy**  The private collection of Mr. C.A. Burden.

(vii)  Two stars each (for MPs) have been added at Lincoln, Stamford, Boston, Grantham; one has been added at Grimsby.

*Barclay's Universal English Dictionary, Newly Revised By Henry W. Dewhurst. London, George Virtue.* [1842].[13]

**Copies**  Leeds UL W 182; Birmingham RL Q 423 BAR; John Rylands UL R 65576.

*A Complete And Universal English Dictionary . . .* [1843].[14]

**Copy**  Leeds RL W 183.

(viii)  *Scale of Miles/16* [= 4.9cm] has been added (Ee, OS).

*(Part 12 . . . Price One Shilling . . . Every Alternate Number Embellished With A County Map, Or Other Illustrative Plate, Beautifully Engraved On Steel. Barclay's Complete*

*And Universal English Dictionary. Illustrated By A Complete Series Of Maps Of Every County . . . London: Published For The Proprietors, By Geo. Henry & Co., 2, Northampton Square . . .* [1844].[15]

**Copies**  The private collection of Mr. T. Burgess. (LAO Scorer Map 42; Louth RL Map A 17).

Issued as a collection of maps, without text. [1844].

**Copy**  The private collection of Mr. T. Burgess.

*Barclay's Universal English Dictionary . . .* [1847].[16]

**Copy**  CUL Syn.4.84.67.

Issued mounted and in a case, with the cover title: *Revised By W. Hughes. Hall's Pocket County Maps, Beautifully Coloured. Containing all the existing Railways . . . Numerous Vignettes . . . Lincolnshire. Price 6d. each, mounted, and in a case . . . London: A. Hall & Co., 24, Paternoster Row.* [1847].[17]

**Copy**  Wisbech Museum TB 11.242.

[Hume's History Of England Div. 7. 7/6. 1849].[18]

**Copy**  The private collection of Mr. C.A. Burden.

(**ix**)  Railways 1, 3–14 and 16 have been added with black dots for stations and with directions, e.g. *From Sheffield. New/Holland* has been added.

*The History Of England, By Hume And Smollett, With a continuation to the Reign of Queen Victoria, By Edward Farr, Esq: London, George Virtue, 26, Ivy lane.* [1849].[19]

**Copy**  The private collection of Mr. C.A. Burden.

(**x**)  Railway 15 has been added.

*A Complete And Universal Dictionary Of The English Language: Comprehending The Explanation, Pronunciation, Origin, And Synopsis Of Each Word . . . By The Rev. James Barclay. A New Edition, Enlarged, Improved . . . By B.B. Woodward . . .* [Vol. III]. *London: George Virtue, 26, Ivy Lane, Paternoster Row.* [1848].[20]

**Copy**  Birmingham RL Q 423 BAR.

(**xi**)  Railways 17, 19–21 have been added. The GNR is named.

*A Complete And Universal Dictionary Of The English Language . . . London: George Virtue . . .* [1852].[21]

**Copies**  Norwich RL R 423. (LRL Map 174).

*A Complete And Universal Dictionary . . . London: James S. Virtue, City Road And Ivy Lane.* [1855].[22]

**Copy**  Leeds UL W 184.

*A Complete And Universal Dictionary . . . London: Virtue & Co., City Road And Ivy Lane.* [1857].[23]

**Copy**  The private collection of Mr. C.A. Burden.

Issued on cards. [1857].[24]

### References

1  Campbell, T. The Original Monthly Numbers Of Moule's 'English Counties'. *Map Collector*, 31 (June, 1985), pp. 26–39 sets out, *inter alia*, a table of the original parts, the map engravers, dates of appearance, etc.

2  Hodson, 1974, p. 135 (item 94 state iii).

3  Harvey and Thorpe, p. 163 (item 95 state ii).

4  Bannister and Moreland, p. 181; a brief but well-illustrated article on Moule is: Zaczek, I. Map-maker Moule. *Art & Antiques Weekly*, Vol. 32, no. 12. July 15, 1978, pp. 14–17.

5  The text between parts is continuously numbered; Lincolnshire being pp. 187–218. The Boston map is included in the complete copies of *The English Counties Delineated* but not all examples of the Dictionary. Loose copies: LAO Dixon 21/4/2/4 and 21/4/2/3; LRL Map 26a; Boston RL Map 3. The first and third of these come from Barclay's Dictionary (i.e. no engraver and plate number 44 added); the Boston RL example is of the Moule issue, state two (engraved by Bingley and trees added in several places); LAO Dixon 21/4/2/3 is state 3, with the engraver's name altered to Schmollinger.

6  Pevsner, N. and Harris, J. Lincolnshire. (London, 1964), p. 610.

7  Todd, pp. 200–201; Kingsley, p. 228.

8  DNB Missing Persons (Oxford, 1993), pp. 41–43 gives details of the Dictionary and the dates for the revisions by Dewhurst (1841) and Woodward (1848).

9  Kingsley, p. 228.

10  In some examples the engraved title-page is dated MDCCCXXXVI; the Guildhall copy (Vol. 1) is one such; the Leeds RL and Chetham's copies (below) are others. Usually the half-titles are undated.

11  An advertisement leaf, before the preliminary pages, lists an item dated 1839.

12  The chronology at the back of the volume ends in Dec., 1834. The Isle of Man map is dated 1839. See note 8.

13  Other commentators have suggested [1842] on the basis of railway data on other county plates.

14  The Isle of Man map has been redated 1843.

15  The cover title is quoted from the part, which included the Lincolnshire map. There were 31 parts and the final number included the title-page for the whole work: *Barclay's Universal English Dictionary* . . . (as [1842]). The Isle of Man map is dated 1843 and appeared in part 10 (i.e. just before the Lincolnshire section).

16  This appears to be a re-set edition of the issue in parts. The county population figure is given as 362, 602 (previously 260,000).

17  A. Hall was a subsidiary of Virtue and used the same address.

18  The title is taken from the spine of a part without title-page. The work is in 8 divisions; at the end of the final part appear three title-pages for the binding up of the intended three volumes. That for the third volume is: *The History Of England From The Accession Of George III To The Tenth Year Of Queen Victoria's Reign. By Edward Farr . . . London: George Virtue.*

19  The preface is dated 1848 and the chronology has been extended to 1848. The railways are those opened by August, 1850. The volume has 'Vol. 2' on the spine and the title quoted is that of the first, lithographed, title-page; the second (letter-press) omits the reference to Farr and the coverage extends only 'To The Death of George II'. Mr. C.A. Burden has, in fact, three full sets of this work, the others being in 3 volumes and 8 volumes.

20  The railways shown suggest a date in the late 1840s; 1848 is given as the date of Woodward's revision – see note 8.

21  Several commentators place this volume at 1852. However, railways 19–21 opened between 1855 and 1857; railways 19 and 21 are wrongly drawn and perhaps represent false anticipation.

22  The text pages are unchanged. The title-page imprint suggests issue in or just after 1855 when James S. Virtue joined the firm.

23  John Virtue joined the firm in 1857 and the new imprint may reflect the new arrangement.

24  Several cards are owned by Mr. C.A. Burden and they correspond with the maps in the [1857] volume. Lincolnshire may have been issued in this form also.

Chubb CCCCLXXI–CCCCLXXIII (1836–1838 editions of Moule only).

## 96

**ROBERT KEARSLEY DAWSON**                                              **1831**

Size: 30.2cm x 25.7cm (from plate mark to plate mark).

Dawson was the son of Robert Dawson (1771–1860), who was a draftsman with the Corps of Royal Military Surveyors & Draftsmen. He had a long career with the Ordnance Survey, finally retiring in 1836.[1] R.K. Dawson was born at Dover in 1798, while his father was working on the first Ordnance Survey of Kent. He was commissioned into the Royal Engineers in 1818 and followed in his father's path with the Ordnance Survey, working under Colby in Scottish and Irish surveys; Colby's name appears on the first O.S. maps for Lincolnshire, finally issued in 1825.[2] In the early 1830s Dawson undertook the immense task of preparing a full set of county plates and maps of all the boroughs of England and Wales, mostly based on fresh surveys.[3] The full range of maps was used, with certain alterations, over the following few years, during which Parliament produced the Reform Bill and undertook a review of the boundaries of all the boroughs involved in the changes to the electoral process instituted in 1832; the final report of the Commissioners affecting the county's towns appeared in 1837. Dawson died at Black-heath in 1861.[4]

The county map, drawn originally on the lithographic stone, is based on Cary (**65**), with the boundaries of the hundreds taken from the larger Cary of 1801 or, perhaps, Rowe. The only places included are the market towns and places of election; the main turnpikes are included but there are no physical features. Dawson also prepared maps of the county's five main towns, which returned members to Parliament.[5]

## EDITIONS

(i)  *LINCOLNSHIRE* (Eb). *Scale of English Miles/10* [= 3.2cm]/ *Engraved by J. Gardner, Regent Street.* (Ee). *Reference to the Hundreds/*31 numbered names (Ab–Ad). *Explanation/*5 lines of symbols, with the signature of *Robt. K. Dawson/Lieut. R.E.* below (Ae). Compass (Ed). There are no frame lines.

*XI. Part III. Parliamentary Representation. Further Return to an Address to His Majesty, dated 12 December 1831; for, Copies of Instructions given by the Secretary Of State for the Home Department with reference to Parliamentary Representation; likewise, Copies of Letters or Reports received by the Secretary of State for the Home Department in answer to such Instructions. Reports From Commissioners On Proposed Division of Counties and Boundaries of Boroughs. Vol. II.- Part I. Ordered by the House of Commons to be Printed, 20 January 1832. 141.*[6]

(ii)  Maltese crosses have been added to mark 18 polling places (not Stamford, however), but including *Lavenby* (sic), i.e. Navenby.

*Accounts And Papers, Eighteen Volumes. – (12.) – Relating To Parliamentary Representation: Boundary Reports, (Kent To Somersetshire.) Session 6 December 1831–16 August 1832. Vol. XXXIX. 1831–2.*

**Copy**   Bod Pp Eng 1831–2/39.

*Plans Of The Cities And Boroughs Of England And Wales: Shewing Their Boundaries As Established By The Boundaries' Act, Passed 11TH July, 1832: Together With Outline Maps, Shewing, The Divisions Of The Counties, The Principal Places Of Election, And The Polling Places, As Established By The Same Act. In Two Volumes. Vol. I. London: Printed By James & Luke G. Hansard & Sons, Near Lincoln's Inn Fields. 1832.*

**Copy**   BL Maps 149.d.28.

(**iii**) *Polling Places* . . . [Maltese cross symbol] has been added to the Explanation.

*Plans Of The Cities And Boroughs . . . 1832.*

**Copies**   CUL Atlas 2.83.2 and 2.83.17; Bod Maps C.17.a.20–21.

*Reports From Commissioners On Proposed Division Of Counties And Boundaries Of Boroughs. Part III. (91b.)*

The cover title is: *Instructions Given By The Secretary Of State* . . . [as above] *In Answer To Such Instructions. Boundaries, Part III. 1832. (91b.)*

**Copy**   BL Maps 145.c.27 (1).

(**iv**)   *Lavenby* and its Maltese Cross have been removed and a Maltese Cross has been added at Leadenham.

*Plans Of The Cities And Boroughs . . . 1832.*

**Copies**   Birmingham RL AF 912.42. (LAO Scorer Map 40).

*Accounts And Papers, Eighteen Volumes. – (12) – Relating To Parliamentary Repre- sentation . . . Vol. XXXIX. 1831–2.*

**Copy**   CUL OP. XXXIX 1831–2.

*Municipal Corporations in England And Wales. Report On The City Of Lincoln, (Lin- colnshire). London: Printed by James & Luke G. Hansard & Sons, near Lincoln's Inn Fields.* [1832].

**Copy**   LAO Monson 7/16/47.

(**v**)   The imprint below the map has been deleted.

*Accounts And Papers . . . Relating To Parliamentary Representation . . . Vol. XXXIX. 1831–2.*

**Copy**   Bod Pp Eng 1831–2/39.

*IX. Part III. Parliamentary Representation. Further Return to an Address to His Majesty . . . 1831 . . . Ordered . . . to be Printed, 20 January 1832. 141.*

**Copies**   Leeds UL W 158; Norwich RL R 328.334.

(**vi**)   As (v) except that there is no line of explanation for the inclusion of the Maltese Crosses.

**Copy**   (LRL Map 772).

### References
1   Hodson, Y. Robert Dawson (1771–1860), Ordnance Surveyor and Draftsman. *Map Collector*, 54 (Spring, 1991), pp. 28–30.
2   DNB. Vol. V. (1886), p. 678.
3   Chubb, pp. 297–304 lists all 277 county and borough plates.
4   Obituary in *The Times*, 1 April, 1861.
5   In most of the works listed below the county map is followed by plans of Lincoln, Great Grimsby, Boston, Grantham and Stamford (in that order). LAO Monson 7/16/47 includes all but Stamford. Several are known in varying states. In *Reports From Commissioners: 1837* (Volumes 6–8) the town maps were issued in fresh lithographs with the latest boundary recommendations shown. BL Official Publications L; CUL OP 1837. XXVI–XXVIII; Bod Pp Eng. 1837/26–28; Sheffield RL 352.042 STQ; Nottingham UL s/JS 3013 E 37; LRL L 352.
6   There are 8 parts divided equally into four volumes. In the case of several counties the electoral

symbols have not been included on the map, nor has the line with the polling place symbol been added to the Explanation. Lincolnshire may exist in this state also.

Chubb CCCCXXXIX (Plans of the Cities . . . 1832 only).


## 97

**WILLIAM COBBETT**                                                                1832

Size: 17.4cm x 10.05cm.

The text for this geographical dictionary was prepared by William Cobbett, but the volume provides no clue to the person responsible for the maps, which face the appropriate county text. The maps in general are of the smallest geographical value; that of Lincolnshire particularly has no merit at all. The county is crudely outlined and 26 market towns are dotted around, without any regard for their spatial relationships. Grantham, for instance is a good way northwest of Boston and Lincoln is north of Wragby, Louth, Horncastle, Alford, Spilsby and Saltfleet. There are no physical features of any sort. The map seems to have appeared once only; although there was a second edition in 1854 copies are very rare and the copies seen do not have any maps.

   Cobbett (1763–1835) was the self-educated son of a labourer, who rose to prominence in many ways, as M.P., writer on many topics, publisher of the *Parliamentary Debates* (taken over by Hansard, the publisher of the map noted above – **96**) and writer/publisher of *Political Register*, a weekly newspaper that ran from 1802 until his death.

### EDITION

(i)   *LINCOLNSHIRE* (in a panel, Ae). *Drawn & Engraved for Cobbett's Geographical Dictionary of England & Wales.* (Be–De, OS).

*A Geographical Dictionary Of England And Wales; Containing The names . . . of all the Counties, with their several Sub-divisions into Hundreds . . . Also, The Names . . . of all the Cities, Boroughs, Market Towns, Villages, Hamlets and Tithings . . . Maps: First, one of the whole country . . . and, then, each County is also preceded by a Map . . . By William Cobbett. London: Published By Wm. Cobbett, 11, Bolt-Court, Fleet-Street . . . 1832.*

**Copies**   BL 577 f 24; CUL S696.c.83.18; Bod 32.418; RGS F 84; Leeds UL W 159; LRL 942.

Chubb CCCCXL–CCCCXLa.


## 98

**JOSHUA ARCHER**                                                                1833

Size: 22.15cm x 16.3cm.

The plates provided by Archer for *The Guide To Knowledge* are quite unique in the record of county map production. The maps were engraved on woodblocks, the lines and letters of the map being incised into the wood, so that, when printed, the resulting impression is of white lines on a black base. The final result is, perhaps, arresting but the process proved unpopular for map work; it was used more for astronomical and town plans but the lack of real clarity proved a barrier. The map is based on Cole and Roper (**63**).

Pinnock, the editor of the work, was noticed above (**77**). Archer was a prolific producer of maps; the others use traditional engraving methods (**111–112** and **119**). Of his life beyond map production nothing seems to be recorded. *The Guide To Knowledge* was a weekly production and the Lincolnshire plate appeared in the issue of 29 June, 1833. The Guide was re-issued again in 1838 and the plates were used again in an atlas produced by Groombridge in association with Shepherd and Sutton, but largely using coloured inks. However, Lincolnshire has not been found in this state. In their final form the plates have been lithographically transferred and the printed effect reversed, that is, black lines appear on a white ground.

## EDITIONS

(**i**)   *LINCOLN/SHIRE* (in a panel)/*London, Edwards, 12 Ave Maria Lane.* (below the panel, Ae). *J. Archer sc.* (Ee). *English Miles/10* [= 2.7cm] (Ea). *No. LXII.] GUIDE TO KNOWLEDGE. SUPPLEMENT./PR. ONE PENNY./* (Aa–Ea, OS). There is no compass. Some of the border is cut away to fit in the map; what remains shows latitude, etc. at 10′ intervals.

*The Guide To Knowledge. Edited By W. Pinnock . . . Vol. 1. – Nos. I. To XCII. London: Printed For The Proprietors, And Published By W. Edwards, 12, Ave-Maria-Lane. 1833.*

**Copies**   BL PP 6000 b; CUL T900.b.1–6; Bod Allen LRO 514 and 3985.d.18–19; Liverpool RL LQ 2. (LAO FL Box L 912; LRL Maps 776 and 776a have both been trimmed and lack the heading).[1]

*The Guide To Knowledge . . . Vol. I London. Printed for the Proprietors And Published At Their Office, 2, Wellington Street, Strand. 1838.*

**Copy**   The private collection of Mr. E. Burden.

## LITHOGRAPHIC TRANSFERS

(**A**)   Size: 23.1cm x 16.1cm. A style of piano key-board border has been added, and the minutes of latitude and longitude placed in circles within the border. The heading above the map has been deleted. *RUTLAND/SH. NORTHAMPTON SH.* and *NORFOLK* have been added; in order to permit the latter *CAMBRIDGE/SH* has been deleted and replaced, in a new site, by *CAMBRIDGE S* and that permits the inclusion of *Cross/Keys* and *R. Nen. Market/Deeping* has been re-engraved in a single line, leaving space for *R. Welland* to be added. Roads leaving the county have been extended to the inside edge of the frame and directions (e.g. *Fr. Nottingham*) added. Place names in adjoining counties have been added: *Kingston upon Hull, Howden, Tuxford, Southwell and Ouse R.* In the latter case the river has been extended to the map frame and *R. Ouse* deleted. Shading has been added to the sea.

*The Guide To Knowledge . . . 1838.*

**Copy**   The private collection of Mr. T. Burgess. (The author has a copy of the loose sheet).

*Descriptive County Atlas Of England And Wales; To Which Is Appended A Brief . . . History Of Ireland, With Map . . . Scotland, With Map . . . Isle Of Wight, With Map . . . London: Published By R. Groombridge, Paternoster Row; And Shepherd & Sutton, Foster Lane, Cheapside. 1844. Stereotyped At The Caxton Foundry, Bishops Court, Old Bailey.*[2]

*The Guide To Knowledge . . . Shepherd and Sutton . . . Vol. 1. 1844.*[3]

(**B**)   Printed with black lines on a white substrate. The imprint and signature have been deleted. Railway i has been added as a solid black line.

*Johnson's Atlas of England; With All The Railways Containing Forty Two Separate Maps Of The Counties And Islands . . . Manchester Published By Tho$^s$. Johnson. 1847. Lizars Sc.*

**Copies**   BL Maps 12.d.2; CUL Atlas 6.84.12; Bod Maps C.17.d.24. (LRL Map 78).

*Johnson's Atlas of England . . . 1863.*

**Copy**   The private collection of Mr. C.A. Burden.

### References

1   A set of two volumes (in CUL T.900.b.1.5–6) was published by Orlando Hodgson in 1837 and MDCCCXXXIX respectively. They were called 'New Series', but did not contain any maps.
2   This work was issued in parts, the Prospectus being dated January, 1844 and the first part February, 1844. There are part sets at CUL (Atlas R.4.84.39), Bod (Maps C.17.b.6) and NLS (Map 11.6.32). From advertisements contained therein the whole was intended to be in 7 parts at 3/6 each, with section '30 Historical Sketch Of Lincolnshire. With Map'. There is also an advertisement for *Shepherd And Sutton's Descriptive County Maps Of England And Wales, Adapted for the pocket . . . Price Ninepence each, tastefully bound'*. Burden, E., Webb, D., and Burgess, T. Pinnock's 'Guide to Knowledge' Maps. *IMCoS Journal*, No. 36 (Spring, 1989), pp. 24–29 discuss the 41 maps found in a private set. Only three counties were missing (unissued?): Lincolnshire, Nottinghamshire and Wiltshire.
3   A new edition was probably intended. However, only one volume appeared (a note in the only copy extant (BL 012216 f 5) suggests six volumes were intended); the single map in volume one is of Germany.

Chubb DXXIII (Johnson's atlas of 1847 only).

**99**

**GRAY and Son**                                                               **1833**

Size: 24.3cm x 18.55cm.

James Bell's *A New and Comprehensive Gazetteer . . .* was published in parts and included several articles that had been taken from Lewis' *Topographical Dictionary of England . . . (***93***)*. In July, 1839 Lewis obtained an injunction against Bell and the work was withdrawn. In its place Fullarton and Co. produced *The Parliamentary Gazetteer*, a new text but the same maps. That work was also issued in parts. The maps are based on Hall (**94**) but a number of mistakes have been introduced, e.g. *Norton* (instead of Nocton) and *Donnington* (instead of Bennington). The production of title-pages seems to have been a Bell speciality and their placing within the four volumes that comprise the sets in both the earlier and the later versions presents problems of bibliographical control; the notes below concerning them are not intended to be comprehensive. The publishers, Fullarton & Co., issued litho. transfers of maps prepared by J. Bartholomew for *Imperial Gazetteer*. In the first volume of that work is a plate *Ports & Harbours On The East Coast Of England*, which includes Grimsby.[1]

### EDITIONS

(i)   *LINCOLNSHIRE.* (Aa). *Pub$^d$. by Arch$^d$. Fullarton & C$^o$. Glasgow. (Ce, OS). Eng$^d$. by Gray & Son* (Ee, OS). *REFERENCE TO THE/ HUNDREDS &c./*31 numbered names (Ea). *The Figures prefixed to the Towns/denote the distance from London/* (Ee). Compass (Ae, above scale). *English Miles/10* [= 3.25cm] (Ae). The inner frame is marked every

5 minutes and numbered every ten. *Meridian of 0 Greenwich* (De, between borders). *Long. 1° West* (Ae, between borders).

*A New And Comprehensive Gazetteer Of England And Wales, Presenting . . . The Population Of The Towns And Parishes, According To The Census Of 1831, And The State Of The Elective Franchise, As Fixed By The Provisions Of The Reform Bill. By James Bell . . . Illustrated By A Series Of Maps, Forming A Complete County Atlas Of England. Vol.* [IV] *Glasgow: A Fullarton & Co., 34, Hutcheson Street; And, 31, South Bridge, Edinburgh. MDCCCXXXIII.*[2]

Issued in an undated atlas, lacking title-page, cover-title and text. [1834].

**Copies**  Leeds UL W 164. (LAO Scorer Map 41).

*A New And Comprehensive Gazetteer . . . Vol. IV . . . MDCCCXXXIV.*

**Copies**  CUL S696.c.83.2–5; Leeds RL 942 B 413.[3]

*A New And Comprehensive Gazetteer . . . Vol. III . . . MDCCCXXXIV.*

**Copy**  NLS Hall 2680.[4]

*A New And Comprehensive Gazetteer . . . Vol. III . . . MDCCCXXXV.*[5]

**Copies**  Bod G.A. Gen. Top. 8° 1686; Glasgow Mitchell L. 914.2003 BEL (B 217526).

*A New And Comprehensive Gazetteer . . . Vol. III, Part I . . . MDCCCXXXVI.*[6]

**Copy**  BL 10358 e 8.

*A New And Comprehensive Gazetteer . . . Vol. IV . . . MDCCCXXXVII.*[7]

**Copies**  CUL S696.c.83.6–9; RGS 510 d.

*The Parliamentary Gazetteer Of England And Wales, Adapted To The New Poor Law, Franchise, Municipal, And Ecclesiastical Arrangements, And Compiled With A Special Reference To The Lines Of Railroad And Canal Communication, As Existing At The Close Of The Year 1839, Illustrated By A Series Of Maps . . . Volume III. Glasgow: A Fullarton & Co., 110, Brunswick Street; 6, Roxburgh Place, Edinburgh; And 12, King's Square, Goswell Street Road, London. 1840.*[8]

**Copies**  CUL S696.b.84.10–13; Bod G.A. Gen. Top. 4° 651; Nottinghamshire AO R 354.42.[9]

*The Parliamentary Gazetteer . . . 1840.* [1843].[10]

**Copy**  London UL [GL] Lc.

*The Parliamentary Gazetteer . . . As Existing In 1840–1. Illustrated . . .* [Volume II E–K] *. . . Glasgow . . . 1842.*[11]

**Copies**  Chester RL 90; Liverpool UL Y 84.5.60–63.

*The Parliamentary Gazetteer . . . Existing In 1840–2 . . . Volume II. E–K . . . 1842.*[12]

**Copies**  Guildhall L. S914/2; Sheffield UL ** 914.2 P.

*The Parliamentary Gazetteer . . . Existing In 1840–3 . . . Illustrated By . . . A Complete Atlas Of England, And By Four Large Maps Of Wales. With An Appendix Containing The Results, In Detail, Of The Census Of 1841. Volume III. L–Q. London, Edinburgh, And Glasgow: A. Fullarton And Co. 1843.*

**Copies**  BL 10347 g 7; RGS 510 g; Liverpool Athenaeum 914.2; Nottingham UL s/DA 625.P2; Leeds UL W 189; Leeds RL 942 P239.[13]

**(ii)**  The signature of Gray & Son has been removed. *Glasgow* in the imprint has been removed, but traces are clearly visible.

*The Parliamentary Gazetteer . . . As Existing In 1840–44 . . . Volume II. E–K . . . 1844.*

**Copy**   Leicester UL H 942 PAR.

*The Parliamentary Gazetteer . . . As Existing In 1840–44 . . .[Part VII] . . . 1844.* [1845].[14]

**Copies**   Leeds UL W 190 and Special Collections. Encyclopaedias. Stack Bibliog. B1.7.

*The Parliamentary Gazetteer . . . As Existing In 1840–44 . . . 1845.*

**Copy**   BL Maps 200.c.5.

**(iii)**   Railways 1, 4, 9–10 and 13 have been added. The imprint has been reduced to: *Arch$^d$. Fullarton & C$^o$.* (The final letter is very faint and has become increasingly so since *Glasgow* was removed).

*The Parliamentary Gazetteer . . . 1840–44 . . . 1845.*[15]

**Copy**   CUL S696.b.84.15–21.

*The Parliamentary Gazetteer . . . 1846.*[16]

**(iv)**   Railways 3, 6–8, 11, 14–15, 17–18, vii (to Spalding only) and viii have been added. An horizontal line above the first two letters of *LINCOLN* may be the wrongly sited railway 16. The imprint has been amended to: *Pub$^d$. by Arch$^d$. Fullarton & C$^o$.*

*The Parliamentary Gazetteer . . . As Existing In 1845–6 . . . Vol. II. E–K . . . 1848.*

**Copies**   CUL Ll.32.31–34; Bod G.A. Gen. Top. 4$^o$ 93–96; NLS Ref. 2.C.17.

**(v)**   The map now only shows railways 1, 3, 7–8 and 11, i.e. only those railways actually open. The horizontal line above Lincoln has been removed.

*The Parliamentary Gazetteer . . . Vol. III . . . 1849.*

**Copy**   Liverpool RL P 59.

### References

1   The map faces p. 813 in Vol. I. Copies include: CUL Ll.20.47–48; Bod G.A. Gen. Top 4$^o$ 188–189.

2   Several commentators quote Chubb (p. 306), who quotes Fordham (1907) for the existence of an 1833 edition. Kingsley, p. 210 cites the same Fordham work in its revised (1914) edition, p. 102. The Leeds RL set has volumes I and II dated 1833, but Lincolnshire is in volume III, dated 1834. The plates were not always placed with the county text; the map often appears in volume two, while the county text is in the third volume.

3   In the CUL set Vol. II is dated 1835. See note 2 for Leeds RL copy.

4   Leicester UL has a set thus, but lacking all maps.

5   In the Bod set volume II is dated 1836 and volume I 1834; the Glasgow set has both volumes I and II dated 1834.

6   The work was issued in parts. In the set belonging to Mr. C.A. Burden Lincolnshire is in Part 7 (equivalent to Vol. IV, part I). The title-pages for the intended 4 volumes appear in part 8 of the BL set.

7   In the CUL set volume III is dated 1836, the rest 1837. In the RGS set volume IV is dated 1838 and the title-page imprint has been changed. Mr. C.A. Burden has a set in three volumes only and the maps are placed together at the end of volume III.

8   This edition was issued in parts, with the Lincolnshire text in part VII (LAC–LON), but the map is in part VI.

9   The Nottinghamshire AO example has the map in the second volume; only the first volume has a title-page.

10   The London UL set is in parts, with the county in part VI. In part XII is an apology from the publishers, dated May 8th, 1843, which refers to the decision to close the work without the promised census figures (later issued as an appendix) – 'the work has been ready for four months but the Census will not be available for 2–3 months . . .' The work in twelve parts or four volumes was, therefore, not complete until 1843. In the Plan (Part I) each part was to be

2s., containing six sheets of letter-press and one map, alternatively with five sheets of letter-press and two maps or plates. The work would comprise '30 Parts or the overplus will be given gratis' and would also be sold in twelve larger portions at 5s. each (the form of the London set). While there was no edition between the issue of part I in 1840 and the final part in 1843 title-pages for 1842 exist (q.v.).

11  Only volume I has a title-page; it indicates coverage from A–D and is dated 1842. As the work's completion was delayed until 1843 impatient owners probably bound up the parts without waiting for the printing of the title-pages for volumes 2–4.

12  The title-page of volume I has 'As Existing in 1842–3' and is dated 1843. The remainder are as transcribed. In the Guildhall set volume IV is dated 1843. Mr. D. Hodson has a set in the original twelve parts with a supplement. Lincolnshire's map is in part VI, while the text is in part VII.

13  In the BL set volume I is dated 1840. The Leeds RL set has a fifth untitled volume, which contains all the maps.

14  Both Leeds sets are in 12 parts, with Prospectuses bound at the front of Part 1. The back of the last part has the title-pages (dated 1845) for binding in volume form.

15  Only volume I has a title-page, quoted in reduced form here. There are 7 parts; at the end there is an offer of the work in 31 parts 'as existing in 1840–6', price 2s. per part, forming four volumes, price £3.10s. in cloth. The county map is in part VI, the end of the planned second volume, while the county text is in part VII.

16  Chubb, p. 342 (item DXI) quotes Fordham, 1914 for this edition.

Chubb CCCCXLV–CCCCXLVa; DX–DXI.


# 100

## HOUSE OF COMMONS                                          1833

Size: 33.9cm x 26.6cm.

A map covering both Lincolnshire and Nottinghamshire was produced to illustrate the Census returns of 1831, finally published in 1833. The map has no physical features and it lacks the names of printer or draughtsman; it was produced by lithography and the passing similarity of the county outline to that of Arrowsmith's 1821 plate for *Valor Ecclesiasticus* (**79**) makes him a possible candidate as printer. The map is divided into nineteen districts (14 in Lincolnshire, 5 in Nottinghamshire); in each section are the population figures for 1801, 1811, 1821 and 1831, foliowed by figures for Baptisms, Burials and Marriages for the same four years. Above these figures are large roman numerals, which corresponds to the key (round the edges of the plate), which gives the names of the hundreds, etc. comprising each district. Two further county tables give totals and averages of the figures in each section. The system and the map appear never to have been used again. The counties are coloured yellow and red (Nottinghamshire and Lincolnshire respectively).

### EDITION

(i)   *COUNTIES OF LINCOLN AND NOTTINGHAM.* (Ba–Da). *Scale of English Statute Miles, Nine = One Inch*/10+40 [10 miles = 2.7cm] (Ce, OS). *TOTALS OF LINCOLN./Population./*4 lines for 1801, 1811, 1821 and 1831/Table of averages of baptisms, burials and marriages for five year periods, ending in 1800, 1810, 1820 and 1830/ (Aa). A similar table headed *TOTALS OF NOTTINGHAM* (Be). *LINCOLN COUNTY./(Coloured Red.)/Square Miles, 2611./Parish-Register Limits./* [the numbers for each district are given, numbered I–VIII under Lindsey, IX–XII under Kesteven and XIII–XIV under Holland] (Ea–Ed). A similar listing headed *NOTTINGHAM*

*COUNTY./(Coloured Yellow.)/Square Miles, 837./* is followed by the five district components, and, finally, the figures for: *V Nottingham Town & County* . . . (Ac–Ad), the space on the map being too small to allow their insertion in the proper space.

*Accounts And Papers, Seventeen Volumes. (14.) Population Of Great Britain. Parish Register Abstract. Session 29 January–29 August 1833. Vol. XXXVIII. 1833.*[1]

**Copies**    BL Official Publications Dept.; CUL Accounts & Papers 1833.7.XXXVIII; Bod Pp Engl. 1833/38.

#### Reference

1    On the page after the title there is the reference: 'Ordered by the House of Commons to be Printed, 2 April, 1833'.

Not in Chubb.

## 101

### JOHN SAUNDERS                                                    1833

Size: 23.95cm x 17.9cm.

The work and life of Saunders have been touched on in connection with Allen's *History of Lincolnshire* (**90**) and the present map first appeared in that work. *Lincolnshire in 1835* was issued in parts and Lincolnshire was intended to be the first section of a much larger work *England In The Nineteenth Century . . . Each View Accompanied By An Explanatory And Illustrated Description* [By Mary Saunders]. There were to be no more than 20 parts for the county, each part costing 2s. The map under consideration was probably prepared almost as a last thought prior to the publication of the last parts of the History. Its purpose is to delineate the new polling places and the changes in parliamentary representation, which resulted from the Reform Bill of 1832. One would normally have expected Saunders to incorporate such details on the plate that accompanied Allen's History. As we have seen, however, that plate seems to have been disposed of, having passed on to Joseph Noble of Hull by 1833. A new map was necessary and was produced by lithography; it draws on Hall for its general outline, but the hachur for the Wolds is drawn differently and Hall's higher ground in the Isle of Ancholme does not appear here. Some roads, hills, rivers and parks of the gentry are marked, with the boundaries of the hundreds, Maltese crosses (marking the election places), the names of the market towns and a few more important villages. The map does not re-appear again after 1836. Since the map appeared at first in Allen's History the full titles, etc. given above (**90**) are not repeated here.

#### EDITIONS

(i)    *LINCOLNSHIRE,/according to the present state/OF/ Parliamentary Representation./* (Ea). *LONDON, J. SAUNDERS, 9 JUN*[R]. *1833.* (Ce, OS). *Explanation/*9 lines of text and symbols (Ae). *Reference to the Hundreds./*31 numbered names (Ee). There is no compass.

*The History Of The County Of Lincoln . . . By Thomas Allen . . . [Vol.1] . . . Leeds . . . Hull . . . London . . . MDCCCXXX.*

**Copies**    LAO FL L 942 ALL; Grimsby RL X000:942 ALL; Scunthorpe RL L 638 942 Q; Stamford RL L 9; Nottingham UL Li 1.D 14ALL oversize.

*The History Of The County Of Lincoln . . . By Thomas Allen . . . London: John Saunders, Junior. MDCCCXXXIII.*

**Copies**   LAO AS 942.51; LRL A9 (SR.); Nottingham UL Li 1.D14 ALL oversize; Louth RL L 9.

*The History Of The County . . . By The Author of . . . Vol. 1. London & Lincoln: John Saunders, Junior. MDCCCXXXIII.*

**Copies**   CUL VIII.2.51; Bod G.A. Top. Gen. 4° 62–63; LRL L 9; Louth RL GL 9; Grimsby RL X000:942 ALL.

*The History Of The County . . . Vol. I . . . MDCCCXXXIV.*

**Copies**   CUL Syn.3.83.1; Bod G.A. Lincs C3–4; Birmingham UL r q DA 670.L69; LAO Sc 942.51; LRL L 9 (SR.) and A 907; Boston RL L 9. (LAO Exley 30/1; LRL Maps 188 and 188a).

(**ii**)   The imprint has been changed to: *London & Lincoln. J. SAUNDERS, JUN^R. 1836.* (Ce, OS).

*Lincolnshire In 1835: with the Rivers Humber, Trent & Witham Displayed In A Series Of Views; Each View Accompanied By An Explanatory And Illustrative Description. Lincoln: Published By John Saunders, Jun. and Sold by all Booksellers in the County. MDCCCCXXXV.*

**Copies**   BL 796 i 12; LAO L 942.08 and FL (PM) L 942.075; Gainsborough RL L 9.

*Lincolnshire In 1836: Displayed In A Series Of Nearly One Hundred Engravings, On Steel And Wood; With Accompanying Descriptions, Statistical And Other Important Information, Maps, &c. Lincoln, Published By John Saunders, Jun. MDCCCXXXVI.*

**Copies**   BL 578 f 32; Bod G.A. Lincs 4° 52; Scunthorpe RL L638 910; LAO Sc 942.52; LRL L 907; Boston RL L 9.

## 102

### CHRISTOPHER and JOHN GREENWOOD                                    1834

Size: 64.1cm x 74.9cm.

The work of the Greenwoods has already been noticed when discussing their large-scale one-inch map (**87**) of which the present map is a reduced version and intended for inclusion in an atlas of all the counties. The process of reduction and correction was begun as early as 1829 (recorded on the 11 plates, which formed part one). The work was advertised in 1828 to appear in four parts at three guineas per part, that is, before all the large scale surveys had been completed, although six counties never were surveyed at the one inch scale.[1] In the prospectus they state that they had been engaged on their surveying work since 1814, costing them 'an average of £5,000 per annum'. There were to be forty-six plates all at one-third of an inch to a mile. The Greenwoods ran into financial difficulties towards the end of the project and the work was finally published by Greenwood & Co. from the address in Burleigh Street, London, which was shared by the engravers of the map plates, Josiah Neele and J. and C. Walker. The Greenwoods had to give up their own premises in Regent Street; they produced no new map plates after 1834. The engraver, Josiah Neele, was the son of the well-known engraver S.J. Neele (**55**, **70** and **77**). Curiously, the engraving of Lincoln Cathedral on the map was entrusted to R. Creighton (**93** and **104**). Although the map is dated 1831 it is not clear if the map was put on sale at once; Lincolnshire was one of twelve plates forming part three, eleven of which were corrected up to 1831; it is probable that the maps were issued as a set in 1832.[2] None of the final twelve plates was prepared until 1834; that is, there was a gap of something over two years between the work on the Lincoln plate and its companions and the completion of the whole work; the delay is largely explained by the need to produce fresh plates for the six counties that they had never surveyed at the larger scale, i.e. three plates to cover the Welsh counties and Yorkshire (on three plates). A later edition (also dated 1834) includes a section of electoral details for the county. The map in state (ii) has no other county equivalent and represents a bibliographical mystery. For detail, general accuracy and balance allied to high quality engraving and colouring this sheet is probably the finest ever single sheet map of the county.

## EDITIONS

(i)  *MAP/OF/THE COUNTY OF/LINCOLN,/FROM AN ACTUAL SURVEY/Made in the Years 1827 & 1828,/BY C. & J. GREENWOOD,/ Published by the Proprietors/Greenwood & Cº.,/ REGENT STREET, PALL MALL,/London./Corrected to the present Period & Published Jan*$^y$*. 26*$^{th}$*. 1831./ENGRAVED BY JOSIAH NEELE 352 STRAND./ (Ea). EXPLANATION/*17 lines of symbols, etc. (Ed). *Scale of Miles/8 Furlongs, 10* [1 mile = 0.8cm] (Ee). *Reference/to the/Wapentakes, Hundreds &c./*33 names divided between Lindsey (18), Kesteven (11) and Holland (4), Boston and Stamford being added to the usual list (Aa–Ab). Compass (Ec). *LINCOLN CATHEDRAL R. Creighton del.* below an engraving (Ae). There is a piano keyboard border, broken at the bottom to allow the map to fit in; minutes are marked by figures at 5′ gaps set in the border, while dots in the inner border mark the single minutes. *Longitude East of Greenwich* (Da). *Longitude West of Greenwich* (Ba).

*Atlas Of The Counties Of England, from Actual Surveys made from the Years 1817 to 1833, by C. & J. Greenwood, Published by the Proprietors Greenwood & Cº. Burleigh Street, Strand, London Engraved by J. & C. Walker. Published April 1*$^{st}$*. 1834.*

**Copies**  BL Maps 150.e.16; CUL Atlas 1.83.3; Bod C.17.a.15 and Allen 97; Leeds UL W 165; Leicester UL FF H942.07 GRE; Hull RL 912.42 Lf. (Bod Map (E) C.17: 39 (26); LRL Maps 121, 122 and 123; Grimsby RL Greenwood 1831 – RM16; Louth RL L912 – 2 copies).[3]

Issued as a boxed set.

**Copy**  Birmingham RL 912.42 Skett 1.

(ii)  Plate number *21* has been entered in letterpress (Ea, OS).

*Atlas Of The Counties . . . 1834.*

**Copy**  (Grimsby RL Greenwood RM 12).

(iii)  *PRINCIPAL PLACES FOR COUNTY ELECTION/*2 columns of place-names and two signs for county divisions and borough boundaries have been added (Ac). *SOUTHERN DIVISION* and *NORTHERN DIVISION* have been added in curving lines on the map. The plate number of state (ii) has been removed.

*Atlas Of The Counties . . . 1834.*

**Copies**  Bod Allen 98; Leeds UL 166. (LRL Maps 768 and 123b).

### References
1    BL MT.6.c.1 and 60.e.15 (6) are copies of the first prospectus; the former with manuscript additions showing the changes that were made between some of the first map proposals and the final product; e.g. the three views of York Minster intended for the Yorkshire plates were

changed to views of Ripon Castle, Beverley Minster and a single view of York Minster. One change that was not noted occurs on the Nottinghamshire plate; the original plan was to include Nottingham Castle but, in the event, Southwell Cathedral was engraved.

2   Watermark evidence is not helpful; the set in the PRO has the map watermarked 1831 and in Leeds UL 1832, but single sheets in the Bodleian and Lincoln RL (Map 122) are watermarked 1829.

3   The BL example is in an intermediate state; Norfolk, for instance, has the addition of the electoral data, while Lincolnshire has not. The loose sheets may have been available separately before the atlas appeared.

Chubb CCCCLVIIIa–CCCCLVIIIb.


# 103

**LONGMAN, REES, ORME, BROWN, GREEN AND LONGMAN        1834**

Size: 7.65cm x 8.95cm.

These small maps accompany a school text-book, which uses the technique of a teacher (Mrs. Rowe), asking questions of two children, George and Anna. The formula teaches geography and history in more or less equal proportions. The maps are simple outlines, with rivers marked with lower case letters, important county areas marked by capital letters and 23 towns marked by numbers; these are all explained at the heading of the book's Lincolnshire section (p. 50). The map was probably prepared by someone in the publisher's office and shares a page with the map of Lancashire occupying the top half. There are neither compass, imprint nor signature. The preface merely reports that the plates were 'executed by one of the best artists in the metropolis'.[1] The author was a writer of school text-books.

## EDITION

(i)   *LINCOLNSHIRE./* (Ca).

*The Geography Of The British Isles, Interspersed With Many Historical Facts And Biographical Sketches; Selected From The Best Authors, And Illustrated With Separate Blank Maps And Explanatory Keys; Showing The Relative Situations, Boundaries, Principal Towns, Rivers, &c. Of Each County. For The Use Of Young Persons And Schools. By Mary Martha Rodwell. In Two Volumes. Vol. I. London: Printed For Longman, Rees, Orme, Brown, Green, And Longman. Paternoster Row. 1834.*[2]

**Copies**   BL 796 b 10; CUL S.18.4–5; Bod 34.295/296.

### References
1   Vol. 1, p. vi.
2   Chubb (p. 320) fails to draw attention in his abbreviated entry to the fact that the work is in two volumes; possibly because the BL copy had been bound as one volume.

Chubb CCCCLVIIIc.

## 104

**R. CREIGHTON**                                                              **1835**

Size: 25.45cm x 18.5cm.

The work of Creighton has been met in the plates he produced for Samuel Lewis (**93**) and in the engraving of the Cathedral used on the Greenwood plate (**102**). The new map is closely based on the Dawson plate (**96**) prepared for the House of Commons' volumes related to the Reform Bill and the subsequent developments in local government boundaries. While Creighton drew the maps the work of engraving was undertaken by the Walkers, whose identity was noted earlier (**81**) and whose largest atlas is recorded below (**105**); they engraved the main part of the plate for the county produced by the Greenwoods (**102**). The map series appeared in harness with the plates issued in Lewis' *Topographical Dictionary* . . . in its 1835 edition and two later issues; there is usually a fifth supplementary volume to the main four volume work, which contains the present plate (LI) and two further plates bearing Grimsby, Stamford, Grantham and Lincoln (all on plate LII) and Boston (plate LIII).

### EDITION

(i)   *LINCOLNSHIRE* (Ea). *Drawn by R. Creighton* (Ae, OS). *Engraved by J. & C. Walker* (Ee, OS). *LI* (Ea, OS). *Scale of Miles/15* [= 4.35cm] (Ee). *Explanation/7* lines of symbols, and text (Ae). Compass (Aa). The map has blue squares at Lincoln and Sleaford for county election places; orange dots inside circles at Lincoln, Boston, Grantham and Stamford and an orange circle at Grimsby. 18 places have Maltese crosses.

*A Topographical Dictionary Of England . . . Third Edition. With A Supplementary Volume, Comprising A Representative History Of England, With Plans Describing The Electoral Divisions Of The Several Counties, And The Former And Present Boundaries Of The Cities And Boroughs. By Samuel Lewis. In Five Volumes. Vol. V. London, Published By S. Lewis And Co., 87, Aldersgate-Street. M.DCCC.XXXV.*

**Copies**   BL Cup 1247 g 1; Leeds W 150; Sheffield RL 914.2 STQ; Nottingham UL s/DA 640.L 4; Liverpool Athenaeum 914.2Q; Grantham RL L 328. (LAO Misc Don. 233a).

*View Of The Representative History Of England, With Engraved Plans, Shewing The Electoral Divisions Of The Several Counties . . . Cities And Boroughs. By Samuel Lewis. London: Published By S. Lewis . . . 1835.*

**Copy**   RGS 7 D 29.

*A Topographical Dictionary Of England . . . Third Edition . . . Vol. V . . . M.DCCC.XXXVII.*

**Copy**   Leicester UL H 942 F LEW.

*A Topographical Dictionary Of England . . . M.DCCC.XXXVIII.*[1]

*View Of The Representative History Of England . . . MDCCCXL.*

**Copy**   CUL Atlas 5.84.20.

### Reference

1   Burden, p. 164 suggests an edition for 1838, but no set has been found, containing this series of maps or those begun in 1831 (**93**).

Chubb CCCLXIV (Lewis' 1835 Dictionary only).

# 105

**JOHN and CHARLES WALKER**                                      **1835**

Size: 38.9cm x 32.1cm.

Some of the biographical record of these two brothers has been mentioned earlier (**81**) as well as some of their other engraving work (**102** and **104**). John (1787–1873) and Charles (died c.1872) were heirs to the engraving and cartographical career of their father, whose chief work had been with the Admiralty. By 1835 they were running a very busy organisation, which was responsible for the final publication of the Greenwoods' atlas, perhaps because the financial difficulties of the Greenwoods stemmed from an inability to settle the bill for the Walkers' engraving work.

The present maps are close neighbours of the maps of Smith (**56**) and Cary (**57**). Although the former was still issuing his atlas, judging by the small survival rate, sales had almost dried up by the mid-1830s. A similar experience was being undergone by Cary and his brother. The new series was probably in mind from 1834, once the Greenwood atlas had been issued and its sales suggested there was still a good market for well-engraved plates of larger size, and its preparation began early in 1835. Plates were issued in pairs initially and Lincolnshire was one of the first pair, dated 1 April, with Gloucestershire. Further pairs of plates were prepared throughout 1835 (none in October but 4 in August), 19 in 1836 but not to a regular pattern, 3 more in January 1837 and the last one, Derbyshire, is dated March 1st, 1837, the date also of the atlas. Curiously the title page had to be re-worked shortly after, since it begins with a dedication to the Duchess of Kent and the Princess Victoria, who, in May, became Queen Victoria; in spite of reworking the upper part of the title-page plate the date remained unaltered. In later editions, only the year was changed. The Walkers adopted a rather cavalier attitude to the compilation of their atlases, with a wide variety of dates on the plates making up any particular issue, thus rendering it impossible to state with any authority that a particular sheet (even though it bears a date) came from a particular atlas. The notes below will also show that close control of the addition of railway lines was primitive; the general intention seems to have been to engrave triple strands to indicate railways, which were later coloured over with a bright red line, but frequently (in haste to be up-to-date, presumably) hand-drawn lines were added before the 'under-engraving' had taken place. It is possible that some of these hand-drawn lines were added by salesmen or owners; confident dating is, therefore, hazardous. The same qualities of engraving are displayed here as in the plates engraved for the Greenwoods.

### Later history
The plates continued in use for the next sixty years, published by the brothers' successors, since they had both died in the early 1870s; John had ceased working due to ill-health in the 1860s and Charles retired some years before his death in c.1872. From 1849 the plates were also used for a series of lithographic transfers, overprinted with fox-hunting material, which continued to c.1880. These fox-hunting atlases were issued by William Colling Hobson. He had produced large-scale surveys of Durham (1839) and Yorkshire (1843) and seems to have retired in 1878, when a few more issues appeared under the name of the Walkers. Also in 1849 the Walker plate was used for a transfer used in Hagar's *Commercial Directory Of . . . Lincolnshire*. After the production of single volume directories of three other counties between 1848–1851 Hagar and Co. left that particular field. The publisher was Stevenson of Nottingham, a firm that had a long history in publishing and engraving in that city, including the issue of a newspaper (variously titled) in the 1850s. Two further lithograph atlases were produced by the firm of Letts.[1] This

firm began in 1809 but went into liquidation in 1884, when the map side was acquired by Mason and Payne[2] and Cassells obtained the diary business, although the diaries appear to this day under Letts' name. The Letts atlas was issued in parts, the first of twelve monthly parts being advertised as being ready on 25th January, 1884. Each part consisted of four maps and cost one shilling. The last atlas using the plates appeared in c.1895. The following list is in two parts; the first deals only with the original engraved plate and its successive states (1835–1895); the second deals with the lithographic transfers from 1849, which appeared in parallel with the Walker atlas. Each transfer is given a separate letter (A–P) and each is subdivided, as cases arise, for additions made to the stone, as opposed to those made to the original engraved plate from which successive transfers were made.

## EDITIONS

(i)   *LINCOLNSHIRE* (in a double framed panel)/*BY J. & C. WALKER/ REFERENCE TO THE WAPENTAKES/HUNDREDS &c./Parts of Lindsey*/18 numbered names (Ea). *Parts of Kesteven*/10 names/ *Parts of Holland*/3 names (Ed). *Lincolnshire contains 2,748 Square Miles/317,244 Inhabitants/returns 4 members to Parliament/for the County and 9 for 5 Boroughs/The Figures to the Towns show the Distance from/London in Miles/along the Roads from Town to Town/* (Ee). *London Published by Longman, Rees, Orme & C⁰. Pater Noster Row April 1ˢᵗ 1835.* (Ce, OS). *BOROUGHS/returning 2 Members/ Lincoln Boston/Grantham Stamford/ 1 Member/Great Grimsby/ PLACES OF ELECTION* . . . /3 lines/*POLLING PLACES*/2 columns of names/2 lines of signs for boundaries (Ae). Compass (Aa). *8 Furlongs 20/English Miles 69.1 = 1 Degree* 20 Miles = 10.05cm] (Reads upwards, Eb, OS). The boundaries of the hundreds, etc. are marked in pink, green, violet and brown; those of the boroughs have fine blue lines. A broad blue line separates north and south Lincolnshire. The border is marked with every minute of latitude, etc. and numbered every ten. *Meridian of Greenwich* (Ad and Ed, between borders).

*To their Royal Highnesses The Duchess of Kent & the Princess Victoria, This British Atlas, Comprising separate Maps of every County in England each Riding in Yorkshire and North and South Wales, Shewing the Roads, Railways, Canals, Parks, Boundaries of Boroughs &c. Compiled from the Maps of the Board of Ordnance and other Trigonometrical Surveys, Is . . . respectfully Dedicated by . . . J. & C. Walker. Published March 1 1837, by Longman, Rees & C⁰. Paternoster Row, and J. & C. Walker, 3, Burleigh Street, Strand.*

**Copies**   BL Maps 196.c.6; CUL Atlas 4.83.3; Bod C.17.b.17; NLS EU 16 Wk; Leeds UL W 170; Peterborough Cathedral S.2.1. (Grimsby RL WALKER; Manchester UL Map C 276).

Issued as a boxed set of maps. [1837].

**Copy**   CUL Maps D.34.63.2/10.

Issued folded and mounted in a small green cover, with a label (also in shades of green), with the title: *WALKER' S/LINCOLNSHIRE/ PRICE 2 . . . 6/* [1837].

**Copies**   BL Maps 2.aa.98; CUL Maps c.34.04.70; LRL Map 204.

*To Her Most Excellent Majesty Queen Victoria And To Her Royal Highness the Duchess of Kent, This British Atlas . . . Is . . . respectfully Dedicated . . . Published March 1 1837* . . .

**Copies**   BL Maps 12.e.32; CUL Atlas 4.83.7.

(ii)   *CASTOR* has been replaced by *CAISTOR/or CASTOR. THORNE* and *BAWTRY*, the roads to them from Gainsborough, and a road between them have been added. The Gainsborough-East Retford road has been extended to the edge of the map as has the Stainforth & Keadby Canal. The five boroughs have fine red lines outlining their boundaries.

. . . *British Atlas . . . 1838 . . .*

**Copy**  The private collection of Mr. C.A. Burden.

(**iii**)  *Tealby* has been added. The following corrections have been made: *N^{th}. Scarle* (formerly *N^{th}. Scale*), *Donna Nook* (*Donna Hook*), and *Ruskington* (*Buskington*).

. . . *British Atlas . . . 1837 . . .* [1839].[3]

**Copy**  Leeds RL RQ 912.42 W152.

(**iv**)  The imprint now reads: *London Published by Longman, Rees, Orme & C^o. Pater Noster Row 1841. Wragby, Alford, Caistor* and *Swinnerby/½ way house* have been added to the list of Polling Places (Ae).

. . . *British Atlas . . . 1841 . . .*

**Copies**  CUL Atlas 4.84.34; RGS 3 H 5.[4]

Issued mounted and folded in a box, with filing tabs. [1856].

**Copy**  BL Maps 15.bb.9.

. . . *British Atlas . . . 1842 . . .*[5]

Issued folded, in a cover with the title: *WALKER'S LINCOLNSHIRE.* [1842].

Issued folded, in a cover with the title: *New Map Of Lincolnshire.* [1842].

Issued folded, in a cover with the title: *Walker's county maps.* [1842].[6]

(**v**)  The population figure (Ee) has been altered to *362, 717*. The imprint now reads: *Published by J. & C. Walker, 9 Castle S^t. Holborn, London, 1843.*

. . . *British Atlas . . . 1843 . . .*[7]

. . . *British Atlas . . . 1865 . . .*

**Copy**  BL Maps 12.e.33.

Issued folded in a cover with the title: *WALKER'S LINCOLNSHIRE/ PRICE 2..6* [1843].[8]

. . . *British Atlas . . . 1845.*

**Copy**  Bod Maps C.17.b.14.[9]

(**vi**)  The year in the imprint now reads *1844*.

. . . *British Atlas . . . 1844 . . .*

**Copy**  (The author's collection).

(**vii**)  The date in the imprint has been altered to *1846*. Additions include: *Breakwater* (off Grimsby), the replacement of *Barrow F^y* by *New Holland, New Dock* (protruding from the coast at Grimsby), *Embankment* added, the land 'reclaimed' and the shading removed at *Cross/Keys Wash*. The Caistor-Croxton-New Holland and Croxton-Melton Ross roads have been upgraded. Engraved lines, coloured over in red have been added for railways i (labelled at the southern end *Cambridge and/Lincoln/Railway*), 1 (*Nottingham, Newark and Lincoln Railway*), 4, 9 and 13 (*Gainsborough and Grimsby Railway*).

Issued folded in a purple cover with the title: *LINCOLNSHIRE.* [1846].

**Copy**   CUL Maps c.34.05.70.[10]

Issued folded in a green cover with the title: *WALKER'S/ LINCOLNSHIRE/PRICE 2..6/* [1846].

**Copy**   The author's collection.

*... British Atlas ... 1846 ...*[11]

*... British Atlas ... 1861 ...*[12]

**Copy**   CUL Atlas 5.86.18.

*... British Atlas ... 1862 ...*[13]

**Copy**   BL Maps 12.e.6.

(**viii**)   The date has been removed from the imprint. Railways have been engraved and are marked by broken lines: ii (starting at Market Deeping), vii (but only to Spalding, where it meets railway (iii), which has no extension to Boston or Lincoln), viii, x (although it does not enter Gainsborough, but joins (v) west of Dunham Bridge), 21 (as a branch from (vii)), 3/6/7, 11, 14 and 16; single lines: 8 (only to Boston) and 17; triple strands: 1, 4, 9, 10 and 13. The Boston-Lincoln section of railway 8 is drawn freehand. Railway 17 is labelled *Lon. & York Rail* at its southern end and *London & York Railway* in Nottinghamshire. All lines are coloured over in red by hand. The hundreds are coloured and the boundaries of the boroughs are blue.

*... British Atlas ... 1848 ...*

**Copy**   Hull RL 912.42.

(**ix**) Railways 3, 6–8, 11 and 14–17 have been engraved as triple-strand lines coloured over in red. Railways vii (to Sutton Bridge only), viii–ix and B have been marked by dotted lines, also coloured over in red. Railway v is marked by a double line. Railways xiii and 16 have been drawn in by hand. Railway (i) has been removed.

*... British Atlas ... Published by Longman, Rees & C°. Paternoster Row, and J. & C. Walker, 9, Castle Street, Holborn, 1849.*

**Copy**   NLS Newman 365.

(**x**)   Railway 12 has been added. The free hand line to Wragby does not appear here. The GNR is now labelled *Great Northern Railway* at its northern end. *East Lincolnshire Railway* has been added, north of Boston. *Ambergate, Nott. & Bos. Rail* has been added, near Bingham, but is still shown by dotted line.

*... British Atlas ... 1851.*

**Copies**   BL Maps 23.e.12; CUL Atlas 4.85.32.

(**xi – a**)   *St., Sta.* or *Station* have been added along all railways properly open, but not in any town except Spalding and *Glamford/Briggs*. Railways ii, iv, vi–vii, 17 remain as dotted lines and v as double lines all uncoloured. Many place-names have been added to the plate (following the addition of meeting places for the hunts to the lithographic transfers prepared for the 1849 hunting atlas – state B below). Among them are: *Three Queens, Hungerton Hill, Bottesford, Elton, Langar, Plungar, Carlby Mill, Scrimshaws/ Mill, Uffington Wood, Kennel* (south of Grimsthorpe Park), *Park* (below Easton), *Ropsley/Rise, T. Bar* (on the Osbournby-Grantham road and at Grayingham), *Coddington/Plantation, Blackmoor/Causeway, Stapleford/ Moor, Langford/Hall, Potter/Hill Bar, Halfway/House, 7th mile stone* (on Foss Road), *Swallow Beck, Drinsey Nook, Thorney/House, Gipple, Aswardby/Hall & Park, Byards Leap, Branston Wood, House* (below Nocton and at Usselby), *by Blankney* (below Martin), *Fisherton Long/Wood* (i.e. Fiskerton), *Wood* (added below Newbold), *Tile House/Beck, Hall* (below Stainfield), *New Park/Wood, Tumby/ Wood, Water/Dyke Ho., Wood* (after Laughton), *Magin/Moor, Grange* (Faldingworth), *Snarford/Bridge, Dunholme/Corner, Linwood/Warren, Cow Pasture* (west of Louth), *Kelsey River Head, Cadney Bridge, Yarborough/Camp, Saxby Mill, Langmere/Furze, Chase/Hill, Ryehill, Riby Cross Roads, Limber Lodge, Lit. Limber Lodge, Fenby Wood, Pelham's/Pillar, Cross Roads* (at Bradley and at Laceby), *Aylesby/Mill, Mill* (below Hatcliffe and beside North/Ormsby), *Houses* (below Grainsby), *Agthorpe Wood,* and *Eastv* (i.e. Eastville). *Wood/End Ho* (east of Fiskerton) has been removed. The maps have been printed back to back and Lincolnshire has Middlesex on the reverse.

. . . *British Atlas . . . 1852.*

**Copy**   CUL Atlas 4.85.10.

(**xi – b**)   The maps are single-sided.

. . . *British Atlas . . . 1853.*

**Copy**   The private collection of Mr. T. Burgess.

(**xii**)   Railway 2 has been added.

. . . *British Atlas . . . 1853.*

**Copies**   The private collection of Mr. T. Burgess. (LRL Map 83).[14]

(**xiii**)   Railway 19 has been added. The eastern portion of railway 15 is coloured red.

. . . *British Atlas . . . 1854.*

**Copy**   BL Maps 23.e.10.

Issued folded in a cover with the title: *LETTS SON & Cº. LINCOLNSHIRE & ROYAL EXCHANGE . . .* [c.1856].[15]

(**xiv – a**)   Railways 1, 3–4, 6–20 and 22 are engraved and coloured. Railways 2, 21 and 24 have been added by hand. The lines at Lincoln have been redrawn; the lines from Market Rasen and to Newark are staggered with the Great Northern instead of crossing in a straight line.

. . . *British Atlas . . . 1858.*

**Copy**   The private collection of Mr. C.A. Burden.

(**xiv – b**)   Railway 2 is not coloured, but railway 23 is coloured over the dotted line. Railway 24 has not been drawn in.

Issued folded, in a cover with the title: *WALKER'S/LINCOLNSHIRE/ PRICE 2 6/* [c.1858].

**Copy**   BL Maps 1.a.61.

(**xv**)   The population figure (Ee) has been changed to *407, 222.* A railway from Willoughby to Mablethorpe has been added by hand, while the manuscript additions of state (xiv) are missed out.

. . . *British Atlas . . . 1856.*

**Copy**   PRO M15.7 PRESS 22.

(**xvi**)   The line to Mablethorpe has not been retained, while the lines of state (xiv) have been added again by hand.

Issued folded, in a cover with title *WALKER'S/LINCOLNSHIRE/2..6/* [1858].[16]

**Copy**   Bod Maps C17: 39 (27).

. . . *British Atlas . . . 1860.*[17]


(**xvii**)   Railway 2 has been added again and now runs parallel to the GNR. Railway 10 joins the MSLR on a curve (previously a T). *G*$^t$*. Northern Rail* has replaced *Lon. & York Rail* at the southern end of the GNR. The railways now underpinned by engraved triple-strand lines are 1–4, 6–17, 19–20, 22 and 25, all coloured in red. Railways 21 and 23 are coloured red over dotted lines. Plain dotted lines continue to mark railways vii (to Spalding only), viii and ix. Railways 26–27 have been drawn in by hand in red. Roads are brown, parks green while North Lincolnshire has its boundary coloured deep pink and South Lincolnshire's is yellow. Borough boundaries remain blue. Railway v is in its previous condition. The junction at Grantham has been re-aligned as has the Edenham line; the latter now curves westwards at its southern end and joins the GNR above the name of *Careby.*

Issued mounted and dissected in a cover with the title: *WALKER'S LINCOLNSHIRE/ PRICE 2 6/* [c.1861].

**Copy**   The author's collection.


(**xviii**)   Railway 21/24 is now engraved and leaves the GNR further north, passing through *24* (the map reference for Winnibriggs and Threo).

. . . *British Atlas . . . 1864.*

**Copy**   The private collection of Mr. D. Hodson.


(**xix**)   Railway 23/27/29 has been extended and forms a T at Kings Lynn with railway H. Railways 26 and 32 have been added. Railway ix is (erroneously) coloured over; the previous dotted line for this railway has been converted to two parallel lines. Dots for stations have been added at Crowle, Thorne (2), Moulton, Whaplode, along the Boston-Grantham line and at three places in Norfolk.

. . . *British Atlas . . . 1867.*

**Copy**   PRO CO 700 Misc 7.


(**xx**)   Railways 30–31 and 33–36 have been added. *St.* has been added to the dots on the Grantham-Boston line. All lines are engraved except railways 32 and 36.

. . . *British Atlas . . . 1869.*

**Copy**   Bod Allen 110.


(**xxi – a**)   Railway 37 has been added by hand (wrongly sited).

. . . *British Atlas . . . 1870.*

**Copy**   BL Maps 33.c.16.


(**xxi – b**)   Railway 39 has been added by hand (wrongly sited).

. . . *British Atlas . . . Published by Longman & C°. 39 Paternoster Row, and J. & C. Walker, 37 Castle Street, Holborn. 1872.*

**Copy**   The private collection of Mr. D. Hodson.

*. . . British Atlas . . . 1873.*

**Copy**    Leeds UL W 171.

(**xxii**)    Railway 38 has been added by hand and railway 37 has been redrawn. Railway 39 is now engraved.

*. . . British Atlas . . . 1877.*

**Copy**    The private collection of Mr. D. Hodson.

(**xxiii**)    Railway 37 has been redrawn and now leaves the ELR below the dot for Little Steeping station. Railway 38/40 has been added but leaves the ELR to the south of Firsby station. Railways 39, 41 and 43 have been added.

[*. . . British Atlas . . . 1878.*]

**Copy**    (The author's collection).[18]

(**xxiv**)    Railway 44 has been added.

*. . . British Atlas . . . 1879.*

**Copies**    Birmingham RL AF 912.42; Brighton RL 912.42 W15.

*. . . British Atlas . . . 1880.*[19]

Issued as a folded map in a cover entitled: *WALKER'S LINCOLNSHIRE.* [1880].[20]

## FIRST LITHOGRAPHIC TRANSFER

(**A**)    The map in state (vi) with a few additions from state (viii) has been transferred. The title is: *MAP/of/Lincolnshire/* (Ea). The Walker imprint has been replaced by: *Engraved at the Offices of Stevenson & Co., Middle Pavement, Nottingham, March 22ᵈ. 1849*. The block of *REFERENCE* has been moved (from Eb to Ed, and its continuation has moved from Ed to Ae). Details of boroughs and election places (formerly at Ae) have been removed. The notes on the county's area, population, etc. have been re-engraved; they now occupy six lines instead of seven. The inner borders are not now marked in minutes by alternating shaded and unshaded sections. The scale (Eb) reads down the map (formerly Eb, OS and reading upwards). Railways 1, B (a short part of its eastern end), 3–4, 6–15, K and V are marked. Dotted lines indicate railway vii (to Spalding only). The Lincoln-Newark line is not named. *Meridian 0 of Greenwich* appears 90 mm. above the inner frame line (Ce) with an arrow pointing to it from its usual place between the frame lines.

*Hagar And Co's Commercial Directory Of the Market Towns In Lincolnshire. Containing An Alphabetical List Of The Gentry, Merchants, Manufacturers, Professions, Trades, Etc. A Classified List . . . Members Of Parliament; Magistrates; Corporation, And Their Offices; Post Office, With The Time Of Arrival And Departure Of Mails; Bankers; Places Of Worship, With The Ministers And Time Of Service; Public And Charitable Institutions; Schools; Coaches; Omnibuses; Carriers By Land And Sea, Etc . . . Price of the Vol. – To Subscribers, 10s.6d.; Non-Subscribers, 14s.6d. Nottingham: Printed By Stevenson, Middle Pavement. 1849.*[21]

**Copies**    BL PP 2507 ebl; LRL L 38; Gainsborough L 929; Scunthorpe RL L638 910. Photocopies of the directory and map were made from LRL's copy and deposited in the libraries at Spalding, Louth, Grantham, Boston. (LAO Scorer Map 45).

## SECOND LITHOGRAPHIC TRANSFER

(**B – i**)    The map described in state (xi) has been overprinted with the outlines of the names of the county hunts: *Mᴿ LUMLEY/See Map 27, LORD YARBOROUGH'S,*

*SOUTHWOLD, BURTON, BELVOIR/ See Map 20; COTTESMORE/See Maps 20 & 30*;
the outlined names have been hand-coloured in royal blue. The boundaries of the hunts
have been marked by broad lines of water colour: pink, lemon, pink, green, orange/brown
and pink respectively. The southern boundary of the Lumley and the western boundaries
of the Lumley, Belvoir and Cottesmore hunts are not coloured. Railways 1 (now named),
3–4, 6–14 are marked and coloured red; railways vii (to Sutton Wash Way only), viii–ix
and B are shown by dotted, uncoloured lines. The places added to the map in state (xi)
are, in most cases, underlined and marked by a large black dot. • *PLACES OF MEETING
OF FOXHOUNDS* has been added below the title (Ea). Plate number *N°. 21.* has been
added (Ea, OS; and Ee, OS sidewards). The 0 meridian appears between the inner and
outer frame lines (Ce).

*Hobson's Fox-Hunting Atlas; Containing Separate Maps Of Every County In England,
And The Three Ridings Of Yorkshire; Comprising Forty-two Maps, Showing The Roads,
Railways, Canals, Parks, Etc., Etc. Compiled From The Maps Of The Board Of Ordnance, And Other Surveys. By J. & C. Walker . . . London: Published By J. & C. Walker,
9, Castle Street, Holborn.* [1849].[22]

**Copies**   CUL Atlas 5.84.12 and 5.85.29. (Grimsby RL RM 12).

(**B – ii**)   The plate numbers have been altered; at the top it now reads *N°. 21* while that
in the bottom corner is *N° 21* and neither has a dot after the number. The hunt names are
coloured in a bright lighter blue; in the case of *LORD YARBOROUGH S* the apostrophe
is faintly outlined only. Railway 2 has been coloured red.

*Hobson's Fox-Hunting Atlas . . .* [1849].

**Copy**   (The author's collection).

(**B – iii**)   The see references below three of the hunt names are outlined but uncoloured.
Railway 17 has been added by hand.

*Hobson's Fox-Hunting Atlas . . .* [1850].[23]

**Copy**   Leeds UL W 196a.

## THIRD LITHOGRAPHIC TRANSFER
(**C – i**)   The references to other plates under the Lumley, Belvoir and Cottesmore hunts
have been removed though traces can still be seen. The Yarborough hunt is mis-spelt
*YAHBOROUGH* and the Lumley is now *M^R. LUMLEY*.

*Hobson's Fox-Hunting Atlas . . .* [1850].

**Copy**   BL Maps 11.c.16.

(**C – ii**)   The H in YARBOROUGH has been changed by hand to R, but the attempt is
emphasised by the use of a different shade of blue. Railway 16 is marked by a dotted
line; the colouring of railway 2 does not appear here.

*Hobson's Fox-Hunting Atlas . . .* [1852].

**Copy**   The private collection of Mr. E. Burden.

(**C – iii**)   Further work on converting the H to R in YARBOROUGH has been tried,
though it differs from the second R, since it lacks the curl at the foot of the right-hand
stroke. Railway 17 has now been engraved with triple strand lines.

*Hobson's Fox-Hunting Atlas . . .* [1852].

**Copy**   The private collection of Mr. T. Burgess.

## FOURTH LITHOGRAPHIC TRANSFER

**(D – i)**   The following places have now been underlined, with black dots added: *Roulston, Digby, Evedon, Aubourn, Mavis Enderby, Revesby, Tumby, Scrivelsby Court, Withcall, Castle Carlton, Fotherby, WRAGBY, Blyton, Laceby Hill, Haltham, Panton/Ho.* and *Lea. Belvoir* (only) has been underlined and has no dot; *SPILSBY* is treated similarly. The Burton hunt has a blue border.

*Hobson's Fox-Hunting Atlas . . .* [1852].

**Copies**   Leeds UL W 196; Leeds RL F799.277 H653.

**(D – ii)**   The following places have been added and underlined: *Mile Stone* (north of Belchford), *Hall* (added below Langton, near Sausthorpe), *North Wood* (above Legsby), *Kennels* (twice – below Ropsley, which is also underlined, and south of Belvoir Castle). The t at Ropsley has been removed.

*Hobson's Fox-Hunting Atlas . . .* [1852].

**Copy**   CUL Atlas 5.85.35.

**(D – iii)**   Railway 19 has been added by hand. While the junction formed by railway 10 with 9 is engraved to form a T at Brigg, a red line is now drawn to form a curve nearer Barnetby.

*Hobson's Fox-Hunting Atlas . . .* [1855**].

**Copy**   The private collection of Mr. C.A. Burden.

## FIFTH LITHOGRAPHIC TRANSFER

**(E – i)**   The transfer is based on a version of state (xiii). The population figure in the panel (Ee) has been changed to *407,222*. The apostrophe in *LORD YARBOROUGHS* has now been removed. Railway 2 is coloured from Tallington to Stamford. The lines at Lincoln have been redrawn so that the lines from Market Rasen and Newark form two separated T junctions with the Gainsborough-Boston line. The plate numbers are now both in the form *N°. 21*.

*Hobson's Fox-Hunting Atlas . . .* [1856].

**Copy**   (LRL Map 97a).

**(E – ii)**   Railways 20–24 have been drawn in, not very accurately.

*Hobson's Fox-Hunting Atlas . . .* [1860**].

**Copy**   CUL Atlas 5.86.27.

## SIXTH LITHOGRAPHIC TRANSFER

**(F – i)**   Mr. Lumley's hunt has now been renamed: *LORD GALWAY*. Railway 21/24 is now drawn in its correct place, leaving the GNR at Barkston. Railways 22–23 and 26–27 have been drawn in. The 'new' lines are drawn very inaccurately, especially railways 26 and 27, the latter following the previous dotted line, which shows the railway north of the Washway road.

*Hobson's Fox-Hunting Atlas . . .* [1860].[24]

**Copy**   The private collection of Mr. D. Hodson. (The author has a loose sheet).

**(F – ii)**   Railways 2 and 17 have been redrawn and run parallel towards [Peterborough]. Railway 25 has been added by hand.

*Hobson's Fox-Hunting Atlas . . .* [1861**].

**Copy**   BL Maps 12.e.26.

(**F – iii**)   Railway 23/27 is drawn in its correct position and its extension (railway 29) added. Railway 31 forms a T at a point half way between Long Sutton and [Sutton Bridge]. Railway 32 has been added. Railway 22 at its southern end joins the GNR on a north-westerly curve, north of Careby.

*Hobson's Fox-Hunting Atlas . . .* [1866?].

**Copy**   (The author's collection).

### SEVENTH LITHOGRAPHIC TRANSFER
(**G – i**)   *GROVE* replaces the Galway hunt name. The plate is transferred from state (xv). The list of polling places and electoral data has ben omitted (Ae). Railways 30 and 33–36 have been added. Railway 31 now joins railway 23/27/29 at Sutton Bridge. The line from Stamford merges with the GNR again. Railway 29 forms a T with railway H at Kings Lynn where railway D has been added. An attempt to remove the double lines from Gainsborough-Bawtry has been made but many traces remain.

*Hobson's Fox-Hunting Atlas . . .* [1867**].

**Copies**   Bod Maps C.17.c.8; Leeds UL W 197.

(**G – ii**)   Railway 37 has been drawn in, but from Little Steeping station.

*Hobson's Fox-Hunting Atlas . . .* [1868**].

**Copy**   CUL Atlas 5.86.21.

### EIGHTH LITHOGRAPHIC TRANSFER
(**H – i**)   Railway 39 has been drawn in but west of its proper line. Railway 37 now curves off the ELR south of Firsby station. *GRIMSBY* (of *Gᵀ./GRIMSBY*) and *LOUTH* have been underlined. There is no imprint; there is no sign that the plate has been closely trimmed.

*Hobson's Fox-Hunting Atlas . . . Published By J. & C. Walker, 37, Castle Street, Holborn.* [1869].

**Copy**   Leeds UL W 198.

### NINTH LITHOGRAPHIC TRANSFER
(**I – i**)   The Burton hunt has been divided to create the *BLANKNEY* hunt (edged in deep green). Roads and railways leading to *MELTON/MOWBRAY* have been added, with the following places (all with dots and underlined): *Saltby, Piper/Hole, Great/Marwood, Langdale/ Lane, Stoneby* and *Waltham/on the Wolds*. Broken lines have been added from *Sewstern* west to Melton Mowbray and then northwards to Bingham (as part of the boundary of the Belvoir hunt). The address in the imprint has been altered to *37 Castle Sᵗ. Holborn. London.* (Ce, OS). Railways 2 and 17 are now close parallel lines. Railway D has been removed. Railway 39 has been redrawn in its correct place.

*Hobson's Fox-Hunting Atlas . . .* [1871].[25]

**Copy**   CUL Atlas 5.87.18.

(**I – ii**)   Railway 37 has been drawn in, again from Little Steeping station, and ending on the north-east side of the town at Spilsby. Railway 41/43 has also been drawn in. Dots for stations have been added on railway 39 but incorrectly sited.

*Hobson's Fox-Hunting Atlas . . .* [1872].[26]

**Copy**   The private collection of Mr. D. Hodson.

## TENTH LITHOGRAPHIC TRANSFER
(**J – i**)   Railway 37 has been drawn in, starting from south of Little Steeping station and ending at its correct site in Spilsby; railway 40 has been added but the line 38/40 starts from north of Little Steeping station. All railways up to 36 with 39 have the lines engraved with triple strands below the red colouring.

*Hobson's Fox-Hunting Atlas . . .* [1873].

**Copy**   Leeds UL W 199.

(**J – ii**)   Railway 38/40 has been redrawn to leave the ELR where the Spilsby line also separates.

*Hobson' Fox-Hunting Atlas . . .* [1875].

**Copy**   BL Maps 10.cc.24.

(**J – iii**)   Railway 44 has been drawn in but incorrectly sited.

*Hobson's Fox-Hunting Atlas . . .* [1877].

**Copy**   Leeds UL W 200.

*Walker's Fox-Hunting Atlas . . .* [1880].

**Copy**   Leeds UL W 201.

Issued folded and mounted in a red cover with the title *FOX HUNTING MAP OF LINCOLNSHIRE* blocked in gold. [1880].

**Copy**   The author's collection.

(**J – iv**)   Railway 41/43 has been redrawn; the clearest effect is seen where the line now passes through the name *South/Willingham* while, formerly, it passed further north through the name *Hainton.*

*Walker's Fox-Hunting Atlas . . .* [1881].[27]

**Copy**   The private collection of Mr. B. Cavalot.

## ELEVENTH LITHOGRAPHIC TRANSFER
(**K – i**)   Railway 41/43 has been engraved with triple strands of line and now joins the Boston-Lincoln line on a south-easterly curve. Railways 45 and 46 have been drawn in but from Pinchbeck.

*Walker's Fox-Hunting Atlas . . .* [1882].

**Copy**   The private collection of Mr. D. Hodson.

## TWELFTH LITHOGRAPHIC TRANSFER
(**L – i**)   The figure for the county population has been changed to *469,994* and the following two lines on the members of parliament have been deleted. The Grove hunt has been renamed *GALWAY. Sta.* has been added along railways 32, 23/27, 17, 35, 30/33, 22 and 44 (now correctly shown, leaving the ELR on a curve south of Louth). The initials *C.G.S.* (Coast Guard Station) have been added at eight places along the coast; *L.B.S.* (Life Boat Station) has been added at two other places. *Lᵗ.Hᵒ.* has been added north of *Stallingborough/Light & Ferry.* A quadrant has been added to the south side of the New Dock at Grimsby. Railways 28 (and a triangular junction formed south-west of Grimsby), O and T/U have been added.

*Walker's Fox-Hunting Atlas . . .* [1883].

**Copy**    The private collection of Mr. C.A. Burden.

## THIRTEENTH LITHOGRAPHIC TRANSFER

(**M** – **i**)    This new transfer is based on state (xxii) minus the names and colouring of the hunts, the dots and underlining of the meeting places but with the additions of K – i. The former title has been replaced by: *LINCOLN* (in a double panel)/*LETTS. SON & Cº. LIMITED./STATISTICS./Population 469,994/Area in Acres 1,767,962/Gross Rental £3,555,684/Inhabited Houses 100,830/Poor Rate £242,426/Paupers 15,987/County Town is Lincoln./* (Ea). The sea is printed in blue with white panels for information boxes; south Lincolnshire is printed green, central Lincolnshire yellow and northern Lincolnshire pink. The boroughs are individually coloured, viz. Grimsby yellow, Stamford and Grantham pink, Lincoln light green and Boston grey/violet. Railways are red, main roads yellow and other roads uncoloured. Post towns have blue triangles and other coloured symbols indicate cathedral cities, places of county courts, towns with quarter sessions, the members of parliament, dangerous hills and market days. A graticule has been added with letters a–e down the sides A–D along the top and bottom. The imprint is: *LETTS, SON & Co. LIMITED. LONDON BRIDGE, E.C.* (Ce, OS). *8 Furlongs + 20/ English Miles 69.1 to a Degree* [20 miles = 10.2cm] (De–Ee, OS). The names Leicestershire and Rutland and the roads and railways leading to Melton Mowbray have been removed. *The Produce of Lincoln is/mainly Agricultural* replaces the material on population (Ee). *LARGE FIGURES in body of Map refer/to WAPENTAKES, HUNDREDS, &c./*18 numbered names (panel, Eb). *Parts of Kesteven/*names numbered 19–28/*Parts of Holland/*3 names numbered 29–31 (in a box, Ed). *EXPLANATION OF SIGNS./* 23 lines and coloured symbols with their meanings (Ae). Railways 1–4, 6–46 are shown; railways B, H and O are the only out-county lines marked. The map is overprinted *NORTHERN/ DIVISION* and *SOUTH/DIVISION*. On the verso of the map *22* appears (top right).

*Lett's Popular County Atlas. Being A Complete Series Of Maps Delineating The Whole Surface Of England And Wales, With Special and Original Features, And A Copious Index of 18,000 Names. Letts, Son & Co. Limited, London Bridge, E.C. 1884.*

**Copies**    BL Maps 33.c.15; CUL Atlas 5.88.7; Bod Maps C.17.c.9; Leeds UL W 225; Leicester UL F.AY; Sheffield RL 912.42 STF.

## FOURTEENTH LITHOGRAPHIC TRANSFER

(**N** – **i**)    The seven new electoral areas are named, in orange in the form: *LOUTH/(EAST LINDSEY)* and coloured violet, pink, yellow, brown or green. Their boundaries have been freshly added as alternating dots and dashes. Grimsby, Boston and Lincoln are coloured brown and Grantham green. The words *WRIT SUSPENDED* (above Boston's name) have been removed. The large red circles to denote the number of county members and the symbol and its meaning in the Explanation panel have been removed. GNR is named *Gᵗ.N.* only at its southern end. The names of the divisions and Lindsey and Kesteven (but not *HOLLAND*) have been removed. The railways are coloured brown. *LONDON.* has been added after the publisher's name, below the title (Ea).

*Lett's Popular County Atlas . . . Mason & Payne, Proprietors And Publishers Of Lett's Atlases, 41, Cornhill, London, E.C. 1887.*

**Copies**    CUL Atlas 5.88.8 and 5.88.31; Bod Maps C.17.c.11.

The Mason and Payne atlas of 1887 was re-issued in 1889 with some plates having the name of Mason and Payne in the maps' imprints. The maps were also issued (for some counties) in covers, marked *GROOM'S NEW MAP OF . . . FORSTER GREEN & Co . . .*, in plain marbled covers, and in covers marked *The Charing Cross Central Depot FOR POPULAR MAPS, 15, CHARING CROSS, S.W.* These three final versions all appeared c.1896. Lincolnshire may exist thus.

## FIFTEENTH LITHOGRAPHIC TRANSFER

(**O – i**)   Based on the eleventh transfer with the following changes: the Burton hunt is not named. Placed well to the west (mostly in Nottinghamshire) $M^R JARVIS$ in a sans-serif type has been stamped in violet. Railways 48–49 have been drawn in but terminate at Alford (instead of Willoughby).

*Walker's Fox-Hunting Atlas . . . 189[2].*[28]

**Copy**   (LRL Map 97).

## SIXTEENTH LITHOGRAPHIC TRANSFER

(**P – 1**)   The Burton hunt is named again and the name of Mr. Jarvis has disappeared. Railways 50–51 have been drawn in. Railway 48 has been redrawn and now joins the ELR at Willoughby.

*Walker's Fox-Hunting Atlas . . . 189[5].*[29]

**Copy**   CUL Atlas 5.89.36.

### References

1   *The romance of the business of a diary publisher*: Charles Letts & Co. Ltd. (London, 1949).
2   While Mason and Payne published the 1887 Atlas (see state 105 N and the note with that atlas), no plate of Lincolnshire with their imprint has been found; it is doubtful if they ever added their names to the county plate.
3   The year on the title-page has been altered by hand to 1839. Ten plates have 1839 in their imprints.
4   While the CUL Atlas has all but nine plates dated 1841 the RGS example has only four. Lincolnshire, in common with most plates, has railways added (by hand, in red) and their state accords with a suggestion that they were added c.1850.
5   CUL Maps BB.53.84.4 is the title-page of such an atlas. The BL catalogue records an example (Maps 23.c.13) but that is missing.
6   Kingsley, p. 218 records Sussex in all the three folding forms with the maps dated 1841 or 1842.
7   In 1991 a dealer broke a copy of the atlas so dated. The county plate was in an earlier state, dated 1835 still. The BL atlas dated 1865 on the title-page has railways up to c.1865 added by hand.
8   John Rylands UL has an example of Nottinghamshire dated 1843 in this format (R 108304).
9   Although the atlas has 1845 on the title-page the range of dates in the volume is 1841–1844.
10   The CUL catalogue suggests this map was issued by Betts & Co., though there is nothing on the map to support this. Such maps were issued by that firm in the mid-1840s (Kingsley records Sussex, dated 1843; Burden notes Berkshire dated 1846 – both in CUL).
11   Whitaker, 1949, no. 531 records an atlas dated 1846.
12   The county plate shows a few minor variations, e.g. the *Cambridge and/Lincoln Railway* is not named and is marked by dots only; most railways up to 1859 have been added by hand.
13   The BL copy only differs from CUL Atlas 5.86.18 in the way railway 21 is extended to Holbeach by hand.
14   The author has a copy that is basically as described but the colouring has been applied differently, e.g. the northern section of the GNR as far south as Retford only has been coloured.
15   No copy has been found but CUL has Buckinghamshire (Maps c.34.06.52) in the same state as in the 1852 atlas, and Middlesex (Maps c.34.06.75) ascribed to 1856.
16   Kingsley records three examples of Sussex dated to c.1860. The cited example has lost most of its original outer cover.
17   The only copy in a public collection – William Salt L. (Stafford) – was missing in 1991. A copy is recorded in the General Printed Catalogue (1912) of the London Library.
18   The map was sold as taken from an 1878 atlas, but no complete atlas has been found. Although the additions have been added by hand the map is intermediate between those set at 1877 and 1879.

19  Mr. E. Burden owns an atlas with a title-page date of 1880; the contents consist of hunting maps issued after Hobson's last issue.

20  Burden, p. 169 records a copy of Berkshire thus (Brunel UL).

21  Hagar & Co. produced directories for Nottinghamshire (1848), Leicestershire (1849) and Durhamshire (sic) in 1851. The LRL copy is signed by its first owner April 14th, 1849.

22  Most commentators have agreed on 1849 as the first date of issue based on the state of railways marked on various county plates. Since all Hobson atlases are undated suggested dates, based on this sort of, often anticipatory, information have to be treated with special caution. The question of the source of Hobson's data on the meeting places has led Mr. B. Cavalot to think 1850 the earliest likely date; he believes the probable source of Hobson's hunting information was *The Fox-Hunter's Guide* by 'Cecil' (London, undated), which lists the places of meets valid for the period from 1st May 1848 to 30th April, 1850 with an appendix to the end of the 1850–1 season. Because of the inclusion of the appendix, this book must have been issued in 1850. If Hobson used this source it is curious that 49 places in Cecil's main lists do not appear on the map, while 7 more appear only in the supplement. 26 names appear on the map at its first appearance, which are not listed in the book at all. Much hunting data was published in *The Sporting Review* or the journal, issued by 'Cecil's' own publishers, *The Sporting Magazine*. These are more likely sources for Hobson's data if he relied solely on published information. Until other research yields more concrete evidence Chubb's suggested date of 1849 has to be accepted.

23  Whitaker suggested 1860 for this volume but the railway data on the county map are a good deal earlier and 1850 is suggested.

24  The 6th Lord Galway resumed the mastership of the hunt from Richard Lumley (later 9th Earl of Scarborough) in 1858. See: Watson, J.N.P. British and Irish Hunts and Huntsmen. Vol. III. (London, 1986), p. [27]. Although the additional railway data are not accurate they suggest a date around 1860. The plates in Mr. Hodson's atlas are all from 1842 or 1843; clearly Hobson was using up old stocks for his transfers. Around this time similar use was made of old sheets – see states (v) and (vii) above.

25  The Burton hunt was divided and the Blankney formed in 1871 when 'Henry Chaplin [found] that hunting six days a week interfered somewhat with his parliamentary duties'. (Olney, R. *Rural Society in Nineteenth Century Lincolnshire*. (Lincoln, 1979), p. 35.).

26  The Louth-Bardney branch was incorporated in 1866 but the decision to proceed was not taken until 1871. (Ludlam, A.J. and Herbert, W.B. *The Louth To Bardney Branch*; 2nd ed. (Oxford, 1987), pp. 7–11).

27  This example is signed by the first owner, 1881.

28  The final digit has been left off for completion by hand. Kingsley, p. 224 records a copy in private hands. The LRL map may have come from a similar source. He also refers to the map of Sussex in identical form in an atlas (also privately owned) ascribed to [1886].

29  Kingsley, p. 225 reports an edition dated 189[4] with the Sussex map in the same state as in 189[5]. Lincolnshire would be the same, presumably.

Chubb CCCCLXXVI–CCCLXXXI (Chubb, of course, records no editions of Walker's atlases after 1870); DXXXII (Chubb only notes the [1849] issue of Hobson's atlas).

# 106

## SAMUEL ARROWSMITH                                                        1836

Size:29.4cm x 31.35cm.

The work of Arrowsmith's father, Aaron, has already been met with in 1815 and 1821 (**72** and **79** respectively). Under the 1815 note the development of the Arrowsmith family firm is set out. The present map includes Nottinghamshire and Leicestershire and is based very loosely on his father's map of 1821; that is to say, it has no geographical features, the coastal outline is inaccurate and Lincoln is the only place named in the three counties. A second map in the volume below shows the southern part of Lincoln diocese.

## EDITIONS

(**i**)  *LINCOLN./(N. Part)/* (Ea). *S. Arrowsmith, Lithog.* (Ee, OS). *11* (Ea, OS). *16 Ordered by The House of Lords to be Printed, 10 March 1836* (De, OS). *A. Archdeaconry of Lincoln./22* numbered names, which form a key to the divisions marked on the map, followed by B. Archdeaconry of Stow (4 names), C. Archdeaconry of Nottingham (5 names) and D. Archdeaconry of Leicester (7 names). (Eb–Ed). *Reference/5* lines of symbols, etc. (De). *Bishops Residence o at Buckden (Huntingdonsh)/Cathedral Church +/* (Ee). The sea is coloured blue, Nottinghamshire pink and Leicestershire green-blue; Lincolnshire is left uncoloured.

*Third Report From His Majesty's Commissioners Appointed To Consider The State Of The Established Church With Reference To Ecclesiastical Duties And Revenues. [Presented by His Majesty's Command.] Ordered, to be Printed, by The House of Commons, 20 May 1836.*

**Copy**   LAO FL 1a C 348.01 Extra Large.

(**ii**)   The imprint has been removed (De, OS).

*Third Report . . . 1836.*

**Copies**   BL Official Publications Dept.; CUL OP. 1836.XXXVI; Bod Pp Engl. 1836/36.

Chubb CCCCLXXa.

**107**

**JAMES STEVENSON**                                             **1838**

Size: 54.5cm x 44.8cm.

In 1838 John Saunders issued a reduced version of Allen's *History of the County of Lincoln* in which the map described below is found. It has already been mentioned (**90**) that Saunders had already disposed of the original large map plate he had used for the earlier editions of Allen's work. The rarity of the new work and loose copies of the map suggest that sales were very poor, though the success of the larger work had no doubt encouraged Saunders to feel that there would be a call for a shorter version.[1]

The map was engraved by James Stevenson and this seems to be his sole attempt at engraving a county plate. Since, as the map freely acknowledges, the detail derives from the Ordnance Survey (O.S.) no great amount of fresh work needed to be put in hand. Nevertheless, the map represents a complete re-engraving of the O.S. sheets while reducing the large sheets to one quarter of their original size. In a second issue J. Addison claims responsibility for the drawing of the map. He is unknown in the world of county map production; he may be related to J. Addison, Junr., who engraved a map of Sussex in 1894 to illustrate Augustus Hare's *Sussex.*[2]

### Later history

In 1839 the map reappeared under the auspices of J. & J. Jackson of Louth. It seems only to have been issued as a loose sheet, perhaps in a cover. Jacksons were booksellers and stationers in Louth for many years. J. Jackson set up in the town in December, 1797, by taking over the premises of Mr. Marsh.[3] By 1827 there were two named J. Jackson operating the business when they published the first work of Alfred Tennyson in *Poems by two brothers* (by Tennyson and, in spite of the title, two of his brothers). The Jacksons gave Alfred and Charles Tennyson £20 in advance in books and cash; they hired a carriage to take them for a day at Mablethorpe to celebrate the publication.[4] The map was, thereafter, used to accompany editions of White's *History, Gazetteer And Directory of Lincolnshire* until that of 1882; the 1892 edition used a transfer provided by Bacon (**120 R**). White (1799–1868) was apprenticed to Edward Baines, who later became the owner of the *Leeds Mercury* and M.P. for Leeds (1834–1841). White formed his own company, using Baines' company to print his directories. He moved from Leeds to Newcastle-upon-Tyne in 1827, but, three years later, he moved to Sheffield, where his firm remained for the rest of its existence.[5]

## EDITIONS

(**i**)   *Map/ of the/COUNTY/OF/LINCOLN/Reduced FROM THE ORDNANCE SUR-VEY./LONDON./J. Saunders Jun$^r$. 7 Dyers Buildings./HOLBORN./1838.* (Aa). *Engraved by Ja$^s$. Stevenson, 2 Southampton Street, Pentonville.* (Ae, OS). *BOROUGHS/Returning 2 Members/*4 names in two columns, followed by a note on Grimsby's single member and two columns of names of *POLLING PLACES* (Ee). *REFERENCE TO THE HUN-DREDS/WAPENTAKES AND SOKES/* 31 numbered names (Ae). *SCALE OF 4 MILES TO AN INCH/12* [= 7.75 cm] (Be). *VIEW OF LINCOLN* (below a pastoral scene with the cathedral seen from the north-east) (Ea). The whole map is framed by a wide leafy border, inside which the minutes of latitude, etc. are numbered at 10′ intervals. *Meridian of Greenwich* (De). *1°. West Longitude.* (Ae). There is no compass.

*History Of The County Of Lincoln With Nearly One Hundred Illustrations On Steel And Wood, Maps, etc. In Two Volumes. Vol. I. London: John Saunders, Junior, 49 Paternoster Row. MDCCCXXXVIII.*

**Copies**   A private Lincolnshire library. (LRL Map 165a).

(**ii**)   The title has been changed to: *Map/of the/COUNTY/OF/ LINCOLN/REDUCED FROM THE ACTUAL SURVEYS/and Carefully Corrected to the Year/1839./PUBLISH-ED BY J. & J. JACKSON,/ LOUTH./* (Ae). *Drawn by J. Addison.* (Ee, OS).

Issued as a loose map (in a plain cover).[6] 1839.

**Copies**   LAO LLHS 15/4; Grimsby X000:912 JAC.

(**iii**)   The title has been changed to: . . . *LINCOLN/REDUCED FROM THE ACTUAL SURVEYS/and Carefully Corrected to the Year/1841./PUBLISHED WITH W. WHITE'S HISTORY/AND/DIRECTORY OF LINCOLNSHIRE./* (Aa).

Issued as a folded map in a green cover with the title: *MAP OF/ LINCOLNSHIRE./* [1842].[7]

**Copies**   LRL Map 213a; Grimsby RL X000:912 WHI; Scunthorpe Museum. LRL Maps 213 and 213b are loose maps without covers.

## LITHOGRAPHIC TRANSFERS

(**A**)   A new title reads: *W. WHITE'S/NEW MAP OF/ LINCOLNSHIRE,/ PUBLISHED WITH HIS/History, Gazetteer, and Directory/OF THE/ COUNTY OF LINCOLN./COL-LEGIATE CRESCENT, SHEFFIELD,/ 1856./* (Ea). The engraving of Lincoln has been replaced by *BOSTON CHURCH* (Aa). The signatures have been removed (Ae, OS and Ee, OS). *Swinderby/Halfway House* have been added to the list of Polling places (Ee). *THE NEW/DOCKS* have been drawn in at Grimsby. *BAWTRY* has been added and *From Bawtry* removed from the Gainsborough road. A compass has been added (Ac). Railways 1–4, 6–17, 19–24, 27 and parts of A, B, C, D have been added (most with the name of the railway); stations are marked by *Station* or *Sta*. Railways 21/24 and 23/27 have *INTENDED RAILWAY* marked on them. The new railways have led to the addition of names, e.g. *Essendine/Station, Aslackton Sta., Elton Station, New/Holland, Burgh Station*. Other places have been added, e.g. *N$^{th}$. Muskham, S$^{th}$. Muskham, Burghley Hall, Misson* (Notts.). Lines delineate *Embankment* (across Cross Keys Wash) and *Proposed/New Outfall* into Boston Deeps from Fosdike and the Witham. The boundaries of the hundreds have, in most copies, been coloured pink, blue, yellow and green, with the boundaries of the boroughs shown with fine blue lines.

Issued as a loose map. 1856.[8]

**Copies**  CUL Maps BB.18.G.30; LAO Dixon 18/5/11 etc.

Issued mounted and folded in a dark blue cover with the cover title: *LINCOLNSHIRE*. 1856.

**Copy**  The author's collection.

Issued in an emerald-green card cover with the title: *WHITE'S NEW MAP OF LINCOLN-SHIRE, WITH ALL THE RAILWAYS, AND OTHER IMPROVEMENTS. PRICE ONE SHILLING. SHEFFIELD: WILLIAM WHITE, 10, BANK STREET. LONDON:– SIMP-KIN, MARSHALL, & CO. AND ALL BOOKSELLERS*. 1856.

**Copy**  LRL Map 206.

Issued mounted on rollers. 1856.[9]

**Copy**  LAO Misc. Don. 96/4/5.

**(B)**  The place and date of publication have been altered to: *18 & 20, BANK STREET,/SHEFFIELD,/1872./* (Ea, below title). *NORTHERN DIVISION* and *SOUTHERN DIVISION* have been engraved across the map. The note (Ee) is now: *BOROUGHS/Returning 2 Members/ LINCOLN,/GRANTHAM BOSTON,/1 Member /GREAT GRIMSBY STAMFORD./PLACES OF ELECTION FOR/*3 columns under the headings for North, Mid and South Lincolnshire, with more places named. *EXPLANATIONS/*4 lines of symbols, etc. (Be, above scale). Railways 18, 25–26, 28–39, F and S have been added. Broken lines indicate the building of railways 41 and 43. Railways 20, 21/24, 22 and 23/27 have all been more accurately drawn. *INTENDED RAILWAY* has been removed from two places. *Station* has been removed from Essendine, while *Station* or *Sta*. have been added to new sites on redrawn lines and new railways.

Issued as a loose map. [1876].[10]

**Copy**  BL Maps 3355 (35).

**(C)**  Below the date in the title a hand stamp (in violet) has been added: *CORRECTED TO/1882*. (Ea). Railways 44–46 have been added and 41/43 converted from broken lines to continuous. The three headings in the lists of polling places are underlined in yellow, blue-violet and reddish-pink respectively and the boundaries of the three divisions have borders similarly coloured. The area, which roughly corresponds to North Kesteven, has its edges (the boundary is not marked) coloured green.

Issued mounted and folded into a dark blue-green cover with the title: *WHITE'S/ LINCOLNSHIRE*. [1882].

**Copy**  LRL Map 229.

**(D)**  The rubber stamp addition does not appear and the date in the title has been changed from 1872 to *1882*. The polling places, listed in three columns, have been removed and the remaining text has been re-engraved (Ee). Railway F has been extended. The map is in a pocket attached to the inside of the book's back cover.

*History, Gazetteer And Directory Of Lincolnshire, Including The City And Diocese Of Lincoln And Comprising A General Survey Of The County And Separate Historical, Statistical, And Topographical Descriptions Of All The Wapentakes, Hundreds, Sokes, Boroughs, Towns . . . Fourth Edition. By William White . . . Price, Cloth, 30s.; Half-bound, 35s. Sheffield: William White, Hoole's Chambers, Bank Street. London: Simpkin, Marshall, & Co. 1882 . . .*

**Copies**   BL P 2506 vi/11; CUL L475.c.18.4; Bod G.A. Lincs. 4° 16; Louth RL L 929; Gainsborough RL L 929; Grimsby RL X000:058 WHI. (LAO Misc. Don. 274; Nottingham UL Li 1: B15.E 82).

### References

1  In the only copy found so far the single volume contains two books; at the end of the volume appears 'End of Vol. 1'; if there is a second volume there would be little difference in size between the original edition and the reduced version.
2  Kingsley, p. 328.
3  *Lincoln, Rutland and Stamford Mercury* 29 December, 1797.
4  Martin, R.B. *Tennyson The Unquiet Heart.* (Oxford, 1980), pp. 44–45.
5  Lincolnshire 1872 . . . A part reproduction comprising Lincoln and eight wapentakes. Introduction [by] Michael Winton. (Kings Lynn, 1988), pp. vii–viii.
6  The LAO copy has signs of an original green cover, while the Grimsby one was purple.
7  Many copies of the 1842 . . . *Directory* . . . have been examined; none contains the map called for in its title.
8  No copy of the 1856 . . . *Directory* . . . has been found containing the map. When David and Charles reprinted the Directory (Newton Abbot, 1969) they did not include a map.
9  On the title-page of the 1856 Directory prices of copies are given and 'With the MAP ON ROLLERS, 4s. extra'.
10  Maps are not called for in the 1872 edition of the Directory. The BL copy was received on 2 DEC 1876. One other copy is known in a private library.

## 108

### JAMES PIGOT                                                              1839

Size: 16.78cm x 10.35cm.

Pigot's work has been discussed when considering the range of directories he produced from the 1820s (**88**). Alongside the directories he had put together the maps that accompanied the directories to produce an atlas from 1830. Perhaps following the union with Isaac Slater in 1839 a new set of reduced maps was prepared to form a new atlas, issued in parts. The first ten parts were issued fortnightly, probably from December, 1838, and they were followed by a further 29 parts at monthly intervals. Part 20 was issued before parts 18 and 19 on 26 November, 1839 and from that the dates of issue are deduced. Lincolnshire was Part 17 and was, presumably, issued in October, 1839. The final part appeared in August, 1841. The map is based on the earlier Pigot map but with the engraving of the Cathedral removed to appear above the text facing the map. The atlas only appeared in two editions.

### EDITIONS

(i)   *LINCOLNSHIRE* (Ca, in a panel set into the upper frame). *Pigot & Slater Engravers, Manch^r.* (Ee, OS). *PUBLISHED BY PIGOT & C^O LONDON AND MANCHESTER.* (Ce, OS). *Scale of Miles/10* [= 1.75cm] (Ce, in a panel set in the lower frame). *Explanation./5* lines of symbols, etc. (Ae). Compass (Ea). The inner frame is marked and numbered at 5′ intervals. *Mer^n. of Greenwich* (De). Railway A is shown.

*No. 17. Lincolnshire. Pigot And Co.'s Pocket Atlas, Topography, And Gazetteer Of England. Publishing In Counties, With Coloured Maps. Price Sixpence. Published Monthly By J. Pigot And Co. At their Directory Offices, 59, Fleet-Street, London, And Fountain-Street, Manchester.* [October, 1839].

**Copy**   Bod G.A. Gen. Top. 8° 479.

*A Pocket Topography And Gazetteer Of England: With Historical And Statistical Descriptions, And Distance, Parochial, Population, & Other Tables. Illustrated By Maps Of The English Counties, And Vignettes Of Cathedrals And Churches. By Pigot & Co . . . In Two Volumes. Vol. 1. London: Pigot & Co. Fleet-Street; Longman & Co. And Sherwood & Co. Paternoster-Row, And Simpkin And Marshall, Stationer's-Court; And Pigot And Slater, Manchester . . . . . . . . . Price, each Volume, 12s.6d.* [1841].[1]

**Copies**   BL 797 e 10; CUL S696.d.84.2–3; Leeds UL W 188a.

*A Pocket Topography . . .* [1850].[2]

**Copy**   BL 10348 c 11.

(ii)   The imprint has been removed. The signature has been redrawn: *I. Slater, Engraver, Manch^r.* (Ee, OS). Railways 1, 3, 5, 9–10 and 13 have been added. A line from Lincoln-Retford is also included.

[The text of the county (pp. [177]–208) preceded by the map and a single-sided leaf containing a mileage chart)]. [1850].[3]

**Copy**   LAO L910.52/8.

### References

1   The full set of maps was only available from August, 1841. There are numerous references in the text to events and data of 1841.
2   Harvey & Thorpe, p. 176 suggested 1850 on the basis of later railway data and this has been adopted by all later commentators, although the Lincolnshire map in state (ii) presents an anomaly. The BL copy was not received until 25 MY 1858. The text is unchanged.
3   Hodson, 1974, p. 152 suggests that the individual counties may have been issued separately without covers, since he recorded three such examples of Hertfordshire. The LAO example is in a home-made folder, with a cover title that suggests the contents come from *Universal Magazine* for May, 1752! The text pages are identical with those of 1841, but the map data suggest it was prepared c.1850.

Chubb CCCCLXII–CCCCLXIII. (Chubb cites an 1835 edition (copy in Bod) as well as one for 1842.

## 109

### W. WRIGHT, junior                                                        1840

Size: 17.2cm x 13.3cm.

Little is known of the engraver or the person responsible for the design. Wright engraved two plates in the book and is not recorded in connection with such work elsewhere. The second map shares the same general characteristics and shows England and Wales with data relating to the early history of the church. The county map may have been based on Arrowsmith's map for *Valor Ecclesiasticus* since the general outline of the east coast is

# LINCOLNSHIRE

showing the number of Parish or Village **CHURCHES** before the Conquest.

R. HUMBER · Barton · Barrow · Killingholme · Habro · Stallingbro · Grimsby · Aukborough · Ferraby · Ulceby · Aylsby · Laceby · Winteringham · Melton · Bradley · Scartho · Burton · Appleby · Barnetby · Waltham · Luddington · Sawcliff · Bottesford · Risby · Bigby · Somerby · Holme · Tonstall · Searby · Grasby · Growle · Scawby · Caistor · Ashby · Wold Newton · Cotes · Blyboro · Kirton · Nettleton · Saltfleet · Corringham · Norton · Claxby · Binbrook · Covenham · Springthorp · Oversby · Harpswell · Cainby · Kirmond · Elkington · Keddington · Fillingham · Normanby · Covenby · Toft · Raisen · Sixhill · Biscathorpe · Gayton · LOUTH · Stow · Frisby · Faldingworth · Linwood · Goldsby · Cukwell · Reston · Gayton in the Marsh · Aistholp · Thorpe · Hackthorn · Beningworth · Fristhorpe · Ranby · Withcall · Tathwell · Witherne · Muckton · Beesby · Scampton · Rand · Scothern · Wragby · Solby · Farforth · Burwell · Well · Bilsby · TORKSEY · N. Carlton · Riseholme · Barlings · Belchford · Tetford · Rigsby · Calceby · Willoughby · Newton · Broxholme · Burton · Willingham · Fiskerton · Ellington · Somersby · Langton · Claxby · Scremby · Dodington · LINCOLN · Greetwell · Bardney · Fulletby · Aswardby · Greetham · Hagworthingham · Skendleby · Welton · Akeley · Washingboro · Branston · Scrivelsby · Lusby · Partney · Ashby · Candlesby · Thorpe · Canwick · Hamworth · Roughton · Hareby · Hundleby · Bratoft · Hykham · Waddington · Harmston · Dunston · Moorby · Bolinbroke · E. Keal · Irby · Stapleford · Boothby · Metheringham · Thorpe · Marcham · W. Keal · Burgh · Bassingham · Wellingore · Blankney · Ruskington · E. Kirkby · Toynton · Steeping · NEWARK · Carlton · Welbourn · R. Witham · Revesby · Sibsford · Louisiham · Branswell · Kirkby · Hough · Normanton · Caythorp · Laythorp · N. Kyme · Dodington · Honington · Wilsford · Rauceby · Sleaford · S. Kyme · Leverton · Berkington · Quarrington · Heckington · Westborough · Syston · Haceby · Haydor · Helpringham · Boston · Belton · Bracely · Osbornby · Sredington · Bicker · Allington · Welby · Aswarby · Frampton · GRANTHAM · Repersley · Sapperton · Pickworth · Newton · Threckingham · Horbling · Barrowsby · Ingoldsby · Walcot · Folkingham · Lowton · Woolsthorpe · Somerby · Aslackby · Pepingale · Lowton · Dowsby · Bichfield · Dunsby · Gosberton · Bassingthorpe · Irnham · Ringsdon · Morton · Stroxton · R. Glen · Old Sea Bank · Spalding · Easton · Bourne · Ketel Abbey in Sutton · Witham · Toft · Fens · Tydd St Mary · Stretton · Braceboro · Baston · Crowland · Greetham · Stow · Uffington · Deeping · STAMFORD · R. Welland · BP Egerine Road · Salt Marshes

W. Wright Jun.r Sculp.t

1 Bartletts Place, Fetter Lane, Holbo

| | | |
|---|---|---|
| Boroughs *as* STAMFORD .............. ⚲ | Sites of Saxon *Monasteries as Crowland* + |
| Markets or Places of Toll as Spalding 8 | Villages *with Churches* .............. *Witham* + |

7. 1840 Map 109 – W. WRIGHT, jnr.

quite similar and, like the earlier map, the outlet of the River Nene is broader, with the eastern side of the river starting from Tidd St. Mary. A further similarity is the ecclesiastical content, purporting to show the villages with churches before the Norman Conquest. The author, Edward Churton (1800–1874), was educated at Christ Church, Oxford (MA in 1824). He became Rector of Crayke in Yorkshire in 1835 and, in 1846, Archdeacon of Cleveland.[1] Two of his sons were prominent churchmen also, one becoming Bishop of Nassau and another Archdeacon of the Bahamas.

### EDITION

(i)   *LINCOLNSHIRE/showing the number/of Parish or Village/ CHURCHES/before the Conquest./* (Ea). *W. Wright Jun^r. Sculp^t.* (Ae, OS). *1 Bartletts Place, Fetter Lane, Holborn.* (Ee, OS). *Boroughs as STAMFORD* [symbol] *Sites of Saxon Monasteries as Crowland +/ Markets or Places of Toll as Spalding 8 Villages with Churches . . . Witham +/* (Ae–Ee, OS). There is no compass.

*The Early English Church. By Edward Churton . . . London: James Burns, 17 Portman Street, Portman Square. M.DCCC.XL.*

**Copy**   BL 1154 d 3; CUL XIX.18.10; Bod 40.68.

### References
1   Crockford's Clerical Directory, 1872.

## 110

**Anonymous**                                                              **c.1840**

Size: 12.5cm x 10cm (approx.).

The British Library possesses maps, which are very similar in size and appearance, of Cornwall, Hertfordshire, Leicestershire, Hampshire and Middlesex.[1] Hodson discusses the map of Hertfordshire, which exists in two states; the first is dated to c.1840 and the second (the BL example) has railway lines opened to 1850.[2] The only features they have in common are the name of the county in an oblong panel, with bevelled corners, and a simple *Scale of Miles*. There are no compasses. The maps were based on the Walker maps of 1835, but the reduction is so great that similarities are not obvious. Railways built are shown by single continuous black lines and those under construction by broken lines. Lincolnshire may exist in one or more forms in a similar style.

### References
1   BL Maps 1820 (19); 2855 (20); 3260 (16); 2550 (26) and 3455 (32) respectively.
2   Hodson, 1974, pp. 152–153.

## 111

**JOSHUA ARCHER**                                                          **1842**

Size: 56cm x 34.6cm.

Archer was, at this time, a prolific engraver of maps. He produced the blocks for the maps used to illustrate *Guide to Knowledge* in 1833–4 (**98**). He also produced plates intended to appear in *The British Magazine, and Monthly Register of Religious and*

*Ecclesiastical Information*; the volumes from 1841–4 contain a series of maps of the dioceses, but after 17 plates, covering the dioceses in alphabetical order as far as Hereford, the whole thing petered out. While he was occupied with that series he was preparing two further sets of county map plates. The present series had only a short life, appearing in two editions, after appearing in parts over a number of years. These plates were his largest county maps and the Lincolnshire plate is sufficiently large for all but the southern tip of Rutland to be included, although Rutlandshire had been included on the very first plate in the series, that for Leicestershire. The map is based on Walker (1835). The maps appeared from early in 1842 from James Gilbert; by June the plan was taken over by Fisher, Son and Co. who saw the project through to completion in 1845. The first ten plates only were engraved by Archer; when Gilbert gave up the new owners soon turned to F.P. Becker to engrave the remaining plates.

In another work in parts a further series was engraved by Archer for *Curiosities of Great Britain* (**112**); he may also have prepared all the plates in a little atlas c.1852 (**119**).

## EDITIONS

(i)   *LINCOLNSHIRE./SCALE OF ENGLISH MILES/2 + 6* [8 = 5.15cm]/*Rectory* [symbol] *Vicarage* [symbol] *Perpetual Curacy* [symbol] Chapel of Ease +/ (Ea). *GILBERT'S COUNTY ATLAS* (Ae, OS). *V/London. Published May 1, 1842 for the Proprietor M. Alleis, by James Gilbert, 49, Paternoster Row.* (Ce, OS). *Drawn & Engraved by J. Archer, Pentonville London.* (Ee, OS). *Longitude West from Greenwich 0° Longitude East from Greenwich* (Ce–Ee, between borders). There is no compass.

*Fisher's County Atlas Of England And Wales. Compiled From Authentic Sources, And Corrected To The Present Time. With A Topographical And Statistical Description Of Each County . . . Fisher, Son & Co., Caxton Press, Angel Street, St. Martin's-Le-Grand, London; Post-Office Place, Liverpool; Piccadilly, Manchester.* [1845].[1]

**Copies**   BL Maps 29.c.17; Bod Allen 106; Leeds RL F 912.42.

(ii)   The imprint has been altered to: *V/Fisher, Son & C⁰. London & Paris.* (Ce, OS).

*Fisher's County Atlas . . .* [1845].

**Copies**   CUL Atlas 5.84.4; Bod Maps C.17.b.3.

*Fisher's County Atlas . . .* [1846].[2]

**Copy**   NLS Map 2 a 49.

### References
1   The whole set was not complete until 1845 (the date on the general map), although the parts were sold as they were ready.
2   The date in the general map has been changed to 1846.

Chubb DIV.

<h2 style="text-align:center">112</h2>

**JOSHUA ARCHER**                                                                          **1842**

Size: 23.45cm x 18.05cm.

Archer's work has been noted earlier (**98** and **111**). These maps were used to illustrate a series of works issued in parts at regular intervals. The title and large parts of the text were used by Thomas Dugdale from 1835 but the map plates used then were those of

Cole and Roper (**63 (v)**). Archer's maps are closely modelled on theirs. The work continued to appear for nearly ten years and then the plates were used to illustrate James Barclay's English dictionary and, finally, a topographical dictionary, edited by E. L. Blanchard, which appeared under several different titles apparently simultaneously. Blanchard (1820–1889) was well-known in the theatrical world, being responsible for the annual pantomime at the Theatre Royal, Drury Lane. He began work on his dictionary in 1852 and supplied Tallis, the publisher, with parts for which he received payments of £2 or £3 and, on one occasion, £4.16.0.[1] On August 16th, 1859 he records payment for part XLI; on September 24th, for part XLII and on December 17th he finished correcting and returned the proofs for the final part.[2]

The Tallis family had taken over the Cole and Roper plates as well as the text for *Curiosities Of Great Britain* in c.1835 and issued the work in two versions, only differing in the maps used during the 1840s. Bibliographical confusion centres on issue in parts, the publisher binding sections in 'Divisions' (Lincolnshire is usually in Division IV) and in volumes (usually three), often with two title-pages.

## EDITIONS

(**i**)  *LINCOLNSHIRE/SCALE/10 Miles* [= 2.8cm] (Ea). *Engraved for Dugdales England and Wales Delineated.* (Ce, OS). *Drawn & Engraved by J. Archer, Pentonville, London* (Ee, OS). *EXPLANATION/*14 lines of symbols, etc. (Ae). *DIVISIONS/1 Lindsey/2 Kesteven/3 Holland* (Ee). There is no compass. *24* (Ea, OS). The inner border is divided and numbered at 10′ intervals. *Meridian of Greenwich* (De, between borders).

*Curiosities of Great Britain. England and Wales Delineated Historical, Entertaining & Commercial. Alphabetically Arranged By Thomas Dugdale, Antiquarian, Assisted by William Burnett. Vol. III. London Published By L. Tallis, 3 Jewin Street, City.* [1842].[3]

**Copies**  CUL 8474.c.39–41 and S474.c.84.8–17; LRL (Pye) PL 9; Nottingham UL s/DA 640.D 8; Birmingham RL T 942.09 DUG; Leicester UL H 942 DUG. (LAO Misc. Dep. 333/1; Louth RL Map A24).

Issued as a collection of maps, without text. [1843 or later].[4]

**Copies**  CUL Atlas 7.84.19; Leeds UL W 185.

*The Universal English Dictionary; Containing An Explanation Of Difficult Words And Technical Terms, Whether In Algebra, Anatomy, Architecture . . . Also A Pronouncing Dictionary . . . An Epitome Of The History Of England . . . A Geographical Description . . . By The Rev. James Barclay. London; J.F. Tallis, 100, St. John Street, Smithfield; And 48, Dundas Street, Glasgow.* [1846].[5]

**Copy**  Manchester RL Q 423 B30.

(**ii**)  The inscription (Ce, OS) has been removed.
*Curiosities of Great Britain . . .* [1848].[6]

**Copies**  BL 010368 p 21; Leeds UL W 186; Liverpool Athenaeum 914.2. (LRL Map 796).

*The Universal English Dictionary . . .* [1848].

**Copy**  CUL S 760.D.84.1.

(**iii**)  Railways 1, 3–4, 6–7, 9–13, 15, 17, x (from Newark to Gainsborough), vii (to Folkingham only), viii and 21 (as a branch of the ANBEJR at Grantham) have been added and the majority are named. Some of the lines that were built are not shown accurately, notably the line to Barton on Humber continues to South Ferriby. A line connects Lincoln with Horncastle, passing south of Cherry Willingham and labelled *Horncastle Branch*.

*Curiosities of Great Britain . . . Vol. IV.* [L. Tallis, London, Edinburgh, Dublin, 1850].[7]

**Copies**   BL 010358 f 34; LRL 942; Nottingham UL s/DA 640.D8; Gloucester RL 913.4; Birmingham RL T 942.09 DUG; Leeds UL W 187. (Grantham RL Map 5; Gainsborough RL Map 3).

*Modern and Popular Geography; Or, A New And Complete History And Description Of All The Empires . . . Of The Habitable World . . . Celebrated Circumnavigators And Travellers . . . In Two Volumes. Vol. 1. By William Goldsmith . . . Embellished with Numerous Elegant Views, Engraved Portraits Maps, &c. Published By J. & F. Tallis, London, Edinburgh, & Dublin.* [1850?].

**Copy**   The private collection of Mr. C.A. Burden.

**(iv)**   *Horncastle Branch* and its line have been removed and, in the process, so has Horsington. Railways 8 (the Boston to Lincoln section of the GNR), 14, vi and xiii have been added.

*Barclay's Universal English Dictionary; Containing The Explanation Of Difficult Words* . . . [1850].

**Copy**   William Salt L. (Stafford) LE 1.4/1 RR.

**(v)**   Railway 16 has been added. Railways vi–viii and xiii, the northern part of x and 24 have been removed and in the processes the mileage *158* at Caistor; Ropsley has been removed also. Several names have also suffered damage in the removal process, most notably Bicker (now *Picker*) and Threekingham (now *Threckingham*). *STA* in sans-serif letters has been added on nearly all lines.

*Dugdale's England And Wales Delineated, Edited By E.L. Blanchard. Vol. II. London: L. Tallis, 21, Warwick Square, Paternoster Row.* [1858].[8]

**Copies**   BL 10351 g 22. (LRL Map 195b; Louth RL Map A25).

**(vi)**   Railway xii has been added.

*The Topographical Dictionary Of England And Wales. Edited By E.L. Blanchard. Volume II. London: L. Tallis, 21, Warwick Square, Paternoster Row.* [1858].[9]

**Copies**   CUL Atlas 6.86.20–21. (LRL Map 195).

**(vii)**   Railways 22–24 have been added; the Sleaford line leaves the GNR at Grantham. Railways 4 and 9 have been redrawn; a triangle of lines at Ulceby results and the extension to South Ferriby has been removed.

*Tallis's Topographical Dictionary Of England And Wales.* [London, Tallis, 1860].[10]

**Copy**   The private collection of Mr. C.A. Burden.

**(viii)**   Railways 2, 20 and 26 have been added.

*The Topographical Dictionary of England & Wales . . .* [1860].

**Copies**   The private collections of Mr. D. Kingsley and Mr. C.A. Burden. (LRL Map 195a).[11]

### References

1  Blanchard, E.L. The Life and Reminiscences. (London, 1891). Vol. 1, p. 217 (for Part XXXIX).
2  ibid., from his diary for those dates in 1859.

3 The work was completed by 1842. On. p. 1580 there is a reference to an event in Feb., 1841. There are minor variations in the punctuation and spelling of the title-pages of apparently identical editions. The sets are usually in 11 volumes. The Leicester set is in 10 coupled in pairs, so the map is in vol. 5/6.

4 Some of the plates (e.g. Yorkshire) show additional railways, suggesting issue after the [1842] volumes above. Whitaker implies that his set (Leeds UL W 185) was issued in 1842; Burden, p. 166 has suggested 1846, but agrees that may be too late.

5 The Chronology in the *Dictionary* ends with Nov. 21, 1846.

6 The receipt date in the BL copy is 13 Nov 48 (i.e. 1848). The title-page of the Leeds copy has no imprint, and the map is in vol. IV (text in vol. VI). The map in the Liverpool copy is also in vol. IV.

7 This volume has no imprint, which is supplied from vol. I. The date is surmised from those railway lines that are accurate.

8 This is one version of the work issued in parts and edited by E.L. Blanchard. The part with the county text and map was received in the BL on 15 JA 58 (i.e. 1858). The work was completed in December, 1859 and published in 1860. Bod G A Gen. Top. 4° 99 is a collection of the first 13 parts only (A–E).

9 This is the same text as *Dugdale's England & Wales delineated.* An engraved title-page *Tallis's Topographical Dictionary of England & Wales* is also included. CUL's set is in 2 volumes, but the set owned by Mr. C.A. Burden is in six, with Lincolnshire's text in volume III.

10 This copy is in one volume and only has a half title as quoted here, i.e. there is no imprint. The spine has *LONDON. L. TALLIS.*

11 Mr. Burden's set is in 9 Divisions; the county is in Division VI, which has the title-page *Curiosities of Great Britain . . .* At the end of Division IX are the title-pages intended for use in binding up the parts *Dugdale's England And Wales Delineated . . .* The title-page of Division I is *Tallis's Topographical Dictionary.* The interchangeability of the various titles for the same texts and maps is thus clearly shown.

Chubb CCCLXIV–CCCLXVII (quoting Fordham for an 1835 edition).


# 113

## VICTOR & BAKER                                                          1843

Size: 20.05cm x 12.8cm.

Victor and Baker were engravers in Newland Street, Lincoln, who also published the first work noted below. In the works noted below there is usually a map of the city of Lincoln, apparently their own work since it shows no debt to Marrat's map of 1817 or Padley's map of the city issued in 1841. The firm of Victor & Baker has proved elusive, not being listed in White's 1842 Directory nor in the early Post Office Directories. The plates must have passed fairly quickly to R.E. Leary, who used them in a series of little guide books, with a variety of titles. In White's *Directory* (1842) Robert Ely Leary's first premises were in Silver Street, Lincoln, but by 1842 he was at 20, Strait (White's Directory); Hagar's *Directory* (1849) has him at 19, Strait and he remained there until his son, also Robert, set up in Saltergate; he worked with him there in the late 1860s for a few more years until his retirement. The plate is based on a variety of sources; perhaps its general outline is closest to Pigot's smaller map (**108**). The inclusion of a *Light* in Boston Deeps is a novelty.

### EDITIONS

(i) *MAP/OF/Lincolnshire,/1843*. (Ae). *Victor & Baker, Engravers.* (Ee, OS). *SCALE/10 Miles.* [= 4.4cm] (Ee). The map shows railway i marked as *Projected Railway* passing from near Market Deeping via Swaton, Timberland, Lincoln to Gainsborough. In the borders are minutes of latitude, etc. numbered at 10′ intervals. *Meridian of Greenwich* (De, between borders). There is no compass.

*The Lincoln Commercial Directory And Private Residence Guide. Lincoln: Printed And Published By Victor And Baker, Engravers, Copper-Plate And Letter-Press Printers, Newland Street. 1843.*

**Copies**    LAO Sc 942.50; LRL LIN 929.

*A Topographical and Historical Description Of The City Of Lincoln, And Its Immediate Vicinity; Containing A Concise . . . Account Of Its Origin And Present State – Ancient Buildings And Relics Of Antiquity – Religious And Benevolent Institutions – Public Charities – Corporation Revenues And Places Of Amusement . . . Lincoln: Printed And Published By R.E. Leary . . . 19, Strait. 1843.*

**Copy**    LRL LIN. 9.LEARY.

(ii)    The date has been removed from the title. Railways 1, 3–4, 6–11, 13–14, 17 and F (extended from *Wisbeach* to join the GNR loop south of Sutterton and avoiding Spalding) have been added. A line from Grantham to [Melton Mowbray] is shown.[1] There are three other lines, which may or not be connected to railways; one leaves the ELR south of Alford, passes through Huttoft and reaches the edge of the frame (in the North Sea); another leaves the Lincoln-Newark line, south of Thorpe on the Hill and reaches westwards to the edge of the frame; the third is from Brigg through Crowle to the left hand edge of the plate.[2]

*The Hand Book Guide To Lincoln, And Business Intelligencer. Lincoln: Printed And published by R.E. Leary, 19, Strait. Also To Be Had Of R. Bulman, 259, High-Street; W.H. Bellatti, 28, Steep-Hill; W. Peck, Hermin Street, Lincoln; J. Noble, Boston; Jacksons, Louth; Bridges, Newark; Oliver, Nottingham; Bemrose, Derby; Creasey, Post-Office, Brigg, Tuxford, Grimsby. [1849].[3]*

**Copies**    Bod G. A. Lincs 8° 48 (1); LAO Sc. P5/2.14 910 Box; LRL N 9; Grimsby RL L400:942 HAM.

(iii)    The lines through Huttoft, from Thorpe on the Hill, from Brigg, from Grantham and railway F have been removed, but traces of them can be seen; some lettering is affected. The imprint has been removed (Ee, OS) but traces are clearly visible.

*The Hand Book Guide To Lincoln, And Business Intelligencer. Second Edition. Lincoln: Printed & Published For W. Doncaster, 45, Silver-St. [1851].[4]*

**Copies**    LAO Sc. P5/2.13 L. Linc. 910 (2 examples).

*The Illustrated Hand Book Guide To Lincoln. Lincoln: Printed And Published By R.E. Leary. [1860].[5]*

**Copies**    BL 10348 ccc 4 (3); Nottingham UL Li 3. D28 ILL.

*Lincoln as it is! What there is to see! And where to see it! Or, The illustrated hand book guide. Lincoln Printed and Published by R.E. Leary, 19, Strait. [1862].[6]*

**Copies**    LAO Sc. P5/2.17; LRL LIN 9.LEAR.

### References

1    A line was proposed in 1845 (the Lincoln-Birmingham direct) to pass from Lincoln, through Grantham, Melton Mowbray to Birmingham; the line marked here is perhaps a part of that proposal but it is curious, if the assumption is correct, that the Lincoln-Grantham section is not included. (Ruddock, J.G. and Pearson, R.E. The Railway History Of Lincoln; second edition. (Lincoln, 1985), p. 92).

2    ibid., pp. 90 and 92 show that a line was projected roughly where the line on the map is shown.

3  The latest date in the text is 1848 and the accurate railway data are consistent with the suggested
   date.
4  On p. 60 'in 1852 there will be an addition of £104' clearly implies issue in 1851. Bod Johnson
   f 1498 is an example of the Third Edition (published by R.E. Leary). It contains a city map
   and an 1826 map of the Fosdyke Navigation only.
5  An advertisement on the back cover refers to *Lincoln Date Book 1859* while a second refers
   to the issue in 1861 of *The original Poems, Tales, Plays . . . By E.G. Kent . . .*
6  On p. 7 the census figures for 1861 are quoted. LRL's copy has the given title only on its
   cover, while the title-page is as [1860].


# 114

## JOHN EMSLIE                                                          1848

Size: 23.68cm x 17.4cm.

Emslie (1813–1875) was apprenticed as an engraver and set up on his own in 1843. He
cooperated with Reynolds in the production of maps, but, when elected to the Royal
Geographical Society (1863), he was described as 'artist'. The maps are based on Hall
(**94**) and Reynolds even uses Hall's title for his first atlases. In 1860 the maps were issued
with colour overprinting illustrating geological strata, using a key at the front of the atlas.
In this form the work continued to be issued until 1927. Of 32 plates only 16 counties
have a plate to themselves. Reynolds was in business from 1825, moving to the Strand
in 1836 where he remained until his death in 1876. He had been elected to the Royal
Geographical Society in 1874, when one of his sponsors was the elder Stanford.[1]

## EDITIONS

(i)   *LINCOLNSHIRE/RAILWAYS/Open* [single line] *Constructing* [double line]/*ENG-
LISH MILES/10* [= 3cm] (Ea). *Drawn & Engraved by John Emslie* (Ae, OS). *Published
by J. Reynolds 174 Strand.* (Ce, OS). *20* (Ee, OS). Railways 1–4 are shown as built and
6–14, 16–17, vii (only to Folkingham), viii and 21 (as a branch of railway vii) are shown
as being under construction. Railway iv is marked but as a branch from Market Deeping
to join the GNR south of [Littleworth]. Compass (Ed).

*Reynolds's Travelling Atlas Of England: With All The Railways And Stations Accurately
Laid Down. Constructed From The Surveys Of The Board Of Ordnance, Railway
Companies, And Other Authorities. London: Simpkin, Marshall & Co. Stationer's Court;
And James Reynolds, 174, Strand. 1848.*

**Copies**   BL Maps 15.b.14; Bod G.A. Gen. Top. 8° 78; NLS Map 6.b.60.

(ii)   Railways 6–7 are now covered in black line to show that they are operational.

*Reynolds's Travelling Atlas . . . Strand.* [1848**].

**Copy**   CUL Atlas 7.84.4.

(iii)   Railways 8–17 are all coloured over in black line.

*Reynolds's Travelling Atlas . . .* [1853**].

**Copies**   CUL Atlas 7.85.3; Leeds UL W 202.

*Reynolds's Travelling Atlas . . .* [1857].[2]

**Copy**   CUL Atlas 7.85.5.

## LITHOGRAPHIC TRANSFERS

**(A)**   Size: 23cm x 17.1cm. The signature and imprint have been removed. + has been added at 16 places; the significance is not clear. Many are at market towns, but several market towns are not marked, while there are marks at West Rasen and Skegness. The outer frame lines have been removed, leaving the inner single line from the earlier plate.

*Portable Travelling Atlas Of England & Wales, Compiled From the Ordnance Survey, &c. Comprising The Following . . . Strand.* [1855**].

**Copy**   CUL Atlas 7.86.13.[3]

**(B)**   Size: 23.68cm x 17.4cm. The imprint, signature and original outer frames have returned. The map is overprinted in brown, yellow, orange, red-brown, light orange, turquoise, green and blue to show the geological strata/soil types; the fen areas are left uncoloured. The boundaries of the coloured areas are defined by fine single lines. Railways 19–26 have been added, though not quite accurately; railways iv, vii–viii, the line to Sleaford from Grantham and the spur joining the Lincoln-Grimsby line to the Grimsby-Gainsborough line at Brigg have been removed, leaving a second spur to Barnetby. Traces can be seen of the deleted lines. The crosses in state (A) have been removed. Plate number *20* has been moved (Ee).

*Reynolds's Geological Atlas Of Great Britain, Comprising A Series Of Maps In Which The Roads, Railways, And Geological Features Of England And Wales Are Accurately Laid Down; With A Geological Map Of Scotland. The Whole Compiled From The Most Authentic Sources. London: Published By James Reynolds, 174, Strand. 1860.*

**Copies**   BL Maps 28.bb.14; Leeds UL W 204; Nottingham UL s/QE 262 (Science L).

*Reynolds's Geological Atlas . . . Map Of Scotland. Preceded By A Description Of The Geological Structure Of Great Britain; Its Mineral Products: And the Agricultural And Water-Bearing Characteristics . . . London: James Reynolds, 174, Strand.* [1860].

**Copies**   Leeds UL Edward Boyle L. Reserve geology A–O. 091 REY; British Geological Survey Y 55 (410) REY 1860; Edinburgh UL SB 912 (42): 55 REY.

**(C)**   The map has reverted to that of state (A) in removal of imprints, frame lines. Railways 26–27 and 29, the latter as a double line (under construction), have been added. There is no geological colouring, although the lines marking the boundaries of the soil types remain.

*Portable Atlas Of England And Wales; With Tourists' Guide To The Principal Places Of Interest, Fishing Streams, Finest Views And Scenery, &c. Of Each County. Thirty Two Maps. London: James Reynolds, 174, Strand, W.C.* [1864].[4]

**Copy**   BL Maps 14.a.59.

*Portable Atlas . . . Reynolds, 174, Strand, W.C. And George Musgrave, Turnham Green, W.* [1864**].

**Copy**   NLS Maps 6.a.29.

**(D)**   The map is coloured for the soil types.

*Reynolds's Geological Atlas . . . A Series Of Maps Geologically Colored, Preceded By A Description Of The Geological Structure Of Great Britain, And The Geological Features Of The Several Counties Of England And Wales, Mineral Products, &c.; With Sections & Views. New Edition. London, Published by James Reynolds, 174, Strand.* [1864].

**Copies**   CUL Atlas 7.86.2; Leeds UL W 205; British Geological Survey Y 55 (410) REY 1864.[5]

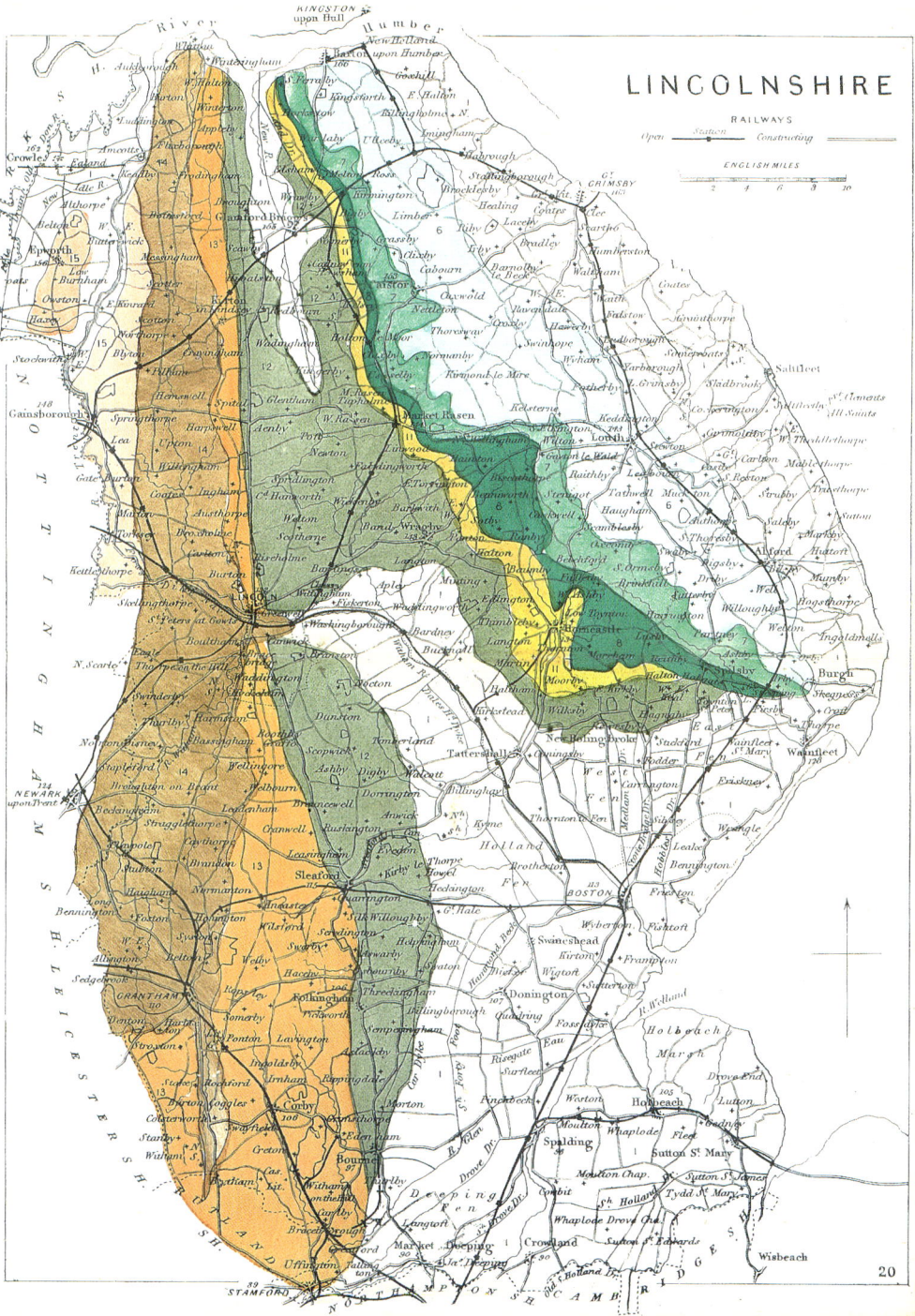

# LINCOLNSHIRE

RAILWAYS

*Open* *Station* *Constructing*

ENGLISH MILES

8.  1864   MAP 114D – JOHN EMSLIE/JAMES REYNOLDS.
By permission of the Map Library, British Library.

*Reynolds's Geological Atlas . . .* [1868].[6]

**Copy**   The private collection of Mr. E. Burden.

**(E)**   Railways 28, 31–35 and 37–38 have been added.

*Reynolds's Geological Atlas . . .* [1875**].

**Copy**   CUL Atlas 7.86.9.

**(F)**   Railways 36, 39–44, xvi and T (with part of U) have been added. The imprint: *London: Published by James Reynolds & Sons, 174, Strand.* has been added (Ce, OS). Notes round the edges of the map refer to geological deposits. A grid has been added and below it *Index to Sheets/of Geol. Ordce. Map* (Ed). A graticule has been added.

*Reynolds's Geological Atlas . . . Sections And Views. Second Edition. London: James Reynolds & Sons, 174, Strand. 1889.*

**Copies**   BL Maps 19.a.11; British Geological Survey AM 111.[7]

The work and maps appeared in new editions in 1903, 1907, 1913 and 1927.

**References**

1   Todd, p. 157; Herbert, F. The Royal Geographical Society's Membership . . . *Imago Mundi.* Vol. 35 (1983), pp. 67–95.
2   There are additions to several county plates, notably Yorkshire; CUL suggests 1857.
3   The numbering allocated to CUL's copy suggests publication in 1860 or later; the grounds are not clear.
4   1864 has been suggested by several commentators; the county railways support the idea.
5   The BGS copy has the title quoted; the others have a second title-page, which shows slight variations in spelling (e.g. Coloured) and lack *New Edition.* Bod Maps C.17.e.3 has no title-page but its contents are similar to the other examples.
6   Several other counties have new railways suggesting 1868.
7   A copy in the Science Library, Nottingham UL (s/QE 262) lacks the Lincolnshire map.

Chubb DXXVI, DXLII–DXLIV.

## 115

**LINCOLNSHIRE RAILWAY GUIDE**                **1849**

Size: 28.7cm x 21.95cm.

This anonymous map is included here for the interest of collectors and those interested in railway history; although it does not quite include the whole county (it excludes the coastal strip from North Somercotes in the north to Wainfleet) it provides a snapshot of the county at an important time in its industrial development. The Lincolnshire Railway Guide, from which the map apparently comes, has not been found; it was presumably a local publication and the map, no doubt, accompanied the timetables then currently in operation in the county.[1] The map has clearly been taken from a larger map, which would have included the coastal strip and the full county name. The map shows the area south to Rothwell (Northants.) and from Tamworth (Staffordshire) eastwards to Lutton.

### EDITION

**(i)** *MAP OF THE RAILWAY COMMUNICATIONS OF THE COUNTY OF LINCOLN,/FOR 1849,/SHOWING THE LINES COMPLETED AND IN PROGRESS;*

*WITH THE ELECTRIC TELEGRAPH LAID DOWN, PRINCIPAL STATIONS, &c./SCALE OF MILES./20* [= 5.2cm] (Aa–Ea, between the upper frame of the map and the outer frame of the plate). *Printed to Illustrate the Lincolnshire Railway Guide.* (Be–De, OS). Railways 1, 3–14, A, C, (part of) G, J and N are marked by single black lines (showing they were operational); broken lines indicate railways 15–17, 23, 27 (with ix) to be built. Similar treatment applies to lines wholly outside the county.

Issued in a volume not yet found.

**Copy** (LRL Map 758).

### Reference

1 Although the files of the *Lincoln, Rutland and Stamford Mercury* for 1849 have been scanned no reference has been found.

## 116

**FRANCIS PAUL BECKER and Co.**                                        **1849**

Size: 28.9cm x 21.3cm.

After Archer lost the contract to supply plates for *Fisher's County Atlas* (**111**) the remaining plates were prepared by F.P. Becker. He also seems to have supplied many of the plates that Kelly's Directories needed as their business burgeoned in the 1840s. Becker's plate of the county for Kelly's bears a great similarity to the Archer map prepared for the Fisher atlas, both taking much of their detail from the Greenwood map of 1834; material taken from Walker has been added also (**105 (vii)**). The maps are all produced by lithography. Becker was the inventor of the Omnigraph in the early 1840s; this was a machine, which speeded up production of large engraved plates. When a plate was passed through the machine various punches could be operated that would stamp out symbols, e.g. for trees, woods, hedges, and decorative devices. The machine was taken up by the Ordnance Survey, after a certain amount of wrangling about the payment Becker should receive. When the O.S. displayed the first map of Lancashire at 6 inches to a mile at the Great Exhibition of 1851 the large plates had been prepared using the newly acquired system. It lasted in use until c.1875 when the costs of retouching the punched symbols began to prove prohibitive.[1]

The first edition is the first map to mention Sutton Bridge. County maps appeared in all nineteenth century editions of the county directory, although they are missing from many of the copies. The plate varied in size by small amounts during the first part of its life, but a noticeably large transfer took place with the 1876 issue; thereafter the size was near to that quoted in the heading. *Lincoln Diocesan Church Calendar* for 1868 (printed by James Williamson of High Street, Lincoln) used Kelly's map instead of the usual illustration of a newly-built church. For the 1889 edition of Kelly's Directory a new transfer was made (**145**).

In 1800 the Post Office produced *New Annual Directory*, the title, in 1801, changing to *Post Office Annual Directory*. It was compiled by two inspectors of the Post Office (Ferguson and Sparke), who were joined by a third, Critchett, in 1803. In 1806 the latter took the whole work over. By 1835, when he sold out to Frederick Kelly, he had established a form of copyright ownership even though questions had been raised officially at his using for private profit Post Office staff to collect his directory information, a practice Kelly maintained until the early 1840s. Although his use of letter carriers was stopped he still kept Post Office in the title of his works, thus gaining an appearance of official status. From 1845 he began a series of county directories and he quickly spread outwards from the Home Counties to cover nearly the whole of the country; his rival Slater now confining himself to the northern counties.[2] After Slater finally closed Kelly's prospered even more; although the last Lincolnshire county directory appeared in 1937 the series of town directories including Lincoln resumed after the 1939–1945 war and only ceased publication in 1975.

## EDITIONS

(**A**)   *POST OFFICE MAP/OF/LINCOLNSHIRE./1849./Scale of Miles/12* [= 4.4cm] */Lincolnshire contains 2,748 Square Miles and 312,717 Inhabitants/* [symbol] *Denotes Polling Places/\* Post Office Money Order Towns/* (Ea). *Kelly & Cᵒ. Post Office Directory Offices, 19 & 20, Old Boswell Court, Temple Bar.* (Ce, OS). *Drawn & Engraved by F.P. Becker & Cᵒ. 11, Stationers Court City.* (Ee, OS). *REFERENCE to the WAPEN-TAKES/HUNDREDS &c/*31 numbered names, divided into the three parts (Ad–Ae). *The County returns to Parliament 4 Members/The Borough of Lincoln 2 Dᵒ/Grantham 2 Dᵒ./Boston 2 Dᵒ/Stamford 2 Dᵒ/Great Grimsby 2 Dᵒ/* (Ee). Railways 1, 3–4, 6–14 are shown by single continuous lines and are named; 2, 15–17, vii (to Spalding only and with 21 as a branch from Grantham) and x are shown by broken lines. Compass (Aa). On the ELR stations are marked and mileages measured from Boston northwards; on the Lincoln-Newark and Lincoln-Boston lines mileages are given, measured from Lincoln.

*Post Office Directory Of Lincolnshire, With Derbyshire, Leicestershire, Notting-hamshire, And Rutlandshire. The Maps Engraved Expressly For The Work London: Printed And Published By W. Kelly & Co., 19 And 20 Old Boswell Court, Temple Bar ...* [1849].

**Copies**   BL PP 2505 ycf; CUL L475.b.15.1; LAO NSR 950.2; LRL L 929; Scunthorpe RL L638 910.

*Post Office Directory Of Lincolnshire, The Map Expressly Engraved ... Subscribers Copy. Price Eight Shillings.* [1849].

**Copies**   Grimsby RL X000:058 KEL; Gainsborough RL L 929.

*Post Office Directory of Lincolnshire ... Price Ten Shillings.* [1849].

**Copy**   BL PP 2505 ycf.

(**B**)   The date in the title has been altered to *1855* and the population figure to *407, 222* (Ea). In the imprint the address has been slightly altered, thus: *19,20, & 21*. The number of MPs for Grimsby has been corrected to 1 (Ee). Railway 17 has been altered to a continuous line and named at its northern end. Stations are marked on the Lincoln-Brigg line. Railways A and C have been added.

*Post Office Directory Of Lincolnshire, With Map Expressly Engraved For The Work And Corrected To The Time Of Publication. London: Printed ... 19, 20 And 21, Old Boswell Court, Temple Bar. 1855 ...*

**Copies**   BL PP 2505 ycf; CUL L475.b.15.2; Bod Dir. Lincoln.d. 1/2; LRL L 929; Grimsby RL X000:058 KEL; Stamford RL R 914.253 KEL.

(**C**)   The date has been removed from the title (Ea). *Printed from Stone by C.F. Cheffins & Son London.* has been added (Ae, OS). Railways 2 and 15–16 are now marked by continuous lines; railways 19–22, 24 and 26 have been added.

*The Post Office Directory Atlas Of England And Wales. London: Published By Kelly And Co., 18 To 21, Old Boswell Court, St. Clement's, Strand; And Sold By ... Price Thirty Shillings.* [1861].[3]

**Copies**   BL Maps 3.d.4; CUL Atlas 5.86.1; Bod Map C.17.c.17; Leeds UL W 203.

**(D)**   The date *1861* has been added in the place of 1855 (Ea).

*Post Office Directory Of Lincolnshire . . . Published By Kelly & Co. 18 To 20, Old Boswell Court, St. Clement's, Strand, W.C. 1861 . . .*

**Copies**   BL PP 2505 ycf; LAO SR. 914.253 and Sc 942.50; LRL L 929; Grantham RL L 929; Scunthorpe Museum. (BL Map 3355 (32)).

*Post Office Directory Of Lincolnshire With Hull And Suburbs. With Map Expressly Engraved . . . 1861 . . .*

**Copy**   Bod Dir. Lincoln.d.1/3.

**(E)**   The year in the title has been changed to *1868*, the acreage has increased to *2,776* and the population to *412, 246*. The signature of Cheffins has been removed (Ae, OS). The reference to Post Office money order towns (Ea) has been removed although the symbols (asterisks) remain on the map. Railways 23, 25, 27, 29–30, 32–33, 35–36 have been added. Several of the new lines have their names added. All broken lines for railways never built have been removed. Railway 16 has been redrawn and now joins the line from Retford-Gainsborough while continuing west from East Retford.

*The Post Office Directory Of Lincolnshire. With Map Engraved Expressly For The Work. Edited By E.R. Kelly . . . London: Printed And Published By Kelly And Co., 12, Carey Street, Lincoln's Inn. W.C. (Removed From Old Boswell Court, St. Clement's, Strand.) MDCCCLXVIII. Price To Subscribers, Twelve Shillings. – Non-Subscribers, Fifteen Shillings.*

**Copies**   BL PP 2505 ycf; CUL L475.b.15.4; Bod Dir. Lincoln.d.1; LAO SR; LRL L 929; Grimsby RL X000:058 KEL.

*The Lincoln Diocesan Church Calendar, Clergy List, And General Almanack, For the Counties Of Lincoln And Nottingham, For The Year Of Our Lord 1869. Published With the Sanction Of The Bishop. Lincoln: Brookes & Vibert, 290, High-Street. And Sold By J. Bell, And C.J. Sisson, Nottingham. London: J. Parker & Co.*

**Copy**   CUL L.125.11.d.L4; Bod Cal. 11126 e 142; Grimsby RL X000:262.3 LIN.[4]

**(F)**   Size: 40.85cm x 29.95 cm. The date in the title has been changed to *1876* and the population figure to *436,599*; the note and symbol for polling places have been deleted (Ea). The symbols for polling places and Post Office money order towns have been removed from the map. The imprint now reads: *Kelly & C⁰. Post Office Directory Office 51 Gᵗ. Queen Street Lincoln's Inn Fields.* (Ce, OS). Becker's signature has been taken out. In the note on MPs Stamford now has 1 Member (Ee). Railways 28, 34, 38–41 and 43 have been added.

*The Post Office Directory Of Lincolnshire . . . Printed And Published By Kelly And Co., 51 Great Queen Street, Lincoln's Inn Fields, W.C . . . MDCCCLXXVI Price To Subscribers Twelve Shillings. – Non-Subscribers, Fifteen Shillings.*

**Copies**   BL PP 2505 ycf; CUL L475.b.15; Bod. Dir. Lincs.d.1; LRL L 929; Grantham RL L 929; Scunthorpe RL L638 910. (LAO Scorer Map 48 and Ex 32/2/160).

*The Post Office Directory Of Lincolnshire And Nottinghamshire. With Maps Expressly Engraved . . . MDCCCLXXVI Price To Subscribers, Eighteen Shillings. – Non-Subscribers, Twenty-Two Shillings.*

**Copies**   LRL L 929; Scunthorpe Museum.

**(G)**   Size: 28.85cm x 21.45cm. The title is: *LINCOLNSHIRE./ 1884./Scale of Miles/12 [= 4.5cm]/Lincolnshire contains 2,776 Square Miles and 469,919 Inhabitants/* (Ea). The imprint is: *Kelly & C⁰. Post Office Directory Office 51, G*ᵗ*. Queen Street London W.C.* (Ce, OS). Railways 31, 42, 44–46 and x have been added. The area around Sutton Bridge has been redrawn; *Cross Keys/Wash* and *Sutton Bridge* have been removed, the road and railway to Kings Lynn are now on reclaimed land and *Breast/Sand* has been added. The railway at Crowle has been re-aligned and is broken by the name of *Tetley*. Re-alignment has also taken place at Bourn and the town is no longer encircled by railways.

*Kelly's Directory Of Lincolnshire With The Port Of Hull And Neighbourhood With Map Of The County. Edited By E.R. Kelly . . . Printed . . . By Kelly & Co . . . MDCCCLXXXV. Price . . . Twenty-Two Shillings. – Non-Subscribers, Twenty-Eight Shillings.*

**Copy**   Bod Lincs.d.2; LRL A 929; Scunthorpe RL L638 910.

*Kelly's Directory Of Lincolnshire. With Map. Edited By E.R. Kelly . . . MDCCCLXXXV. . . . Fifteen Shillings. – Non-Subscribers, Eighteen Shillings.*

**Copies**   Grantham RL L 929; Gainsborough RL L 929; Peterborough RL Dir.

**References**

1   Wakeman, G. Aspects Of Victorian Lithography. (Wymondham, 1970), pp. 39–40.
2   Norton, op. cit., pp. 145–146; Shaw and Tipper, op. cit., pp. 9–10 and 51–55; Todd, p. 111; Brown, p. 104; Smith, D. pp. 141–142.
3   The county map is identical in all essentials to state (D); an advertisement leaf refers to several items issued in 1861.
4   The LRL copy (L 052) has Becker's map of Nottinghamshire instead of Lincolnshire.

Chubb DXLVI (the 1861 atlas).

**1i7**

**JOHN MURRAY**                                                          **1850**

Size: 15.85 x 11.1cm.

During the years 1843–1870 the Royal Agricultural Society of England published maps of most counties in its *Journal*. 12 counties were not dealt with and there seems to have been no pattern to those that were printed. Very few of the maps, which are geological in nature, are signed and they contain no clue to their origins. The map consists of strata shown by various types of shading, rivers, numbers (for the towns) and letters (for the types of soils). The county map illustrates the prize article *On the Farming of Lincoln-shire*, which the author published separately in the following year. The author wrote on agricultural topics from 1852 to c.1867.[1]

**EDITION**

**(i)**  *GEOLOGICAL   MAP/OF/LINCOLNSHIRE./EXHIBITING   THE   DIFFERENT DRAINAGES/AND THE COURSES OF THE PRINCIPAL/STREAMS, TOGETHER WITH THE/CANALS, RAILROADS, ETC./ CAREFULLY CONSTRUCTED/IN THE YEAR/1850./* (Ea). Compass (Ee). *Numbers and corresponding Towns on Map of Lin-colnshire./*53 numbered names in 6 columns, followed by 15 shaded rectangles, lettered A to O indicating different soils and rocks; *Canals, Rivers, &c., marked thus* [single black line] *Railways* [broken lines] *Embankments* [alternating dots and dashes] (in an unframed area below the map).

*The Journal Of The Royal Agricultural Society Of England. Volume The Twelfth. Part 1 . . . No. XXVII–1851. London: John Murray, Albemarle Street.*

**Copies**   BL Ac 3485; CUL P 440.c.29.12; Bod Radcliffe Soc 19195 e 298/12.

*On The Farming Of Lincolnshire. By John Algernon Clarke. Prize Report. London: Printed By William Clowes And Sons, Stamford Street. 1852.*

**Copies**   BL 7076 c 12; Bod G.A. Lincs 8° 17; LAO R.L 631 CLA; LRL L 63; Grimsby RL X000:630 CLA; Louth RL GL 63.

**Reference**

1   These are the dates attached to his works in the British Museum Printed Catalogue.

Chubb DXII.


## 118

**WILLIAM ROCK**                                                    **1850**

Size: 24.2cm x 18.3cm.

William and Henry Rock formed a firm in 1834, which specialised in Christmas cards and topographical steel engravings. The series of prints runs into thousands; Stamford and Lincoln were the subjects of such collections.[1] Other views in the county also appeared. The firm continued in business until the end of the century.[2] Some of the quality of the firm's work can be gauged from the maps they produced in the early 1850s.[3] The maps are closely related to Hall (**94**) but the towns are indicated by tiny views in circles only 80mm. across, each one distinctive and, in some cases, hand coloured. There are thirteen such in the county, plus Newark. The firm also engraved a map of London for James Reynolds in 1860.[4] The identity of the author of *A Farming Tour* . . . has not been established; the book is not listed in any of the standard dictionaries of anonymous and pseudonymous literature.

## EDITIONS

**(i)**   *LINCOLNSHIRE./Scale of Miles/12* [= 3.6cm] (Ea). *Rock & Co. London, April 1850.* (Ae, OS). The inner border is marked in two minute sections and numbered every 20. *West of Greenwich* (Ae, between borders). *East of Greenwich* (De, between borders). There is no compass. Railways 1, 3–4, 6–14 are shown.

Issued in cover with the title *MAP OF LINCOLNSHIRE. ILLUSTRATED. 1/–* [1850].[5]

**Copies**   CUL Maps d.18.G.76; Grimsby RL X000:912 ROC.

**(ii)**   Railways 2, 15 and 17 have been marked by single lines. Broken lines mark a projected railway from Grantham-Sleaford.

*A Farming Tour, Or Hand Book On The Farming Of Lincolnshire. By A Lindsey Yeoman. London:– Simpkin, Marshall, And Co. Stationer's Court. Market Rasen:– J.G. Caborn, Bookseller, Market-Place. 1854.*

**Copies**   BL 7075 d 33; Bod G.A. Lincs 8° 17 and 23(9); LAO Ex 942.51; LRL A 63; Nottingham UL Li 1.P14 FAR; Grantham RL L 63.

**References**

1   Elvin, L. Lincoln As It Was, Volume III. (Nelson, 1979), p. [4] illustrates three of Rock's views of Lincoln from an album in LRL.
2   Adams, B. London Illustrated, 1604–1851. (London, 1983), pp. 478–479 and 496–497.
3   Three other county maps are known so far: Sussex, Kent and Derbyshire.

4   Hyde, R. Printed Maps of Victorian London, 1851–1900. (Folkestone, 1975), item 54–4.
5   The cover of the CUL copy is black and Grimsby's is dark green; the copy of Derbyshire
    (Derby RL. Local Studies Dept. 8307) has a pale yellowish mock-vellum cover. It was
    published by Rock Brothers & Byre at 6d.

# 119

## JOSHUA ARCHER                                         c.1852

Size: 7.6cm x 5.35cm.

The work of Collins, the publisher, has been encountered when making lithographic
transfers from earlier plates, during the period 1848–1858 (approx.) (**62A, 71A–C, 92A**).
Although the map has no sign of its origin the series has been assigned to Joshua Archer,
on the basis of his signature on the general map. The county maps have been individually
transferred from a large map of England and Wales, which appeared in at least two atlases
in the 1850s. For Archer's earlier maps see **98, 111–112**. The Lincolnshire plate also
includes Rutland and areas as far south as Mildenhall (Suffolk) and west to [Re]tford.

## EDITION

(i)   *Lincoln./17*. (Ca, OS). There is no scale or compass. Railways 1–13, A, C, D, H, I,
J, K and N are shown by thick continuous lines; 14–16 by a thinner line and 17 by a quite
fine line. Railways in other counties are also similarly delineated.

*Collins' Pocket Ordnance Railway Atlas of Great Britain. H.G. Collins, 22, Paternoster
Row, London.* [1852].[1]

**Copy**   BL Maps 1.aa.6.

### Reference
1   Chubb suggested 1852, which is generally accepted. Lincolnshire railway data could support
    this, only if the way the three types of line used to delineate railways descend from thick for
    open lines, thinner for lines being built to fine lines for planned lines. A better suggestion
    might be c.1850.

Chubb DXXXV.

# 120

## JOHN DOWER                                            1858

Size: 43cm x 31.2cm.

The maps for this atlas were originally produced directly on the stone and there was no
intaglio version. *The Weekly Dispatch* included a series of county maps between 1856
and 1862, the artists of the majority of the plates being B.R. Davies, Edward Weller and
John Dower, who prepared the map of Lincolnshire. The county map, which is modelled
on those of the Ordnance Survey, first appeared in the issue for August 15th, 1858; when
the series was complete (there were very detailed maps of railways, including two plates
delineating the route of the Great Northern Railway) they all appeared as an atlas in 1863.

The plates soon passed to the publishers, Cassell, Petter and Galpin. John Cassell (1817–1865) came into publishing rather oddly. After a very deprived childhood and youth in the industrial north he became a keen supporter of the teetotal movement. After marrying Mary Abbott, whom he met in Lincolnshire during one of his tours lecturing against the evils of drink and who provided the money, he founded a tea and coffee business (the positive alternative to alcohol). Then he moved into publishing to pursue his reforming activities, mostly by means of weekly magazines. He was joined as his partners by George William Petter and Thomas Dixon Galpin in 1858 and, as Cassell & Co., the firm has remained in publishing until the present day.[1] Petter and Galpin were one of the firms of printers used by Cassell during the 1850s. The father of the engraver, John Dower (1825–c.1888), was also an engraver (John Crane Dower), who had worked for Greenwood, Teesdale, Moule and Bradshaw (famous for his railway timetables) and published his own *A New General Atlas of the World* in 1831. The son engraved the plates for *Lewis's Atlas . . . of Ireland . . .* (1837, etc.), the equivalent to Lewis's *Topographical Dictionary of England . . .* (**93**), Padley's map of Lincoln (1841, the plate being used for editions until 1883) and *PLAN of the BOROUGH OF STAMFORD . . .* by James A. Knipe (1833). He also prepared plates for *Bacon's Illustrated Complete Atlas* (1871). He was elected to the Royal Geographical Society in 1854 but his name was removed in 1867.

Within a few more years the plates had passed to G.W. Bacon and Co. This well-known firm of mapsellers was formed in 1863 by George Washington Bacon (1830–1922), who, not surprisingly, began with maps associated with the Americas. Edward Stanford proposed him as a Fellow of the Royal Geographical Society in 1866; the following year he went bankrupt but by 1869 he was back in business and produced his first county atlas.[2] The plates remained in use well into the present century and were as much used for touring, cycling and motoring folding sheets as they were for atlases. His whole efforts concentrated on the production of the same plates in as many ways as possible; apart from the addition of new railways many of his basic plates remained in use for over forty years.

**EDITIONS**

(**A**) A drawing of Hermes, above a hemisphere, balanced on a banner, which bears the words: *THE DISPATCH ATLAS/LINCOLNSHIRE./* and below *British Statute Miles./10* [= 5.3cm] (Ae). *Weekly Dispatch Atlas 139 Fleet Street.* (Ae, OS). *Drawn & Engraved by John Dower. Pentonville London.* (Ee, OS). *Day & Son Lith^{rs} to The Queen* (Ce, OS). *Lincolnshire contains 2729 Square Miles/returns 2 Members to Parliament for each/ Division of the County and 2 each for/the Boroughs of Lincoln, Boston, Grantham,/ Stamford & 1 member for Great Grimsby./3* lines of text and symbols concerning the county boundary, divisions of the county, railways and stations (Ea). Compass (Eb). Between the inner and outer frames (Ee) appear numbers, which are noted in brackets after the locations below in a distinctive type. Their significance has not yet been determined. The inner border is marked with each minute of latitude, numbered every 10. *Meridian of Greenwich* (De, between borders). Railways 1–4, 6–22, A, C, I and M are marked.

Issued as a loose sheet with the newspaper stamp *WEEKLY DISPATCH/OF/SUN-DAY/AUG^{ST}. 15TH/1858/* the letters appearing as white on pink. (Aa, OS).

**Copy** The author's collection (*3*).

*The Dispatch Atlas. London: Published At The 'Weekly Dispatch' Office, 139, Fleet Street, London. And Sold By . . . 1863.*

**Copies** BL Maps 43.f.5 (*2*); CUL Atlas 3.86.6 (*3*); Leeds UL W 206 (*3*); LAO HD 67/27 (*3*)[3]; Birmingham RL AE 912.42 (*9*); RGS 1 A 55 (*10*). (LAO FL Box L 912 (*3*); the author (*7*)).

*Cassell's Complete County Atlas Containing Two Hundred and Sixty Folio Maps, (Beautifully Engraved And Coloured,) Presenting A Full And Most Accurate Survey Of The World's Surface. London: Cassell, Petter, And Galpin, La Belle Sauvage Yard, Ludgate Hill, E.C.* [1865].[4]

**Copy**   CUL Atlas 3.86.17 (*11*).[5]

(**B**)   *24* has been added (Ea, OS).

*Cassell's County Atlas: Comprising Fifty Beautifully Engraved and Coloured Maps Of The Counties Of England And Wales. Including Maps Of Liverpool, Manchester, And Birmingham. London: Cassell, Petter, And Galpin, La Belle Sauvage Yard, Ludgate, E.C.* [1866].[6]

**Copy**   NMM 912.44 (42) ″18″ B 5553 (*6*).

(**C**)   The Hermes and hemisphere symbols have been removed and *BY JOHN DOWER F.R.G.S.* added after *LINCOLNSHIRE* (Ae). The note on the lithographer has been replaced by *LONDON, PUBLISHED BY CASSELL, PETTER & GALPIN, LA BELLE SAUVAGE YARD, LUDGATE HILL. E.C.* (De–Ee, OS). The signature (Ae, OS) has been removed. *20* (Ee, OS to the right) has replaced the plate number *24* (Ea, OS). *AREA.–1,775,457 Acres, or 2,774 Square Miles./ POPULATION (at Census of 1861), 412,246; increase in decennial period 1851–61, 5,024/or 1 per cent . . . / . . . /* followed by tables for marriages, births and deaths, paragraphs on physical geography, Divisions, Parliamentary representation, population of principal towns, education and crime (Ea–Eb). The compass and check digits have been removed. Railways 23–24, 26–27 and L have been added. Yorkshire east of *Winestead*, the name *Donna Nook*, the ends of several other place-names and the descriptive material have been removed to allow the inclusion of the above tables (Ea–Eb).

*Cassell's Illustrated Family Paper. Vol. XIV.– New Series. London: Cassell . . . 1864.*

**Copy**   BL PP 6004 da.

Issued in Johnson & Co.'s Lincolnshire Directory 1864.[7]

**Copy**   Grimsby RL X000:058 JOH.

Issued as an atlas without title-page. [c.1865].

**Copies**   Leeds UL W 208. (LRL Map 69; Grimsby RL 1864–7 Cas; Nottingham UL Li 1).

Issued in a cover with the title *PRICE FOURPENCE CASSELL'S COUNTY MAPS LINCOLNSHIRE ROAD AND RAIL. LONDON: CASSELL, PETTER, AND GALPIN . . .*[8]

Issued in a cover with the title *CASSELL'S SHILLING COUNTY MAPS LINCOLNSHIRE. MOUNTED ON LINEN LONDON. CASSELL, PETTER & GALPIN LA BELLE SAUVAGE YARD, LUDGATE HILL, E.C.*[9]

*Cassell's British Atlas: Consisting Of The Counties Of England, With Large Divisional Maps Of Scotland, Ireland, And Wales; Copious Maps Of All The Principal Routes Of Railway Throughout The Country, With indications of every object of importance and interest to the Traveller along the Lines; Separate Maps Of Cities, Towns, And Places Of Importance; The Great Map Of London . . . A Facsimile Of Ralph Aggas's Map of Old London, As It Was In the Time Of Queen Elizabeth. London: Cassell, Petter, And Galpin, La Belle Sauvage Yard, Ludgate Hill, E.C.* [1867].

**Copy**   CUL Atlas 3.86.7.

(**D**)  The imprint has been resited; it now has the first word *LONDON* starting below the fourth minute west of O°; previously the first word commenced below the sixth minute west.

*Cassell's British Atlas . . .* [1867].[10]

**Copies**  BL Maps 4.e.11; Bod Maps C.17.a.2; Leeds RL F 912.42 C272; Manchester UL AMGS C46.

*Mercer And Crocker's General, Topographical, and Historical Directory And Gazetteer For Lincolnshire With Hull, &c. Comprising Classified Lists Of The Nobility, Clergy, Gentry, Bankers, Merchants . . . with an . . . Account Of Each Town And Village . . . Subscriber's Copy, Price 15s. Non-Subscriber's, Price 21s. Hull: Mercer And Crocker, Printers And Publishers, 1870.*

**Copy**  LRL L 929.

(**E**)  Size: 59.2cm x 43.1cm approx. The title now is: *BACON'S/MAP OF/LINCOLN-SHIRE/BY JOHN DOWER F.R.G.S./British Statute Miles./10* [= 7.4 cm] / *Lincolnshire Contains 2729 Square Miles/ returns 2 Members to Parliament for each/Division of the County and 2 each for/the Boroughs of Lincoln, Boston, Grantham,/Stamford, & 1 Member for Great Grimsby./* (Ea). 3 lines on the boundary markings and signs for railways (Eb). *G.W. BACON & C⁰. 337, STRAND, LONDON.* (Ce, OS). The map is coloured – the North Division is pink, Mid Division yellow and South Division green. Railways 25, 28–37 are marked in blue (already operating), 39 in red (under construction) and xiv as double uncoloured lines. *Donna Nook* and the Yorkshire coast to Spurn Head have been added again and the place-names, which were curtailed now re-appear in full (e.g. *SALTFLEET*).

*Bacon's County Atlas: Comprising Forty two Beautifully Engraved and Coloured Maps Of The Counties Of England And Wales. Also Including Maps Of Liverpool, Manchester, And Birmingham. Contents . . . London: G.W. Bacon & Co. 337, Strand, Opposite Somerset House. 1869.*

**Copy**  CUL Atlas 3.86.10.

Issued mounted on linen on a hanging rod, with the boundaries of the *NORTHERN DIVISION* and *SOUTHERN DIVISION* edged in purple and green respectively; the boundary of *MIDDLE DIVISION* is not coloured. Railways are marked in black-violet, but blue if under construction (35, 39 and xiv). There are marks at 10′ intervals round the entire inner frame (previously the top frame had no marks). [1869].

**Copy**  LAO 3 Cragg 1/24.

Issued in a card cover with the title *BACON'S NEW TOURIST MAP OF LINCOLNSHIRE FROM THE ORDNANCE SURVEY, WITH THE NAME OF EVERY RAILWAY STATION Cloth Case 1s* [Paper] *Price 6d On Cloth 2s. London. G.W. Bacon & C⁰. Ordnance Map Agents, 127, Strand . . .*[11]

(**F**)  Size: 60.3cm x 43.2 cm. The title has been moved to (Ea). Below the title *Spurn Head* is shown, separated from the rest of Yorkshire, and, below that, *Boundary of County* [symbol]/further lines of explanation and symbols. Imprint: *G.W. BACON & C⁰. 337, STRAND, LONDON.* (Ce, OS). The overprint in central Lincolnshire is *MID DIVISION*; the north division is now coloured pink, the mid yellow and the south blue-green; in each case the boundaries are coloured in deeper shades. The sea (along the coast) is coloured blue. The railways are coloured red, except a portion of the Barkston-Sleaford line at Normanton. There is a spur at Bourn pointing northwards (railway 39 is shown as two parallel lines). The engraving of railway 23/27/29 is broken at Sutton Bridge by the River Nene. Railway has been added; railway xiv has been removed.

*Bacon's Large Print County Atlas: Comprising Forty Beautifully Engraved and Coloured Maps of the Counties of England And Wales. Including A Large Scale Map Of The Environs Of London. Contents . . . London: G.W. Bacon & Co., Agents . . . 127, Strand, 20 Doors West Of Somerset House. 1870.*

**Copy**  (The author's collection.)[12]

**(G)**  Size: 60.9cm x 42.8cm. The material below the scale has been moved to (Ae) and replaced by a view, in an oval, of Lincoln cathedral from the castle wall. Railways 38–39 are shown as open and a Louth-Lincoln line, which joins the GNR between Washing-borough and Lincoln is shown by two parallel lines. There is no imprint. The Yorkshire eastern coast has reverted to that of state (C).

Issued mounted on rollers. [c.1870].[13]

**Copy**  Scunthorpe Museum.

Issued in a card cover, with the title: *Bacon's New Large-Print Map Of Lincolnshire, Reduced From The Ordnance Survey Showing The Parliamentary Divisions. Also, all the Railways . . . Price: Coloured, on Roller, Varnished, Cloth Back 4s.0d. . . . on Cloth, in Case 4s.0d. Folded in Paper boards 1s.0d. London: G.W. Bacon & Co . . . 127, Strand. 1870.*[14]

**(H)**  The imprint has been added again in the form: *G.W. Bacon & Cº. 127, STRAND, LONDON.* The map is otherwise as state (F).

*Bacon's Large Print County Atlas . . . 1870.*

**Copy**  The private collection of Mr. T. Burgess.

Between c.1871 and 1875 maps of other counties are known in a variety of covers, with altered titles and other changes. If Lincolnshire existed in such formats, the cover titles could have included: *LINCOLNSHIRE FROM THE ORDNANCE SURVEY*[15], *BACON'S NEW POCKET MAP OF LINCOLNSHIRE FROM THE ORDNANCE SURVEY . . . In Cloth case 6d.– On cloth 1ˢ.*[16] and/or *BACON'S NEW TOURIST'S MAP OF LINCOLN-SHIRE FROM THE ORDNANCE SURVEY.*[17]

**(I)**  Size: 42.9cm x 30.92cm. The title is *BACON'S/MAP OF/ LINCOLNSHIRE./BY JOHN DOWER F.R.G.S./* (Ae, with the scale below). *Note – The official names of Railway Stations are all engraved/in a special character, thus Ludgate, except where indicated/by the name of the town being in bold letters and identical with/the name of the station/Lin-colnshire Contains 2729 Square Miles/returns 2 Members . . ./*5 further lines on MPs, 3 of symbols for boundaries and 1 on railways and stations (Ea). Compass (Eb). Because a smaller type has been used for the notes the whole coastline of Yorkshire is now shown. *Bacon's Map Establishment, 127 Strand, London* (Ae, OS). Many place-names are now in the distinctive sans-serif type to indicate the presence of a railway station and new stations have been indicated for the first time. Railway 40 is shown, but the double lines for railway 41/43 have been removed. Railway P has been removed. Railway 23/27/29 is now continuous from Spalding to Kings Lynn. Page number *32* is stamped (Ea, OS). The boroughs are coloured orange-brown, and the Northern, Mid and Southern Divisions green, pink and yellow respectively.

*Bacon's New Quarto County Atlas: Comprising 55 Beautifully Engraved and Coloured maps Of The Counties Of England And Wales. Contents . . . London: G.W. Bacon & Co., Agents . . . 127, Strand, W.C. (20 doors west of Somerset House.)* [1876].[18]

**Copy**  BL Maps 12.e.7.

Issued in a cover with the title *BACON'S NEW POCKET MAP OF LINCOLNSHIRE.*[19]

**(J)**   Size: 45.7cm x 31.1cm. The map has been newly transferred and shows more of Yorkshire, north to *BEVERLEY*. In extending northwards the extra mark between the frame lines (Ea) is shown as *20'* (it should be 50'). The title is *LINCOLNSHIRE./ REDUCED FROM THE ORDNANCE SURVEY./Divided into 5 Mile squares/* (Ae, with the scale below). There is no imprint. *EXPLANATION./*3 lines on boundary markings, 1 on railways, and 4 lines, beginning *Note – The official names of Railway Stations . . .* (Ea). A graticule has been added with letters A–P down the sides and numbers 1–11 along the top and bottom, all between the inner and outer frames. Railways 41–46, P, R, S, T, U and V have been added. Many place-names have been added, mostly in Yorkshire or on new railways outside the county. *26* has been stamped (Ee, OS on its side, and on the verso). North and South Divisions are coloured pink and the Mid Division yellow; the boroughs are red and adjoining counties ivory.

*New Large Scale Ordnance Atlas Of The British Isles With Plans Of Towns, Copious Letterpress Descriptions, Alphabetical Indexes And Census Tables. London: Edited and Published by George W. Bacon . . . 127, Strand . . .* [1883].[20]

**Copies**   BL Maps 11.c.30; Leeds UL W 222; Leeds RL Q 912.42 B133. (LAO FL 912).

*New Large Scale Ordnance Atlas . . . 1884.*

**Copy**   The private collection of Mr. T. Burgess.

Issued in a cover with the title: *BACON'S New COUNTY GUIDE AND MAP OF LINCOLNSHIRE FROM THE ORDNANCE SURVEY WITH ALPHABETICAL INDEX . . .*[21]

**(K)**   *(Parl'y)* has been added to the first line of the *Explanation*.

*New Large Scale Ordnance Atlas . . . 1884.*

**Copies**   Bod Allen 111; RGS 7 E 17; Birmingham RL AF 912.42 BAC. (LRL Map 235b; Grimsby RL RM 210 and RM 11).

*New Large Scale Ordnance Atlas . . .* [1885].[22]

**Copies**   BL Maps 37.e.31; CUL Atlas 5.88.14; Bod Map C.15.c.1; NLS Map 1.b.2.

**(L)**   Size: 46.2cm x 31.15cm. A new transfer has added a small area at the top of the map. The Explanation has been replaced by *Parliamentary Divisions/West Lindsey or Gainsborough, colored thus* [a box 5mm. long x 2 mm. deep, filled with colour]/7 further lines of parliamentary divisions and 6 coloured boxes, the seventh being represented by an irregular shape coloured pink/a line on railways and stations/4 lines, beginning *Note* (Ea). The seven divisions are coloured yellow, brown, pink and green with the boroughs red.

*New Large Scale Ordnance Atlas . . .* [1886].[23]

**Copies**   The private collection of Mr. T. Burgess. (LRL Maps 235c and 98).

**(M)**   The boxes in the notes on the parliamentary divisions are now 7.5mm long and are no longer separate but form a larger rectangle with 7 internal divisions. Railway X has been added. The erroneous *20'* (Ea, between the inner and outer frame) has been replaced by an inverted 40 (a double error).

*New Large Scale Ordnance Atlas Of The British Isles . . .* [1886].

**Copy**   The private collection of Mr. T. Burgess.

*New Large Scale Ordnance Atlas . . .* [1888].[24]

**Copies**   Leeds UL W 223. (Grimsby RL BACON RM 10).

Issued mounted on linen in a cover entitled: *BACON'S County Guide AND MAP OF LINCOLN FROM THE ORDNANCE SURVEY. Coloured on Cloth, 1s. LONDON: G.W. BACON & CO., 127 STRAND* . . .

**Copy**   The author's collection.

(**N**)   The information on the parliamentary divisions has been moved down the map by 22mm. (the coloured boxes and the words *colored thus* have been removed) and is followed by the 3 lines on boundaries and railways (Ea). The title is *MAP OF/LINCOLN-SHIRE./* (Ae), with scale below. *PRINTED FOR THE LINCOLN & DISTRICT DIREC-TORY.* (Ae, OS). *PUBLISHED BY AKRILL, RUDDOCK & KEYWORTH, LINCOLN.* (Ee, OS). The inverted *40* has been corrected to *50* (Ea, between inner and outer frames). Railway 48 has been added.

*Directory Of The City Of Lincoln, With The Surrounding District, Containing Street, Court Guide, Alphabetical & Classified Trades Lists, With Map Of County (Engraved expressly for this Work). Price: Subscribers, 4/6; Non-Subscribers, 5/6. Akrill, Ruddock, & Keyworth, High Street & Silver Street, Lincoln. Compiled by Wells & Manton. MDCCCLXXXVIII* . . .

**Copies**   LAO Ex 942.55. (LAO Ex 942.50).

(**O**)   A new transfer, which is basically as state (M). *Parliamentary Divisions/West Lindsey or Gainsborough, colored* . . ./a large rectangle, with seven internal divisions, now 1.3cm across, the remaining notes as in states L and M (Ea). There is no imprint. The title is again as in states (K)–(M). *LINCOLN.26.* has been added (Ee, OS, reading down the right frame line). *26* appears on the verso. Railway 49 has been added. Railway 22 has been removed. The inverted 40 reappears (Ea, between the frames).

*New Large Scale Ordnance Atlas* . . . [1889].[25]

**Copies**   BL Maps 19.c.2; CUL Atlas 3.88.12; Bod Map C.15.c.3; NLS Map 1.b.4. (Louth RL Map A 10).

(**P**)   Size: 45.5cm x 31.2cm. In the transfer part of Yorkshire does not appear, including *BEVERLEY*. The title is *LINCOLNSHIRE,/* (Ae), with the scale below. The imprint *HARRISON & CO., Limited, 3, Liverpool Street, LONDON, E.C.* has been added (Ce, OS) and *Copyright* (Ee, OS). The Wisbech-Peterborough line has been restored but it now crosses the Spalding-March line. All railways and stations outside the county are now named, including two new ones on railway X. The mouth of the Nene is enclosed north of *King Johns/Ho* and the Norfolk coastline is slightly further north. *SUTTON/S*$^T$ *MARYS* has been changed to *LONG/SUTTON, the O in FOLKINGHAM* to *A* and *Naconby* has been changed to *Hacconby*. The graticule and the key letters and numbers have been removed. Below the note on the colouring of parliamentary divisions, which has moved 65mm. lower, (the word *thus* re-appears in the first line) are three lines only, two referring to the marking of boundaries, one to railways (Ea). Railway G extends to the inner frame (previously it was shown extending to the outer frame-line). The railways at Wisbech are shown as two separate lines, with their own stations. Railway 26 has been redrawn; this results in *Toft* losing its final letter (previously obscured by the railway). A tongue of reclaimed land partially obscures *River Witham* as it enters the Wash; the name Fosdyke is spelt thus in three cases. Railway 47 has been added. The numbers for degrees along the top frame have been removed. The corrected 50′ mark reappears (Ea, between borders).

Issued in an unidentified book. [1889].[26]

**Copy**   (LRL Map 70).

**(Q)** This transfer is basically state (N) to which have been added: the railways of the intervening states, three large black dots for stations on railway 27 and three on railway T/U in Leicestershire and Nottinghamshire. The Wisbech-[Peterborough] line does not cross railway 34 but joins it near the 1' west mark on the map's inner frame.

*New Large-Scale Atlas Of The British Isles . . . And Tables Of Population. Edited By G. W. Bacon, F.R.G.S. London: Published By G.W. Bacon & Co., Limited, 127, Strand . . . 1890.*

**Copy** Hull RL 912.42.

*New Large-Scale Atlas Of The British Isles . . . 1891.*

**Copies** PRO M7 PRESS 20; Liverpool RL Gf 119.

*New Large-Scale Atlas Of The British Isles . . . Tables Of Population, Census Of 1891. Edited By G.W. Bacon . . . 1891.*

**Copy** BL Maps 33.c.2.

*New Large-Scale Atlas Of The British Isles . . . 1892.*

**Copy** The private collection of Mr. T. Burgess.

**(R)** Size: 66.95cm x 45.35 cm. The transfer is basically state (O), but with the following differences. The added names for railway stations in the adjoining counties and the northern extension in Yorkshire to include Beverley re-appear. The title now reads: *WHITE'S/COUNTY MAP OF/LINCOLNSHIRE/SHOWING ALL THE/RAILWAYS and NAMES of STATIONS,/ALSO THE VILLAGES, TURNPIKE ROADS, GENTLEMEN'S SEATS, &c./REDUCED FROM THE ORDNANCE SURVEY./*with *Roads* and symbol (Ea). The imprint: *PUBLISHED BY WM. WHITE, LTD., 9, ST. JAMES' ROW, SHEFFIELD.* has been added (Ce, OS). Below the scale (Ae) *Lancashire, Derbyshire & East Coast Railway thus* – – – has been added. There is a graticule with letters and figures between the frame lines. The names of railways in adjoining counties have been removed. Railway 52 is marked by a broken line as far as the ELR at Alford. *FALKINGHAM* has reverted to its earlier spelling. The map should be found in a pocket at the back of the volume, but is frequently missing (e.g. two copies in LRL and one in Scunthorpe RL).

*History, Gazetteer And Directory Of Lincolnshire, Including The Diocese Of Lincoln; And Comprising A General Survey Of The County . . . Fifth Edition. By William White, Limited . . . Price:– Cloth, 30s.; Half-Bound, 35s. Sheffield: William White Limited, 9, St. James' Row. London: Simpkin, Marshall, Hamilton, Kent & Co . . . 1892–3 . . .*

**Copies** Nottingham UL LH Li.1.B15 E 92. (LAO Misc Don 461 is an example of the map, but mounted on a roller).

**(S)** The transfer and size are as state (P) but the Wisbech-Peterborough section of railway 31 has been deleted. The colouring of East Lindsey is now pale yellow (previously brown).

*New Large-Scale Atlas Of The British Isles . . . And Gazetteer, With Census Of 1891. Edited By G.W. Bacon . . . 1893.*

**Copy** The private collection of Mr. T. Burgess.

**(T)** Size: 46.4cm x 31.1cm. This transfer is based on state (P), with the title, scale, graticule and the Yorkshire extension to *BEVERLEY*; railways 1–4, 6–21, 23–49 are marked here as well as railways A, C, G, I, J, L, M, P, R, S, T, U and X. The L.D.E.C.R (52) is marked by broken lines as far as Mumby, but the note recording this has been removed (Ae). The title is *LINCOLNSHIRE/REVISED THROUGHOUT BY THE ORD-NANCE SURVEY/Divided into 5 mile squares/Scale of English Miles/10* [= 5.25cm] (Ae). A note on *ADJOINING/ COUNTIES/* 5 names and numbers (Ee, OS at right angles to the map). *Bacon's Geographical Establishment* (Ae, OS). *REFERENCE TO COLOURS./ Parliamentary Divisions & Boroughs. Colours. Populations./1881. 1891*/a table of 11 names with the colours used on the map and the census figures for 1881 and 1891, followed by county totals; *Note – The official names of Railway Stations* . . . on four lines, and *Explanation*/8 lines of text and symbols (Ea). The letters and numbers to the graticule are now 1–11 along the top and bottom and A–Q down the sides. The compass has been removed. The divisions are coloured green, yellow, pink and brown, the boroughs purple and the adjoining counties pale lemon. The numbers at 10′ intervals return along the upper frame, between the inner and outer borders.

*Commercial And Library Atlas Of The British Isles . . . 1895.*

**Copies**  BL Maps 33.b.32; CUL Atlas 5.89.51; Leicester UL F912.42; Sheffield RL 912.42 STQ; Liverpool RL GF 119.

*Commercial And Library Atlas . . . 1896.*

**Copies**  BL Maps 49.d.40; Leeds UL W 224; PRO CO 700 Misc.14.

(**U**)  Size: 62.9cm x 45.7cm (approx.). Although a larger transfer, there is no part of Yorkshire further north than *Burton/Pidsea*. The *REFERENCE TO COLOURS* has been moved to (Ae). The title is preceded by *BACON'S/MAP OF/* and moved to (Ea), followed by the scale in *British Statute Miles/Explanation*/8 lines/*Note* . . . on 4 lines. The compass has been reinstated (Eb). *London: G.W. Bacon & Co. Ltd 127 Strand.* (Ee, OS). The signature (Ae, OS) has been removed. Railways 50–51 have been added. The figures and letters to the graticule revert to those states before state T.

Issued in a cover with the title *BACON'S COUNTY MAP & GUIDE LINCOLN In Cloth Case, 6d. On Cloth, 1/– THE ROVER* [below an illustration of a woman and a man cycling] *FOR CYCLISTS & TOURISTS FROM THE ORDNANCE SURVEY G.W. BACON & CO. LTD. 127, STRAND, LONDON.* [1896].[27]

**Copies**  BL Maps 3355 (38); LRL Map 850.

Issued mounted, lacquered and on rollers.

**Copy**  LAO 4BM 16.

(**V**)  Size: 46.4cm x 31.1cm. The transfer is that of state (T), with the addition of railway 52.

*Commercial And Library Atlas . . . 1897.*

**Copy**  The private collection of Mr. T. Burgess.

(**W – i**)  Size: 62.5cm x 45.6cm with layout arranged as state (U), i.e. title, etc. (Ea), table of population, colours, etc. (Ae), and imprint (Ee, OS). Code number *6.7.G.12+* has been added (Ae, OS).

Issued in a blue cover with the title *BACON'S COUNTY MAP & GUIDE LINCOLN* [illustration of a man and woman holding their cycles, above the name of their machines] *ROVER CYCLES FOR CYCLISTS AND TOURISTS FROM THE ORDNANCE SURVEY. LONDON* . . . [1897**].[28]

**Copy**  Grimsby RL X000:912 BAC (pamphlet).

(**W – ii**)   As in state (W – i) but with the code number 98G12+ (Ae, OS).

Issued in a buff cover; the wording of the cover title is as in the previous state, except that *AND* replaces the ampersand and *SHOWING DANGER HILLS* has been added at the top.

**Copy**   LAO L 912.

(**X – i**)   The map is as state T. Code number *38M10* has been substituted (Ee, OS).

*Commercial And Library Atlas . . . 1898.*

**Copy**   Leeds UL W 224a.

(**X – ii**) The code number is *18M10* (De, OS).

*Commercial And Library Atlas . . . 1899.*

**Copy**   BL Maps 18.c.19.

*Commercial And Library Atlas . . . 1900.*

**Copies**   PRO M7 Press 20; Liverpool RL Gf 119; Leeds RL 912.42 B 133.

(**Y – i**)   Size: 72.5cm x 45.3cm; the layout is as state (U). *Hills to be ridden with caution . . .* [symbol]/*Hills dangerous* [symbol] have been added (Ce, OS, in orange). The symbols have been added to the map. The table (Ae) and the note (Ad) on dangerous hills, etc. have been removed and the space is occupied by details of Rutland, which means that railway 5 is now seen. Railway 52 has been added, with stations; an extension is marked by broken lines across mid-Lincolnshire and joining the Mablethorpe line at Mumby. There is no code number. The roads are coloured orange-brown, with the border of the county edged in brown and the danger symbols orange; no other colour.

Issued in a dull buff cover with the title: [symbol] *SHOWING DANGER HILLS* [symbol] *BACON'S COUNTY MAP AND GUIDE LINCOLN WITH PARTS OF ADJOINING COUNTIES Cloth Case, 6d. On Cloth, 1s.* [Illustration] *THE ROVER CYCLES FOR CYCLISTS AND TOURISTS BASED ON THE NEW ORDNANCE SURVEY. G.W. BACON & CO., Ltd., 127, STRAND, LONDON.*

**Copies**   BL Maps 3363 (1); LAO Ex 942.50; Nottingham UL 11 B8.E 95.

(**Y – ii**)   The map is as in (Y – i). The cover is identical except that *BASED ON THE NEW ORDNANCE SURVEY* has been omitted.

**Copy**   The author's collection.

(**Y – iii**)   As in (Y – i), except that (a) the cover is overprinted with red and blue; (b) the price is shown in the form: *On Cloth, In Case, 1s.;* (c) at the foot of the cover *WITH ROUTE GUIDE* has been added; and, (d) *BASED ON THE NEW ORDNANCE SURVEY* has been added above the imprint.

**Copy**   LRL Map 999.

The issue of maps in separate covers continued well into the 1920s. Railways were added to the maps and other features were provided, such as a guide to suitable cycling routes. The wording of the cover titles and the illustrations were varied. At first the lady cyclist was retained with an early motor car in the background. By c. 1910 two cyclists are shown, vainly pursuing an open car. In the 1920s the cover picture shows a couple consulting a map, with a car in the distant open road. By 1910 Bacon & Co., also provided alternative covers, with the names of local firms as the apparent publishers. The author has examples issued by J.W. Ruddock of Lincoln and Leayton & Eden of Grantham.

The atlas with plates in very much the same transfer as issued in 1900 (the addition of railways and changes to the census figures are the key indicators of their later vintage) appeared until c.1932. Parts of the plates were also transferred to create *Bacon's 'Excelsior' Post Cards* (c.1910), still being used as late as 1927.

### References

1   Nowell-Smith, S. *The House of Cassell*. (London, 1958), *passim*; the volume has no references to *The Weekly Dispatch* or to the Cassell atlases; Brown, p. 34.
2   Smith, D. George Washington Bacon 1862–c.1900. *Map Collector*. No. 65. (Winter, 1993), pp. 10–15.
3   The LAO copy is very defective and has no maps before that for Devonshire in the alphabetical sequence.
4   The plates passed to Cassell in 1864; the reproduction of the Aggas map of London is dated 1865.
5   A privately owned copy has the county plate showing check digit 6.
6   The slight change may have been effected before the new atlas of 1867 appeared and 1866 is suggested. A privately owned copy shows the check digit 8.
7   This work lacks a title-page and the title is taken from the spine.
8   CUL has Berkshire in this series and its cover records Lincolnshire offered in this style.
9   Mr. E. Burden owns a copy of Sussex in this series and there is a copy of the Derbyshire map in Derby RL (Local Studies Centre), item 4507; the covers show that all counties were issued.
10  The BL copy consists of the monthly parts with original covers dated as received from Feb., 1864 to Aug., 1865 (at 1s. each), followed by 5 supplementary parts from Sept., 1866–Jan., 1867. The Lincolnshire part should have been issued in May, 1864 but it arrived with a collection of parts, all received 2 JU 1864.
11  Kingsley, p. 249 dates to c.1870 an example of Sussex. The author has a copy of Buckinghamshire, which can also be dated to c.1870. An example of Yorkshire is held in CUL (Maps c.93.88.1).
12  The loose map is believed to come from such an atlas.
13  The railways in operation suggest 1871–2; the Louth & Lincoln Railway Co. was incorporated in 1866 to build a line as far as Five Mile Houses station; the route is roughly that shown on the map. However, by 1867 building difficulties forced a route change, given royal assent in 1872. Ludlam & Herbert, op. cit., pp. 7–11.
14  Mr. E. Burden has examples of Sussex, Huntingdonshire and Norfolk. The map at Scunthorpe is almost certainly an example of the same series; the alternative formats (referred to in the title) have not been found.
15  Lancashire is held by John Rylands UL (R 143995).
16  CUL Maps 88.86.1 is an example of Sussex.
17  Reading RL (LMV 149) is an example of Berkshire; examples of Sussex and Somerset are held by CUL.
18  The BL copy was received 30 AUG 1876. The county railways are correct to July, 1873 only. Mr. T. Burgess has an example, differing only in not having the page numbers; he dates his copy to 1873.
19  CUL Maps c.34.013 consists of 9 county maps in such covers. Lincolnshire may exist thus.
20  The BL copy was received in 1883.
21  The county map has not been found. The author owns a copy of Shropshire, the inside cover of which records the availability of a series of 40 uniform county maps. Kingsley, p. 254 notes a copy of Sussex in Brighton RL and Burden, p. 197 owns a copy of Somerset.
22  The Bod copy was received 14 Feb 85 (i.e. 1885).
23  The date is assumed from the inclusion of the new parliamentary boundaries, following the Redistribution of Seats Act, 1885.
24  Ascribed by Whitaker to 1888 in his catalogue of the Leeds UL collection, p. 101.
25  The BL receipt stamp is 11 NOV 1889; the Bod copy contains notes on the claiming of their copy from Bacon's in Sept., 1889.

26  The publisher of this map only occurs as the publisher of one other map – in the 'Finger Post' guides (**57M**). It is possible that there was another issue of that guide, containing the present map; the LRL copy has been laid down on card and it is not possible to see any sign of original inclusion in a book. Ironically, this was the first map to show the Alford Tramway, which closed in 1889. Harrison, the publisher, is not noted for producing any other county map and is only recorded in the Post Office Directory of London for 1889, hence the suggested date.

27  The BL receipt stamp is 29 APR 1896.

28  This, and the other issues in covers, described below were issued in 1897 or just after; perhaps after 1900 even. The Goole-Reedness railway (opened 1900) is not shown on the plate, but Bacon may have delayed its addition to the plate. The listing of the variety of wordings shown in the cover titles is probably not complete.

Chubb DXL; DLI; DLIX.


# 121

**EDWARD STANFORD**                                                        **1861**

Size: 30.6cm x 13.2cm.

This map was transferred by the firm of Stanford's from plates which covered the whole of the country, and is used to illustrate a topographical book by Mackenzie Edward Charles Walcott, which covered all the east coast, the material of which appeared under different titles. Walcott (1821–1880) was the son of an M.P., who, after Winchester, went into the church and was, from 1863 until his death, Precentor of Chichester Cathedral. He wrote on a wide number of topographical and antiquarian topics, including *A guide to the cathedrals of England and Wales* (1858) and *Memorials Of Lincoln and the Cathedral* (Lincoln, 1866).[1]

The publisher of the map, Edward Stanford (1827–1904), set up in business in 1852, after serving an apprenticeship with a printer in Malmesbury. When his master died he returned to London, his birth-place, and worked for Trelawny Saunders whose map business he took over fully in 1853 after their partnership had lasted only ten months.[2] Stanford was elected F.R.G.S. in 1853;[3] in 1858 he took over the plates formerly in the possession of H.G. Collins (**62A, 71A–C, 92A** and **119**) and the firm prospered; further work is discussed later (**135–136, 138**). When Stanford retired in 1881 his son, also Edward (1856–1917), took over; the firm was taken over by George Philip in 1946, although the name still survives in the map world. The present plate was only used in the two works by Walcott; there is a close family relationship between this and the later Stanford maps. It shows the area from just north of Whitby in Yorkshire as far south as Upwell; the adjoining counties are shown in as much detail as Lincolnshire.

## EDITION

(i)  *COUNTIES OF YORK & LINCOLN* (Ba–Da, OS). *Stanford's Geographical Establishment, 6, Charing Cross*. (Ee, OS). Compass (Eb). *LINCOLN* (across the centre of the county). *English Statute Miles/10 + 10* [10 miles = 2.35cm] (Ce, OS, below the level of the imprint). Railways 1–17, 19–23, A, C, F, G, H, I, L, M and N are shown, as well as railways in north Yorkshire. The county's boundaries are coloured lemon and those of Yorkshire pink; the names of several sea-side resorts (Cleethorpe (sic), Mablethorpe and Frieston in Lincolnshire) are underlined; in some copies the black underlining is superimposed with pink.

*A Guide To The Coasts Of Lincolnshire & Yorkshire Descriptive Of Natural Scenery Historical, Legendary, And Archaeological By Mackenzie E.C. Walcott. London Edward Stanford, 6 Charing Cross, S.W. 1861.*[4]

**Copies**   BL 10368 aa 71; LRL L 91; Grimsby RL X000:914 WAL; Louth RL GL 9.

*The East Coast Of England From The Thames To The Tweed Descriptive Of Natural Scenery Historical, Archaeological, and Legendary. By . . . London Edward Stanford . . . 1861.*

**Copies**   BL 10358 b 17; CUL D.38.4; Bod G.A. Gen. Top. 8° 61; LAO; LRL L 91; Gainsborough RL L 551.46.

### References
1   DNB. Vol. LIX (1899), pp. 11–12.
2   Stanford, E. jun. *Edward Stanford with a note on the history of the firm from 1852.* (London, 1902), pp. [1]–4; Hyde, 1975 op. cit., pp. 11–13.
3   Herbert, F. 1983, op. cit., pp. 67–95.
4   Lincolnshire occupies pp. [121]–157 in both works noted here and the map is placed before the first page of the county text.


## 122

### EDWARD WELLER                                                        1861

Size: 15.5cm x 13cm.

Weller was one of the engravers responsible for some of the plates in the *Weekly Dispatch Atlas* (**120**), though not Lincolnshire. He also prepared the plates for the series used in *Philips' Atlas Of The Counties . . .* (**123**) and the smaller plates also issued by Philips (**130**). He remained active until about 1875; his son, F.S. Weller, provided the plates for Brabner's *Comprehensive Gazetteer . . .* (1894) (**149**) and a later map in 1898 (**151**). The present map simply outlines the county, emphasising the rivers and hills; only the main towns are marked and a few physical features. Stamford is notably misplaced, being sited where Market Deeping might have been. A novelty is a north-south cross-section of the county down the left side of the plate.

### EDITION

(i)   *LINCOLNSHIRE.* (Ea). *E. Weller* (Ee, OS). *Scale of/Feet/*reading up from 100 to 500 (Ab, at right angles to the frame). 7 (Ee, OS at right angles to the right edge). There is no compass or general scale. The county is coloured green, the hills buff and the coast has a blue edge.

*Physical Atlas Of Great Britain And Ireland With Illustrative Letterpress By Walter M'Leod . . . London Longman, Green, Longman, And Roberts 1861.*

**Copies**   BL Maps 15.b.10; CUL Atlas R.7.86.3; Bod G.A. Gen. Top. 8° 624; NLS Map 1.1.14.

### Reproduction
This map is shown in Beresiner, p. 152.

Chubb DXLVIII.

## 123

EDWARD WELLER                                                    c.1862

Size: 42.5cm x 33.75cm.

These large plates were first issued as loose sheets in covers and, perhaps, as separates. They made their first appearance in an atlas in 1865 and remained in use until c.1905.[1] The maps are based on the Ordnance Survey and have a more than passing resemblance to those of the *Weekly Dispatch Atlas*, though Weller was more selective in the detail he showed, compared with Dower, and produced a map with perhaps greater clarity. The plate was produced by lithography and was never used for intaglio printing.

George Philip (1800–1882), born in Aberdeenshire, left in 1819 to join his brother, who was a minister in Liverpool. After obtaining work with a bookseller he opened his own shop in 1834 and began to publish educational books; his success is measured by his moving to larger premises in South Castle Street within a year. His only son, also George (1823–1902) joined him in 1848, two years after they had introduced litho printing to the firm's letterpress capability and a year or two after opening a London office (in John Murray's first office in Fleet Street; it was his ship motif that Philip used for his colophon). It was only around 1860 the firm undertook its own preparation of map plates, plates having been made for them by John Bartholomew, A. Petermann and W. Hughes (**126**).[2] The use of Rowe's plates of 1813 has been noted above (**71E–F**). At one time Philip employed more than 80 people as map colourists. Philip, senior, retired in 1880, the firm being run by his son and a nephew, who had joined in 1851.

The author of Murray's *Handbook For Lincolnshire* (state **N**) was G.E. Jeans. George Edward Jeans (b.1848) was born at Tetney, the son of a cleric, and educated at Pembroke College, Oxford. Ordained in 1875 he spent 4 years as Assistant Master at Haileybury before taking up clerical posts in the Isle of Wight. He also prepared Murray guides to Hampshire and the Isle of Wight. It is recorded that Jeans sold the copyright to Murray for £141.15.0. and his friend, Morgan George Watkins, was paid 2 guineas for corrections.[3] Watkins (b.1835) came from a clerical family and, after education at Exeter College, Oxford, took holy orders in 1858; after a period teaching in Devonshire he became Rector of Barnetby-le-Beck from 1861 to 1885. He was Rural Dean of Great Grimsby and Diocesan Inspector of Schools, before becoming Vicar of Kentchurch in Herefordshire.[4] Among his many works is *The Worthies of Lincolnshire* (1883).

### EDITIONS

(**A**)   *LINCOLNSHIRE./BY EDW^D. WELLER, F.R.G.S./English Miles/10* [= 5.6cm]/ symbols for Railways, Canals, Roads (on 3 lines) (Ea). *GEORGE PHILIP & SON, LONDON & LIVERPOOL* (Ce, OS). There is no compass. Railways 1, 3–27, A, C, D, F, G, I, L and M are marked and stations shown by dots with *Sta*. The map is coloured: *NORTHERN/DIVISION* grey-violet, *SOUTHERN/DIVISION* light buff and the sea pale blue.

Issued in a cover with the title *ONE SHILLING. PHILIPS' NEW SERIES LINCOLN FROM THE ORDNANCE SURVEY. OF COUNTY MAPS LONDON, GEORGE PHILIP & SON, 32, FLEET STREET, AND CAXTON BUILDINGS, LIVERPOOL.* [1862].[5]

**Copy**   NLS Maps M 51.2.

Issued as a loose sheet.

**Copies**   BL Maps 3355 (33); Bod Maps C.17: 39 (5); LRL Map 762.

(**B**)   *22* appears on the map verso.

*Philips' Atlas Of The Counties Of England, Reduced From The Ordnance Survey. By Edward Weller, F.R.G.S. London: George Philip And Son, 32, Fleet Street; Liverpool: Caxton Buildings, South John Street, And 51, South Castle Street. 1865.*

**Copy**   Liverpool RL O.R. 21–2.

Issued in a cover with the title as in state A. [1866].[6]

*Philips' Atlas . . . 1866.*[7]

**(C)**   Railways 30–35, 37 and O have been added. Not all the lines are marked accurately, e.g. the line to Wisbech is drawn from *SUTTON/S$^T$ MARYS* instead of [Sutton Bridge].

*Philips' Atlas . . . 1868.*

**Copy**   The private collection of Mr. T. Burgess.

**(D)**   Below the title, etc. *The Colouring represents the Parliamentary Divisions/& Parliamentary Boroughs* has been added (Ea). A graticule has been added, with letters A–I added along the top and bottom and 1–8 down both sides; numbering of degrees of latitude and longitude appears at 10′ intervals between the inner and outer frames. *Meridian of Greenwich* (De, between borders). The plate number on the map's verso no longer appears. Railways 36, 38–40, P and R have been added; railway 31 (to *WIS-BEACH*) is now drawn correctly. Railway xvii is marked by a double line; railway xiv is similarly marked and continues south to Sleaford. The Nene has been embanked and the coast line redrawn at *Lutton/Sluice*; the words *Cross Keys/Wash* have been deleted. *22* has been added (Ee, OS, sidewards on). North Lincs. is coloured yellow, Mid Lincs. pink and South Lincs. green; Grimsby, Grantham and Stamford pink; Lincoln and Boston yellow.

Issued in a green cover with the letters embossed (the county name is in gold); the cover title is *PHILIPS' NEW SERIES LINCOLNSHIRE FROM THE ORDNANCE SURVEY OF COUNTY MAPS* [1874**].

**Copy**   Grimsby RL X000: 912 WEL.

*Philips' Atlas Of The Counties . . . Weller, F.R.G.S. New Edition, Improved And Corrected To The Present Time. London . . . Liverpool . . . And 49 & 51, South Castle Street. 1874.*

**Copy**   A private Lincolnshire library.

**(E)**   The plate number is a different, sans serif figure *22*. *Railways Constructing* [two parallel lines] has been added below the other explanatory notes (Ea).

*Phillips' Atlas . . . New Edition, With A Complete Consulting Index, By John Bartholomew, F.R.G.S. London . . . Liverpool . . . 1875.*

**Copies**   BL Maps 10.cc.12; Leeds UL W 216; RGS 7 G 6; Liverpool RL GQ 153.

*Philips' Atlas . . . 1876.*

**Copy**   BL Maps 10.cc.25.

*Philips' Atlas . . . 1880.*

**Copy**   Leeds UL W 217.

**(F)**   In two lines above the top *SUPPLEMENT TO THE PICTORIAL WORLD, September 27th, 1879.* (Aa–Ac, OS) and *THE 'PICTORIAL WORLD' MAP OF LINCOLN-SHIRE.* (Aa–Ea, OS, below the first heading) have been added. The plate number has been erased. The title, engraver's name and reference to colouring have been erased (Ea), leaving the scale and four lines of symbols and their meanings. Their removal has led to the breaking of the graticule in that area. The imprint has been removed. Railways 41,

43 and xvi (there is only one branch – to Saltfleet Haven) have been added. Railways 45–46 are shown by double parallel lines. The county area is coloured pink and the sea blue. The large map was folded into the periodical; space on the verso is used for two pages of text on the county.

*The Pictorial World – an illustrated weekly newspaper.* 1879.

**Copies**   (BL Map 3355 (52); LAO Scorer Map 50; LRL Map 205).

A collection of the loose sheets bound as an atlas.

**Copy**   Leeds UL W 220.

**(G)**   The two lines of heading have been removed.

Issued as a loose sheet, with the labels of *JAMES WYLD, Geographer to the Queen,/11 & 12, Charing Cross, S.W.* stuck in the margin below the map and down the right side.

**Copy**   The private collection of Mr. E. Burden.

**(H)**   The map, in state E, has railways 39, 41–46, T and U added.

*Philips' Atlas . . . 1883.*

**Copy**   Middlesbrough RL 912.42 (No. 961).

Issued in a green cover with embossed lettering, the county name being in gold, with the title: *PHILIPS' NEW SERIES LINCOLNSHIRE FROM THE ORDNANCE SURVEY OF COUNTY MAPS.* [1883].

**Copy**   The author's collection.

**(I)**   There is a broken line to mark railways 48–49. The & has been replaced by *and* in the last line of the notes (Ea). Compared with the colouring of states (D) and (E) South Lincs. is now yellow. *LINCOLN* has been added on the map's verso.

*Philips' Atlas . . . 1885.*

**Copy**   Bod Allen 112.

**(J)**   *Each Parliamentary Division & Borough/returns 1 member* has been added below the title, scale and explanatory notes (Ea). The engraver's name (below the county name) has been omitted. The names of the county's divisions have been overprinted in black, e.g. *N. LINDSEY/OR BRIGG.* The divisions and boroughs are coloured pink, yellow, green, violet and brown, with parks bright green and the sea blue. The area around Crowland is coloured pink as part of South Kesteven. *Sta.* has been added to new lines, e.g. at Bourn and along the line to Sleaford; some names have been re-engraved, e.g. *Navenby/Sta* (previously *Sta.* only and east of the line); *Sta.* has been resited, e.g. Reepham, Leadenham; the village name has been added above, e.g. *Goxhill/Sta.* Railway 35 previously made a fork on joining the Sleaford line at Honington; the eastern curve has been removed. Railway 31 at Wisbech is no longer joined to the other lines there but there is only one station still. The broken lines for railways 48–49 are not shown.

*Philips' Atlas Of The Counties Of England, Including Maps Of North & South Wales, The Channel Islands, And The Isle Of Man. Reduced From The Ordnance Survey, And Coloured To Shew The New Political Divisions, According To The Redistribution Bill, 1885. New And Revised Edition, With A Complete Consulting Index. London . . . Liverpool . . . And 45 To 51 South Castle Street. 1885.*

**Copies**   BL Maps 10.a.33; CUL Atlas 4.88.14; RGS 7 G 7; Liverpool RL GQ 153; Brighton RL F 912.42 P 53; Hull RL 912.42.

Issued in a cover with the title: *PHILIPS' POPULAR MAP OF LINCOLNSHIRE* 6ᵈ. FROM THE ORDNANCE SURVEY. ON CLOTH & IN CASE 1/– GEORGE PHILIP & SON, LONDON & LIVERPOOL. [1885].[8]

Issued in a red card cover, mounted on linen, with the series title *PHILIPS' NEW SERIES OF COUNTY MAPS.* embossed in black and split between the top of the cover and the bottom with, embossed in gold in the centre, *LINCOLNSHIRE FROM THE ORDNANCE SURVEY CLOTH EDITION.* [1885].[9]

**Copy**   The author's collection.

**(K)**   The explanatory notes, below the scale, now read *Main Roads Coloured Brown* [symbol]/ Cross Roads [symbol]/Railways [symbol]/ (Ea). Red initial letters *H, H.C.X., CX, H.C., C,* and arrows to mark steep hills have been added; the initials refer to types of stopping places, recommended by the Cyclists' Touring Club. The parliamentary divisions are not named. Apart from the roads and the sea (blue) there are no colours.

Issued in a cover with the title *PHILIPS' CYCLISTS' MAP OF THE COUNTY OF LINCOLN SHEWING THE MAIN ROADS DISTINCTLY COLOURED REDUCED FROM THE ORDNANCE SURVEY PRICE ONE SHILLING. GEORGE PHILIP & SON. 32, FLEET STREET, LONDON: AND LIVERPOOL.* [1885].[10]

**Copy**   BL Maps 10.a.23; CUL Maps c.34.08.70; Bod Map C.17: 39 (6); NLS Map M3 (24).

**(L)**   As state (J) but with broken lines for railways 48–49.

*Philips' Atlas Of The Counties . . . 1889.*

**Copy**   Leeds UL W 218.

**(M)**   The broken line indicating railways 48–49 has been redrawn; at the southern end it passes to the left of Mumby (previously it passed through the initial letter of the village name).

*Philips' Atlas Of The Counties . . .* [1890].[11]

**Copy**   The private collection of Mr. D. Hodson.

**(N)**   The map is basically as in state M but railways 48–49 are now marked with the correct symbol to show they are open. Railway 22 has been deleted. The imprint now reads: *London. John Murray, Albemarle Street.* (Ce, OS). There is no plate number. Lindsey is coloured pink, Kesteven green and Holland yellow, with the sea blue. *Sutton* has been changed to *Sutton/on Sea.* The map is in a pocket at the back of the volume and is frequently missing (as in the BL copy) or substituted with the wrong map (as in Bodleian, which has the quite different map from the second edition).

*Handbook For Lincolnshire, With Map And Plans. London, John Murray, Albemarle Street. 1890.*

**Copies**   Nottingham UL Li a.D28 MUR; LAO Wright 942.51, etc.; LRL L 91; Louth RL GL 910; Scunthorpe RL L 638 910; Stamford RL L 91.

**(O)**   As state (K) but with railways 48–51 and X added. The letters for hotels and coffee 'taverns' have been greatly increased and many of the existing letters have also been altered. The arrows to indicate hills have been made larger. Many place-names have been added, e.g. *Sutton Bridge, Clipsham Quarries, Tuxford.* Below Sutton has been added *on Sea*; *Pier* has been added at Skegness and Cleethorpes; *Hotel* has been added between Trusthorpe and Sutton/on Sea; *Sta.* has been added at Grimsby Basin, on some of the newly added railways and between Grimsby and Cleethorpes. The names of the parliamentary divisions from state (J) are shown, but the colours of state (L) apply.

Issued in a green cloth cover with gold lettering; the title is *PHILIPS' CYCLISTS' MAP OF THE COUNTY OF LINCOLN 1/6* [1896].[12]

**Copy**   Grimsby RL X000: 912 PHI Pamphlet.

**(P)**   As state (J) with the addition of the railways and place-names up to state (M).
*Philips' Atlas Of The Counties . . . 1896.*[13]

**(Q)**   Railway 52 has been added; all railways are now marked by solid single black lines. Scunthorpe is shown as a town (previously marked in 'hamlet' size lettering).
*Philips' Atlas Of The Counties . . . 1899.*[14]

*Philips' Atlas Of The Counties . . . 1900.*

**Copy**   Leeds UL W 219.

### References
1   Smith p. 149.
2   Philip, G. The Story Of The Last Hundred Years. (London, 1934), pp. 19–48.
3   Lister, W.B.C. A Bibliography of Murray's Handbooks. (Dereham, 1993), pp. 90 and 137.
4   ibid., p. 179.
5   Other commentators have placed their county maps at 1862. The county railways support this.
6   Sussex is known thus (in the author's collection); Lincolnshire could exist thus.
7   Smith, p. 148 postulates an edition of 1866. Kingsley, p. 270 notes a loose sheet of Sussex in a state between the atlases of 1865 and 1868, which he dated to 1866 or 1867.
8   Mr. E. Burden owns a copy of Cornwall and the list on the back cover shows that Lincolnshire was also issued in this series.
9   The leaf inside the cover records that this series was issued in three versions: 'in a cover for 6d.'; the copy described here for 1s.; and, 'Mounted on cloth, superior style, in case 1s. 6d. each'.
10  The BL copy's receipt stamp reads 21 MAY 1885. The Bod copy may not have had a cover.
11  It is assumed the minor alteration was made soon after state J (1889) was issued.
12  Similar maps for other counties have the suggested date 1898. 1896 is suggested based on Lincolnshire railway data.
13  Mr. D. Kingsley owned a copy of the atlas, since sold. It is assumed that the county plate was in the 1890 state, with additional railway data as suggested by Smith, p. 149.
14  Such an atlas is listed in Whitaker, 1948 but no example is known. The county plate should be the same as the 1900 edition.

## 124

**W.J. SACKETT**                                        **1864**

Size: 42.65cm x 34.32cm.

Nothing is recorded of the lithographer, who produced the maps constructed by James Thomas Law, who had prepared them to illustrate a series of lectures on the various divisions of the Kingdom into Provinces and Dioceses delivered at Lichfield Theological College. Law (1790–1876) was the eldest son of the Bishop of Bath and Wells and, after Christ's College, Cambridge, he became Prebend of Chester Cathedral in 1818 but, only three months later, became Prebend at Lichfield and Chancellor in 1821. He spent the greater part of the rest of his life there and gave lectures on ecclesiastical law in 1861, from which the work below derived.[1] When his eyesight faded William Francis helped to see the work below through the press.[2] The map, which was never used for intaglio printing, shows places, rivers and the deaneries, numbered in accordance with the lists in the Reference. The map includes Nottinghamshire, which is coloured mustard-yellow, Lincoln Archdeaconry being salmon pink and Stow bright green. The source of the map is not clear, but one suspects that Arrowsmith (**79**) was nearby; it is very inaccurate in its general outline. The maps were issued in parts and Lincoln was no. 13.[3] For the second map in the volume see below (**125**).

## EDITION

(**A**)   *Lincoln.* (Aa). *W.J. Sackett, Lithographer & Printer, 11, Bull S^t. Birm^m.* (Ce, OS). *Reference*/a four column list of the deaneries arranged under the names of the Archdea-conries of Lincoln, Stow and Nottingham. The upper half of the verso is stuck to a leaf, which bears a list of parochial names; the verso of the lower half is blank.

*A New Set Of Diocesan Maps. By James Thos. Law . . . And William F. Francis . . .* [1864].[4]

**Copies**   BL Maps 3.d.8; CUL Atlas 5.86.17; Leeds UL W 207; Birmingham RL AF 912.42.

### Reference
1   DNB., Vol. XXXII, (London, 1892).
2   Smith, D., p. 111.
3   The CUL atlas retains the covers of the original parts.
4   There is no imprint or colophon. The preface is dated Nov. 1st., 1864.

## 125

**W.J. SACKETT**                                                                    **1864**

Size: 7.9cm x 12.7cm. (approx.).

The Lincoln section of the work noted above (**124**) consists of six 'pages'; the first leaf has the title of the section; on its verso is the beginning of a list of the archdeaconries, subdivided into deaneries, under which are listed the names of the parishes; the third 'page' continues the listing of parishes and its verso is stuck to the top half of the map's verso, the map thus forming 'pages' four and five; the sixth 'page' is blank. On [2], above the list of deaneries, etc. is a smaller key map; apart from a few rivers it has no geographical features. It has the deanery boundaries, which are numbered in accordance with a key printed to the left of the map; additionally the three archdeaconries of Lincolnshire and Nottinghamshire are lettered: A for Lincoln, B for Stow and C for Nottingham. The upper panel above the lists of deaneries and parishes has *Lincoln* in a gothic script at the left, the central space is taken up with the *REFERENCE* listing the names of the deaneries and the right hand portion contains the map. There are no divisions between the three sections. The only features named on the map are: *River Humber, Wash, Fossdyke Wash.* Outside the map area are the names of the four surrounding dioceses, e.g. *DIO. OF/YORK.*

The title of the work and the locations are as in **124**.

## 126

**WILLIAM HUGHES**                                              **1865**

Size: 31.1cm x 24.2cm.

*The National Gazetteer* was issued in parts from 1863 until 1868. The prospectus announced that the work 'would be completed in about Thirty-five Parts of 80 pages each, imperial octavo, or in about 12 Divisions at 7s.6d . . . The price of each Part will be Two Shillings, and the whole, when completed, will form three handsome volumes of 800 or 1,000 pages each . . .' The Lincolnshire map appeared as Part 20 in the sixth division, which came out in 1865;[1] the text for the county was in the seventh division. There was no order in the appearance of the maps, Lincolnshire appearing with those of Cornwall, Derbyshire, Durham, Hebrides & Sutherland, and Herefordshire. The maps were gathered up to form an atlas in 1873 and in c.1886.

William Hughes (1817–1876) was styled Professor of Geography in King's College, London from 1863 to 1875. In the history of Philip, Son and Nephew the W. Hughes recorded as a supplier of plates before c.1860 is equated with the Hughes who was Professor of Geography.[2] Hughes was a cataloguer in the British Museum from 1841–1843, working on the geography books and there is a long list of geographical works assigned to W. Hughes from 1840 to 1870. It seems likely that he provided the drawings of the county maps, which are based on the maps of the Ordnance Survey and were presumably engraved in the publisher's office. William Hughes was elected a Fellow of the Royal Geographical Society.[3] W. Hughes (1803–1861) has been considered as a possible candidate[4] since he was a noted wood engraver, but he died two years before the project began.

The publishing firm was founded by George Virtue (c.1793–1868) and was noted as the publisher of the dictionaries, in which the Schmollinger/Bingley plates made their final appearances (**95 (vi–xi)**) in the 1840s and 1850s. His son, James Sprent Virtue (1829–1892), expanded the firm greatly, opening up in the United States and developing close links with some of the literary giants of the time. He had taken more or less complete control of the firm of Alexander Strahan, who had serialised Trollope's *Rachel Ray* (1863), published *Phineas Finn* (1869) and *He Knew He Was Right* (1869), the last two serialised in *St. Pauls Magazine*, which Virtue founded and Trollope edited from 1867–1870. When Virtue took over, Strahan had run up large debts, not least by giving Tennyson £5000 a year from 1868 for five years to republish his old works.[5]

**EDITIONS**

(**A**)   *LINCOLNSHIRE./English Miles/10* [= 3.9cm]/*Boundary between North & South divisions . . . . . . ./Railways* [symbol]/ (Ae). *LONDON JAMES S. VIRTUE* (Ce, OS). W. Hughes. (Ee, OS). Lindsey is coloured yellow, Kesteven green, Holland pink, the sea blue. Railways 1–4, 6–17, 19–24, 26–27, A, C, D, F, H, I, M, N, and O are marked with the names of the places of the lines' origin, e.g. *from Rugby* at Stamford. Latitude and longitude are numbered at 10′ intervals between the borders. *West of Greenwich 0⁰ East of Greenwich* (De, between borders). There is no compass.

*Sold To Subscribers Only. The National Gazetteer Of Great Britain And Ireland. London: Virtue & Co., City Road And Ivy Lane.* [1865].[6]

Second title-page: *The National Gazetteer Of Great Britain And Ireland. London, James Virtue, City Road.*

**Copy**   BL 10348 i 7.

*The National Gazetteer: A Topographical Dictionary of the British Islands. Compiled*

*From The Latest And Best Sources, And Illustrated With A Complete County Atlas, And Numerous Maps. Vol. II. Faccombe – Myton-upon-Swale. London: Virtue & Co., City Road And Ivy Lane. 1868.*

**Copies**   CUL 8500.b.5 and S 696.b.86.15; Bod G.A. Gen Top 4° 278–280. (Grimsby RL Hughes RM 4).[7]

**(B)**   The imprint is *LONDON VIRTUE & CO*. Railways 25, 28–33 have been added, although railway 31 curves off between *SUTTON/S^T MARY* and *Sutton Br.*

*The National Gazetteer . . . 1868.*

**Copy**   Leicester UL H 942 HAM.

**(C)**   Railways 34–35 have been added.

*The National Gazetteer . . . 1868.*

**Copy**   (LAO Scorer Map 47).

**(D)**   *St.* has been added to the railway symbol (Ae). Railways 36–37 and S have been added.

*A New County Atlas Of Great Britain And Ireland Containing Sixty-Eight Coloured Maps By W. Hughes . . . London Virtue* [1873].[8]

**Copies**   BL Maps 11.c.11; CUL Atlas 5.87.6; Bod Maps C.15.c.2; NLS Map 12.a.9; Leeds UL W 208a. (Grimsby RL Hughes RM 3).

*The National Gazetteer of Great Britain And Ireland. London, James S. Virtue, City Road.* [Division VI. 1875].[9]

**Copy**   Leeds UL W 215.

**(E)**   The imprint is *LONDON, J.S. VIRTUE & C°. LIMITED* (Ce, OS). Railways 38–41, 43–44 have been added; the Mablethorpe branch and Louth-Bardney line are wrongly routed; there is also a link from Grimsby to Little Coats. The county is coloured buff and is overprinted with the boundaries and names of seven parliamentary divisions (e.g. *BRIGG*) in red; the sea is coloured green.

*A New Parliamentary And County Atlas Of Great Britain And Ireland Containing Seventy-Two Coloured Maps By W. Hughes . . . And Others Edited By Professor A.H. Keane . . . London J.S. Virtue & Co., Limited, City Road.* [1886].[10]

**Copies**   BL Maps 11.c.34; Leeds RL Q 912.42 H 874; Manchester RL F 912.42 K 1; Manchester UL AMGS B 33; Hull RL 912.42; Bristol RL EN 1.

**References**

1   The date of receipt in the BL set is, in fact, 20 NO 1865. Several parts were deposited the same day but Part 20 was certainly issued in the autumn of that year.
2   Philip, G. op. cit., p. 38.
3   *Proceedings of the Royal Geographical Society*, Vol. 21 (1876–7), p. 429 contains Hughes' obituary.
4   Kingsley, p. 281.
5   Hall, N.J. *Trollope*. (Oxford, 1991), pp. 306–307, 337–338, 348–355; Martin, R.B. op. cit., pp. 407 and 502. Hall says the sum was £4000 p.a.
6   The assumed date of the issue in parts. The title quoted is that of the first engraved title-page (of two, the second being letterpress) in part 1 of the original parts issue.
7   The map in CUL S 696.b.86.15 has no imprint; it was probably trimmed in binding.
8   1873 has been suggested by several commentators; there are no railways after 1868 on the county map.
9   This set was issued in parts and the first has only the original type of engraved title-page.

Division VI is taken from the spine. At the end of the last part are three letterpress title-pages for the final intended binding and the Poor Law returns for 1875.

10  The date is assumed from the inclusion of the boundaries of the divisions created by the Redistribution of Seats Act of 1885.

Chubb DLXII.

## 127

### PARSONS, R.M.                                                    1868

Size: 48cm x 33.3cm.

The map was included in the Report of the Boundary Commissioners of 1868, required under the provisions of the Representation of the People Act of the previous year. The map was prepared in the Ordnance Survey offices under the direction of Capt. R.M. Parsons. It is a simple outline map, showing the boundaries and names of the wapentakes, key roads, a few rivers and the railways, with their names. Only the towns and a few important villages are named. The map is signed by Col. Henry James, the famous director of the Ordnance Survey. He had become Director-General in 1845, having started as a second Lieutenant in the service in 1826. While in Paris in 1855 he had looked at the possibility of combining lithography with the new craft of photography; the camera he was commissioned to obtain for the Ordnance Survey was 18″ square and 3′ long. In 1859 A. de C. Scott introduced the use of photozincography, the process which James ensured became the standard for the production of O.S. maps, making possible the reduction of standard large-scale maps to smaller scales. He was appointed a Fellow of the Royal Society in 1848, knighted in 1860 and became a commander of the order of Queen Isabella the Catholic by the Queen of Spain in 1863. He retired from the O.S. in 1875 and died in 1877.[1]

The report also included maps of the towns, which were to be affected by the parliamentary boundary changes. Maps of Lincoln[2] and Great Grimsby[3] thus appeared as a result.

Eyre and Spottiswoode were the official government printers in the nineteenth century. William Spottiswoode (1825–1883) succeeded his father in 1846 as head of the firm but combined it with his career as a mathematician (he lectured at Balliol) and original work on the polarization of light and electrical discharge in rarefied gases. The firm continues to thrive although it has lost its earlier privileges.

### EDITION

(A)  *LINCOLNSHIRE/(NEW DIVISIONS OF COUNTY)/* (Ea, in a rectangular panel). *Zincographed at the Ordnance Survey Office, Southampton under the superintendence of Capt^n. R.M. Parsons R.E. F.R.A.S. Col. Sir H. James . . . Director./1868/* (Be–De, OS). *Scale – 5 Miles to an Inch/Miles 5 + 20 Miles* [25 miles = 13cm] (Ba–Da, in a panel below the map but above the imprint). *REFERENCE/*5 lines of colours for the boundaries and election places (Ab–Ad in a panel). *Henry James./Colonel Royal Engineers.* (signature, Db–Dd in a panel). Railways 1, 3–4, 6–27, 29, 32, 34, A, D and H are shown by single continuous lines. Red and blue dots represent proposed and temporary places of election respectively; red and blue lines mark proposed and present parliamentary boundaries respectively and yellow those of the hundreds.

*The Representation Of The People Act, 1867. (30 & 31 Vict. Cap. 102.) Boundary Commission. Report . . . For England And Wales. 1868. Presented to both Houses of*

*Parliament by Command of Her Majesty. London: Printed By George Edward Eyre And William Spottiswoode, Printers To The Queen's Most Excellent Majesty. For Her Majesty's Stationery Office. 1868.*

**Copies**   BL Official Publications Dept. BS 18/94 (3); Bod G.A. Lincs. C1 and Pp Eng 1867–8/20. (LAO Scorer Map 46).

**References**

1   Twyman, op. cit., pp. 43–49; DNB. Vol. XXIX (1892), pp. 210–213.
2   LRL LIN 352 is a copy of the report as it affected Lincoln with the map; LRL Map 951 is a copy of the map.
3   LRL Map 220a is an example of the map of Grimsby.

## 128

## SEELEY, JACKSON AND CO. AND S. W. PARTRIDGE & CO.         1869

Size: 5.1cm x 3.9cm.

This outline map only marks 15 towns, *HUMBER R., NORTHERN OCEAN* and *THE WASH* and the names of 4 of the adjoining counties. The only physical features are two unnamed rivers. The map appears as part of a children's puzzle picture in a periodical and made a second appearance some 14 years later in a different paper. The origins of the map can not be established and the name of the artist of the woodblock on which the whole page is engraved is not known. Earlier maps in the same series bear the initials J.P.

## EDITIONS

**(i)**   *LINCOLNSHIRE.* (Ce, in a rectangular panel). The page is headed with the page number *148* (Aa, OS) and *[The Children's Friend, October 1, 1869* (Ea, OS). *HIERO-GLYPHICAL READINGS.* (Aa–Ba). *No. XX. LINCOLNSHIRE* (Da–Ea). The coat of arms of Lincoln (Ca). The map appears in the top left corner, followed by text and pictures, e.g. [Map, i.e. Lincolnshire] *is principally/remarkable/for the im-/ mense flocks/of/*[picture of fowl with a decoy]*/which resort/to it. They are/mostly caught in/* . . . The map, etc. appears on the verso of p. 147, which contains the end of a story [Jack the Conqueror] and two short anecdotes, etc.

*No. CVI. October 1, 1869 The Children's Friend. Price One Penny.* [Top of outside cover]. [London, Seeley, Jackson And Co. and S.W. Partridge And Co. 1869].

**Copy**   BL PP 1163.[1]

**(ii)**   The map and text are unchanged. Page number *144* (Aa, OS). *PUZZLE-PICTURES. No. XXI.* (Ca, OS). Page 143 on the back of the map page has the continuation of an article on James Clerk Maxwell, a poem *NOTHING TO DO* and the answer to picture-puzzle No. XX.– Leicestershire.

*Early Days. Sept., 1884. Price One Penny.*

**Copies**   BL PP 183 b; Bod Per. 1419 d 268/1884–5.

**References**

1   The volume in Bod (Per. 1419 d 297/1869) lacks October; the answer to the puzzle appears in the November issue (p. 175).

## 129

### JOHN BARTHOLOMEW                                                1872

Size: 44.6cm x 34.45cm.

George Bartholomew (1784–1871) was an engraver in Edinburgh but it was his son John (1805–1861), who really founded the firm, which became famous for its maps. John undertook his apprenticeship in 1820 but specialised in map engraving from 1826.[1] In 1839 he obtained the right (held for a further 99 years) to print the maps in *Encyclopaedia Britannia*. Until 1860 he only undertook map work for other firms and this remained one of the firm's strong suits. His son, also John (1831–1893), took control in 1859, introducing lithography. He was elected to the Royal Geographical Society in 1857, one of his sponsors being George Everest. He refused an offer to join with George Philip in Liverpool or open jointly in a London venture in 1879. His son, J.G. Bartholomew (1860–1920), took over in 1888; he was instrumental in developing the lithographic process and exhibited at the Paris Exhibition in 1878 the pioneering technique of contour layer colouring. The first book incorporating maps using the technique was Baddeley's *Guide to the Lake District* (1880). He financed the first lectureship in geography in Edinburgh University and was a founder-member of the Scottish Geographical Society in 1884. The family tradition has been maintained and since 1910 the firm has been Geographer and Cartographer to the King. The firm is still, of course, in business and is particularly well-known for its county series at a scale of two miles to an inch. The first series of these is distinguished by the cover title: *BARTHOLOMEW'S REDUCED ORDNANCE SURVEY* . . . These sheets first appeared in c.1892. It is not known when the sheets covering Lincolnshire first came out; sheet 14 covered the southern part of the county and may have been issued in 1896. Sheet 10 covered the northern section – Lincoln was on the exact divide of the two sheets – and was certainly issued later as one of the last sheets to come out, perhaps after 1900. After pressure from the Ordnance Survey Bartholomew removed the reference to that body on the covers, rewording the title as *BARTHOLOMEW'S NEW REDUCED SURVEY* . . .; these maps have the badge of the Cyclists' Touring Club and The Edinburgh Geographical Institute (founded in 1902).

The first publisher of the present map was W.H. Smith. The business was started by two brothers in c.1820, Henry Edward and William Henry (1792–1865). The elder brother resigned in 1829 and the more energetic, W.H., ran the firm until 1846, when his son, also William Henry (1825–1891) joined him and 'Son' was added to the firm's name. In 1848 they obtained the concession from the London & North Western Railway to establish a newsagent's shop at Euston Station. Soon they were establishing themselves along the main railway routes, selling cheaply produced novels, papers, periodicals and maps to the intending traveller. The firm became the largest newspaper distributor as well as developing a network of subscription lending libraries that rivalled Mudie's. The elder Smith retired in 1858 and the younger turned to other activities; he became an M.P. in 1868, held a number of important posts, including becoming the First Lord Of The Admiralty in 1877, and inspiring Gilbert and Sullivan to satirise him in *H.M.S. Pinafore*. He was appointed Lord Warden of the Cinque Ports in 1890, the year before his death. The firm continues to flourish.[2]

The map was prepared for lithographic use from *The Imperial Map Of England & Wales According To The Ordnance Survey* . . . *By John Bartholomew*, a multi-sheet work at 4 miles to the inch. The plates remained in service until just before the 1939–1945 war. Its first appearance must have occurred in 1872, if the railway data are up-to-date and accurate, in a folding map for railway users. A slightly earlier transfer from the same source was issued by Houlston and Wright whose partnership ended with Houlston's death in 1869.[3] Entitled *ENVIRONS OF LINCOLN* it covered only the central part of the county, extending west beyond Worksop but not reaching the east coast, the Humber or further south than Belton Park.[4] Although probably issued c.1870 the map includes the Lincoln-Grantham line, which opened in 1872. It was also listed on the inside cover of the county map by W.H. Smith as number 23 in their series. This map of central Lincolnshire was used again to illustrate the notes on places visited when the Royal Archaeological Institute of Great Britain held its annual meeting in Lincoln in the summer of 1880.[5] Further transfers from the same source include maps entitled: *ENVIRONS OF NOTTINGHAM* and *HULL & DISTRICT*.[6] They both slightly overlap the Lincoln area map. A further map from the same source, but somewhat reduced, appeared in c.1879, put out by the firm of R. Allen of Nottingham. The map, *ALLEN'S GUIDE MAP OF THE LINCOLNSHIRE COAST*, omits the very northern fringes of the county and all of the area south of Grantham and Spalding.[7]

## EDITIONS

**(A)**  *LINCOLNSHIRE/Scale of Miles/8* [= 5cm] (Ea). *London. W.H. Smith & Son, 186, Strand.* (Ce, OS). *J. Bartholomew, Edin[r].* (Ee, OS). *LINCOLN* is printed across the map in black. The map is coloured; the county is pink, the sea blue and adjoining counties yellow, green or brown. Railways 1, 3–39, F, I, L, M and N are marked by a thick line between two thinner lines; railway xvii is marked by 3 thin strands. There is no compass. A graticule covers the map but there are no marginal numbers or letters; nor are there are marks for latitude, etc.

Issued in a card cover with the title: *W.H. SMITH & SON'S REDUCED ORDNANCE MAP OF LINCOLN AND ENVIRONS. Scale 4 Miles to an Inch. PRICE ONE SHILLING Mounted on Cloth. LONDON W.H. SMITH & SON, 186 STRAND. AND ALL RAILWAY STATIONS.* [1872**].

**Copy**  Bod Johnson Maps 139.

**(B)**  Size: 46.9 cm x 34.3 cm. The county name in the title has been moved (the final letter is now 3mm. from the border instead of 1mm.). The whole of the county is confined within the map's inner frame; previously the line of the Humber projected into the area between the inner and outer frames. The W.H. Smith imprint has been replaced by *John Bartholomew & Co., Edin[r].* (Ee, OS). *Copyright* has been added (Ae, OS). Railways 40–46 and 48 have been added, and railway xvii removed. The outlet of the Nene has been redrawn and the words *Reclaimed* and *VICTORIA* removed. The name *RUTLAND* has been removed. The letters *N, M,* and *S* (denoting North, Mid and South) have been removed. All the major roads in the county have been redrawn; previously there was no distinction between the roads. The county boundary is edged in pink, the coast line blue and the boundaries of adjoining counties marked in olive, green or brown (Rutland and Northamptonshire are not marked separately); main roads are brown.

Issued in a cover with the title: *SMITH & SONS SERIES OF REDUCED ORDNANCE MAPS FOR TOURISTS BY J. BARTHOLOMEW MAP OF ENVIRONS OF LINCOLN LONDON W.H. SMITH & SON, 186. STRAND & ALL RAILWAY BOOKSTALLS. MOUNTED ON CLOTH.* The words *ENVIRONS OF LINCOLN* are on a label stuck across the centre of the cover. [1887**].

**Copies**  Nottingham UL Li 1. B8 E 90 (Sir Francis Hill's copy) and Li 1.B8 F.

Issued in a cover with the title *PRICE ONE SHILLING W.H. SMITH & SONS SERIES ... LINCOLN JOHN BARTHOLOMEW & CO. EDINBURGH LONDON W.H. SMITH ... [1888].*[8]

**Copy**   CUL Maps 34.058.70.

**(C)**   The title is *LINCOLNSHIRE/BY J. BARTHOLOMEW, F.R.G.S.* (Ea). *SCALE 4 MILES TO AN INCH* (Ce, OS). Railway 49 has been added.

Issued in a cover with the title as in State C. [1888**].

**Copy**   Grimsby RL X000: 912 BAR.

Similar maps, divided into two sheets, one for the southern and one for the northern sections of the county, were issued as part of the Bartholomew 4 Miles to an inch series after 1900; they appeared thus as part of *Royal Atlas Of England And Wales Reduced From The Ordnance Survey*. Originally issued in 20 parts at 6d. each from March, 1899 to May, 1900, there were 31 regional maps transferred from *The Imperial Map Of England & Wales* . . . Lincolnshire appeared as part of plates 27 and 31 at 4 inches to a mile, and on plate XI at 10 inches to a mile. The parts appeared in atlas form in 1900 or 1901.[9]

The maps also appeared in small sections on postcards. They were issued by John Walker & Co. Ltd. in their *Geographical Series*, with a small picture in the corner of the map relating to the area covered by the map. No. 545 covers the area within the triangle Bourn, Sleaford and Grantham, with a picture of Grantham Church; No. 547 has Horncastle at the western edge and reaches Skegness, whose Clock Tower and Parade are depicted; No. 548 shows Skegness, Burgh Le Marsh and Wainfleet, the illustration of The Pier, Skegness occupying the whole of the lower portion. They were available by 1904.[10]

A further appearance (with changes) occurs in *Lincolnshire* by J. Charles Cox in Methuen's Little Guides Series, (London, 1916).

### References
1  Gardiner, L. Bartholomew 150 years. (London, 1976).
2  Pocklington, G.R. The story of W.H. Smith and Son. (London, 1921); Wilson, C. First with the news: the history of W.H. Smith, 1792–1972. (London, 1985), *passim*.
3  Boase, F. Modern English biography. (London, 1965).
4  CUL Maps D.70.87.1 is an example.
5  LAO Sc. P2/16; LRL LIN 9 ROY; Grimsby RL X000: 720 ROY. An example of the map only is Bod Map C17:39 (9).
6  NLS has copies of both (Maps s.52.35), the Hull map being numbered 51 in the series.
7  The only known copy (LAO Scorer Map 51) is a loose sheet; it is not known from which publication the map was taken.
8  It is suggested that the map was reissued ready for the summer season, 1888, before the railway in state C was completed.
9  The atlas made up from the original parts with their covers is in Bod (Maps C17.b.11); CUL (Atlas 5.90.2) and Nottingham UL Dept. of Geography have copies of the final atlas form.
10 The author owns the cards described, which were postally used in 1904; no. 546 in this series must relate to the county and others in the series probably display other areas of the county.

## 130

**JOHN BARTHOLOMEW**                                         **1872**

Size: 20.3cm x 15.2cm.

Bartholomew provided the plates for George Philip and Son; based on the Ordnance Survey maps they are a reduced and simplified version of **129**. John Pincher Faunthorpe (1844?–1924), the author of the text, for which the plates were first prepared, produced many county geographies during the early 1870s for Philip. After graduating from London University in 1865 he was ordained priest in 1868 and, as the title-page tells us, became Vice-Principal of St. John's College, Battersea at the time of writing. In 1874 he became Principal of Whitelands Training College and remained there until his retirement in 1907. He was a Fellow of the Royal Geographical Society. The *Handy Atlas* appeared almost annually into the present century. Edward Adrian Woodruffe-Peacock (1858–1922), the author of the short text on Lincolnshire's natural history divisions (state P), was a son of Edward Peacock, 'the leading authority on the dialect of north-west Lincolnshire'.[1] The son was born at Bottesford Manor Farm, studied at St. John's College, Cambridge, but transferred to Durham to train for the church in 1881. By 1891 he was Vicar of Cadney cum Howsham; an early member of the Lincolnshire Naturalists' Union he was secretary from 1893 and President 1905–1906. He was also a Fellow of both the Linnean Society and the Geological Society and acted as Curator of the Lincolnshire County Herbarium. With two of his siblings he prepared the draft material for the unpublished *Lincolnshire Place-Name and Dialect Dictionary* and acted as Editor-in-chief.[2] He was inducted into the living at Grayingham in 1920. The map with Woodruffe-Peacock's article was prepared by a fellow geologist, Alfred John Jukes-Brown. Jukes-Brown (1851–1914), after taking his degree at St. John's, Cambridge, was employed by the Geological Survey from 1874 and spent 10 years working in Lincolnshire and several of the adjoining counties. He later worked in the south-west and retired early on grounds of ill-health, the year after winning the Murchison medal in 1901.[3]

Versions of the map were also provided for the firm of W.J. Cook of Boston; they produced directories for several towns in the county; not all the editions have been found and they may also contain county maps, the Lincoln directory for 1895 being a case in point – (state (Q)). The firm of W.J. Cook (states N and Q) seems to have issued in Boston directories with maps from 1894. In all contemporary directories covering Boston the only publisher thus named is variously given as Joseph Cook or, more often, Cooke, proprietor of the *Boston Guardian*. By 1899 the firm had moved to Derby and its directory publications covered many towns in the East Midlands and Hull. In its 1899 edition of the Hull directory were listed directories for Boston & District (1898), Gainsborough & District (1897), Gainsborough and Retford combined (1897), Grimsby & Cleethorpes (1898), Grantham area (1896) and Lincoln & District (1897). None of these has yet been found and it is not known, therefore, if maps were included.

The examples of atlases recorded in these notes as in the ownership of Mr. D. Kingsley were sold in 1994; their present owners are not known. See also Addenda, pp. 411–12.

## EDITIONS

(**A**)   *THE COUNTY OF/LINCOLN/English Miles/6* [= 1.6cm]/*1. NORTH/2. MID/3. SOUTH/Railways/Roads/Canals* [indicated by a solid single; two thin lines and one thin respectively] (Ea). *PHILIPS' EDUCATIONAL SERIES OF COUNTY MAPS./* (Ba–Da, OS). *GEORGE PHILIP & SON LONDON & LIVERPOOL* (Ce, OS). Latitude and longitude are numbered at 30′ intervals with intervening marks at 10′ gaps. *Meridian of Greenwich* (De, between borders). Railways 1, 3–39, A, C, L, M and S are marked and, mostly, named. The map is coloured; pink for the north and south sections and yellow for mid Lincs., the sea blue and the adjoining counties cream-pink. There is no compass.

*The Geography Of Lincolnshire For Use In Schools. By Rev. J.P. Faunthorpe . . . London: George Philip And Son, 32 Fleet Street; Liverpool: Caxton Buildings, South John Street, And 49 And 51 South Castle Street. 1872 . . .*

**Copies**   BL 10048 aaa 27; CUL Maps c.34.016.70; Bod Gough Adds. Lincolnshire 8° 23 (2); LAO Sc P2/19 (L 914.2); LRL L 9; Louth RL GL 91.

Collected into an atlas without text or title-page.[4]

**Copy**   Bod Maps C17.d.15.

*The Geography Of Lincolnshire* . . . has two other maps of the county, which do not justify separate places in this catalogue. On p. 6 a map (8.35cm x 8.7cm) shows the main physical features, and below is a note to show that the figures on the map represent the depth of the sea and (in brackets) the rise of tides. On p. 14 a map (8.3cm x 7.45cm) shows the geological divisions by various shading. Both have text on the versos.

**(B)**   The heading above the map has been removed. Plate number *20* has been added (Ee, OS sideways and on the reverse). A graticule covers the map and letters A–I have been added (between borders) along the top and bottom, with numbers 1–7 down both sides.

*Philips' Handy Atlas Of The Counties Of England, By John Bartholomew . . . London . . . Liverpool . . . 49 & 51 South Castle Street. 1873.*

**Copies**   BL Maps 14.a.36; CUL Atlas 7.87.1; Bod Maps C17.e.5; Leeds UL W 209.

**(C)**   The plate number on the verso has been removed. Railway 40 has been added.

*Philips' Handy Atlas . . . 1874.*

**Copy**   Leeds UL W 210.

**(D)**   The plate number is in larger figures and has been moved 2mm. downwards. A fine red line outlines the county borders. The adjoining counties are no longer coloured. A small spur at Grimsby links the MSLR and ELR. Stations are indicated by short lines drawn at right angles across the line. *Sta.* has been added to the symbol for railways (Ea).

*Philips' Handy Atlas . . . Of England: New And Revised Edition, With A Consulting Index. By John Bartholomew . . . 1876.*

**Copies**   BL Maps 14.a.56; CUL Atlas 7.87.4; Bod Maps C17.e.6; NLS Map 2.e.17; Leeds UL W 211.

*Philips' Handy Atlas . . . 1877.*

**Copies**   The private collections of Mr. D. Kingsley and Mr. E. Burden.

**(E)**   The colour of mid Lincs. is orange-brown instead of yellow.

*[Philips' Handy Atlas . . . 1878.]*[5]

**Copy**   The private collection of Mr. D. Kingsley.

*Philips' Handy Atlas . . . 1879.*

**Copy**   The private collection of Mr. E. Burden.

**(F)**   The colour of mid Lincs. reverts to its pre-1878 yellow. Railways 41–44, I and T/U have been added.

*Philips' Handy Atlas . . . 1879.*

**Copy**   The private collection of Mr. D. Kingsley.

*Philips' Handy Atlas . . . By John Bartholomew . . . New And Revised Edition, With Consulting Index . . . 1880.*

**Copy**   Norwich RL R 912.42.

**(G)**   *GLANFORD BRIGG* has been reduced to *BRIGG*.

*Philips' Handy Atlas . . . 1881.*

THE COUNTY OF
# LINCOLN
English Miles
0 1 2 3 4 5 6

The Colouring represents the
Parliamentary Divisions,
each returning 1 member:
Railways
Roads
Canals

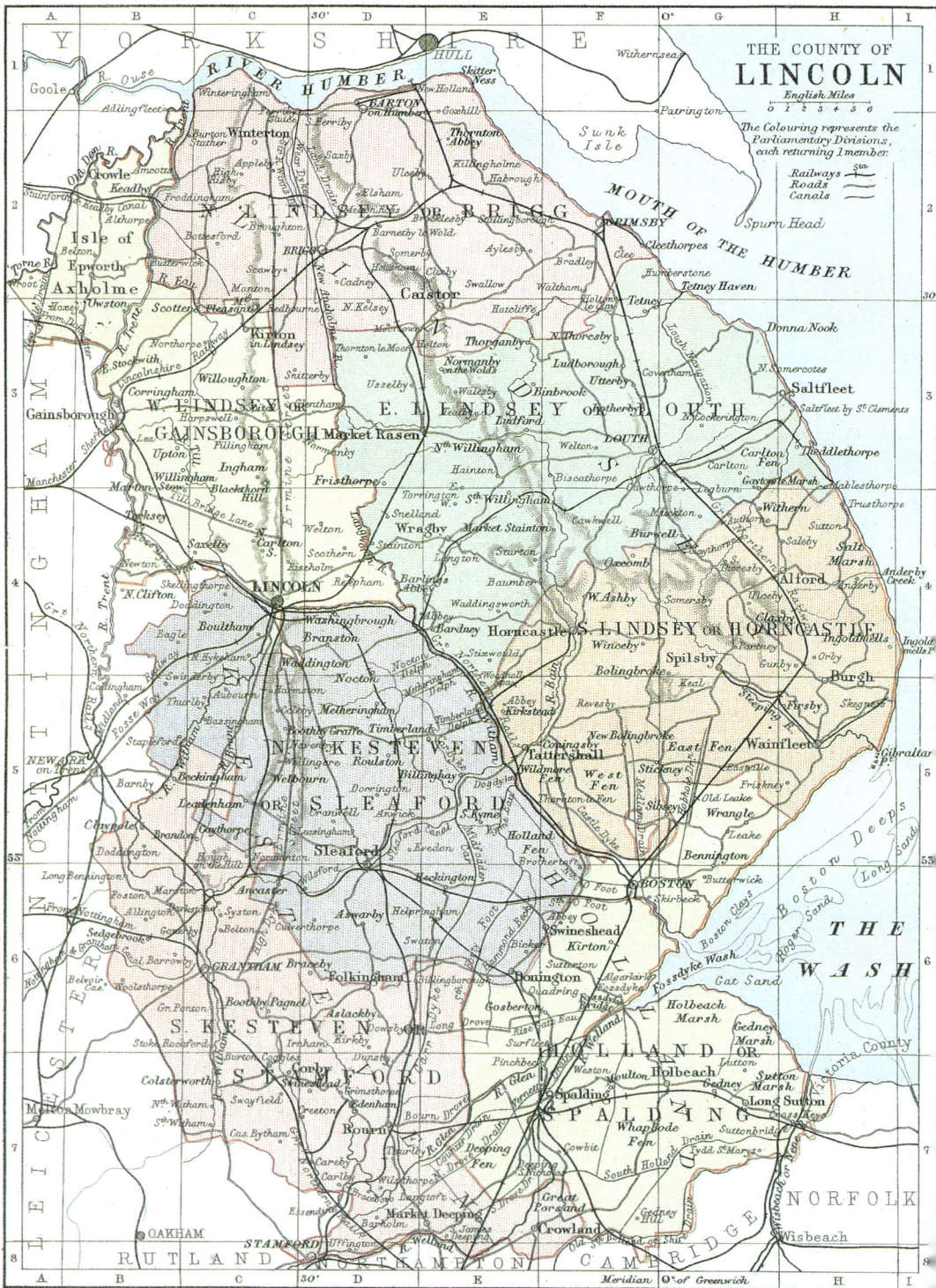

9. 1887    MAP 130K – GEORGE PHILIP & SON.

**Copy**  The private collection of Mr. E. Burden.

*Philips' Handy Atlas . . . 1882.*

**Copy**  The private collection of Mr. D. Hodson.

(**H**)  Railways 45–46 have been added. Stations are marked on all railways. The Bardney-Louth line has been redrawn, passing south of Biscathorpe, etc. The Mablethorpe line has also been redrawn, passing north of Theddlethorpe.

*Philips' Handy Atlas . . . 1882.*

**Copy**  Leeds UL W 212.

*Philips' Handy Atlas . . . 1884.*

**Copy**  The private collection of Mr. E. Burden.

(**I**)  The map is overprinted with the names of the new parliamentary divisions, e.g. *N. LINDSEY OR BRIGG.* Below the scale *The Colouring represents the/Parliamentary Divisions,/each returning 1 member* replaces the previous note. The divisions are coloured pink, brown, yellow, green, violet-grey, the sea is blue and the remainder uncoloured. The borough boundaries are marked by fine red lines. The Sleaford avoiding line has been added. The line from Doncaster to Gainsborough is now a curve from the MSLR; previously the line to Doncaster was an extension of the line from Lincoln, with a spur southwards to form a link with the MSLR. A new cut links the *Fossdyke Nav* to Torksey. A 'blob' on the plate, on the name of Kirton/in Lindsey has been removed; the missing 't' in Kirton has been replaced and two minor roads added in the same area.

*Philips' Handy Atlas Of . . . England Including Maps Of North & South Wales, The Channel Islands, And The Isle Of Man. Reduced From The Ordnance Survey, And Coloured To Shew The New Parliamentary Divisions, According To The Redistribution Bill, 1885. New And EERevised Edition. With Consulting Index. London . . . Liverpool . . . 45 To 51 South Castle Street. 1885.*

**Copies**  BL Maps 14.a.60; RGS 8 A 2.

(**J**)  *LINCOLN* has been printed on the map verso.

*Philips' Handy Atlas . . . 1886.*

**Copies**  BL Maps 15.a.44; NLS Map 2.e.18.

(**K**)  *Stow* (south of Willingham) has been added.

*Philips' Handy Atlas . . . 1887.*

**Copies**  BL Maps 16.a.42; CUL Atlas 7.88.10.

*Philips' Handy Atlas . . . 1888.*

**Copy**  The private collection of Mr. E. Burden.[6]

(**L**)  Railways 48–49 have been added, but from Alford.

*Philips' Handy Atlas . . . Of England . . . And The Isle Of Man. New And Revised Edition, With Consulting Index. London . . . Liverpool . . . 1891.*

**Copy**  The private collection of Mr. C.A. Burden.

(**M**)  The title has been changed to *LINCOLN/comprising/Lindsey, Holland & Kesteven./English Miles/6* [= 1.6cm]/*The Colouring represents the/Parliamentary Divisions/ and Boroughs./*the symbols for *Railways/ Roads/Canals/* (Ea). Railway X has been added and appears twice (at Hull and at Goole), the central portion being 'off the map'. The boroughs are now coloured orange-mustard, the divisions yellow, pink, brown, green or slate-blue; the county is edged with a broad band of brown; the sea is blue.

*Philips' Handy Atlas . . . 1893.*

**Copy**   Leeds UL W 212a.

**(N)**   Down the right side of the map *J.H. SMALL & SON, Drapers,* has been added. Down the left side *15, MARKET PLACE, Boston.* has been added; all the lettering is outside the frame of the map. *COOK'S DIRECTORY.]* (Ae, OS). *[BOSTON AND DISTRICT.* (Ee, OS). *W.J. Cook & Co., Boston, Lincolnshire.* (Ce, OS). The plate number, the county name on the verso and Philip's imprint have been removed. The map plate remains as in state M.

*Cook's Directories! . . . Boston And District Directory, Containing A Description Of The Dock And Town . . . By W.J. Cook & Co., Directory Publishing Offices:– 13, Strait Bargate, Boston. Second Issue. Price 4/6 Nett. Wing & Broughton, Printers, Boston. 1894 . . .*

**Copies**   Grimsby RL B608: 058 COO; Boston RL L. BOST. 929.[7]

**(O)**   The map is basically as state (M), with the imprint, plate number and county name (on verso) restored and the additions of state (N) removed. Railways 50–51 have been added. Railway 48 has been redrawn to leave the ELR south of Alford. *Note. Railway Stations marked/'Sta.' have the same name as their/named town or village./* has been added below the scale; the symbols are now to the east of Spurn Head (Ea). Villages, which have railway stations have been named, e.g. *Willoughby Sta.* or *Sta.* has been added to place-names. There are numerous examples of both additions. The area around Crowland is differently coloured from Holland. The boroughs are now bright yellow. The broad band of brown around the county border has been removed. *Old/Don R.* has been re-engraved (previously on one line).

*Philips' Handy Atlas Of . . . England, Including Maps Of The County Of London, North And South Wales, The Channel Islands, The Isle Of Man; And Plans On An Enlarged Scale Of The Environs Of Six Important Towns. New And Revised Edition, Shewing Every Railway Station In England And Wales. With Consulting Index. London: George Philip & Son, 32 Fleet Street, E.C. Liverpool: 45 To 51 South Castle Street. 1895.*

**Copies**   BL Maps 32.a.60; CUL Atlas 7.89.6; Bod Map C17.e.12; Leeds UL W 213; NLS Map 2.e.30.

*Philips' Handy Atlas . . . South Castle Street.* [1896].

**Copy**   The private collection of Mr. E. Burden.[8]

**(P)**   The plate number has been removed. The title now reads: *A SKETCH MAP OF THE SOILS OF/LINCOLNSHIRE,/By A.J. Jukes-Brown, B.A., F.G.S./Brown Alluvium. Peat, and Fen Soils./Blue Clay Soils./Pink Gravel and Sand Soils./Green Chalky Soils./Yellow Limestone Soils./ NATURAL HISTORY DIVISIONS,/By Rev. E. Adrian Woodruffe-Peacock,/Shown with Red Boundaries & Numbers/* (Ea). This involves the removal of all East Yorkshire, including Spurn Head. *The Naturalist – Oct. 1895.* (reads down Ea, OS). *To be inserted between pp. 288 and 289.* (reads down Ee, OS, 3.5cm from the map's frame). The map is as state (O), with the colour overprinting now related to the soils and natural history divisions, the boundaries of which are marked by broken lines. The various natural history divisions have numbers and letters in red. A line across the county is marked *H.C. Watson's Line of Demarcation/between Lincs N and Lincs S.*

twice; the line of grey-blue is broken on the west side but continuous when the mouth of the Witham is reached.

*Sketch Map Of The Soils & Natural History Divisions Of Lincolnshire. A.J. Jukes-Brown . . . and Rev. E. Adrian Woodruffe-Peacock. Price One Shilling. October, 1895. Reprinted from 'The Naturalist' for the Lincolnshire Naturalists' Union.*

**Copies**   Bod G A Lincolnshire 8° 253 (2); LAO Sc P2/20 and P2/21; LRL L. A53 and Pye L 553; Louth RL GL 5; Gainsborough RL L 553.

*The Naturalist: A Monthly Journal of Natural History For The North Of England. Edited By Wm. Denison Roebuck . . . 1895. London: Lovell Reeve & Co., 5, Henrietta Street, Covent Garden, E.C. McCorquodale & Co. Limited, Cardington Street, Euston; and Leeds: Basinghall Street, 1895.* [Issue for October 1895].

**Copies**   BL Ac 3016; Bod Radcliffe Per 19138.d.4.

**(Q)**   The title, etc. again reads: *THE COUNTY OF/ LINCOLN/ English Miles/6* [= 1.6cm]*/ The Colouring represents the/ Parliamentary Divisions,/ each returning 1 member./* the symbols of *Railways/ Roads/ Canals/* (Ea). There is no plate number. Railways 48–51 and X are now shown. The colours of the divisions are as states I–L.

*Cook's Directories . . . Grimsby And Cleethorpes Directory, Containing: Historical Account Of The Town; Places Of Worship . . . W.J. Cook & Co., Local Directory Offices, 13, Bargate, Boston. Second Edition. Price 6/6 Nett. Grimsby: Richard H. Eden . . . 4, Victoria Street and 12a, Flottergate. 1896.*

**Copy**   LRL (Pye) PL Grim 929.

*Cook's Directories . . . City Of Lincoln and District (Embracing The Surrounding Villages Within Ten Miles Radius) Directory Containing: Historical Account Of The City . . . Map, Etc., Etc. By W.J. Cook & Co., Local Directory Publishers, 13, Strait Bargate, Boston. Price 5/6 Nett. Lincoln: Clifford Thomas . . . 202, High St. 1895.*[9]

**(R)**   The title is as in state (M). *Railways Roads Canals* are now in one line separated by their symbols, below the note on railway station names. The plate number has been restored (Ee, OS, sidewards). Railway 52 has been added. *LINCOLN.* has been added on the map's verso.

*Philips' Handy Atlas . . . Liverpool: Philip, Son & Nephew, 45 To 51, South Castle Street.* [1897].

**Copy**   The private collection of Mr. D. Hodson.

*Philips' Handy Atlas . . . 1898.*

**Copy**   The private collection of Mr. D. Kingsley.

*Philips' Handy Atlas . . .* [1900].[10]

**Copy**   Birmingham RL A 912.42.

After 1901 maps in this series may be distinguished by the addition of *London Geographical Institute* to the maps. The Handy Atlas continued to appear for several years until, at least, 1910, while the maps also appeared in series designed for school use. One such item is E. Mansel Sympson's *Lincolnshire* (Cambridge, 1913 and 1914); on the end papers are two maps of the county, one entitled *Physical Map Of/Lincoln* and the other *Geological Map Of/Lincoln*. Both use the basic map with new railways, overprinted with colours for heights above (and below) sea-level and for geological strata respectively.

## References

1   Elder, E. The Peacocks of North-West Lincolnshire . . . *Lincolnshire History and Archaeology*, Vol. 24 (1989), p. 29.

2   ibid., p. 39; Elder, E. Edward Peacock And His Family, in *Some Historians Of Lincolnshire*, ed. by C. Sturman. (Lincoln, 1992), pp. 73–77 (portrait on p. 72), with a useful note on his writings (pp. 80–81), which refers to: Seaward, M.R.D., Biographical and Bibliographical Notes on the Rev. Edward Adrian Woodruff-Peacock in: *Lincolnshire History and Archaeology*, Vol. 6 (1971), pp. 113–124; *Lincolnshire At the Opening of the Twentieth Century*. (Brighton, 1907), p. 210; a family tree of the Peacock Family appears in: Leach, T.R. and Pacey, R. Lost Lincolnshire Houses, Vol. 4. (Burgh le Marsh, 1993), p. 56.

3   *Who Was Who, 1897–1915.*

4   The spine title suggests that the plates cover a period from 1870 to 1889. Maps for some counties are in later states and do not show the features common to the first issue.

5   The only copy lacks a title-page. Most counties are printed in brighter colours and a railway added to the Cornwall plate indicates a later issue than the 1877.

6   A copy in Bod (Maps C17.e.8) lacks a title-page. It is ascribed to 1887 but could equally be 1888.

7   The Grimsby copy lacks a title-page, but the spine is dated 1894.

8   Kingsley, p. 298 recorded a copy in the Sussex Archaeological Society Library, but only the Sussex map from this work is held there.

9   The only copy (LRL LIN 912) lacks all but a fragment of a map, which forms the frontispiece. The map may be of Lincoln.

10  Burden, p. 218 suggests 1900, but 1899 is possible, as the Goole-Reedness line (opened in 1900) is not marked.

131

## EDWARD WELLER                                              1872

Size: 20.9cm x 16.2cm.

This small map is another based on the Ordnance Survey and it is closely related, therefore, to Bartholomew and Philip (**129** and **130**); the lettering of the county name has the same 'type-face' as in the earlier of those two, while the remainder of the title area has the appearance of the latter. The map was first used in a school primer written as part of a series published by William Collins and sold for twopence a copy. This firm, which is still going strongly, was founded in Glasgow by William Collins (1789–1853) and, on his death, taken forward by his son, also William (1817–1895). A Provost of Glasgow he was knighted in 1881. The author of the little primer was Thomas Archbold (1835–1902), who, at the time of issue, was Headmaster of the Diocesan Middle Class School at Burgh le Marsh. After obtaining his M.A. at St. John's College, Cambridge and being Vice-Principal at the Diocesan College at Culham from 1864–1867 he came to Lincolnshire in 1869. In 1875 he became Principal of the Diocesan Training College in Norwich, retiring in 1895, when he became Rector of Burgate, where he died.[1]

## EDITIONS

(**A**)   *LINCOLNSHIRE/English Miles/10* [= 2.75cm]/*Railways/Canals/Roads/* [with three symbols]/*1 North Division/2 Middle D°./3 South D°./* (Ea). *Edw^d. Weller* (Ee, OS). Railways 1–4, 6–17, 19–27, 30, 32–33, 35–36, A, C, D, F, G, H, I, J and L are shown. The map is coloured: North Lincs. pink, Middle Lincs. green, South Lincs. buff-brown and the sea blue. The map lacks compass, imprint, marks for latitude, etc. and indication of the meridian.

*Collins' County Geographies. Edited By W. Lawson, F.R.G.S. Geography Of Lincoln-shire. Adapted To The New Code. By Rev. Thomas Archbold . . . With Illustrations. London: William Collins, Sons, & Co.,* (sic). [1872].[2]

**Copy**   The author's collection.

(**B**)   An imprint: *William Collins, Sons & Co. London & Glasgow.* has been added (Ce, OS).

*Geography Of Lincolnshire . . . Collins, Sons, & Co., 17 Warwick Square, Paternoster Row.* [1872].

**Copies**   BL 10348 bbbb 20; CUL Maps c.34.017.70; Bod G.A. Lincs. 8° 23 (3). (LRL Map 167).[3]

(**C**)   The signature has been removed. Plate number 7 has been added. (Aa, OS, and on the verso).

*Collins' Series Of Atlases. Atlas Of England And Wales, Containing Maps Of All The Counties, Coloured Into The Parliamentary Divisions, And Showing Railways, Roads, And Canals; Together With A Railway Map Of England. London: William Collins, Sons & Company.* [1877].[4]

**Copies**   BL Maps 3.a.5; CUL Atlas 7.87.3; Bod Maps C17.e.9.

*Collins' . . . Atlas Of England . . . 1877.*

**Copy**   CUL Atlas R.7.87.69.

(**D**)   The scale, imprint, notes and symbols have been removed. The map is printed in brown on buff coloured card as part of the back cover of the booklet; Lincolnshire is on the inside and the map of Norfolk on the verso (i.e. the outside).

*The Eastern Counties Atlas. Consisting Of Sixteen Maps, Full Colored constructed & engraved by John Bartholomew . . . Together With Maps Of The Eastern Counties Of England . . . William Collins, Sons & C°. London & Glasgow Price Sixpence* [1878].[5]

**Copy**   CUL Atlas R.6.87.32; Bod 2027.d.49 (28).

(**E**)   Size: 25.9cm x 20.1cm. The map in state D is printed in violet on buff.

*The Eastern Counties Atlas . . .* [1878].[6]

**Copy**   BL Maps 40.c.32.

(**F**)   The map is in the same state as B, i.e. with imprint, symbols and explanation and plate number. Railways 34, 37–38, 40 and 44–46 have been added.

*Collins' Series Of Atlases. Atlas Of England . . . London: William Collins, Sons & Co.* [1882**].

**Copy**   Leeds UL W 221.

### References
1   Crockford's Clerical Directory, 1880.
2   On p. 10 the fishing figures for Grimsby for 1871 are given, implying issue in 1872. The text refers to the branch railways at Firsby to Spilsby and to Wainfleet, i.e. the Skegness extension (1873) was not in evidence. Neither of these lines is on the map. The address for the publisher given on the cover differs from that given on the title-page of state (B). It is: Bridewell Place, New Bridge Street.
3   Grimsby RL X000: 914 ARC is a copy of the book, but lacks the map.
4   The preface is dated 1877; BL received its copy on 14 JUN 1877.

5   It is assumed the different arrangements of the maps were issued more or less simultaneously.
6   The BL receipt stamp is dated 6 NOV 1878.


## 132

**ANDREW REID**                                                    **1875**

Size: 21.18cm x 20.1cm.

It is not certain whether the writers of the article, which this map illustrates, J. Daglish and R. Howse, drew the map as well. The map is fairly simple and has close similarities to the basic outline and some of the content of Philip's map of 1872 (**130**). That it was taken from a larger map (almost certainly Ordnance Survey) is suggested by the inclusion of much of West Norfolk. The map shows geological strata. Only the main towns and a few villages are marked. Although the title would suggest that it only covers the north of the county the map shows the whole county, though the strata are not separated in the south as they are in the north. The text of the article covers pp. 23–33, with the map being placed between pages 32 and 33. Andrew Reid was an engraver, artist as well as a lithographer who entered the field of publishing in 1845; from 1849 he published *Reid's Monthly Timetable*, a rival to Bradshaw's Timetables, which survived for over 100 years. He was an early pioneer in the issue of folding strip maps showing the then main railway routes and illustrated with his own engravings.[1] The firm was also responsible for the production of other non-railway maps, most notably a plan of Newcastle (1896) and of the River Wear (1898).[2]

### EDITION

(**A**)   *GENERAL GEOLOGICAL MAP OF NORTH LINCOLNSHIRE./ Scale of Miles./10* [= 2.8cm] (Ba–Da, OS). *Vol. XXIV PLATE VIII.* (Ea, OS). *To illustrate Mess^{rs}. Daglish & Howse's paper on 'The North Lincolnshire Ironstone Field'.* (Ba–Da, OS above the map title and below the note on the volume). *And^w Reid Newcastle* (Ee, OS). Below the map's lower neat line are 9 boxes in two lines, preceded by *Faults* [and its symbol]. The boxes are coloured, and bear the names of different strata/rocks; the colours (green, yellow, orange, rust-red, violet-pink with grey stipple) are also applied to the map. Railways 1, 3–21, 23–27, 29–36, 39, A, D, H, L, M, O, Q and S are marked.

*North Of England Institute Of Mining And Mechanical Engineers. Transactions. Vol. XXIV. 1874–5. Newcastle-Upon-Tyne: A Reid, Printing Court Building, Akenside Hill. 1875.*

**Copies**   BL Ac 3232; CUL Q 432.3.c.2.24.

#### References
1   *Railway Magazine*, Vol. 95, No. 582 (July–August, 1949), pp. 266–267: an anonymous article celebrating the centenary of the timetable.
2   Tooley, op. cit., p. 532.

# 133

## JOHN BARTHOLOMEW                                              c.1878

Size: 140.6cm x 84.6cm approx.

This very large map was intended for school use and is based on Ordnance Survey models. In spite of its size (it includes the area as far south as Peterborough) the map does not carry all the detail offered by such a canvas. The villages are all shown and some of the more grand country houses, together with the usual geographical features; however, the many small groups of houses and individual properties noted, for instance, on Bryant or Greenwood are not to be found here and perhaps such detail was considered irrelevant to its instructional purpose. The names of the wapentakes are given but the boundaries are not delineated. The map is found in two sheets as well as in a single linen-mounted form. It is possible that the map was issued as a rival to a series of similarly large sheets also intended for school use but put out by Collins. With the title: *Collins' New School Room Map Of* . . . examples are recorded for a number of counties but none for Lincolnshire has yet been found.[1]

## EDITION

(A)  *PHILIPS' NEW MAP/OF/LINCOLNSHIRE/FROM THE ORDNANCE SUR-VEY./By J. BARTHOLOMEW, F.R.G.S./SCALE OF MILES/10* [= 15cm]/*LONDON, PUBLISHED BY GEORGE PHILIP & SON, 32. FLEET STREET,/LIVERPOOL, CAXTON BUILDINGS, AND 51. SOUTH CASTLE STREET./* (Ae). Railways 1, 3–41, 43–44, A, B, C, F, I, J, K, L, M, N, P, R and S are shown and mostly named, with directions added, e.g. *from Doncaster.* There is no compass. The map is overprinted *NORTH DIVISION, MID DIVISION* and *SOUTH DIVISION.* The map is coloured: North and South Divisions and Lincoln borough pink, Mid Division, Grantham and Stamford green, Boston and Grimsby yellow and the sea varying shades of blue to indicate the depths of water. The boundaries of all the divisions and boroughs are overlaid with a band of colour in a deeper shade. The inner border is marked with every minute of latitude, etc. and numbered every 10 minutes; a graticule covers the map based on the 10′ markings. *Meridian 0 of Greenwich* (De, between borders).

Issued as a separate map, in one or two sheets, dissected on cloth. [1878**].

**Copies**   BL Maps 3355 (37); CUL Maps 70.87.2; Bod Maps C17: 39 (10); NLS Map 1.23.31.

### Reference
1   BL has copies for Cumberland, Derbyshire, Durham, Gloucestershire and Hampshire. The series was edited by W. Lawton.

# 134

## ANONYMOUS                                                    1880

Size: 2.5cm x 2.5cm approx. There is no frame.

This tiny map formed part of a puzzle issued in a children's periodical to illustrate the life of John Wesley. The magazine *Early Days* began life in 1824 as *Child's Magazine and Sunday Scholar's Companion* and took the title *Early Days* in 1846. From 1917 until

1958, when it closed, it was called *Kiddies' Magazine*.[1] The map marks Lincoln and Boston only; the single other named feature is *The Wash*; the Witham and the Wolds are the only physical features marked. The map was probably prepared in the periodical's office from an unidentified source; the whole page of text and illustrations was cut into a single wood-block. Only 14 of these biographical puzzles were prepared and only four county maps formed a part, including Lincolnshire. The Wesley puzzle also has maps of England and the Americas. The key to the puzzle appeared on p. 175 (Nov., 1880). The magazine featured a different map of the county in another puzzle in 1884 (**128 (ii)**).

## EDITION

(i)   The map forms part of a pictorial life of John Wesley and lacks all usual map features, e.g. scale, title, compass, provenance. The page is headed *160* (Aa, OS); *BIOGRAPHI-CAL PUZZLE.– No. X.* (Ca, OS). The first line of the text reads: [picture of John Wesley] *was/born/in* [Map of Lincolnshire] *When/a/little* [boy with a hoop] *he was/rescued/from a/* [picture of a house on fire] . . .

*Early Days. 1880. London: 'Early Days' Office, 2, Castle-Street, City-Road, And 66, Paternoster-Row.* [The issue for October, 1880. Price One Penny].

**Copy**   BL PP 183 b.[2]

### Reproduction
(i)   The whole page was reproduced in *Lincolnshire Past & Present*, No. 3 (Spring, 1991), p. 10.

### References
1   Mr. Jim English kindly provided the periodical's previous history.
2   Bod Per 1419 d.297 is another set but lacking the puzzle picture.

## 135

### EDWARD STANFORD                                                  1880

Size: 27.6cm x 20.1cm.

Stanford's prepared this map (which is related to **93**) to accompany a guide written by Sir Charles Anderson of Lea. The map is transferred from the firm's own larger maps of England and Wales, but based on the Ordnance Survey Old Series one-inch maps. It extends as far south as Peterborough. Stanford has already been noted, when he acquired the use of Ebden's plates in c.1871 (**85 C**); the firm became a prolific producer of maps used to illustrate the many guide-books of the 1880s and 1890s designed to satisfy the demands of cyclists and other travellers. Other work is noted below (**136 and 138**). The author, Sir Charles Henry John Anderson (1802–1891), was the ninth baronet and came from an old established Lindsey family. After Oriel College, Oxford (B.A. in 1826, M.A. in 1829) he settled down to life in Lea as squire, sitting on the bench for more than sixty years and undertaking many other acts of local service; he was High Sheriff in 1851.[1] The guide book to the county first appeared in 1874, but this had no map. *A Short Guide To The County Of Lincoln* (Gainsborough, 1847) was, perhaps, an even earlier attempt, but it has no maps and nor are they called for.[2] The edition of 1880 included the map as did the so-called second edition of 1881; after Anderson's death a further edition was prepared by A.R. Maddison in 1892. Arthur Roland Maddison (1843–1912) was educated at Rugby and Merton College, Oxford. After ordination he was curate in several

Lincolnshire parishes before becoming vicar of Cherry Willingham from 1898 to 1901; in 1907 he was presented to the living of Burton by Lincoln. He was Cathedral Librarian in 1877 and Succentor from 1879–1903.[3] He became a prominent student of local history and is particularly noted for his two series of *Lincolnshire Wills* (1888 and 1891) and four volumes prepared for the Harleian Society of *Lincolnshire Pedigrees* (1902–1906).[4]

The *Lincoln Diocesan Church Calendar* made use of the map on five occasions from 1889; the 1869 edition used Becker's map (**116 (v)**). A further appearance of the map occurs in the second edition of *Murray's Handbook of Lincolnshire* (1903); the first edition (1890) had a map drawn by Edward Weller (**123 N**).

## EDITIONS

(**A**)   *MAP TO ACCOMPANY THE GUIDE TO LINCOLN.* (Ba–Da, OS). *London: Published by Edward Stanford, 55, Charing Cross, S.W.* (Be–De, OS). *Stanford's Geog[l]. Estab[l].* (Ee, OS). *Scale of Miles/10* [= 3.3cm] (Ea). *LINCOLN* (across the map from north-west to south-east). There is no compass. Railways 1–41, 43–44, A, B, C, H, I, J, L, N, R and X are shown.

*The Lincoln Pocket Guide : Being A Short Account Of The Churches And Antiquities Of The County, And Of The Cathedral Of The Blessed Virgin Mary Of Lincoln, Commonly Called The Minster. By Sir C.H.J. Anderson . . . With Map, and Plan and Illustrations of the Cathedral. London: Edward Stanford, 55, Charing Cross, S.W. 1880.*

**Copies**   BL 10352 aaa 15; CUL 8474.d.73; Bod G.A. Lincs 8° 62; LAO Sc P 5/2.21 and L. Linc 910; LRL LINC 9; Louth RL L 9. (LAO Scorer Maps 32 and 49).

*The Lincoln Pocket Guide . . . With Map and Illustrations. Second Edition. London . . . 1881.*

**Copies**   LRL LINC 9; Louth RL L 9.

(**B**)   The imprint has been removed.

*The Lincoln Diocesan Church Calendar, Clergy List, And General Almanack, For The County Of Lincoln, For The Year Of Our Lord, 1889. Published With The Sanction Of The Bishop. Lincoln: James Williamson, 290, High Street; London: James Parker & Co., Southampton Street.*

**Copy**   LAO FL; LRL L 052.

(**C**)   The imprint *London: Edward Stanford, 26 & 27 Cockspur St., Charing Cross, S.W.* has been added (Be–De, OS). Railways 45–46 and 48–49 have been added.

*The Lincoln Pocket Guide . . . By Sir C.H.J. Anderson . . . Third Edition, Edited And Revised By The Rev. A.R. Maddison . . . London: Edward Stanford, 26 & 27, Cockspur Street, Charing Cross, S.W. 1892.*

**Copies**   BL 10351 aa 58; CUL 8474.d.75; Bod G.A. Lincs 16° 3; LAO Ex 942.55; LRL LINC 9; Gainsborough RL L 9.

(**D**)   The heading has been altered to *MAP OF THE DIOCESE OF LINCOLN.* The imprint has been removed. Below the map, under the heading *Deaneries.*, are five columns recording the numbers and names of the deaneries in the archdeaconries of Lincoln and Stow. The numbers are printed on the map in red and the boundaries of the deaneries are also marked in red.

*The Lincoln Diocesan Church Calendar . . . 1895 . . . Lincoln . . . London . . .*

**Copies**   CUL L.125.11.d.L30; Bod Cal. 11126 e 42; LAO FL; LRL L 052; Grimsby RL X000: 262.3 LIN.

(E)    The map is unchanged; the lists below the map have been reset; the only difference being that, in the fifth column, *D* is set on the line below *40* . . . in the adjacent column; (previously D was half-way between the levels of 39 and 40 in the fourth column).

*The Lincoln Diocesan Church Calendar . . . 1896 . . .*

**Copy**    CUL L.125.11.d.L31; Bod Cal. 11126 e 42; LAO FL; LRL L 052.

(F)    A thick red line has been printed to mark the boundaries of the north and south of the county. The lists below the map have been reset in four columns, with references to the pages in the text, which refer to that deanery, with, below them, a note about the red line.

*The Lincoln Diocesan Church Calendar . . . 1899 . . .*

**Copy**    CUL L125.11.d.L34; Bod Cal. 11126 e 42; LAO FL; LRL L 052.

*The Lincoln Diocesan Church Calendar . . . 1900 . . .*

**Copy**    CUL L125.11.d.L35; Bod Cal. 11126 e 42; LAO FL; LRL L 052.

The map re-appeared in *Handbook for Lincolnshire . . . Second Edition, thoroughly revised, with map and 3 plans. London: Edward Stanford, 12, 13 and 14 Long Acre, W.C. 1903*. The map now has a double frame, the inner one being marked with degrees of latitude. The county name is given (Ca, OS), imprint (Ce, OS), scale (Ae, OS) and signature (Ee, OS). It is coloured to indicate heights above sea-level, an explanation of which appears (Ea).[5]

### References

1    Olney, R. op. cit., pp. 39–41; *The Times*, obituary on 10 October, 1891; *The History Of Lea . . . by Oxoniensis . . .* (Gainsburgh (sic), 1902), pp. [17–18].
2    There is a copy in BL (10360 aa 8) of this work, printed and published by Alfred Smith of Gainsborough.
3    *Lincolnshire At the Opening of the Twentieth Century.* (Brighton, 1907), p. 182.
4    Bennett, N. A.R. Maddison and the Development of Local History in Lincolnshire. In: Sturman, C. (ed.). *Some Historians of Lincolnshire.* (Lincoln, 1992), pp. 27–31.
5    The Bod copy of the 1890 edition of Murray's Handbook of Lincolnshire has, erroneously, this map in the pocket at the back of the volume.

## 136

### EDWARD STANFORD                                                    1881

Size: 2 plates each 8.7cm x 13.8cm.

This little map was prepared for a series of school primers on geography; the northern part of the county as far south as Horncastle appears on one page, while the continuation southwards is on the facing leaf. The basis for the map is the Ordnance Survey one-inch series but there is very little similarity with the Stanford transfers of 1880 (**135**). The emphasis is on physical features: the Wolds, the 'Cliff' and the sandbanks in the Wash are especially prominent. Only the main towns and large villages are marked; there are no roads but railways are included. The map extends as far north as Beverley and south to Peterborough. A series of letters and figures against the five main towns are a novel feature, not explained on the map itself; +C (at Lincoln) means there is a cathedral; B 1 or 2 refer to borough status and the number of MPs returned.

    In 1881 Edward Stanford, senior, retired and his son, also Edward, took over the running of the business. The son was elected to the Fellowship of the Royal Geographical Society in 1877; among his proposers was J.P. Faunthorpe. Under his control the firm continued to prosper and, in 1893, it was appointed Geographer to the Queen, an office renewed by Edward VII and again by George V.[1] Charlotte Maria Shaw Mason (1842–1923) was a noted educationist, who worked her way up from teaching in a London infants' school to become a lecturer in Education. She retired because of ill-health in 1878 and went to live in the Lake District where she founded the House of Education, and wrote many textbooks for the new National schools; Charlotte Mason College, Ambleside is named after her and now forms part of Manchester University.[2]

## EDITIONS

(**A**)  *LIN/COLN* (down the centre and split between the two leaves). *20 ENGLISH MILES* [= 3.7cm] (Ae, OS, of southern plate). *Stanford's Geogl. Estabt. London.* (Ee, OS, of southern plate). There is no compass. There is a graticule. Railways 1–44, A, C, D, G, H, I, J, K, L, M, N, O, P, Q, R, S, T, U and V are shown. Although unnumbered themselves the maps have the texts of pages 161 and 164 on their versos.

*The London Geographical Series. Geographical Readers For Elementary Schools. By Charlotte M. Mason . . . Book III. for Standard IV. The Counties Of England. With Thirty-six Maps. London: Edward Stanford, 55, Charing Cross, S.W. 1881.*

**Copies**   BL 10001 bbb 14/11; CUL 1881 (6) 332; Bod 2017 f 55.

(**B**)  *LINCOLN, NORTH* runs down the right side of the northern plate; *35* (Ee, OS and sidewards on the northern plate). *LINCOLN, SOUTH* and *36* appear in the same positions on the southern plate. The areas of Lincoln, Grantham, Boston and Grimsby are now marked out. The figures for MPs have been reduced to 1 at Boston, Lincoln, Grantham and Kings Lynn; at Newark, Doncaster, Retford and Stamford the figure has been eliminated. The names of the new parliamentary constituencies have been added; e.g. *HOLLAND/OR SPALDING*. Railways 45–46 have been added. The map is printed in dull blue; the divisions lemon or pink with their names in deep pink; the four boroughs are also pink. The Isle of Axholme is excluded from the colour process.

*Stanford's Handy Atlas And Poll Book Of The Electoral Divisions Of Great Britain And Ireland, With Synopsis Of The Representation Of People Act. Statistical Tables Of Each County, And List Of Members, Indicating The Supporters Of Mr. Gladstone's Home Rule Bill, The Unionist Liberals, And The Conservatives. London: Edward Stanford, 55, Charing Cross, S.W. 1886.* [The preface is dated June, 1886].

**Copies**   BL Maps 4.a.49; CUL Atlas 7.88.1.

(**C**)  *1886.* has been added below the title on both plates. The map is printed in dull blue, with N. Lindsey pink, Grimsby violet, Lincoln, Boston and Grantham mustard and the rest of the county lemon. The lemon colour signifies places returning conservative members and the pink liberal MPs. Hachur is grey blue, the divisional boundaries red and their names pink.

*Stanford's Handy Atlas . . . Of Each County, And An Alphabetical List Of Members, Indicating The Political Party With Which They Are Associated, And The Constituencies By Which They Have Been Returned. Second Edition. London . . . 1886.* [The preface is dated October, 1886].

**Copies**   BL Maps 4.a.51; Birmingham RL A 912.42.

**References**

1   Herbert, F. op. cit., pp. 67–95.
2   DNB Missing Persons. (Oxford, 1993), pp. 450–451; Cholmondeley, E. The Story Of Charlotte Mason. (London, 1960).

## 137

### JOHN BARTHOLOMEW                                               1882

Size: 23.35cm x 16.6cm.

In outline this series, prepared for the ninth edition of *Encyclopaedia Britannica*, bears the expected resemblance to the maps Bartholomew designed for use by Philip in the *Handy Atlas . . .* (**130**). There is, however, a greater amount of detail in the present case, but no special distinguishing features. Apart from its use in the various re-issues of the *Encyclopaedia Britannica*, which appeared until c.1903, there is one use of the plates in an atlas of 1892, issued by the well-known Edinburgh publishing firm of Adam and Charles Black. Adam Black (1784–1871) set up his bookselling business in Edinburgh after a short time with Lackington, Allen and Co., who published Luffman's little pocket atlas of 1806 (**59**). He took his nephew Charles into partnership and they acquired, in 1827, the copyright of *Encyclopaedia Britannica*. John Bartholomew acquired the privilege to print the maps in that work in 1839, a right exercised until 1938. The business thrived to the extent that they paid £27,000 for the copyright of the works of Sir Walter Scott. The firm was carried on by Adam Black's two sons after his death and still thrives.

### EDITIONS

(**A**)   *LINCOLN* (Ca, OS). *VOL. XIV* (Aa, OS). *PLATE VIII* (Ea, OS). *J. Bartholomew Edin^r.* (Ee, OS). *ENCYCLOPÆDIA BRITANNICA, NINTH EDITION* (Ce, OS). *Scale of Miles/10* [= 3cm]/*Note/N. North Lincoln/M. Mid Lincoln/S. South Lincoln* (Ea). At 10′ intervals are marks and numbering for latitude, etc. between the borders. *Meridian of Greenwich* (De, between borders). Railways 1, 3–46, A, C, F, I, K, L, N, P, R and T/U are shown. The county is overprinted in brown with the sea blue; a fine red line marks the borough boundaries and divides north, south and mid Lincs. There is no compass.

*The Encyclopedia Britannica A Dictionary Of Arts, Sciences, And General Literature Ninth Edition Volume XIV Edinburgh: Adam And Charles Black MDCCCLXXXII . . .*

**Copies**   BL 12214 a 1; CUL Ant.b.108.365; Bod 399.a.705.

(**B**)   The title now reads *LINCOLN.* (Ca, OS) and is repeated on the verso, in brown ink. The volume, plate and Encyclopædia references have been removed. The signatures now read: *John Bartholomew & Co.* (Ee, OS) and *Published by A. & C. Black, London.* (Ce, OS). *The Edinburgh Geographical Institute* (Ae, OS). *Black's Handy Atlas of England & Wales* (Ea, OS, reading down the side of the plate). *Plate 29.* (Ee, OS, sidewards on). *Parliamentary Divisions 1885* heads a list of 7 numbered areas, followed by *Boroughs* heading a list of four boroughs, numbered 8–11 (Ea). The numbers appear on the corresponding areas of the map. *PLATE 29* (on the verso, in brown). *NORTH/SEA* has been engraved east of Tetney Haven (previously it read up the right-hand side of the plate in one line). *Sutton/on Sea* has been altered from *Sutton in/the Marsh*. Railways 48–49 have been added. The boroughs are coloured pink; their boundaries are outlined with fine red lines, while those of the divisions are edged in lemon, lime-green and orange-brown with the sea blue.

*Black's Handy Atlas Of England & Wales A Series Of County Maps And Plans With Descriptive Index And Statistical Notes Edited By John Bartholomew . . . London Adam & Charles Black 1892.*

**Copies**   BL Maps 33.a.5; CUL Atlas 7.89.1; Bod Maps C17.e.11; NLS Map 3.e.34; PRO M 12.

**(C)**   *Page 653* has been added (Ae, OS). The imprint has been altered to *W. & A.K. Johnston Limited.* (Ee, OS). The references to volume, plate and *ENCYCLOPEDIA BRITANNIA, NINTH EDITION* of state A have been restored, while the additions of state B have been removed. Only the scale remains (Ea). Railways 50–51 are marked by dotted lines (to suggest that they were under construction). The county is overprinted pink with the sea blue and the rest uncoloured.

*The Encyclopædia Britannica . . . 1882.* [1892].[1]

**Copy**   Lincoln Cathedral L.

**(D)**   Railways 50–51 are now marked in the same way as other lines.

*The Encyclopædia Britannica . . . 1882* [1895].[2]

**Copy**   Leicester UL REF. 032 ENC.

**(E)**   The imprint now reads: *W. & A.K. Johnston.* (Ee, OS).

*The Encyclopædia Britannica . . . MDCCCLXXII.* [1899].

**Copies**   Guildhall L. S 032; Nottingham UL Enc. AE 5.E 6.[3]

**References**

1   On the verso of the title-page of the Guildhall copy (see 1899) is a printing history, which shows that there were reprints in 1890, 1892, 1895, 1896 and 1898. Only the 1892 matches the map's state.
2   The 1895 (or 1896) reprint is the likeliest date – see note 1.
3   See note 1 re Guildhall's copy; the Nottingham set appears to be basically the same but there are 11 supplementary volumes, issued in 1903. The 1899 issue was specially put out by *The Times*, which claimed to have sold 'more than 15,000 copies . . . between March 23rd, 1898, and March 23rd, 1899' (advert in Mr. E. Burden's set).

## 138

**EDWARD STANFORD**                                                      **1885**

Size: 2 plates each 17.9cm x 24cm.

This map is, like that of 1881, split between two plates printed face to face (**136**). There is, however, considerably more detail in the present map in terms of names of places, equally applied to the adjoining counties. The emphasis in the earlier on hachur for the Wolds and the 'Cliff' is notably absent here; although the sandbanks in the Wash area are equally pronounced in both series here there is no attempt to indicate the higher ground. The map is a transfer from *STANFORD'S LIBRARY MAP OF ENGLAND AND WALES*, first issued about 1881.

**EDITION**

**(A)**   *LINCOLN. (PARTS OF LINDSEY.)* (Ca, OS of the 'upper' plate). *48* (Ea, OS, above

*VII.30a.* on the 'upper' plate). *London: Edward Stanford, 55 Charing Cross.* (Ce, OS, on both plates). *Stanford's Geographical Estab*ᵗ. (Ee, OS, on both plates). *SCALE OF MILES* (Ae, on both plates between the inner and outer frame lines. 5 miles = 1.8cm). *LINCOLN. (KESTEVEN & HOLLAND)* (Ca, OS, on the 'lower' plate). *49* (Ea, OS, above *VII. 30b.* on the 'lower' plate). *LIN/COLN* (across the two plates). The inside of the outer frame is marked off in 'notches' with numbering, which represents miles, every five divisions; along the top and bottom the numbers are 5–70 and up the sides of each half map 5–50. Railways 1–46, A, B, D, F, H, I, J, K, L, M, N, O, Q, R, S, T, U and V are marked. The divisions are named, e.g. *SOUTH KESTEVEN/OR STAMFORD* in red and each is coloured yellow, green, purple or brown with boundaries marked in red; the boroughs (4) are coloured with pink diagonal stripes and the coastal areas blue. There is no compass or indication of latitude, etc.

*Stanford's Parliamentary County Atlas And Handbook Of England And Wales Containing Also Geological And Orographical Maps Of Great Britain, And Physical, Statistical, And Administrative Maps Of England And Wales With Lists Of Parishes, Petty Sessional Divisions And Unions, Population Tables, And Other Particulars Relating To County Statistics, Local Administration, And The New Parliamentary Constituencies London Edward Stanford 55 Charing Cross, S.W. 1885.*

**Copies**   BL Maps 19.a.12; CUL Atlas 7.88.11; Bod 22774 e 7; Leeds UL W 228; NLS Map 1.h.10; Hull RL 912.42.

**139**

**ORDNANCE SURVEY**                                                          **1885**

Size: 49.6cm x 39.5cm.

In 1885 the O.S. issued its own map to show the new arrangements determined by the Boundary Commission and consequent upon the Redistribution of Seats Act of that year. Further issues followed using a newly prepared map to cover a variety of official purposes, one being an index to the O.S. map series; all emphasise the parish names and boundaries (**143**). One oddity of this new series is the failure to include any railway after 1868.

**EDITIONS**

**(A)**  *LINCOLNSHIRE/NEW DIVISIONS OF COUNTY./Scale: Four Miles to an Inch/4 + 12* [12 = 7.7cm] (Ea). *Zincographed at the Ordnance Survey Office, Southampton. 1885.* (Ce, OS). *REFERENCE./5* lines and 3 symbols denoting the proposed parliamentary boundaries and those for petty sessional divisions and boroughs. Below is the signature of Lt. Col. R. Owen Jones R.E. (Ae, in a separate box from the References). Railways 1, 3–20, 22, 25–26, 30, 33, 35 (only to Navenby), 37, A, C, F, I, J, K, L and N are marked. The names of the divisions and their boundaries are printed in red and the boroughs are overprinted in dull violet. There is a graticule, but no compass nor indication of latitude, etc. The map is not a regular rectangle; a square is cut out of the lower right-hand corner and the spread of the east coast of the county has lead to the frame in that area being extended.

*Boundary Commission (England And Wales). Report Of The Boundary Commissioners For England And Wales. 1885. Part 1. Counties. Presented to both Houses of Parliament by command of Her Majesty. London, Printed By Eyre and Spottiswoode . . . 1885. [C.–4287.]*

**Copies**   BL Official Publications Dept. BS 18/94 (4) and Maps 10.cc.3; CUL OP. 1884–5 XIX; Bod Pp Eng. 1884–5/19; Norwich RL R 328.334; Leeds UL W 226; Manchester RL Parl F324.42 B8. (Nottingham UL Li 1.F 19 BOA).

**(B)**   The title and scale (Ea) and the table of references, with signature (Ae) are printed in smaller type, thus allowing more of the map (over which they are printed) to be seen.

*Report Of The Boundary Commissioners . . . 1885 . . .*

**Copy**   The private collection of Mr. T. Burgess.

**(C)**   The panels surrounding the title and scale, the References and signature have been removed. Nothing of the adjoining counties is shown. The divisions now appear in the form *EAST LINDSEY/OR/ LOUTH* (previously *LOUTH* only). The boundaries of the 7 divisions are red, the 4 boroughs are overprinted in violet and the divisional and petty sessional division names in large black letters. The names of the sub-divisions have been redrawn; the letters of *ELLOE* measure 1.4cm. (3.2cm in state A). Their boundaries are marked by thin black lines. Railway 5 and all those outside the county do not appear here. The wording of the *REFERENCES* is in larger type.

*Redistribution Of Seats Act, 1885 (Contents of County Divisions). Return to an Address of the Honourable The House of Commons, dated 7 July, 1885; – for, Return showing with respect to each of the Several Counties of England, Scotland, and Ireland, divided by 'The Redistribution of Seats Act, 1885' . . . Home Office, July, 1885. County Of Lincoln. (Sir Charles Dilke.) Ordered . . . to be Printed, 8 July 1885. London: Printed By Henry Hansard And Son . . . 258 – (25.)* The work forms part of Vol. XIX of *Accounts and papers for the session 23 October 1884 – 14 August 1885.*

**Copies**   BL Official Publications Dept. and Maps 14.d.3; CUL OP. 1884–5 LXIII; Bod Pp Eng. 1884–5/63. (Nottingham UL Li 1.F 19 BOU).

Collected as an atlas. The spine title is *Redistribution of Seats 1885.*

**Copy**   Leeds UL W 227.


## 140

## RAPHAEL TUCK and SONS                                      1885

Size: 8.4cm x 11.2cm.

Although these cards are strictly speaking outside the definition for inclusion their rarity and interest decrees otherwise. The firm of Raphael Tuck has been a large-scale producer of picture cards for well over a century and still continues to flourish. They had, at the time the cards were produced, offices in New York, Paris and Berlin. The maps show Leicestershire, Rutland, Nottinghamshire, Derbyshire as well as Lincolnshire. The main towns are shown with the railways and roads and, as space permits, some of the villages, not all of them the most important. A year or two after their first issue the cards were reproduced in Paris by a firm called Bognard. They were used in this country as reward cards for diligent scholars; there is space on the back for school use.

### EDITIONS

**(A)**   *LINCOLN* (across the county and below an engraving of the cathedral (from the south) (Ae). There are sepia illustrations of *SUDBURY HALL DERBYSHIRE* (Ba), *ENTRANCE LODGE/OF BURGHLEY/HOUSE* (Ea), *NEW UNIVERSITY – NOTTING-*

*HAM.* (Ce–Ee) and *NEWARK* (Aa–Ca). *REGISTERED* (Ed, reading down the side of the card). *RAPHAEL TUCK & SONS* (Ae, below the engraving of the cathedral). *London & New York* (Ee). Railways 1, 3–17, 19–21, 23–27, 30, 32–47, K, N and T are marked. The map is coloured pink, blue and yellow. On the verso are topographical notes on the five counties.

*Educational Series. The United Kingdom. The Whole Of The Counties Of England, Scotland, Ireland & Wales. With Leading Places Of Interest In Each. Illustrated In 22 Pictorial Maps . . . With descriptive information on each Map. One Shilling Complete Raphael Tuck & Sons . . .* [1885].[1]

**Copy**  The private collection of Mr. T. Burgess.

**(B)**  The references to Tuck & Sons, their offices in London & New York and the word *REGISTERED* have been removed. *BOGNARD, PARIS.* has been added (Ad, reading up the card).

Issued as a reward card for the School Board of London, by the Parisian firm named on the card. Above the topographical notes (on the verso) a panel permits the entry of the name of the school, name of the pupil, his standard, the signature of the Head Teacher and the date (in the 1880s) of the award for 'Punctual and Regular Attendance during the School quarter . . .'[2]

**References**
1  The title is taken from the cover of the holder containing the cards. The date is based on the county railway data and is agreed by other commentators.
2  The card containing Lincolnshire has not been found. Mr. T. Burgess has two cards, which conform to the description given here. They were probably issued a year or two after the Tuck issue since they were signed in June, 1888 and January, 1889.

# 141

## ORDNANCE SURVEY                                                  c.1886

Size: 68.1cm x 58.15cm (map only).

This map was prepared to accompany a large chart, which tabulates, parish by parish, their areas and the amounts within each of land, water, salt marsh, foreshore and tidal water. The map takes up the left hand half of a single larger sheet. The only clues to its dating are the legend at the foot of the chart stating that the names of the Parliamentary divisions and the marking of their boundaries accord with those of the act of 1885 and the inclusion of a label below the adjoining chart of the firm of Edward Stanford as agents of the O.S. They had obtained the sole agency in 1884. It is assumed that the map predates that of 1887 (**143**).

## EDITION

**(A)**  *ENGLAND AND WALES/ORDNANCE SURVEY/OF/ LINCOLNSHIRE/* (Ea). *All rights of reproduction reserved.* (Ae, OS). *Scale of this Index/2 + 8 Miles/Three Miles to One Inch* [8 miles = 4.8cm.] (Ee). *Price Two Shillings and Sixpence uncoloured/Three Shillings coloured* (Ee, OS). *CHARACTERISTICS*/above a chart showing the signs used for types of land, boundaries and the distinctive lettering used for parish and district names (Ee, above the scale). *REFERENCE BY LETTER TO THE SMALL PARISHES/ AND DETACHED PARTS OF PARISHES/* above a chart divided into two columns (Eb).

*Enlarged sketch of Part of the City of Lincoln*/above a map of Lincoln and its parishes (Ae). *Enlarged Sketch of the portion marked b*/above a map of the area centred on Langriville parish, and below the map, *REFERENCE TO THE PARISHES &c., SHEWN ON/ ENLARGED SKETCH MARKED b/* (Ac). *Enlarged Sketch of the Town of Stamford*/above a very small map of Stamford and its parishes (Be). Lindsey is coloured brown, Holland green and Kesteven pink. Railways 1–4, 6–47 are shown. No areas outside the county are included.

Issued in a cover with the label of Edward Stanford; the county name is added by hand.

**Copy**   LRL Map 238.


## 142

**TRÜBNER & Co.**                                                    **1886**

Size: 17.35cm x 11cm.

It is assumed that this map was prepared by the publisher's staff, probably working from drawings provided by the English Dialect Society (E.D.S.), since it appears in two volumes by two local authors, published three years apart both published by Trübner under the auspices of the E.D.S. Nicholas Trübner (1817–1888) was born in Heidelberg but came to England in 1843 and began his publishing business in 1852. The firm specialised in the publication of academic works, particularly oriental texts, a field in which he was an expert; later he obtained the contract to print the works of the Early English Text Society as well as the volumes of the E.D.S. The year after his death the firm amalgamated firstly with Trench to become Kegan Paul, Trench, Trübner & Co; as Routledge the firm still exists, being well known for its sociology and psychology lists. The map has no great distinction, merely noting the sites of 14 towns in the county plus Newark, the areas of the three wapentakes studied by Cole and Peacock, the *Foss Way* (from Newark to Lincoln) and *Ermine Street.*

Robert Eden George Cole (c.1831–1921) was Rector of Doddington for 48 years (retiring in 1909), Rural-Dean of Boothby Graffoe and a canon of Lincoln cathedral. At the time of his death in January, 1921 he was 89 years old. In 1878 Cole published plans to collect lists of Lincolnshire dialect words; from the responses he received (first published in his Deanery magazine) came the publication in 1886 of the *Glossary Of Words Used In South-West Lincolnshire* (a misnomer as Elder has pointed out, in that the glossary is confined to Boothby Graffoe deanery or wapentake, a small area 11 by 8 miles and a good way from the south-west of the county).[1] The map re-appeared in Edward Peacock's equivalent work for Manley and Corringham wapentakes, published in 1889, the first edition of which (1877) does not include a map. Peacock (1831–1915), the somewhat eccentric squire of Bottesford and father of Edward Adrian Woodruffe-Peacock (**130**), was a noted collector of dialect words and phrases. His major work in this field was the work recorded below.[2]

## EDITIONS

(**A**)   *SKETCH MAP OF LINCOLNSHIRE.* (Be–De, OS). *ENGLISH DIALECT SOCIETY.* (Ca, OS). No scale or compass. The boundaries of Manley and Corringham Wapentakes are marked and *Graffoe/Wapentake* is coloured pink.

*A Glossary Of Words Used In South-West Lincolnshire. (Wapentake of Graffoe). By The*

*Rev. R.E.G. Cole . . . London: Published For The English Dialect Society. By Trübner & Co., Ludgate Hill. 1886.*

**Copies**   BL Ac 1934/20; CUL 768.c.87.29; Bod Soc. 30205 e 2 (36); LAO R L 427.003 COL; LRL L 427; Grimsby RL X206:427 COL.

**(B)**   The colouring of *Graffoe/Wapentake* has been removed. The lettering of *Graffoe/ Wapentake*, *ISLE OF/AXHOLME* and *Ermine Street* (in two cases) has been redrawn. The Manley and Corringham wapentakes are coloured brown-orange.

*A Glossary Of Words Used In The Wapentakes Of Manley And Corringham, Lincolnshire. Second Edition. Revised And Considerably Enlarged By Edward Peacock . . . Vol. 1. London . . . 1889.*

**Copies**   BL Ac 1934/26; CUL 768.c.87.34; LAO R L 427.003 PEA; LRL L. Man 427; Grimsby RL X000: 427 PEA; Boston RL L. MANL. 427.

**References**
1   Elder, E. The Peacocks of North-West Lincolnshire. *Lincolnshire History and Archaeology.* Vol. 24 (1989), pp. 34–35 and 39 (notes 48 and 52).
2   ibid., pp. 29–39; Elder, E. Edward Peacock and his Family. In: Sturman, C. (ed.) *Some Historians of Lincolnshire.* (Lincoln, 1992), pp. 70–81.

## 143

### ORDNANCE SURVEY                                                        1887

Size: 54.1cm x 45.6cm (measured from the outer borders of the main map area; two small maps are outside the borders; one other, which breaks the lower frame, is not included in the measurement).

In 1887 the Ordnance Survey produced an index map to its own county series at various scales; this map, which shows and names all the individual parishes, was then used as the base map for a series, which offered material of an official or administrative nature. Although updated in terms of railways, compared to the 1885 map, they always lag a few years behind in the incorporation of later changes. There are no details of the adjoining counties.

### EDITIONS

**(A)**   *INDEX/to the/ORDNANCE SURVEY/of/LINCOLNSHIRE/Shewing /CIVIL PAR- ISHES/Scale of this Index 4 Miles to 1 Inch/N.B. The 6″ Scale Map of the County is being published/by Quarter Sheets . . . It is also/being published on the Scale of 1/2500 . . . / . . .* (Ea, OS). *Photozincographed and Published at the Ordnance Survey Office, Southampton, 1887./Price 2d/All rights of reproduction reserved./* (Ce, OS). *References/* 11 lines of symbols and their meanings (Ae, OS, in a separate box). *Enlarged Sketch of the/City & County of the City of Lincoln/above a map with, to the right, a box headed REFERENCE,* which provides a key to the features on the map (Ec, OS). *Enlarged Sketch of portion marked__a/above a map* of the area centred on Langriville, with a box headed *REFERENCE* to the right (Ee, OS). *Enlarged Sketch of portion marked b/above a map* of the parishes of Stamford, with a 3-line key to the right (Ce, in a box straddling the lower neat line). *Quarter sheets are composed of/four 1/2500 plans, thus:/above a* diagram, with 3 more lines of text below (Ad). The map is divided into large numbered rectangles, subdivided into 9 smaller numbered rectangles, each further divided into 12 more un-numbered rectangles. The boundaries of parishes are marked and their names given in small upper case letters. The only physical features are the sandbanks of the Wash and the coastal strip. *LINCOLNSHIRE* is printed in black across the centre of the map. Railways 1–4, 6–46 and K are marked.

Issued as a separate sheet.

**Copy**   CUL Map Library.

(**B**)   The frame lines, the numbered rectangles, the diagram of the Quarter sheet arrangement and the main Reference block have been removed. The title is *DIAGRAM/of the/SANITARY DISTRICTS/in/ LINCOLNSHIRE/Shewing also/CIVIL PARISHES/ Scale of this Diagram 4 Miles to 1 Inch/* (Ea). The date in the imprint has been altered to *1888.* Below the imprint is *Price 6d./All rights . . . reserved./* (Ce, OS). *REFERENCE TO COLOURS/*5 lines of text and symbols (Eb). The sketch of Lincoln has *CITY & COUNTY OF THE/CITY OF LINCOLN* added below the box of references (Ab). . . . *sketch . . . a* has been moved (Ec). *References/*9 lines of description and symbols (Ad, in a box). The map is overprinted with the boundaries of the unions and their names in red. Lincoln is coloured red; Louth, Stamford, Boston, Grimsby and Grantham are coloured brown; light blue is used for urban sanitary districts, but three, Long Sutton, Winterton and Weelsby, are in a darker blue. The boundaries of the three parts of the county and their names are green.

Issued as a separate sheet. 1888.

**Copies**   BL Maps 3364 (13); CUL Maps 70.88.1; LAO Kesteven Papers KCC 1/1.

(**C**)   The overprinted colours, the title below the sketch map of Lincoln, the imprint, etc., and the reference to colours have been removed. The title is *LOCAL GOVERNMENT BOUNDARIES COMMISSION./ DIAGRAM/of the/ALTERATIONS PROPOSED/BY THE BOUNDARY COMMISSION, /In the COUNTY of/LINCOLN/ Showing/*4 lines of colours and their meanings (Ea). *DANGERFIELD LITH 22 BEDFORD S$^T$. COVENT GARDEN 17195.* (Ee). 6 Parishes, from Misterton in the north to West Burton in the south plus an area west of Crowle, are coloured red to show that they are proposed for transfer from Nottinghamshire to Lincolnshire. Stamford is coloured green to signify the proposal to transfer it to Rutland, while Uffington, Tallington and West Deeping are also coloured green to show their proposed transfer to Northamptonshire. The old county boundary is marked in green and the suggested new one in red.

Issued as a separate sheet. [1888].

**Copy**   LAO Kesteven Papers KCC 1/1.

(**D**)   The title, altered from state C, is: . . . */In/LINCOLN,/Showing/ CIVIL PARISHES ADDED TO THE COUNTY – – – Hatched Red* [a box coloured red]/7 further lines relating to boundaries and their colours/ *Scale of this Diagram 4 Miles to 1 Inch./*signature of *Rob Owen Jones.* (Ea). The colours are as state C. *RUTLANDSHIRE* has been altered to *RUTLAND.*

*Report Of The Boundary Commissioners Of England And Wales, 1888. Vol. 1. Ordered to be printed 11th August 1888. Printed By Eyre And Spottiswoode, Printer To The Queen's Most Excellent Majesty. And to be purchased . . . from Eyre and Spottiswoode . . . or Adam and Charles Black, 6, North Bridge, Edinburgh; or Hodges, Figgis, & Co., 104, Grafton Street, Dublin. (279.) [Price 25s.]*

**Copies**   BL Maps 14.d.5; Leeds UL W 230.

(**E**)   The colours have been varied; Lincoln is grey with a red boundary line; outer

Boston's boundaries are red and green; the boundaries in the sketch map b are red and green. *PARTS OF LINDSEY*, etc. are in large green letters; the names of the unions are green. Many towns and large villages are shaded grey; Stamford and North Scarle are coloured green, while the area to the west of Crowle and that between Skellingthorpe and Saxilby are red.

*Report Of The Boundary Commissioners . . . 1888 . . . Dublin. 360.*

**Copies**   CUL OP. LI; Bod Pp Eng. 1888/51; Norwich RL R 328.334.

(**F**)   The title is: *DIAGRAM/of the/ORDNANCE SURVEY/of/ LINCOLNSHIRE/ Shewing/CIVIL PARISHES/Scale of this Diagram 4 Miles to 1 Inch/* (Ea). The imprint, price, etc. are as in state A with the date changed to 1888. There is no colouring.

Issued as a separate sheet. 1888.

**Copies**   BL Maps 3364 (14); LAO Scorer Maps 55–70 (includes 17 copies of the same map).

(**G**)   The title and map are as state A, with the date in the imprint altered to *1890*. The sketch of area a has been moved (Ce, above the imprint). The sketch of area b is now resited (Ad), and the diagram on Quarter sheets has dropped down (Ae), with *References*/11 lines of text and symbols (Be). The sketch of Lincoln has been moved, under the heading *CITY AND COUNTY OF THE CITY OF LINCOLN* (Ee). The tables of reference attached to the sketch maps have gone. Railway 48 has been added and 22 removed. There are no colours.

Issued as a separate sheet. 1890.

**Copy**   CUL Maps 70.89.1.

(**H**)   The title has been altered: *ENGLAND AND WALES/ COMBINED INDEX/ SHEW- ING/CIVIL PARISHES AND THE ORDNANCE SURVEY/ OF/ LINCOLNSHIRE/ON THE 1-INCH, 6-INCH, AND 25-INCH SCALES./Surveyed in 1883–87./* (Ca, OS). The imprint has been altered to *Photozincographed . . . Southampton. 1893./Price of this Index 2d./ Rights . . . reserved.* (Ce, OS). The map of Lincoln appears under the heading *COUNTY OF THE CITY OF LINCOLN* (Ed). *Scale of this Index Four Miles to One Inch/Miles 2+12 Miles* [12 = 7.5cm] (Ce, OS, between the imprint and the sketch map centred on Langriville). The *References* table has only 8 lines of symbols in a box, and a single line below the box (Be). *Enlarged Sketch of portion marked b* above a map of an area, east of Baston (Ad), with, below it, a note on the way Six-Inch Quarter Sheets are numbered and are made up of 4 1/2500 sheets. *HUMBER MOUTH* has been added (previously *MOUTH OF THE HUMBER*). Several areas of coast-line have extra shading (notably in squares 14 and 23) and all areas south from Wainfleet, which were previously stippled, are now shaded with sloping lines.

Issued as a single sheet. 1893.

**Copy**   BL Maps. Index Boxes.

(**I**)   The title is *ENGLAND AND WALES/ORDNANCE SURVEY/OF/ LINCOLNSHIRE/ SHEWING/CIVIL PARISHES/* (Ea). The imprint is *Photozincographed . . . 1894. Price 2d . . .* (Ce). The box of *References* has 9 lines of symbols, etc. All the large rectangles and their subdivisions have been removed. The map and the sketch maps are all unframed. *LINCOLNSHIRE* is printed across the centre of the map; outside the county map area, in large capital letters, are the names of *CAISTOR, LOUTH, BOSTON*, and *SPALDING* (at the right) and *BARTON, LINCOLN* and *GRANTHAM* (at the left). The notes on the arrangement of Six-inch Quarter Sheets (Ae) have been removed.

Issued as a single sheet. 1894.

**Copy**    Boston RL Map 18.

## 144

**JOHN CARY**                                                                 **1887**

Size: 94.3cm x 73cm (approx.).

In 1794 Cary produced *Cary's New Map Of England and Wales with Part of Scotland* . . .; published in volume form Lincolnshire is found, wholly or in part, on sheets 33–34, 42–43 and 51–52.[1] It was reprinted at intervals[2] until a second edition appeared in 1816.[3] A corrected re-issue appeared in 1822.[4] A complete revision, under the title, *Cary's Improved Map Of England And Wales, With A Considerable Portion Of Scotland* . . . appeared on 65 sheets in 1832.[5] It was also issued as a boxed set[6] or mounted with each of the book pages occupying a separate card[7] and also, in Lincolnshire's case, mounted as a single sheet made up of the part pages and issued by J.A. Knipe of Stamford in 1833.[8] Subsequently the plates passed to Cruchley, who issued them separately and as an atlas in c.1868.[9] A number of cases are recorded of the use of the plates to create cycling maps, centred on important towns.[10] In some cases, the maps had been issued by Bacon or, under licence, by local promoters. Following Cruchley's death the Edinburgh firm of Gall & Inglis owned the plates and was responsible for a variety of issues from the 1890s until the 1914–1918 war, when, it is believed, the metal was melted down.

The present map was prepared by Gall and Inglis for issue by the firm of James Williamson of Lincoln. Williamson had a shop in the upper part of High Street (no. 290) and published his popular *Guide through Lincoln* (2nd edition, 1881; 7th edition 1901).[11] He also published *Williamson's Lincoln & Lincolnshire Railway Guide* from September, 1891.[12] Williamson's rival, Clifford Thomas, had similar maps prepared for issue in the mid 1890s; one map includes the area around Lincoln to a forty mile radius.[13] A very similar map, based on the same Cary source, covered the area sixty miles around Lincoln. There were several editions of that map, which, it is presumed, appeared largely in the 1890s also.[14] One other use of the plates must be referred to. The firm of A.H. Swiss of Devonport produced a series of hunting maps in the late 1890s, of which no. 15 covered the northern part of the county, reaching nearly as far south as Grantham.[15] The southern part of the county with the Belvoir and Cottesmore hunts may not have been included in any of these maps. See also Addenda, pp. 411–12.

The Cary original was a very detailed map and the revision of 1832 owed much to the maps of Bryant and the brothers Greenwood. There is considerable detail also in the coastal areas with numerous figures for depths of water; the wapentakes are named with the boundaries indicated by dotted lines.

## EDITIONS

(A)   *WILLIAMSON'S MAP OF THE COUNTY OF LINCOLN.* (Ba–Da, OS). *Gall & Inglis, Edinburgh* (Ee, OS). *PUBLISHED BY JAMES WILLIAMSON, BOOKSELLER, 290 HIGH STREET, LINCOLN.* (Ae, OS). Railways 1, 3–43, F, I, J, K, N, R and S are marked by solid black lines; railways xvi (continued to Alford), T and U are marked by broken lines. All stations are marked and named. The roads are coloured buff, the county is pale brown with the boundary edged in pink; the adjoining counties are coloured yellow, blue, brown and green, with their boundaries shown by light shades of the same tone; parklands are green and the coastline blue. There is no compass, nor reference to latitude, etc.

Issued as a loose sheet. [1887**].[16]

**Copy**   CUL Maps c.18.G.66.

(**B**)   Size: 72cm. x 51cm. A new title reads: *SAVORY'S 'ECLIPSE' SERIES/ OF/COUNTY MAPS./LINCOLN/AND PARTS OF /.ADJOINING COUNTIES/ Prepared From The Ordnance Surveys/Published by E.W. SAVORY, STEAM PRESS, CIRENCESTER./LONDON: SIMPKIN, MARSHALL & CO.* (Ea, in a box). *Gall & Inglis, Edinburgh.* (Ce, OS). The map is coloured blue, brown, green, yellow, etc., with roads brown. *Scale of English Miles/20 Miles* (Ae–Ce, OS). Along the top and bottom, outside the area of the map, are a number of advertisements, printed from type. On the reverse there are 24 'panels' (4 rows, each with 6), which carry further advertisements.

Issued in a red cover, with black lettering; the title is very similar to that on the map, but the price (SIXPENCE) is added. [1895].[17]

(**C**)   The map is the same size and format as state **A**, but without the imprint (Ae, OS). The map is overprinted with the names of the divisions in two lines, e.g. *GAINSBOR-OUGH/DIVISION*. The divisions are coloured brown, pink, green, yellow, with their boundaries in a deeper shade of the same colour. Lincoln and Boston are coloured yellow and Grantham and Grimsby pink, parks are green, the sea is blue and the adjoining counties have boundaries coloured green, lemon or violet. Railways are named. *WRYDE STA.* has been removed (on Wisbech-[Peterborough] line. Railways 44–46, 48–51, L, T and U have been added as single solid lines; railway xvi has been deleted. Railway 52 and the proposed line from Lincoln to Belchford and Mumby are shown as 'building'. *ALGARKIRK &/SUTTERTON STA.* has been altered from *SUTTERTON STA*; *SCRED-INGTON/STA* has been replaced by *ASWARBY/STA*.[18]

Issued as a sheet on rollers. [1897**].

**Copy**   LAO FL Maps 9 (2 copies).

The county from Navenby to Braceborough is shown on a transfer used to illustrate T.F. Dale's *The History Of The Belvoir Hunt* . . . (London, 1899).[19]

**References**

1   CUL Atlas 5.79.2, etc.; British Geological Survey ZF 912 (410) CAR 1794; Chesterfield RL 912.42.

2   CUL Atlas 5.80.11 has plates watermarked 1804.

3   BL Maps 10.c.19; CUL Atlas 5.81.12; PRO PRESS 23/M 13.

4   BL Maps 15.c.12; CUL Atlas 5.82.3.

5   CUL Atlas 1.83.9.

6   Nottingham UL G 5512.

7   BL 1130 (1).

8   BL Maps 3425 (1); CUL Maps c.36.83.1, which has a label stuck on the bottom corner reading: Published By The Proprietor J.A. Knipe, Stamford, 1833. In this form the map includes much of East Riding and other adjoining counties; it measures 121cm x 94cm.

9   In atlas form: CUL Atlas 2.86.13. As a separate sheet: Glasgow Mitchell L. Sheet 39 (CR 66), issued in c.1854; an example, using sheet 39 and covering the south-east corner of the county, was used to accompany *An Act Proposed: Norwich and Spalding Railway . . . Engineer, Peter Bruff. Nov., 1852* (in Cambridge AO Q/RU m 28). An example of sheet 44 is held in Nottinghamshire AO (Map N 9 L); this covers the north-west section of the county, as far south as Waddington and eastwards to Usselby; it was issued by Cruchley, in c.1867, judged by railway data.

10   Kingsley's 145C (p. 321), issued c.1890; Burden's 142 (p. 216), issued c. 1899.

11   LRL L. Linc 91; Grimsby RL X400: 914 WIL (1881 edition); LRL LIN 9 (1901 edition); several intervening editions are also in LRL.

12   Copy of the first issue (only) in BL (PP 2500 fn). It cost one penny or 1/6 p.a. (in advance) by post.

13   Copy in the author's collection; it was issued in an orange card cover with the title: *THOMAS' NEW LARGE SCALE CYCLING, TOURING, HUNTING, AND DRIVING ROAD MAP OF 40 MILES ABOUT LINCOLN . . .*

14   It measures 158cm x 84cm. but only includes the county area as far south as Sutterton. *THOMAS' NEW LARGE SCALE CYCLING TOURING & DRIVING ROAD MAP 60 MILES ABOUT LINCOLN.* (Aa–Ea, OS). *CLIFFORD THOMAS, BOOKSELLER, 202 HIGH STREET, LINCOLN.* (Ae–Be, OS). *Gall & Inglis, Edinburgh* (Ee, OS). *Scale/10* [= 129cm] (De, OS). The roads are coloured brown. The map is mounted and folded to produce a booklet 59.3cm x 74.5cm with the following cover title: *THIRD EDITION. THOMAS' NEW LARGE SCALE Cycling, Touring, and Driving ROAD MAP OF 60 MILES ABOUT LINCOLN WITH ONE MILE CIRCLES. Showing at a glance the Distances and Roads from Lincoln to any part, including the 'Dukeries.' SCALE: HALF AN INCH TO A MILE. CLIFFORD THOMAS, Bookseller and Stationer, 202 HIGH STREET, LINCOLN.* Dating is only tentative. The latest railway shown is the Lincoln-Spalding line, opened in 1882. The map must have been issued much later than that but before 1900. A copy in LAO (Ex 942.50) is signed on the cover by its (first?) owner, C.L. Exley, 1900.

Clifford Thomas is noted in local directories at the above address in 1896 and 1900; he sold the site in 1900 and the map must have appeared in that year or just before. The original shop (now occupied by Lloyds Bank) is illustrated in Elvin, L. Lincoln as it was, Vol. III. (Nelson, 1979), p. [23]. Thomas re-opened at 206, High Street, the site now occupied by Marks & Spencers.

15   BL Maps 4.a.44 is bound with a printed cover; it sold for 5s. or, as a wall map, for 10s., which may account for the map's rarity. It is illustrated, in colour, in Smith, D., plate V.

16   Dating is very tentative; the map may well have appeared earlier.

17   No example has been found of Lincolnshire in this form. The description here is based on that of Derbyshire (BL Maps 31.b.4) and the description of the Warwickshire map in Harvey and Thorpe, p. 240. BL has these two counties and Shropshire. CUL has 5 examples: the same 3 as those in the BL plus Somerset and Staffordshire. They were received in CUL in 1895 or 1896.

18   The station was properly called Scredington and Aswarby.

19   Copies in LRL L 799.2; Grimsby RL X000:799.25 DAL; Grantham RL L 799.2. See also Addenda, pp. 411–12.

## 145

**KELLY & Co.**                                                           **1889**

Size: 40.4cm x 30.3cm.

After the 1885 editions of the county directory Kelly & Co. ceased using the Becker plates, which had given service for over 30 years (**116**). Their own map is based on Ordnance Survey maps and has some resemblance to Weller's 1862 map (**123**), probably because they share the same source. The new map is larger than that used in the 1885 directory and it allows more detail to be added. The series continued until the final version of the county directory in 1937. Many copies of the directories lack the map.

### EDITIONS

(**A**)   *KELLY' S/MAP OF/LINCOLNSHIRE/1889/Scale of Miles/10* [= 5.1cm]/Compass/ (Ea). *London – Kelly & Cᵒ. Directory Offices, 51 Great Queen St. Lincolns Inn Fields, W.C.* (Ce, OS). *Drawn & Engraved by Kelly & Cᵒ. 51 Great Queen Street W.C.* (Ee, OS). Railways 1, 3–4, 6–27, 30, 32–46, 48–49, A, C, K and T are marked.

*Kelly's Directory Of Lincolnshire, With The Town Of Hull And Neighbourhood. With A New Map Of The County Engraved Expressly For the Work. London . . . Kelly And Co.,*

*51, Great Queen Street, Lincoln's Inn Fields, W.C. Branch Offices: Birmingham . . .*
*Manchester . . . Sheffield . . . MDCCCLXXXIX. Price To Subscribers, Twenty-Two Shil-*
*lings And Sixpence. – Non-Subscribers, Thirty Shillings.*

**Copies**   LRL (Pye) P.L. 929; Stamford RL L 929; Peterborough RL Dir.

*Kelly's Directory Of Lincolnshire. With New Map . . . MDCCCLXXXIX. Price To Sub-*
*scribers, Fifteen Shillings.– Non-Subscribers, Eighteen Shillings.*

**Copies**   CUL L475.b.15.6; Bod Dir. Lincoln.d.2; LAO SR; Grimsby RL X000: 058
KEL; Grantham RL L 929; Stamford RL L 929.

(**B**)   The date in the title has been changed to *1892*. The imprint now reads (partly):
*London – Kelly & C°. L^{td}. Directory Offices* . . . Railways 2, 28–29, 31 have been added;
railway 52 (labelled *LANCASHIRE, DERBYSHIRE & EAST COAST/ (Constructing)/* is
marked by a broken line east to Mumby. Railway 22 has been deleted. *STA* has been
added at more places with the village name, where necessary, e.g. on the GNR and the
Spalding-Norwich line; the name of the railway has been added in several places, e.g.
*G.N.R.* (near Newark) and *M.S.& L. RAIL^{Y}.* (west of Torksey). *SUTTON BRIDGE* has
been added and the boundary with Norfolk marked. *Sutton on Sea* replaces *Sutton in/the*
*Marsh.*

*Kelly's Directory Of Lincolnshire. With Map Engraved Expressly For The Work. London*
*. . . Kelly & Co., Limited . . . Lincoln's Inn Fields, W.C. (Removing to 182, 183 & 184,*
*High Holborn, W.C.) Branch Offices:– Birmingham . . . Sheffield . . . Paris . . .*
*MDCCCXCII. Price To Subscribers, Sixteen Shillings; Non-Subscribers, Twenty Shil-*
*lings.*

**Copies**   LRL L 929; Boston RL L 929; Gainsborough RL L 929; Grantham RL L 929;
Scunthorpe RL L638 910.

*Kelly's Directory Of Lincolnshire With The Town Of Hull And Neighbourhood. With Map*
*Of The County Engraved Expressly . . . MDCCCXCII. Price To Subscribers, Twenty-Two*
*Shillings And Sixpence; Non-Subscribers, Thirty Shillings.*

**Copies**   CUL L 475.b.18.4; Scunthorpe Museum; Nottingham UL Li 1.B15 E 92.

*Kelly's Directory Of Lincolnshire With The Town Of Hull . . . MDCCCXCII. Price Thirty*
*Shillings.*

**Copy**   Hull RL (H 1717113).[1]

(**C**)   The date in the title has been changed to *1896*; the imprint now reads *London.–*
*Kelly & C°. Ltd. Directory Offices, 182 to 184 High Holborn, W.C.* (Ce, OS). The
signature has been altered to *Drawn . . . Kelly & C°. Ltd. 182 to 184 High Holborn, W.C.*
(Ee, OS). Railways 5, 50–52 have been added and extend to the left-hand frame; 52 is
now named *LANCS. DERBYSH. & EAST COAST RAILWAY* and extended to *Sutton on*
*Sea/Harbour & Docks. Melton Mowbray* and *Saxby* have been added in Leicestershire.

*Kelly's Directory Of Lincolnshire. With Map . . . Engraved . . . For The Work. Kelly &*
*Co., Limited. Branch Offices:–Birmingham . . . Sheffield . . . Liverpool . . . Leeds . . .*
*Manchester . . . Glasgow . . . Paris . . . Hamburg . . . MDCCCXCVI Price . . . Sixteen*
*Shillings; Non-Subscribers, Twenty Shillings.*

**Copies**   LRL L 929; Grantham RL L 929; Scunthorpe Museum.

*Kelly's Directory Of Lincolnshire, With . . . Hull . . . Kelly & Co., Limited, London: 182,*
*183 And 184, High Holborn, W.C. Branch Offices . . . MDCCCXCVI. Price Thirty*
*Shillings.*

**Copies**  CUL L 475.b.18.5; Bod Dir. Lincoln.d.1/1896; Hull RL (H1717114); Scunthorpe RL L638 910.

(**D**)  There is no date in the title area. The imprint and signature are *Kelly's Directories Limited* and *Kelly's Directories Ltd* respectively. The map has been freshly transferred and the detail shown in the county mapping is equalled in the adjoining counties. Parks have been added to form a prominent feature. The name of the L.D.E.C.R. has been removed. *The County for Parliamentary purposes is divided/into 7 Divisions, each returning 1 Member/1 West Lindsey 4 South Lindsey/2 North Lindsey 5 North Kesteven/3 East Lindsey 6 South Kesteven/7 Holland* (Ee, in a box; the names are arranged in two columns, separated by a vertical line). The 7 divisions are coloured pink, green, yellow and dull blue; the coastline is blue and the parklands bright green; the boundaries between each of the adjoining counties are pink. Railways J, L, M, N, R, S, U and X are now marked.

*Kelly's Directory Of Lincolnshire. With Colored Map Of The County. Kelly's Directories Limited, London . . . City Depot: 50, Leadenhall Street. Branch Offices: Birmingham . . . Glasgow . . . And At Hull, Etc., Etc. Paris . . . Hamburg . . . New York . . . Toronto . . . MDCCCC Price To Subscribers Sixteen Shillings; Non-Subscribers Twenty Shillings.*

**Copies**  Birmingham RL BQ 942.53; Gainsborough RL L 929; Stamford RL L 929; Grantham RL L 929; Peterborough RL Dir. (LRL Map 201).

*Kelly's Directory Of Lincolnshire With The City Of Hull. With Colored Map . . . MDCCCC. Price . . . Twenty-Two Shillings And Sixpence; Non-Subscribers, Thirty Shillings.*

**Copies**  CUL L 475.b.18.6; Bod Dir. Lincolnshire d.2/1900; Grantham RL L 929; Scunthorpe RL L638 910.

*Kelly's Directory Of Lincolnshire With The City Of Hull . . . MDCCCC. Price Thirty Shillings.*

**Copy**  Hull RL (H1717115).

### Reference

1  This is included to indicate the work's existence. However, the only copy seen, in Hull RL, lacks the map (and the section on Hull).

## 146

**W. & A.K. JOHNSTON**                                                              **1889**

Size: 33.2cm x 24.4cm.

This firm was founded in 1825 by Sir William Johnston (1802–1888), who was joined the following year by his brother Alexander Keith (1804–1871). Initially their Edinburgh business concentrated on engraving and printing but soon expanded into map production, since the younger brother was a prominent cartographer, who exhibited the first globe showing physical characteristics; he also showed up the deficiencies in the Ordnance Survey's work in Scotland and this led to improvements. The firm's work was continued by the sons of the founders and they were especially prominent in producing a variety of school atlases. Their use of the plates in *Encyclopædia Britannica* has been noted (**137C**). The new plates were used in one atlas only, although the plates for some counties were used in guide books. The map is based on the one-inch series of the Ordnance Survey, but is transferred from the large *Stanford's Modern Map of England and Wales* of 1889.

One innovation is the inclusion of the distances from Hull and Grimsby to places abroad served by ship.

## EDITION

(A)  *LINCOLN/English Miles/10* [= 5.5cm] (Ea). *W. & A.K. Johnston, Edinburgh & London.* (Ee, OS). Railways 1–21, 23–46, 48–51, A, C, D, F–V, X and other lines in Northants. and Cambridgeshire are shown. On the verso is the page number 29. The boundary of the county is coloured light violet and the coastline is blue. A graticule covers the county; along the top are letters A–D and down the sides a–d. There is no compass and no indication of latitude, etc.

*The Modern County Atlas Of England & Wales Comprised In Fifty Seven Maps, All On One Scale Arranged Alphabetically With Complete Index W. and A.K. Johnston Edinburgh And London 1889.*

**Copies**  BL Maps 19.b.2; CUL Atlas 6.88.18; Bod Map C17.d.5; RGS 7 D 25; Birmingham RL AQ 912.42.

## 147

## J. JAQUES & SON                                                    c.1890

Size: 7.6cm x 5.8cm.

The firm of Jaques was founded in 1795 and produced chessmen and billiard balls. In 1861 the firm became John Jaques and Son and developed a wider range of indoor and outdoor games. In c.1890 the firm adopted the above name and issued a card game, which used maps of the English counties; the boxed game contains two sets of cards; on one set appear outline county maps with few features; on the cards in the second set the same outlines are filled in to form figures, which are used to illustrate comic verses. The firm's premises in London were bombed during the 1939–1945 war with the loss of the archives. Jaques And Co., Ltd. continues to make specialist sporting equipment at premises in Thornton Heath in Surrey.[1] Perhaps their most famous line has been the production of the chess pieces named after Howard Staunton, the world champion from 1843–1851. They were designed by Nathaniel Cook, chess correspondent of *Illustrated London News*, in 1843 and Jaques soon acquired the copyright.

## EDITIONS

(A)  *LINCOLNSHIRE* (Be–De, below the map; there is no frame). *10* (Aa). The map is a simple outline, with 6 rivers and 30 circles for town sites. There is no compass, scale or name of any feature.

*Skits A Game Of The Shires A Game Both Instructive & Pleasant. It Tickles The Fancy & Sharpens The Wits. Published by Jaques & Son. 102. Hatton Garden. London.* [c.1890].[2]

**Copies**  The private collections of Mr. E. Burden and Mr. D. Ollis.

(B)  The same outline map of the county as in state A, with colours and shading to provide a caricature of a monk, facing left and carrying a lamp. Above the map the first two lines of the verse appear and the last three are below; the verse is '*Tis a Lincolnshire monk with a light,/Who is faring alone through the night,/The fens are around,/ And dangers abound,/God help the poor venturous wight!/*

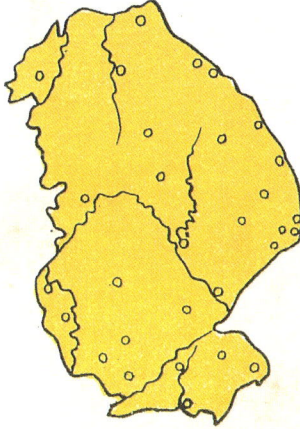

LINCOLNSHIRE

'Tis a Lincolnshire monk with a light,
Who is faring alone through the night.

The fens are around,
And dangers abound ;
God help the poor venturous wight !

10.  1890?   MAP 147 – J. JAQUES & SON.
By permission of Mr. E. Burden.

*Skits . . .* [c.1890].

**Copies**   The private collections of Mr. E. Burden and Mr. D. Ollis.

**References**

1   I am grateful to Mr. E. Burden for some of this information, for drawing my attention to the existence of the cards and allowing use of the county card as an illustration. Mr. Ben Jaques also provided useful facts.
2   Evidence for the date of issue is lacking; the game must have been issued between 1890, when the firm adopted the title in the heading, and 1899, when the firm became a limited company. The owners of the cards have suggested 1890.

## 148

**ORDNANCE SURVEY**                                          **1891**

Size: 67.7cm x 57.7cm.

This O.S. map differs from earlier transfers made between 1885 and 1887 (**139, 141** and **143** respectively) in the inclusion of a much larger amount of geographical detail. The 1885 had less detail and the prominent feature was the overprinted information on the new parliamentary divisions; the 1888 had no important topographical detail, since it lacked all roads, had few rivers and the emphasis was entirely on the parishes and their boundaries. The new transfer includes the definition and naming of the parishes combined with greater geographical detail, the numbers of the O.S. sheets and, included in the transfer (as opposed to their superimposition in a second printing operation), the names of the new parliamentary divisions. The result is a rather flat presentation of a wealth of information; because each parish is coloured the numbers of the O.S. sheets are so obscured to be almost invisible. A later transfer of the basic map was issued in 1899 (**152**).

## EDITION

(**A**)   *INDEX/to the/ORDNANCE SURVEY/of/LINCOLNSHIRE/ON THE SCALE OF SIX INCHES TO ONE MILE/* (Ea). *Reference/Numbers of . . . Sheets are shown thus e.g. XXXIV, N.E./Price of each Quarter Sheet —— One Shilling/Scale of this Index/Miles 2 + 8 Miles* [8 = 6.9cm.]/*Three Miles to One Inch* (Ee). *Photozincographed at Ordnance Survey Office Southampton, 1891./Price of this Index 6d./* (Ee, below the Reference). *REFERENCE BY LETTER TO THE SMALLER/AND DETACHED PARTS OF PAR-ISHES/*above a box of letters and descriptions (Eb). *Enlarged Sketch of the portion marked . . .b* above a map in a box of the area around Langriville, and, below it, a table headed *REFERENCE TO THE PARISHES/ . . ./* (Ab). *Note.– The names and Boundaries . . ./ . . . are those of the Act of 1885./* (Ac). *Enlarged Sketch of part of the City of Lincoln/* above a map of Lincoln (Ae). *Enlarged Sketch of the Town of Stamford/ST. GEORGE/*above the map (Be). *CHARACTERISTICS/*above a box depicting the various physical features and their depiction (Ed). Railways 1–4, 6–46 and 48 are shown, stations are marked *Sta.* and lines are named. The parishes are coloured yellow, violet, green, pink and brown; the parliamentary divisions have coloured boundaries of brick red; the numbers of the O.S. sheets are pink; the inset maps are also coloured. The map is attached to an almost equally large chart, listing the parishes and their physical composition in acres in terms of land, water, salt-marsh, foreshore and tidal waters.

Issued as double sheets of map, with attached chart. 1891.

**Copies**   BL Map Dept. (Open shelves); CUL Map Dept.

**149**

**F.S. WELLER**                                                    **c.1894**

Size: 28cm x 21.67cm.

Francis Sidney Weller is noted as the cartographer of a map of London included in *Cassell's Gazetteer Of Great Britain and Ireland* . . . This work was published in three volumes, plus an atlas volume, in 1893. It includes regional maps by W. & A.K. Johnston and Lincolnshire is spread over parts of sheets VIII, IX and XII.[1] An enlarged edition in six volumes was issued between 1894 and 1898[2] and in 1898–1900.[3] In the work below Weller was responsible for all the county plates. F.S. Weller was the son of Edward Weller, who prepared the maps for George Philip and Collins (**123** and **131**, respectively). The map is based on Ordnance Survey sources. The publisher William Mackenzie was active from 1866 until 1895.

**EDITION**

(**A**)   *LINCOLN*/*Scale of Miles*/*10* [= 3.45cm]/*Railways . . . Roads . . . / Canals . . . Drains* . . . [and their symbols]/(Ea). *WILLIAM MACKENZIE, LONDON, EDINBURGH & GLASGOW* (Ce, OS). *F.S. Weller, F.R.G.S.* (Ee, OS). There is no compass. Railways 1–21, 23–46, 48–51, A, F, H, I, J, K, L, M, N, R, S, T and U are shown; stations are marked. The parliamentary divisions are named and coloured green, pink, brown, yellow and violet. The boroughs are pink, the sea blue, the county boundary red-brown and adjoining counties pale lemon. Latitude and longitude are marked every 20' (between borders) and provide the basis for a graticule. *Meridian of Greenwich* (De, between borders).

*The Comprehensive Gazetteer Of England And Wales. Edited By J.H.F. Brabner* . . . [Volume IV]. *London: William Mackenzie, 69 Ludgate Hill, Edinburgh And Dublin.* [1894].[4]

**Copies**   BL 2367 d 3; CUL NA.1.56–61; Bod G.A. Gen. Top. 4° 174 = R.Top. 502; Leicester UL H 942 BRA; Liverpool UL Dept. of Geography.

**References**

1   Set in LAO SR. 950.2.
2   Sets in CUL S696.b.89.14–18 (bound as 5 volumes although there are 6 title-pages); Bod G.A. Gen. Top. 4° 171; Northampton AO NRO Lib. No. 1451/1–6; Birmingham UL s/DA 640.
3   Leicester UL H 942 CAS.
4   The introductory volume refers to the availability of the 1891 census figures in 1893 and, it is assumed, the fourth volume was issued in 1894; CUL's volume bears the receipt date 14 JUL 94.

**150**

**EDWARD STANFORD**                                               **1897**

Size: 26.35cm x 15.9cm.

Stanford's prepared this plain map to illustrate a pair of ecclesiastical works, issued in 1897. One work was written by Edmund Venables and completed after his death by G.G. Perry. Venables (1819–1895), ordained by the Bishop of Chichester in 1844, completed the requirements for his Oxford M.A. in 1845. He remained in Sussex until his appointment as examining chaplain at Lincoln cathedral. In 1865 he became prebend, of Carlton with Thurlby and, in 1867, Precentor and canon-residentiary. He edited the fourth edition of Murray's *Handbook for Wiltshire, Dorset and Somerset* (1882). He began work on his study of Lincoln in the series *Diocesan Histories* issued by S.P.C.K. but died on 5 March, 1895; his wife died the day after and they were buried together in the cathedral cloisters.[1] George Gresley Perry, M.A., who completed the work, was Rector of Waddington from 1852 and, at the time of publication, Archdeacon of Stow. The map also appeared in *Historical Church Atlas*, which dealt with the organisation of the Anglican church world-wide. Its author, Edmund McClure, educated at Trinity College, Dublin, spent 40 years as Secretary of the Society for the Promotion of Christian Knowledge from 1875. At the time of his death in 1922 he was an honorary Canon of Bristol Cathedral.[2] The map is a simple outline map of the Lincoln diocese, including parts of other counties as far south as the River Thames. The map makes reference to *Valor Ecclesiasticus* and this concerns the spelling of place-names, hundreds, etc. The atlas, prepared by McClure, has other plates, which either show part of the county (that of York, plate 68) or all of the county (the plate for Peterborough, plate 62). The latter map includes few place-names but emphasises the physical features in a way not included on the Lincoln diocesan map, which only has rivers, and a few towns or villages. It was probably transferred, in outline only, from the Stanford *LIBRARY MAP* with the ecclesiastical detail added from type.

## EDITIONS

(**A**)  *A MAP/TO ILLUSTRATE THE/ANNALS OF LINCOLN/English Miles/10* [= 3.2cm]/*REFERENCE/5* lines on various boundaries and their markings and 2 on *Churches in 1541* and Abbeys, etc. with symbols (Aa). *From Valor Ecclesiasticus*. (Ae, OS). *Stanford's Geog*$^l$*. Estab*$^t$*. London*. (Ee, OS). There is no compass. Lincolnshire is coloured pink, with the diocesan boundary marked in a deeper shade of pink. There is a graticule at one degree intervals (marked between borders). *Meridian of Greenwich* (De, between borders).

*Diocesan Histories Lincoln By The Late Edmund Venables . . . And George G. Perry . . . With Map Published Under The Direction Of The Tract Committee London Society For Promoting Christian Knowledge Northumberland Avenue, W.C. 43, Queen Victoria Street, E.C. Brighton: 129, North Street New York: E. & J.B. Young & Co. 1897.*

**Copies**  BL 2210 a 9; CUL XXVII.102.58; Bod 111 f 55; LAO Sc 942.50; LRL LINC 9 VEN and L 283; Louth RL GL 9 and L LINC 910.

(**B**)  The title is now *A MAP/TO ILLUSTRATE/LINCOLN/*followed by the same scale and references (Aa). There is no colour.

*Historical Church Atlas. Consisting of Eighteen Coloured Maps and Fifty Sketch-maps in the Text. Illustrating the History of Eastern and Western Christendom until the Reformation, and that of the Anglican Community until the present day. By Edmund McClure . . . Published Under . . . Tract Committee. London: Society For Promoting Christian Knowledge . . . Brighton . . . New York . . . 1897.*

**Copies**  BL Maps 48.c.27; CUL Atlas 6.98.17; RGS 12 G 16; Chethams L. q 270.084 Ma 173.

### References
1   DNB; *Lincoln Gazette*, 9 March 1895.
2   Who Was Who, 1916–1928, which fails to provide a date of birth.

## 151

### F.S. WELLER                                                        1898

Size: 18.6cm x 11.9cm.

Weller's work was noted above (**149**) and the life of the author, E.A. Woodruffe-Peacock, recorded in connection with his earlier work (1895) on the county's natural history divisions. The map is a simple outline affair, with 7 towns named and the emphasis placed on the 18 divisions that the author uses for his classification of the county's natural history. Like the map used in his work of 1895 (**130 P**) H.C. Watson's Line of Demarcation is marked on the map.

### EDITION

(**A**)   *Sketch map of the/NATURAL HISTORY DIVISIONS/of/ LINCOLNSHIRE./By Rev. E. Adrian Woodruffe-/Peacock./English Miles/5* [= 2.05cm] (Ea). *F.S. Weller. litho.* (Ee, OS). There is no compass. The county is divided into 18 numbered divisions in the form *ALFORD//AND/BURGH/11*. Across the map *H.C. Watson's Line of Demarcation/* Between Linc. N. & Linc. S./ roughly follows the line of the Witham with the wording repeated at the western and eastern ends. There is no colour. A graticule covers the county at 30′ intervals (marked in the borders). *Longitude West of Greenwich* (Ce, between borders).

*The Natural History Of Lincolnshire; Being The Natural History Section Of Lincolnshire Notes & Queries, From January 1896, to October 1897. Edited By The Rev. E. Adrian Woodruffe-Peacock . . . Horncastle: W.K. Morton. High Street. 1898.*

**Copies**   LAO Sc 942.52; LRL L 574; Grimsby RL X000:574 WOO; Grantham RL L 574; Gainsborough RL L 574.

*Lincolnshire Notes And Queries. Vol. V. No. 33. January, 1896. Natural History Section.*

**Copy**   CUL Q 475.c.9.6.

*The Naturalist: A Monthly Journal Of Natural History For The North Of England, Edited by Wm. Denison Roebuck . . . London, 1895.*

**Copy**   CUL P382.c.22.23.

## 152

### ORDNANCE SURVEY                                                  1899

Size: 85.3cm x 68.5cm (there is no frame).

The map described here is basically the same map as the 1891 (**148**), but it is overlaid with a network of rectangles giving the 6″ series of Ordnance Survey sheet numbers and overprinted (twice) with names of unions and sanitary districts. Since the surrounding tables and sketch maps differ from the 1891 this map is treated separately as a new transfer. While the map here is a re-issue dated 1899 the inference is that there was an issue in 1897; no copy of such a map has yet been found. In revised forms it was re-used again in 1902 (simply showing the civil parishes and index of O.S. maps up to the 25″ scale)[1] and in 1908.[2]

## EDITION

(**A**) *ENGLAND & WALES./DIAGRAM/OF THE ADMINISTRATIVE COUNTIES OF/PARTS OF HOLLAND, PARTS OF KESTEVEN,/& PARTS OF LINDSEY,/Shewing Unions, Sanitary Districts, Boroughs, and Civil Parishes;/AND THE SHEET LINES OF THE ORDNANCE SURVEY MAPS OF/LINCOLNSHIRE,/ON THE SCALE OF 25.344 INCHES TO 1 MILE (1/2500)/Surveyed in 1883–87.* (Ea). *Scale of this Index/Miles 2 + 8 Miles* [8 = 6.8cm]/*Photozincographed at the Ordnance Survey, Southampton 1897./Price 3s./* . . . (Ce). *REFERENCE*/6 lines of symbols for boundaries and boroughs/ *N.B. The Boundaries on this Diagram are revised up to date (26–9–99).* (Ce, above the scale, etc). *REFERENCE BY LETTER TO THE SMALL PARISHES/AND DETACHED PARTS OF PARISHES./*above a table, in three columns, the third giving the O.S. sheet number (in previous versions the table lacked the column of sheet numbers) (Ec). *CHARACTERISTICS*/above a box containing symbols used on the map (Ed). A 3-line note refers to the numbering and price, etc. of 1/2500 sheets (Dd). *Enlarged Sketch of the portion marked . . .b*/above a map of the area centred on Sibsey (Ac). *REFERENCE TO THE PARISHES &c. SHEWN ON/ENLARGED SKETCH MARKED b./* above a table of three columns (Ad). *Enlarged Sketch of Part of the City of Lincoln*/above a coloured map of south Lincoln (Ae). *Enlarged Sketch of the Town of Stamford*/above a map (Bd). The boundaries of the unions are coloured pink, with those of the Rural District Councils blue, the 3 administrative counties green, Urban District Councils brown, boroughs blue and county boroughs purple, with their boundaries blue. The names of the administrative units are in capital letters in the appropriate colours. The map is overlaid with a grid, each section of which bears its 25″ sheet number. Railways 1–4, 6–48 are marked.

Issued as separate sheets. 1899.

**Copies**   BL Maps 3364 (4); CUL Maps 70.90.1.

### References
1   CUL Map Library.
2   PRO M29 PRESS 28.

## 153

**GALL & INGLIS**                                             **1900**

Size: 33.1cm x 27.6cm.

This map was transferred from a large map of England and Wales; Gall & Inglis were, at this time, putting out county maps, based on transfers from Cary's *Improved Map of England And Wales* of 1832 (see **144**), but this is taken from Ordnance Survey sources, as the map acknowledges. It appeared in a county guide book prepared for cyclists by G.J. Wilkinson, who was employed by the *Lincoln Gazette* and also *Lincolnshire Echo* as the cycling correspondent. It is a useful and reasonably detailed map, but, oddly, lacks any representation of hills.

## EDITION

(**A**) *LINCOLN/SHIRE* (printed across the centre of the county). *Gall & Inglis, Edinburgh.* (Ee, OS). *Prepared for WILKINSON'S Illustrated Guide to LINCOLNSHIRE.* (Ae–Be, OS). *By kind permission of the Controller of H.M. Stationery Office.* (Ce, OS). *Scale 10* [= 4.2cm], the numbers run from right to left (De, OS). Railways 1, 3–46, 48–52, A, F, H–N, S–U and X are marked. On the verso are six panels of advertisements for R.M. Wright & Co. Lincoln. There is a graticule but it lacks marginal numbers or indications of latitude, etc.

*Illustrated Guide to Lincolnshire, what to see and how to see it. The Tourists' Handbook. An Illustrated, Historic, and Descriptive Guide to the beautiful and interesting district in and around Lincolnshire. By G. J. Wilkinson . . . Published by the Author At 18, York Avenue, Lincoln, And At The 'Gazette' and 'Echo' Offices, Lincoln. London:– Messrs. Iliffe, Sons & Sturmy, Ltd., 3, St. Bride Street, E.C. Lincoln: Printed at the 'Gazette' and 'Echo' Offices, 1900.*

**Copies**   CUL 9474.d.1109; LAO Wright 942.51; Grimsby RL XOOO:914 WIL; Louth RL L 9 BL; Gainsborough RL L 91.

## 154

## ORDNANCE SURVEY                                             c.1900

Size: 34.8cm x 26.8cm.

This map is very similar to that provided by the O.S. for Gall & Inglis to use in 1900 (**153**). There are differences, however; in the present map *Kirton/in Lindsey* differs from *KIRTON/in Lindsey* in the Gall & Inglis; *LONG SUTTON* compares now with the former's *SUTTON S^T MARY*, and *CLEETHORPES* in the present case differs from *Clee/thorpe*. The map was issued by *Lincolnshire Chronicle*, the reason being unclear; the key to the map must have appeared in the newspaper, since the only copy of the map found so far lacks that feature.

## EDITION

(**A**)   *MAP/OF THE/DIOCESE OF LINCOLN./PRINTED AND PUBLISHED BY/Lincolnshire Chronicle, Ltd.,/LINCOLN./*above a crest surmounted by a mitre and imposed on two crossed keys and a bishop's staff, the whole being surrounded by a double circle around which *DIOECESIS* (sic) *LINCOLNENSIS.* appears (Ea). *Reproduced from the Ordnance Survey Map with the sanction of the Controller of H.M. Stationery Office.* (Ee, OS). There is neither compass nor scale. The map is overprinted, in red, with the boundaries of the deaneries in the county and each is numbered, also in red. Railways 1, 3–21, 23–52, A, F, I–N, R–U and X are marked. There is a graticule, but there are no marks for longitude or latitude.

Issued as a separate sheet. [1900?].[1]

**Copy**   Grantham RL.

### Reference
1   The date is conjectural. The railways shown were all open by March, 1897; the next line to be built in the county, from Goole to Reedness, opened in January, 1900 but is not shown here.

APPENDIX

ROAD BOOKS 1675 – 1900

1

JOHN OGILBY                                                               1675

Size: 33.4cm x 44cm (approx.).

Ogilby (1600–1676) came to cartography when most men would have settled for the
retired life. Born in Edinburgh he came to London as a very young boy and was
apprenticed as a dancing master and through that skill he became part of the Earl of
Strafford's household in Ireland. He became eventually Master of Revels in Ireland and
built a theatre in Dublin. He lost everything in the Civil War but after the return of Charles
II he returned to London and was able to set up as a publisher.[1] However, he again lost
everything in the Great Fire, including a stock valued at £3,000. With three others he was
given the task of surveying property in the City in order to settle disputes between the
owners of premises lost or damaged in the Great Fire. In 1669 he projected an *English
Atlas*, a work to be published in five volumes and intended to cover the world.[2] In fact,
the idea had apparently occurred the year before and Ogilby advertised frequently his
'Lottery of Books' meant to raise money for his project.[3] The first volume, Africa,
appeared in 1670 and the second, dealing with America, in 1671.[4] Ogilby received a
Royal Warrant in 1671 which authorised him to make the necessary surveys and enquiries
by way of preparation for his *Britannia*. The warrant referred to Ogilby as Royal
Cosmographer, a title he may already have been awarded earlier.[5] In 1673 a further part
dealing with Asia appeared. More lotteries were held in order to defray the expenses of
the whole project, which were estimated at £14,000; the King subscribed as did other
royalty and the City of London (£1900).[6] Ogilby also received relief from tax on paper
imported for the work and hoped to obtain patronage from those willing to pay to have
plates dedicated to them.[7]

   Britannia was intended initially to be in three parts, covering the roads, followed by
25 town plans and, finally, a county atlas. In the event only the road book was finished
and three counties surveyed and issued.[8] Kent and Middlesex were published in 1672–3,
but Essex did not appear until 1678, after Ogilby's death, and the survey of Surrey,
undertaken by John Aubrey in 1673, was never finished. Ogilby announced his intention
to issue *Britannia* (his intended fifth part) before the third and fourth (Asia and Europe).
The survey of the roads was already in progress by early in 1672; Richard Shortgrave
and Gregory King (and, perhaps other surveyors) had begun the measurement of the
roads, which was completed in 1675. The *London Gazette* for November 18–22, 1675
carried the announcement of its completion. In the following January the alternative title
*Itinerarium Angliae* was also advertised.[9] The work was also offered in sheets at 6d. each.
At the end of 1676 it was announced that the final sections of *English Atlas* were to be
completed by Ogilby's relation, William Morgan. Apart from the Essex map the only
other Ogilby work to be issued was the survey in 20 sheets of London at the scale of '100
Foot in an Inch'.[10] Sales of Ogilby's books and the map of London were still being
announced in 1678 in order to pay for the costs of completing the *English Atlas* but the
larger work remained incomplete.

   Ogilby adopted the mile of 1760 yards as opposed to the 'computed' mile of some
2640 yards or the old British mile of 2428 yards; and in his preface referred to the fact
that everywhere the local miles were longer than the statute mile.[11] His work was the
major factor in the adoption of the mile as legally defined in 1593 (35 Elizabeth). Analysis
of Ogilby's plates showed that, on average, the old mile was equal to 1.3 statute miles.[12]
Interestingly, the mileages are given in the form 7'3, etc., quite similar in style to that still

used in Germany, i.e. 'Seven comma three'. Apart from the work of Ogilby himself and his surveyors the sources at hand for Ogilby were very few. At that time no complete series of county maps included roads. It has been seen that Saxton's maps did not include roads as a general feature and that all later county map series were entirely derived from Saxton for their geographical material. However, there had been various attempts to make maps of several counties, which had included roads, among other 'new' features. John Norden had produced several maps of counties, which included roads; although not printed by Norden himself it has been inferred that several maps drawn by William Smith and engraved by Jodocus Hondius at Amsterdam for publication by Hans Woutneel in 1602–1603 are based on Norden's work.[13] The map plates turned up later in the stock of Peter Stent (with another Norden map – of Hampshire) and passed from him to Overton, father and son, who were issuing them in the eighteenth century. It will be recalled that the mileage charts, that appear on the same plate as Van Langeren's thumb-nail county maps (**10**), are derived from Norden's *England: An Intended Guyde*. Speed only showed a small portion of road on one map. Robert Walton, one of the most prolific and successful of sheet map publishers in the mid-seventeenth century, began to include roads on his maps of England a few years before Ogilby's work appeared.[14] Holinshed's *Chronicles of England, Scotland and Ireland* (1577) had been the chief source of information on coach routes, since it contained tables of roads and the stage distances. These features had been copied by others, until Norden refined the mileages in his charts.

Ogilby and his fellow workers had some basis, therefore, for the material that appeared on the hundred plates of *Britannia*. It was only in 1640 that the first stage-coach route, from London to Oxford, made its appearance. From that time a very great increase in coach travel developed on roads, which were newly being converted from mud tracks to something on which conveyance of goods and people became possible. This phenomenon was being repeated in Western Europe since roads were not only of strategic importance but also facilitated the exchange of goods and ideas, important to the individual. The need for a set of maps that would furnish the traveller with all the information needed in relation to journeys from the centre of all commercial activity, London, seemed obvious to all informed opinion. The success of Ogilby's efforts is shown in the pattern he set that was repeated, without much alteration, for over a hundred years and provided a source that had a great influence on the subsequent makers of county maps. The plates made one further appearance in 1698 in two versions; one from Swalle and Morden and the other, quite rare, by Swalle alone.

## EDITIONS

Each plate comprises 6 or 7 strips on scrolls; they are read upwards, starting from the bottom left corner and ending in the upper right corner. Some plates may contain more than one road. There are no plate numbers. There are compasses on each strip.

(**i**)  [Plate 6]. *The Continuation of the Road/from/LONDON to BARWICK/Beginning at Stilton and extending to/Tuxford/Plate the 2ᵈ Containing 62 Miles/From Stilton to Stamford 13¾ to Grantham/ 21½ to Newark 14 to Tuxford 12¾/* (decorated panel, at the top of strips 3–4) There are 6 strips on separate scrolls.

[Plate 36]. *The Road from LONDON to BOSTON in Lincolnsh.:/By JOHN OGILBY . . . /Containing 114 Miles 3 Furlongs vizᵗ./From the Standard in Cornhil, LONDON, to Stilton in the Barwick Road 69.M. 1 Furl:/to PETERBOROUGH 7'3. to Wellington 3'6. to Crowland 7'6. to Spalding10'0. to Setherton 10'2/and to BOSTON 6M.1 Furl:/ Continued to the City of LINCOLN 37 M.4 F. vizᵗ./From BOSTON to Heckington. 13'3. to Sleford 6'1 and to LINCOLN 18'0./* (panel surrounded by pastoral scenes, top of strips 3–5). There are 7 strips on separate scrolls.

[Plate 41]. *The Road from/LONDON to FLAMBOVROVGH HEAD in Cō* Ebor/By IOHN OGILBY . . . /Containing 212 mile 6 Furlongs viz$^t$./*From the Standard in Cornhill LONDON to Stilton in the Barwick Road.69./1. to PETERBOROW 7.1 to Market Deeping 10.6 to Born .6.4.*(sic) *to Morton.2.3/to Sleeford.14.5. to LINCOLN 18/thence to Flambourough in plate the 2$^d$..84.1/*otherwise 218 mile 4 Furlongs thus/From the Standard in Cornhill LONDON to Temsford in y$^e$. S$^t$. Neots road/52 to Eaton.3.6 to Bugdon 5 7 to Stilton aforesaid. 13.3./ thence to LIN COLN vt supra./ (in a decorative panel, above strips 3–5). There are 7 strips on separate scrolls.

[Plate 42]. *The continuation of the Road from/LONDON to FLAMBRUGH. Comebor.* (sic)/*By JOHN OGILBY Esq$^r$. his Ma$^{ties}$. Cosmographer/Containing 84. Miles 2. Furlongs viz$^t$./From Lincoln to Redbourn 19.0. to Glamford Bridges. 5.5./to Barton 10.1. to Hull 6.4. to Beverley 9.2 to Beseck 7.0./to Kilham 12 0 to Burlington 7 4 to Flambourough 5 0/& to Flambourough head. 1.2./* (panel surrounded by nautical scenes, top of strips 3–5). There are 7 strips on separate scrolls.

[Plate 78]. *The Road from/NOTTINGHAM to GRIMSBY/in the Cōūn̄t* (sic) *of Lincolne./By JOHN OGILBY . . . /Containing 67 Miles 2 furl. viz$^t$./From Nottingham to N. Wark.* (i.e. Newark) *17.1. to the City of/Lincolne 14.6. to Wilton 6.4. to Markit Raisin 9.4/to Stanton in the Hole 5.5. to Briggesley 7.4/And to Grimsby 5.4./* (in a decorative panel at the top of columns 3–4). There are 6 strips on separate scrolls.

*Britannia, Volume the First: Or, An Illustration Of The Kingdom Of England And Dominion of Wales: By a Geographical and Historical Description Of The Principal Roads thereof. Actually Admeasured and Delineated in a Century of Whole-Sheet Copper-Sculps. Accomodated With the Ichnography of the several Cities and Capital Towns; And Completed By an Accurate Account of the more Remarkable Passages of Antiquity, Together with a Novel Discourse of the Present State. By John Ogilby Esq; His Majesty's Cosmographer, and Master of His Majesty's Revels in the Kingdom of Ireland. London, Printed by the Author at his House in White-Fryers. M.DC.LXXV.*

**Copies**  BL 192 f 1; CUL Atlas 4.67.6; NLS Newman 388.

(ii)  The following alterations have been made to the plates:

[*6*] In column 1: *to* has been added at Stibbington and at Stilton; in column 3, *a Rill* has been added (three times), i.e. above mileage 97, at Easton and below the 100 mile mark; in column 4, *a Rill* has been added twice near *Gunnerby*; in column 5 *to/Dodding* has been corrected by the addition of the final *ton* (just below mileage 114).

[*36*] At the top of column 7 *Foss fluv:* has been added.

[*41*] In column 3 the river now reaches the left side of the scroll and *Nine fluv.* has been added; in column 6, *or open Arrable* has been added to the right of *Corn/fields* below mileage 113 (on the Sleaford-Lincoln road).

[*42*] At Lincoln (column 1) *Foss fluv* has been added to the west and *Witham fluv* to the east (where *Fos fluv* appeared originally); column 2 now has *Glamford bridges* (formerly *Glamford brides*); in column 3, at mileage 159 and also between mileages 161 and 162 a line has been added above the m in *comon*; in column 5, between *Wood* and *Brook*, by mileage 182, an ampersand has been inserted .

[*78*] In column 1 *Torent fluv* (i.e. Trent) has been added, east of Nottingham and the river extended northwards; in column 3 *Witham fluv* has been added, at *North Hikeham*. The Trent river (columns 1–2) and the Witham in column 3 have now been outlined with stronger lines along their 'banks'.

*Britannia . . . M.DC.LXXV.*[15]

**Copies**  BL Maps C.6.d.8; Bod Vet A3 b 10; Leeds UL W 240; RGS 264 H 13;

Birmingham RL F094/1675/20. (LRL Maps 55 and 56 (both plate 42); Grimsby RL RM 44 (plate 36), RM 43 (plate 6), RM 42 (plate 41), RM 121 (plate 78); Nottingham UL EM.B8.C 75 (plate 78)). All the plates in the BL example are marked with double red lines to form a frame; the lines in the titles of the maps, the title-page and text leaves are also treated similarly; the copies in Grimsby RL come from a similar source.

*Itinerarium Angliae: Or, A Book Of Roads, Wherein are Contain'd The Principal Road-Ways Of His Majesty's Kingdom Of England And Wales: Actually Admeasured and Delineated in . . . Copper-Sculps, And Illustrated with the Ichnography of the several Cities and Capital Towns. By John Ogilby . . . London . . . M.DC.LXXV.*

**Copy**   Liverpool RL Fo. F 247.

**(iii)**   Plate numbers have been added (Ee, OS). Plate 78 has its number immediately above a poorly obliterated number *51*.

*Britannia . . . M.DC.LXXV.*

**Copies**   Nottingham UL DA 615.04; NLS EU 18 Rd.

*Itinerarium Angliae . . . M.DC.LXXV.*

**Copies**   BL 457 f 1; Bod Gough Maps 100; NLS Newman 389.

*Britannia: Or, The Kingdom Of England And Dominion of Wales, Actually Survey'd: With A Geographical and Historical Description Of The Principal Roads; Explain'd by One Hundred Maps on Copper-Plates. With The Ichnography, or Draught of the several Cities, &c. And a View of the Churches, Houses, and Places of Note on the Road. Also An Account of the most remarkable Passages of Antiquity relating to them, and of their present State. By John Ogilby . . . London: Printed for Abel Swall, at the Unicorn in Paternoster-row, MDCXCVIII.*

**Copies**   Birmingham UL **F 914.2 (O); Liverpool Athenaeum 910.3F; Hull RL 912.42. (LRL Map 773 (plate 6), Map 799 (plate 36), Map 800 (plate 41); Boston RL Map 13 (plate 36); Gainsborough RL Map 11 (plate 41).[16]

*Britannia . . . London: Printed for Abel Swall, at the Unicorn in Pater-noster-row and Robert Morden, at the Atlas in Corn-hill, M DC XCVIII.*

**Copies**   BL Maps C.25.c.4; CUL Ely a 119 and Peterborough 0.6.8; Bod Gough Maps 101 and Vet A3 b 31; Leeds UL W 241; Sheffield UL **F 914.2 (O); RGS 209 H 1 and F 256.

### Reproductions
(i) A reduced format version of the maps was published by Alexander Duckham in 1939; the maps have pairs of double lines drawn in red around the plates, a feature to be found in some of the above examples. (ii) *John Ogilby: Britannia 1675 With An Introduction . . . By J. B. Harley.* (Amsterdam, 1970). (iii) *Ogilby's Road Maps Of England and Wales from Ogilby's 'Britannia', 1675.* (Reading, 1971). Both of these editions have plate numbers. (iv) Plate 36, on pp. [34–35] and plate 78 (reduced) on p. 49 illustrate the piece: Famous Early Maps in Lincolnshire John Ogilby, 1675 by John E. Holehouse in *Lincolnshire Life*, Vol. 13, no. 12 (January, 1974). Plate 36 is in state (ii) and plate 78 in state (iii). (v) Molyneux, F.H. and Wright, N.R. An atlas of Boston. (Boston, 1974), p. 11 includes a facsimile of the heading of plate 36 and the three columns immediately below it.

### References
1  DNB, Vol. XLII (1895); Tyacke, pp. 129–130; Van Eerde, K.S. John Ogilby And The Taste of his Time. (Chatham, 1976).

2  Harley, J.B. John Ogilby: Britannia, 1675. (Amsterdam, 1970), pp. v–xxxi.
3  Tyacke, pp. 129–130; and p. 3 (items 1–4) for advertisements in 1668; Fordham, F.G. John Ogilby, His Britannia And The British Itineraries Of The Eighteenth Century. *Library*. Fourth Series, Vol. VI, pp. [157]–178 also quotes an advertisement on 10 May 1669 (Bod. Wood 658 f 792) and the Term Catalogue (Michaelmas, 1670).
4  Tyacke, p. 5 (item no. 10) gives an announcement from *London Gazette* for 23–26 October, 1671 that it would be available from 3 November, 1671.
5  Bod MS Aubrey 4, f 220, dated 24 August, 1671; reprinted in Fordham (see note 3).
6  Bod Wood 658 f 786, cited in Clapp, S.L.C. op. cit., pp. 365–379.
7  Letter of 27 February, 1671/2, cited by Clapp (note 6).
8  Skelton, pp. 185–186.
9  Tyacke, p. 13 (items nos. 48–49); further advertisements appeared in 1676.
10  Part of the announcement in *London Gazette* for 25–29 January, 1676/7, cited by Tyacke, p. 18 (item no. 4).
11  Seebohm, F. Customary acres and their historical importance. Part II: The Old British Mile. (London, 1914); Close, Sir Charles C. The Old English Mile. *Geographical Journal.* Vol. LXXVI (1930), pp. 338–342.
12  Close, ibid., citing Seebohm (above) and Petrie, Sir Flinders. The Old English Mile. *Proceedings of the Royal Society of Edinburgh*, Vol. 12 (1883–4), p. 254.
13  Skelton, pp. 19–21 discusses the basis of Smith's 12 maps.
14  Skelton, p. 135.
15  In the BL example the plate number 78 is engraved above plate number *51*; in all other examples an attempt has been made to scratch out this number.
16  All these loose copies could have come from any of the state (iii) works. The same applies to loose copies listed above in state (ii).

Chubb C–CIIa.

# 2

## HERMAN MOLL                                                          1718

Size: 27.4cm x 29.2cm.

The work and life of Moll have already been discussed in connection with his atlas of 1724 (**24**). Around 1717–1718 there developed a great rivalry amongst a number of dealers in maps and prints to produce a pocket-sized version of Ogilby's large sheets. Some of this has been discussed in connection with Bowen's *Britannia Depicta* (**23**). Moll seems not to have been a serious rival in the event. By the end of 1718 the scheme Moll had begun and in which he was quickly joined by Thomas Bowles and others was aborted. Eleven strip maps had been issued, all of them rare and, one assumes, that that fact is an indication of the small success of the operation. In 1718 Bowen was already advertising *Britannia Depicta* and Bowles joined that project, possibly because the creation of pocket-size versions of Ogilby was a better idea than Moll's large sheets of small scale maps.[1]

### EDITION

(**i**)  *An Actuall Survey of the/NORTH ROAD from LONDON/by York to BERWICK. And from/Ferry Bridg by Wetherby to Borou:Bridg./The Black Spots markt in the Road thus [symbol]/are Statute Measured Miles, Commencing from yᵉ/Standard in Corn-hill. each mile containing 5280 F./The Computed or Post Miles, viz From London to/Waltham 12. Ware 8. Royston 13. Caxton 8. Hun/tingdon 9. Stilton 9. Stamford 12. Post* (sic) *Witham*

*8./Grantham 8. Newark 10. Tuxford 10. Bawtry 12./7* further lines/*Made Useful for ye Pocket By H: Moll Geog. 1718./* (Ea, in a rectangular box). *Sold by H. Moll between Temple Bar and S^t. Clements Church in the Strand* (De–Ee, OS). *Sold by J. Hyde at ye Statinors* (sic) *Arms behind/the Royal Exchange. T. Bowles near to ye Chap:/ter house in S^t. Pauls Church Yard. and P. Overton/near S^t. Dunstans Church Fleetstreet./* (Db, reading up the plate in column 4). There are 5 strips, with London in the lower left corner and Berwick (Eb) in column 5, with (above it) *The Road from/Ferry Bridge/by Wetherby to/ Borough Bridg./* (Ec, in a separate box). The Lincolnshire portion occupies the lower part of column 2.

Issued as a separate sheet.

**Copies**   BL Kings Top. Coll. 5. V93; London UL [G.L.]. 1730; NLS EMGB s 6.[2]

### References

1   Hodson, 1984, pp. 149–150 summarises the background to the issue of Bowen's work and the activities of his main rivals.
2   The NLS example lacks the lower half of column 5.

Not in Chubb.

# 3

## THOMAS GARDNER                                                         1719

Size: 16.8cm x 25.5cm (approx.)

Thomas Gardner is virtually unknown; only the roadbook below is recorded as evidence of activity in the map world. Yet, by a very short head, he was the first to produce a pocket sized version of Ogilby. Bowen first advertised the preparation of *Britannia Depicta* on 29 December, 1718;[1] in the following day's issue Gardner announced that his book was ready[2] and two days later, on January 1st 1719 the first volume of Senex' pair of volumes appeared.[3] After the single edition Gardner and his book seem to have been overtaken by the success gained, firstly by Senex and then, from 1720, by Bowen.

Of the publishers, Jacob Tonson (1656–1736), was a well-known figure. He was active from 1678, having taken up his freedom in the previous year. In 1679 he published, with Abel Swalle, Dryden's *Troilus and Cressida* and Dryden edited for publication in 1684 *Tonson's Miscellany.* He also published Addison and Steele. On retiring in c.1720 he handed over to his nephew.[4] John Watts (c.1678–1763) had one of the most important printing houses of the time, since he was renowned for the quality of his typesetting. He encouraged the early development of Caslon, whose typefaces are still in use and, at one time, Benjamin Franklin worked for him as a compositor.[5]

### EDITION

Each strip has a compass.

(i)   6 (Ee, OS). *The continuation of the Road from LONDON to BARWICK./Beginning at Stilton and extending to Tuxford./Plate the 2^d. Containing 62 Miles Viz./From Stilton to Stamford 13¾. to Grantham 21½. to/Newark 14 to Tuxford 12¾./Humbly Inscrib'd to the Honourable/FRANCIS WILLOUGHBY, Esq./* (Ca, above the top of columns 3 and 4). There are 6 strips on scrolls and closely imitative of the Ogilby originals.

*36 (Ee, OS). The Road from LONDON to BOSTON in LINCOLNSHIRE/BY Tho^s GARDNER/Containing 114 Miles 3 Furlongs Viz^t./ From the Standard in Cornhill*

*LONDON to Stilton in the Barwick Road 69'1. to/PETERBOROUGH 7'3 to Wellington 3'6 to Crowland 7'6 to Spalding 10'0 to Setherton 10'2/& to Boston. 6'1. Continued to the City of LINCOLN 37'4.Viz$^t$./From Boston to Heckington 13'3. to Sleford 6'1. & to LINCOLN 18'0./ Humbly Inscrib'd to the Honourable/RICHARD GRANTHAM, Esq;/* (Ca, above the top of columns 3–5). There are 7 strips on scrolls.

*41* (Ee, OS). *The Road from LONDON to FLAMBOUROUGH in COM: EBOR/By Tho: Gardner/Containing 212 Mile* (sic) *6 Furlongs Viz$^t$/ From the Standard in Cornhill LONDON to Stilton in the BARWICK Road 69'1 to PETERBORO'/7'1 to Market Deeping; 10'6. to Bourn 6'4 to Morton 2'3 to Sleeford 14'5. to LINCOLN 16. thence to/ Flambourough in Plate 2$^d$. 84'2./Otherwise 218 Mile 4 Furlongs Viz$^t$./From the Standard in Cornhill LONDON to Temsford in the S$^t$. Neots Road 52 to Eaton 3'6 to Bugdon/5'7. to Stilton a. fores$^d$* (sic) *13'3 thence to LINCOLN ut Supra./Humbly Inscrib'd to the Honourable/Sir IOHN BROWNLOW, Bar$^t$./* (Ca, at the top of columns 3–5). There are 7 strips on scrolls.

*42* (Ee, OS). *The Continuation of the Road from LONDON to FLAMBOUROUGH Com ebor/By Tho$^s$ Gardner/containing 84 Miles 2 Furlongs Viz./From Lincoln to Redbourn 19.0 to Glamford Bridges 5.5. to Barton 10'1 to/Hull 6'4. to Beverley 9'2 to Beseck 7'0. to Kilham 12'0 to Burlington 7'4 to/Flambourough 5'0 & to Flambourough head 1'2/Humbly Inscrib'd to the Honourable/Sir ARTHUR KAY, Bart./* (Ca, across the top of columns of 3–5). There are 7 strips on scrolls.

*78* (Ee, OS). *The Road from NOTTINGHAM to GRIMSBY./in the Com. of Lincoln./By THO$^S$. GARDNER./Containing 67 Miles 2 Furl. Viz$^t$./ From Nottingham to Newark 17'1. to y$^e$ City of Lin-/coln 14'6. to Wilton 6'4. to Market Raison 9'4. to Stan/ton in y$^e$ Hole 5'5 to Briggesley 7'4. & to Grimsby 5'4./Humbly Inscrib'd to the Honourable/Sir GEORGE MARKHAM, Bar$^t$.* (Ca, at the top of columns 3–4). There are 6 strips on scrolls.

*A Pocket-Guide To The English Traveller: Being a Compleat Survey And Admeasurement Of all the Principal Roads and most Considerable Cross-Roads in England and Wales. In One Hundred Copper-Plates. London: Printed for J. Tonson at Shakespear's Head over-against Katherine-Street in the Strand, and J. Watts at the Printing-Office in Wild-Court near Lincoln's-Inn Fields. MDCCXIX.*

**Copies**   BL Maps C.22.a.22; Bod Gough Maps 112. These copies are bound as oblong quartos. Other copies are known bound as two volumes as a result of the plates being folded down the centre, thus creating octavo size pocket books; examples are BL 195 a 21 (Vol. [1] only) and NLS Newman 840 (Vol. [2] only) and EME b.4.40. For the two volume versions a second title-page, identical to that for the first, was included.

**References**
1   *Daily Courant*; cited by Hodson, 1984, p. 80.
2   *Daily Courant* for 30 December, 1718 and *Evening Post* for 27–30 December, 1718, cited by Hodson, 1984, p. 81.
3   *Daily Courant* for 1 January, 1719, cited by Hodson, 1984, p. 81.
4   DNB; Plomer, 1922, pp. 291–292.
5   Plomer, 1922, p. 304.

Chubb CXXXVII.

**4**

## JOHN SENEX                                                                    **1719**

Size: 14.9cm x 19.7cm.

John Senex (*fl.*1702, died 1740) was apprenticed to the Stationers' Company in 1695, being freed in 1705. In 1706 he was joined in partnership by Charles Price and advertised that year 'A New Pair of Globes'.[1] In the following year, they began the issue of a *New Sett of Correct Mapps*, which in 1710 were being sold exclusively by Willdey, for whom Price was now working. In 1715 Senex published Halley's account of the total eclipse of the sun[2] and, in 1718, was one of a number of undertakers in an enterprise to issue a general atlas, based on data communicated by the Royal Society. He was made a Fellow of the Royal Society (1728). He died in 1740 but, in his will, he left all to his wife, Mary, who carried on the business for some years and saw the second, third and fourth editions of his road book through the press.[3] The plates were passed on to John Bowles and Son in 1748 or very soon after; two editions, which only differ from the change to the title-page imprint, were probably issued quite close together. The plates remained with the Bowles family for twenty years. R. Wilkinson came into possession of them and issued a very rare edition in 1780. The odd appearance of the plates with the title- page altered by hand to suggest a late issue in 1792 remains unexplained.

The plates prepared by Gardner are Ogilby miniaturised and every effort had been made to retain the original's appearance on the page. Senex, however, has tried to reduce the scrolls and eschewed entirely the ornate titles and the decorative engravings around the titles. The result is a plain but practical pocket volume.

### EDITIONS

All strips have simple pointers.

(i)   6 (Ee, OS). There is no title; the plate shows the continuation of the London to Berwick road, from Stilton to Babworth in Nottinghamshire (miles 69–138). There are 6 strips on separate scrolls.

*36* (Ee, OS). *The ROAD from LONDON to BOSTON in Lincoln sh./ With a Branch from Boston to Lincoln./Commencing at Stilton in the Barwick Road/Stilton 69/Peterborough 76½/Widrington 80/Crowland 88./Spalding 98/Setherton 108¼/Boston 114/And from thence to/Kekington* (sic) *13½/Sleaford 19½/Lincoln 37½/* (the places and mileages are arranged in two columns, the second beginning with Boston (Ca, at the top of columns 3–5). There are 7 strips on scrolls.

*41* (Ee, OS). *The ROAD from LONDON to FLAMBOROUGH in YORK-SH/Commencing at Stilton in the Barwick Road:/And Shewing the Way to Stilton through Tamesford./ Stilton 69/PETERBOROUGH 76¼/Market Deeping 87/Born* (sic) *93½/Morton 96/Sleaford 110½/LINCOLN 128½/Redbourn 147½/Glamford Bridges 153/ Barton 163¼/Hull 169¾/Beverley 179/Beseck 186/Kilham 198/ Burlington 205½/Flamborough 210½/ Flamborough/Head 212¾/* (Ca, across the top of columns 3–5). The place-names and mileages are arranged in three columns, with Stilton, LINCOLN and Beseck at their heads. There are 7 strips on scrolls showing the route from Tamesford (i.e. Tempsford) to Lincoln (52–128 miles).

*42* (Ee, OS). There is no title; the plate has 7 columns and continues the route of plate 41, starting at Lincoln and ending at Flamborough.

*78* (Ee, OS). *The Road from NOTINGHAM* (sic) *to GRIMSBY./in LINCOLN SHIRE/Containing from Notingham to/NEWARK 17/ LINCOLN 31¾/Walton 38½/Market Raising*

*48½/Stanton 54¼/ Briggesly 61¾/Grimsby 67/* (Ca, at the top of columns 3–4). The place-names are arranged in two columns, headed by Newark and Stanton. There are 6 strips on scrolls.

*An Actual Survey Of all the Principal Roads of England and Wales; Described by One Hundred Maps from Copper Plates. On which are delineated All the Cities, Towns, Villages, Churches, Houses, and Places of Note throughout each Road. As Also Directions to the Curious Traveller what is worth observing throughout his Journey. The whole described in the most easy and intelligible Manner. First perform'd and publish'd by John Ogilby, Esq; And now improved, very much corrected, and made portable by John Senex. Vol. I. Containing all the Direct Roads from London through England and Wales In 54 Plates. London: Printed for and sold by J. Senex at the Globe in Salisbury-Court, Fleetstreet. 1719.* Part II has a similar title-page, with 46 plates, although the two parts are often bound together.

**Copies** BL Maps C.24.b.9; CUL Atlas 6.71.1; Leeds UL W 243; Guildhall L. S 388/1. (LRL Map 52 (plates 36 and 78).

**(ii)** On plate 6, the erroneous 30 (mileage at the top of column 1) has been corrected to *80.*

*An Actual Survey . . .* [1719].

**Copies** Bod Gough Maps 120[4] and Johnson Maps 242; Leeds UL W 242; Stafford, William Salt L W.K.2.14.

*An Actual Survey . . . Vol. I . . . London: Printed for and sold by J. SENEX at the Globe in Salisbury-Court, Fleetstreet, and W. TAYLOR at the Ship in Paternoster-Row. 1719.*

**Copy** Norwich RL R 912.42 (S.C. 3.20).

*An Actual Survey . . . By John Senex, F.R.S. In 2 Vol. The Second Edition London: Printed for & sold by M. Senex at the Globe ag^st. S^t. Dunstan's Church Fleetstreet.* [1742].[5]

**Copy** CUL Atlas 6.74.3.

*An Actual Survey . . . In 2 Vol. The Third Edition London: Printed for and sold by M. Senex . . .* [after 1742].[6]

**Copies** RGS 10 A 31; Northamptonshire AO GK 1109; NLS Newman 629.

*An Actual Survey . . . In 2 Vol. The Fourth Edition London: Printed for & Sold by M. Senex at the Globe ag^st. S^t. Dunstans Church Fleetstreet.* [1748].[7]

**Copy** The private collection of Mr. D. Webb.

*An Actual Survey . . . The Fourth Edition London Sold by John Bowles and Son at the Black Horse in Cornhill* [1753].[8]

**Copy** The private collection of Mr. D. Webb.

**(iii)** All plates are double-sided.

*6 (Aa and Ea, OS). The Miles shown on this Itinerary are those through Huntingdon continued. But the Mile Stones placed upon the Road are measured through Hatfield & Bugden* has been added (Ae–Ee, OS). The following changes have been made to the plate: column 1: *Silton* has been corrected to *Stilton,* its mileage changed from 69 to *71,* (and all other marks increased by two); to the east of the village *to Peterborough/& Boston Pl. 37./to Lincoln, Hull &c/see Pl. 43 & 44.* has been added; below Stilton 4 lines have been added about mileage variations from London. Column 3: At *Coltsworth* the words *the birth Place/of S^r Isaac/Newton/* have replaced *to Corbey;* at the by-road at *little*

*Panton* (sic) *to Ancaster &/Lincoln/of late a much/resorted way/* has been added. Column 5: *to Nottingham/see Pl 86./* has been added at mileage 120; below the name *Newark, to Lincoln/see Pl. 86/* has been added.

*37* (Aa and Ea, OS, instead of [plate] 36). The mileages on the map have been increased by 2 and in the chart below the title, but only in the road to Boston. Column 4: *Horizontal/distance 95/miles/* has been added above Boston. Column 7: South-east of Lincoln *Pl. 86/to/* has been added above the word *Newark*; above Lincoln *to Barton &/Hull Pl. 44/* and *to Market Raisin/and Grimsby/Pl. 86./* have been added.

*43* (Aa and Ea, OS, instead of [plate] 41). *Another Road to Lincoln is through Panton Pl. 6. The Stage Coach goes thro' Grantham.* has been added to the right of column 7, reading up the plate. The mileages from *Stilton* to Peterborough have been increased by 2 on the road and in the chart (Ca), and, thereby, the mileage south of Peterborough has been corrected (formerly 70 instead of 76); mileages after Peterborough have been increased by 1 as far as Lincoln. Column 1: *Plate 47 shows the Road/from Barnet to Temsford/* has been added at the foot. At *Temsford* has been added *a bridge*; at St. Neots *to Cam/bridge/Pl. 65/* has been added. Column 2: *to York Pl. 6* has been added at the top (Stilton). Column 4: *Pl. 37* has been added below *to/Crowland* at Peakirk.

*44* (Aa and Ea, OS, instead of [plate] 42). All mileages have been increased by 1 throughout the plate. Column 1: *Plate 86 shows from/Lincoln to Market/Raising and Grimsby./* has been added at the foot; *to/Nottingham/Pl. 86/* and *to Grimsby/Pl. 86/* have been added south-west and north-east of Lincoln respectively. Column 4: a ship has been drawn in the Humber. Column 6: to the direction at Kilham *& Scarborough* has been added. Column 7: *The Road for Scarbo-/rough turns off at Dros-/-field* (sic, i.e. Driffield) *over the pleasant/Woulds through Seamer/to Scarborough./* has been added at the top.

*86* (Aa and Ea, OS, instead of [plate] 78). Column 1: *to/London/Pl. 42* has been added to the east of Nottingham. Column 6: Below Grimsby *56 direct/Horizontal/distance/* has been added.

*The Roads Through England delineated or, Ogilby's Survey, Revised, Improved, and Reduced to a Size portable for the Pocket By John Senex F,R,S. Being an Actual Survey of all the Principal Roads of England And Wales, Distinctly laid down on one hundred & one Copper Plates on which are delineated all the Cities, Towns, Villages, Hills, Rivers, Brooks Churches, capital Seats, and every place worthy of Note, throughout each Road; With The addition of some Roads newly drawn, which were omitted by M$^r$. Ogilby, and several necessary Corrections made in others: Together with a great number of explana-tory references, by which this Edition of the Roads is render'd of more general Use to Travellers. Printed for John Bowles and Son at the Black Horse in Cornhill London 1757.*

**Copy**  RGS 262 B 24.

(iv) Plate 44: *Direct Horizontal distance/168 miles.* has been added (Ec, reading upwards in the 7th strip).

*The Roads Through England delineated . . . 1757.*

**Copies**  BL Maps C.27.a.6; CUL Atlas 6.75.8; Bod Map C.17.e.34; Sheffield UL **912.42 (O); NLS Newman 630; Hull RL 080.91242.

(v) On plate 6, column 1 *N.B. Stilton is/* has been added above the four lines at the foot. The legend below the plate has been extended by the addition of *& are 4 miles more.* Plate 37: *Plate 43 shews from Peterborough to Lincoln.* has been added below columns 1 and 2. On Plate 43: Column 2: *71.m. by the way of/Huntingdon Pl. 5./* has been added at Stilton. Column 3: *75 miles by way/of Bugdon* has been added at Stilton. From Peter-borough to Lincoln all mileages have been increased by 1, ranging from 79 to 130 as a result. Column 6: *Pl.37* has been added at Sleaford below the reference to the Boston road. Plate 44: all mileages have been increased by 1, ranging from 130 to 214 (there is no 131) as a result.

*The Roads Through England . . . Printed for John Bowles and Son at the Black Horse in Cornhill London 1759 Price 7<sup>s</sup>.6<sup>d</sup>*

**Copies** CUL Atlas 7.75.6; NLS EME b.1.4; Leeds RL R912.42 OG4; Norwich RL R 912.42 (S.C.3.19).[9]

*The Roads Through England . . . Printed for John Bowles at the Black Horse in Cornhill London 1762 Price 7<sup>s</sup>.6<sup>d</sup>*

**Copies** BL Maps C.24.b.17; CUL Atlas 6.76.9; Bod Vet A5 d 169; William Salt L. (Stafford) W.K.2.14.

*The Roads Through England . . . Printed for John Bowles at Number 13 in Cornhill London. Price 7<sup>s</sup>.6<sup>d</sup>* [1768?].[10]

**Copy** Bod Gough Maps 130.

*The Roads through England delineated . . . Publish'd as the Act directs 1<sup>st</sup>. May, 1780, by R. Wilkinson, at Number 58, in Cornhill, London. Price 7<sup>s</sup>.6<sup>d</sup>.*

**Copy** The private collection of Mr. D. Webb.

*The Roads through England . . . London 1792 Price 7<sup>s</sup> 6<sup>d</sup>* [11]

**Copy** RGS Fordham 87.

### References

1 *London Gazette*, 6–9 May, 1706, given in full in Tyacke, p. 89.
2 Advertised twice in *London Gazette* in March and April, 1715 (Tyacke, p. 103, items 402 and 403).
3 Tyacke, p. 142.
4 Bod Gough Maps 120 has 1719 very faintly visible on the title-page of Vol. II.
5 John Senex died in 1740 (Plomer, 1932, p. 224). Volume II has its own title-page and this is dated 1742. The work sold for 5s.6d. – advertised on a broadsheet held in CUL (Maps bb.999.74.1).
6 Only the title-page of [Part 1] has 'The Third Edition'. The two parts are bound together in all cases; the title-pages of Part II are identical to that of the second edition dated 1742. The work must have been printed after 1742, but how much after is not clear. Webb, D. Further Notes on the Senex Road Books. *IMCoS Newsletter*, No. 34 (Autumn, 1988), pp. 25–27 suggests 1742–1748, the latter being the suggested date of the Fourth Edition (q.v.).
7 The general map of England and Wales is dated 1748.
8 While the general map is still dated 1748 the date 1753 is suggested. John Bowles did not advertise as John Bowles and Son (i.e. Carington Bowles) until 22 Feb., 1753 in *Public Advertiser*. (Cited by Hodson, 1984, p. 188). It is not known exactly when Bowles acquired the Senex plates.
9 The Norwich copy has been differently imposed in the press; the effect of this is that plate 6 is the verso of plate 5 and faces plate 7; in other examples plate 6 has plate 7 on its verso and so on throughout.
10 The numbering of houses began in London during the 1760s. Hodson, 1984, p. 187 notes that Bowles first used the newly numbered property in *Gazetteer and New Daily Advertiser* for 1 July, 1768. Advertisements placed earlier in that year used the former type of address.
11 The last two digits in the date have been changed by hand.

Chubb CXXXVIII–CXLV (including editions now shown to be by others).

**5**

**EMANUEL BOWEN**                                                    **1720**

Size: 18.2cm x 11.65cm (approx.; much material is outside the frame-lines).

The basic details of this work by Bowen have been discussed above (**23**). The full details of titles, imprints and copies already provided are not repeated here. It is important to remember that copies of this work vary a great deal; the way examples were made up in the print shop was, even for its time, haphazard. Thus, volumes may include plates, which are in a revised state, while other plates are in an earlier version, even though a later state appeared to be already available. Only the examples noted below against each edition have the plates in the state being described. The copies noted below include the changes described in that issue; no assumptions should be made that the copies so noted also incorporate all the earlier alterations.

**EDITIONS**

Each strip has a compass.

(i) *14* (Aa, OS). The road from Stilton to milestone 100 (5 miles south of Grantham on the Great North Road) is shown in three strips. Down the left side are notes on *Harlax/ton, Stamford* and *Burley/Hall* (i.e. Burghley); under the three strips are notes on *Hartford* (sic) with *The ARMS of HARTFORD* (Ed).

*15* (Ea, OS). Three further strips continue the road on plate 14 as far as Tuxford. There are notes on *Belvoir/Castle* (Ec) and *Grantham* below the three strips, with *The ARMS of GRANTHA$^M$* (Ad).

*87* (Ea, OS) *The Roads from/LONDON to BOSTON &c Com̃encing at Stilton (See Page/Containing* . . . heads two columns of town names with their mileages from Stilton to Boston and on to Lincoln, via Sleaford. Below is *The DESCRIPTION of LONDON – continued.* The strips for the above roads begin with:

*88* (Aa, OS). There are three strips; down the left side is a note on *Crowland or/Croyland;* along the bottom is a note on *PETERBORO'* with *Y$^e$ ARMS of PETERBORO* (Ed). The strips show the roads from Stilton to milestone 107 (south of the present day Sutterton).

*89* (Ea, OS). There are four strips, the major part of the first showing the end of the road from Spalding to Boston (Ab); the top of the first and the remaining three strips show the continuation from Boston to Lincoln through Heckington and Sleaford. Notes appear on *Spalding* (Ea, below the page number), *Peakirk* (Ec–Ed), *Sleaford* (Ae) and Boston (the foot of strips 2–4, with the final part in column 1 (Ad) and *y$^e$ Arms of BOSTON* (Bd).

*103* (Ea, OS). Four strips show the road from milestone 81 (2 miles south of Peakirk) to Lincoln through *Born* and Sleaford. There are notes on *Sleaford* (Eb), *Born* (Ec), *Stow* (Ed) and *LINCOLN* (below the four strips) with *y$^e$ Arms of LINCOLN* (Bd). The note on Stow mistakes the village a mile north of Market Deeping for Stow, west of Lincoln.

*104* (Aa, OS). Three strips show the continuation of plate 103 from Lincoln to Barton. Notes appear on *Barton* (Ab), *Glamford-Br.* (Ac) and *Kingston/upon Hull* (begins at Ac and takes up all the lower part of the plate, with *the Arms of KINGSTON/upon HULL* (Ed) and the arms of the Duke of Kingston (Aa, below the page number).

*206* (Aa, OS) contains the map of Lincolnshire and describes the contents of the next plates: *The ROAD from/NOTTINGHAM TO GRIMSBY IN/LINCOLN SHIRE . . ./*

*207* (Ea, OS). Three strips show the road from Nottingham to Lincoln; there are notes

11. 1720 APPENDIX MAP 5 – EMANUEL BOWEN, PLATE 103

on *Bridgford* (Eb), *Shelford* (Ec), *The BISHOPRICK of LINCOLN* (below the three strips and a note below that on titles associated with Lincoln). The arms of the Bishop (Dd) and of the Earls of Lincoln (Ee) are shown.

*208* (Aa, OS). Three strips show the road from Lincoln to Grimsby. There are notes on *Market/Rasin* (Aa, below the page number), *Grimsby* (Ac–Ae and continued with two final lines along the bottom of the plate), *The DEANERY-/of LINCOLN* (Bd, below the arms of the Dean) and *Stoake* (sic – near Newark, Dd–Ed). *y$^e$ Arms of GRIMSBY* appear above the note on the town (Ab).

*Britannia Depicta . . . 1720.*

**Copies**   BL Maps C.27.a.12; CUL Atlas 7.72.13; Bod G.A. Eng. Roads 8° 92; RGS 260 d 24; Nottingham UL s/G 5514; Guildhall L S 388/1. (LRL Map 169 is pp. 207–208; LRL Map 198a is pp. 104–105).

**(ii)**   *87* At the end of the third line *13* has been added, as the conclusion of the words *see Page* on the previous line.

*104* The mileage south of Barton has been corrected to *159* (formerly 150).

*Britannia Depicta . . . 1720.*

**Copies**   BL Maps C.27.a.13; CUL Atlas 7.72.8; LAO 950.2; Northampton AO NRO Lib 53.

**(iii)**   *14* The mileages have been increased by 2 throughout; they run continuously from 71–102 (previously they were 69–100). Mileage mark 97 is now below *Lopthorp*; (previously 95 was at the site of two side roads, which have largely disappeared in the removal of 95).

*15* The mileages have been increased by 2 throughout; they run from 103–133 (previously they were 101–131).

*Britannia Depicta . . . 1736.*

**Copy** Bod Godw. 8° 853.[1]

*Britannia Depicta . . . 1753.*

**Copy**   BL Maps C.27.a.22.[2]

**(iv)**   *15* Below Newark *to Nottingham/See Pl. 206/* has been added in column 2 and, opposite, *To Lincoln/see Pl. 206./* has been added.

*87* In line 3 the page reference has been altered to *11*.

*88* In column 2 *To Market/Deepi$^{ng}$ See Pl 101/* has been added, south-east of mileage 84.

*89* In column 4 (top) *To Barton & Hull/See Pl.101* has been added, the words being divided by *LN* of *LINCOLN*. Below Lincoln *To Market/Raising & Grims=/by see Pl. 206* has been added.

*103* Mileages have been increased by one throughout columns 1–4 as far as mileage 124; 124 then appears twice, followed by 125 & 126; there is no 127 and the next mark is 128, followed by 129 (where there was no mark previously). North-east of Lincoln has been added +; in the margin + *Another/Road to Lin:/coln, is thro' /Panton See/Pl. 12. The/Stage Coach/goes thro'/Grantham./*

*104* Mileages have been increased by 1 throughout and now read from 130 to 166 (in the Humber). In column 1 *to Nottingh-/am. See Pl./206./* has been added (south-west of Lincoln). In column 3 (below Barton) *131 computed* has been added.

*207* In column 1 above the last part of *Nottingham* has been added *to London* and below

*S P. 274*. Below Newark has been added *to London/see Pl. 5 &c/*. In column 3, *See Pl. 101* has been added below the note *Road to/London*.

*Britannia Depicta . . . 1759.*

**Copies** BL Maps C.27.a.23; Bod Vet A5 e 2499.

(**v**) *14* Mileages have been increased by 2 and now run from 71–102; most of the numbers have been re-engraved and show marked stylistic differences in their engraving (e.g. 94 compared with the previous 92). The references to mileages in the accompanying text have been altered: *106* (below the piece on Harlaxton), *84¾* (Stamford) and *84* (Burley/Hall).

*15 (at 106½).* has been added in the bottom line of the text on Grantham. The mileages have been increased by 2 and now run from 103–133, the replacement figures not always sited as before.

*87* The figure in the title: *116* (in line 3) replaces the previous 114′3. The figures in the table below have been altered in the second column to accord with the new calculation in the heading, viz. *71′2, 78′4, 82′2, 90, 99′4, 110′2, 116* (instead of *69′1, 76′4, 80′2, 88, 98, 108′2, 114′3* respectively). The mileages from Boston to Lincoln remain unaltered.

*88* In line 2 of the text on Crowland the mileage reference has been increased from 88 to 90. Mileages have been re-engraved and increased by 2 throughout, running now from 72 to 109. Above the text for *PETERBORO* has been added *at 78½*.

*89* Mileages have been re-engraved and increased by 2 in that part of column 1, which refers to the road to Boston. The section on Sleaford (Ad) has been deleted. The mileage at the end of the text on Spalding has been changed to: *(at 99)/P: 88* (previously (at 98) only). At the end of the text for Crowland a similar change has altered the mileage to 85, with the same page reference added.

*103* The mileages have been re-engraved, increasing by 2 throughout, running now from 83–130. The mileages attached to the marginal notes have been altered to *112½, 95½* and *90½*, from 111, 93 and 88 (*Sleaford, Born, Stow* respectively).

*104* Mileages have been re-engraved and increased throughout by 2, running from 130–166 (166 has been added, in the Humber). The mileages attached to the notes on *Barton* and *Glamford-Br.* have also been increased to 165 and 155 respectively. At the end of the piece on Kingston upon Hull *[at 172 P. 105]*. has been added.

*206* In the heading the figure 67½ in the third line has been reduced to *67* and a space left where the fraction has been removed. To make the table of mileages (below) match this alteration all the figures in the second column have been revised to: *17, 31′6, 38′6, 48′4, 54′2, 61′6, 67*.

*207* The error in mileage near Boultham (20) has been corrected to *29*.

*208* At the end of the marginal note on *Stoak* the reference has been altered to: *[at 14 P. 207]*. (formerly [at 14]).

*Britannia Depicta . . . 1764.*

**Copies** BL G 4697; CUL Atlas 7.76.4; Bod Vet A5 e 5417; Grimsby RL 914.2 OWE; Manchester John Rylands UL L 942.006 Og4 (R 85119).

*Britannia Depicta . . .* [1767].

**Copy** BL Maps C.27b.68.

**References**
1  No other example dated 1736 has this revision; the other copies all continue to include the

plates in their unrevised form. The unrevised plates were still being incorporated in copies as late as 1764 (the Guildhall copy).

2  Two copies dated 1753, in CUL and Leeds RL, contain unrevised leaves.

Chubb CXLVI–CLVI.

6

**S. BEE**                                                                                    **1756**

Size: 70.6cm x 63.1cm

Nothing is recorded of S. Bee, whose name is given in the title, nor of William Rower of Boston, who drew the map and engraved it. The map acknowledges the use of Morden's map of 1695 in the preparation of the eastern part. At half an inch to a mile there is plenty of space to add detail but only the positions of towns and villages have been included while the roads connecting them to each other are not shown, and nor are other physical features, such as the main drains. The publisher is Robert Sayer, who was advertising from the address in Fetter-lane from as early as 1748.[1] It is presumed that the map was prepared soon after, if not in support of, the turnpike's creation; the Lincoln Heath to Peterborough trust was created in 1756.[2] If that is the case the map was no longer listed in Sayer's stock in his 1766 catalogue. The dedication is to Peregrine Bertie, who died in August, 1778.

**EDITION**

(i)  *A/SURVEY/and/PLAN/of/The Road from the City/of/Lincoln/ Over the Heath, thro Dunsby-Lane,/Sleaford, Folkingham, Bourn, and/Market-Deeping to the City of/PETER-BOROUGH./And also/From Bourn thro' Edenham,/Grimsthorp-Park, by Swinstead,/ and Corby, to/COLSTERWORTH./By/S. Bee./* (Aa, in a rococo cartouche). *A Scale of Miles/12* [= 14.7cm] (in a panel)/*N.B. All the Towns and Roads Eastward of the /said Road from Lincoln to Peterborough/are laid down from MORDEN'S Map/of Lincolnshire By the same SCALE./William Bower in Boston delin. et scrip$^t$./Sold by R. Sayer, opposite Fetter-Lane, Fleet-Street, Price 6.$^s$* (Ce). Compass (Ec). *A TABLE of References, Shewing All the/Noblemen and Gentlemens Names that are/Owners of the several Parishes . . . / . . . setting forth each Parish's/Distance in Miles from Peterborough & London./* (Ab–Ae, with a small break to allow the road to Colsterworth (at the edge of the map) to be shown). In the table letters A–Z, a–z (both less J/j) and numbers 1–16 are applied to the names of owners of parishes, etc., and provide a key to the letters or numbers placed on the map. *TO/The Most Noble/Peregrine Bertie/Duke of Ancaster & Kesteven/Lord Great Chamberlain of England, &c./Lord Lieutenant and Custos Rotulorum of the County of/LINCOLN./And to the . . . Promoters of,/and Subscribers to the Repair of these Roads/This PLAN is Humbly Dedicated by His/Grace's, and their/Most Obedient Servant,/Robert Austin./* (Ee, in a rococo cartouche). A note refers to the existence of 'near Thirty Miles' of country from the Humber to Lincoln (Ca). The area shown is from Lincoln to Stilton and from Colsterworth to the Lincolnshire coast.

Sold separately.

**Copy**   Bod Gough Maps Lincolnshire 13.

**References**

1  Hodson, 1984, p. 74.
2  Wright, N.R. 1982, op. cit., pp. 37 and [259].

# 7

## Anonymous c.1758

Size: 29.75cm x 23cm.

The Donington Trust was set up in 1758 and controlled a network of roads reaching as far as Boston.[1] Two versions of a map concerned with the setting up operation are known. It is assumed the earlier version was prepared to accompany the proposal for an act to facilitate the creation of the trust. The second version incorporates the changes made to carry a road from Boston to Langrick Ferry and thus obtain access to the gravel bed at Amber Hill. Who drew the maps and was responsible for their execution is not known. Although there is no scale the map is quite accurately drawn at one inch to a mile.

### EDITIONS

(i)   *A Plan of the Roads from the Borough of Boston in the County of LINCOLN/To the Eight Miles Stone in the Road to Spalding . . . 8:0:0/ And also of a New proposed Road from Boston to Swinshead* (sic) *North end 5:6:0/And also of the Road from thence to Hale Barr in the Road to Lincoln 1:6:0/And also of the Road from thence thrô. Swineshead to Donnington High Bridge 6:2:0/* (Aa–Ea). Compass (Ce) – the map is drawn with north at the bottom.

**Copies**   RGS 1 C 88 (88); LRL Map 2.

(ii)   An extra line has been engraved below the title, etc. It reads: *And also of the Road from Langrick Ferry to Kirkton Holme . . . 3:0:0/.* A drawing has been added and is labelled: *Double Pointing Doors/for the ends of y^e Drains/* (Ab). Among the many additions are: *Sutterton Drain, North Forty Foot, Brothertoft, Gill Syke, Witham R.* (twice). Two new cuts and the word *Drain* have been added, west of *Donnington. New Cut* has also been added twice, south of Boston.

**Copy**   LRL Map 3.

### Reference
1   Wright, N.R. 1982, op. cit., pp. 38 and [259].

# 8

## LE ROUGE, GEORGE LOUIS 1759

Size: 15.6cm x 20.2cm.

Le Rouge was a military engineer by profession but, after taking up cartography in c.1740, produced a large body of work, specialising in town plans, maps of military battles and fortifications. He also produced general atlases, including several of American subjects, *Recueil des Villes, Ports d' Angleterre* (1759) and *Topographie des Chemins d' Angleterre* (1760).[1] The maps in the work below are based closely on those of Senex, but with directions and notes in French. The influence of Senex may be readily seen in the use of Notingham (sic) on Plate 86, compared with Plate 78 in Senex' first edition. The plates seem to have passed to Louis Charles Desnos, who followed le Rouge in presenting them with a dual language title-page. Desnos flourished from c.1750 to 1790; he was appointed as globe-maker to the King of Denmark, but spent the largest part of his working life in Paris. He produced roadbooks of France also.[2] The volume of 1767 noted below consists of three parts; the first is an atlas of England (dated on its title-page 1760), in which Lincolnshire appears on parts of three plates (the west of the county on 83, the east and north on 84 and the south on 81); the second part contains the road strips and the third engravings of ports and the coast-line usually as seen from an off-shore vessel, the aforesaid *Recueil des Villes* . . .[3]

## EDITIONS

Each strip has a simple pointer.

(i)  6 (Ea, OS). There is no title but the map continues plate 5, the road from London to Berwick, starting at Stilton and continuing to Babworth (Notts.). There are six separate rectangular strips. *Les Miles marqués dans cet Itinéraire sont oeux* (sic) *qui sont Continues par Huntingdon, mais les bornes placées sur le Terrain sont mesurées par Hatfield et Bugden* (Ae–Ee, OS). Other notes in French appear on the plate especially at the foot of column 1.

*37* (Aa, OS). *de Londres à Boston/ et de Boston à Lincoln/ Commancant à Stilton/Route de Barwick pl: 5./* (Ca, in a box across the top of columns 3–5). There are seven strips, each being separate rectangles; Stilton to Boston finishes near the top of the (shortened) fourth column; the rest is taken up with the Boston-Lincoln road.

*43* (Aa and Ea, OS). *de Londres à Flamborough/Pr^{ce}. de York/ Commence à Stilton/Route de Barwick pl. 5./de plus le Chemin de Stilton par Temsford/* (Ca, in a box across the top of columns 3–5). There are seven separate rectangular strips for the route from Stilton to Lincoln through Peterborough and Sleaford. *Autre Route à Lincoln Parpantpon* (sic; i.e. Little Ponton) *Pl. 6. le coche passe par Grantham.* (Ec–Ea, OS, reading up the plate).

*44* (Ea, OS). There is no title. The seven separate rectangular strips show the way from Lincoln to Flamborough. Notes in French appear particularly at the foot of column 1 and the top of column 7.

*86* (Ea, OS). *de Notingham* (sic) *à Grimsby/en Lincoln/* (Ca, in a box at the top of columns 3–4). There are six separate rectangular strips.

*Itineraire De Toutes Les Routes De L'Angleterre Revuës Corrigé^{e}s (sic) augmentées, & réduites, Par Senex en 101 Cartes. Contenant un détail exact* (a letter has been partially removed – probably an erroneous e – at the end of the word) *de toutes les Villes, Bourgs, Villages, & Montagnes, Rivieres, Ponts, Eglises, Maisons Seigneuriales, et de tous les Lieux remarquables. Bowles a ajouté en 1757 plusieurs nouvelles Routes à cet ouvrage, plusieurs renvois et Corrections necessaires. Cet ouvrage a été traduit de l'Anglois Par le S^r. le Rouge Ing?* (sic) *Géographe du Roy, et se vend A Paris Chez le même Rue des grands Augustins. M.DCC.LIX. Prix 9.^{#}*

The French language title occupies the left half of the leaf; an English language version occupies the right hand half: *The Roads through England or, Ogilby's Survey; Revised, Improved, and Reduced By Senex Distinctly laid down on one hundred & one Plates With The addition of some Roads newly drawn and Several Corrections of more general Use to Travellers. Printed at Paris. For le Rouge Geographer great Augustin Street. 1759.*

**Copy**  RGS 260 g 19.

*Les Routes D'Ogilby Par L'Angleterre Revues . . . M.DCC.LIX. Prix 9.^{#}*

**Copies**  CUL Atlas 6.75.7 (7.75.7 on the title-page); NLS Newman 595; Edinburgh UL EB 911 (42) Ogi.

*Nouvel Atlas D'Angleterre. Divisé En ses 52. Comtés Avec Toutes les Routes Levées Topographiquement par ordre de S/M Britannique et les Plans des Villes et Ports de ce*

*Royaume. A Paris Chez le Sieur Desnos Ingénieur Géographe pour les Globes et Spheres Rue S<sup>t</sup>. Jacques. 1767. N<sup>d</sup>. On trouve chez le même toutes les Routes differentes de l'Europe.*

**Copy** BL Maps C.22.bb.6; Bod Gough Maps 105; Leeds UL Special Collections. Anglo-French.[4]

*Nouvel Atlas . . . S<sup>t</sup>. Jacques. N<sup>d</sup>. On se trouve . . .*

**Copies** Leeds UL Special Collections. Anglo-French.[5]

**References**

1 Moreland and Bannister, p.134.
2 ibid., p. 232.
3 The third title-page *Recueil des Villes Ports D'Angleterre Tiré des Grands Plans de Rocque et . . . du Belin* precedes plates showing towns, etc. and appears in all copies of *Nouvel Atlas* dated 1766.
4 The 'double' title-pages, one in French with another in English to its right, follow the exact wording of the edition by le Rouge. The imprint reads: *A Paris, Chez le S<sup>r</sup>. Desnos, Ingénieur geographe pour les Globes et Sphéres, Rue S<sup>t</sup>. Jacques à l'Enseigne du Globe et de la Sphére 1766. on se trouve . . .* Some confusion has arisen over the years, arising from Fordham's belief that Desnos had copied from Senex directly (Fordham, F.G. Studies in Carto-bibliography, p. 41); he does not refer at all to Le Rouge.
5 On the atlas title-page there is a 'thin' where the date formerly stood and it may have been issued after those noted above. However, the second title-page (*Itineraire De Toute Les Routes D'Angleterre . . .*) and the third (*Recueil des Villes . . .*) are both dated 1766. Normally the plates are single-sided, but copies of loose sheets printed on both sides have been reported. (R. Lintott in a letter in *Map Collector*, No. 51 (Summer, 1990), pp. 53–54).

Chubb CXLIII (the 1767 edition by Desnos only, ascribed to Senex).

**9**

**JOHN HINTON**                                              **1766**

Size: each plate varies; see below.

Hinton's work has been noted in connection with the early plates for *The Large English Atlas* (**34**), including that for Lincolnshire of 1751. Between 1765 and 1773 Hinton issued 39 plates, covering the same ground as Ogilby but on sheets several times larger and, in some cases, displaying several roads not especially related to each other on the same plate. The road strips appeared in *The Universal Magazine of Knowledge and Pleasure . . .* Hinton used both Kitchin and Bowen for map engraving, but there is no evidence for a firm attribution for the engraving of the road strip plates.

**EDITION**

There are simple pointers in each strip.

(i) *Plate IV.* (Ea, OS). Size: 29.2cm x 36.3cm. *The ROAD from LONDON/to the 145.<sup>th</sup> Mile-stone in the way to/BERWICK:/Actually Survey'd from Cornhill to/Waltham 12/Ware 21½/Royston 38/ Huntingdon 58/Stilton 71¼/Stamford 85/Grantham 106½/ Newark 120½/Tuxford 133¼/ and/Tarworth 145/NB. This Road will be continued in Pl. V. to Berwick.* (Da, in a panel at the top of columns 5–6). The places and their mileages in the panel are arranged in 3 columns, headed by Waltham, Stilton and Tuxford. There are 8 strips, separated by vertical single rules; the first and second (lower part only) strips relate to the road from Bristol to Huntspil (Somerset); the rest of the second strip shows the route from Bath to Bristol. The road from London northwards begins at the foot of the third and reaches Tarworth at the top of the eighth strip. The road to Berwick continues in Plate V (July, 1766) but contains no part of Lincolnshire.

*The Universal Magazine of Knowledge and Pleasure: Containing News Letters Debates . . . Geography Voyages . . . And other Arts and Sciences . . . Vol. XXXVIII. Publish'd Monthly . . . By John Hinton, at the King's-Arms in Paternoster Row, near Warwick Lane London Price Six Pence.* [Issue for June, 1766].

**Copies**   BL PP 5439. (RGS Portfolio 202).

*Plate XV.* (Ea, OS). Size: 30.4cm x 34.3cm. *A SURVEY of the ROAD from/LONDON to BOSTON,/in LINCOLNSHIRE,/Commencing at Stilton in the Berwick Road Plate IV./Stilton 71/Peterborough 78½/Crowland 90/Spalding 100/Setherton 110¼/ Boston 116/With a Branch from/BOSTON to LINCOLN,/Boston Sleaford 19½ Lincoln 37½/* (Aa, in a box at the top of columns 2–3; the places and their mileages in the first list are arranged in three pairs). The road to Boston from Stilton occupies the first two columns and the road to Lincoln the third and most of the fourth; the remainder of the plate contains: *A SURVEY of the ROAD from/LONDON to BURTON,/* . . . (in a box at the top of columns 6–7). The road shown starts at Darlaston and ends at Burton on the Westmoreland border. There are 8 columns.

*The Universal Magazine . . . Vol. XLI . . .* [Issue for December, 1767].

**Copies**   BL PP 5439. (Grimsby RL; LRL Map 780 is the portion of the plate showing the roads to Boston and Lincoln, i.e. the first 3 strips and all the fourth but for a small part at the top).

*Plate XVI.* (Ea, OS). Size: 30.2cm x 35.6cm. *A SURVEY of the ROAD from LONDON/to FLAMBOROUGH HEAD, YORK-SHIRE,/ commencing at Stilton in the Berwick Road Plate IV./Stilton 71/ PETERBOROUGH 78¼/Market Deeping 89/Born 95½/Sleaford 112½/LINCOLN 129½/Redbourn 149½/Glanford Bridg$^s$. 155/ Barton 164¾/Hull 170¾/Beverley 180/Kilham 200/Burlington 207/Flamborough 212/Flamborough H$^d$. 214½/* (Da, in a panel at the top of column 6, with small portions in columns 5 and 7). The place-names and mileages in the heading are in three columns, headed by Stilton, Lincoln and Beverley. The first 2 strips show the road from 12 miles south of Kendal to Carlisle, being a continuation from plate 15. The main strip starts at Stilton (milestone 71), a third of the way up the third strip and reaches Flamborough Head at the top of the eighth strip.

*The Universal Magazine . . . Vol. XLII . . .* [Issue for February, 1768].

**Copies**   BL PP 5439. (RGS Portfolio 202).

*Plate XXXIII.* (Ea, OS). *A Survey of the ROAD from/NOTTINGHAM/to/ GRIMSBY,/IN LINCOLNSHIRE./Commencing in Plate 32./* (Aa, in a box at the top of the first strip, with a small overlap into the second strip). The map shows the road from *Stoak* (2 miles south of Newark) to Grimsby, which fills the first two and a quarter strips. There are 8 strips; the rest of the third strip, columns 4–5 and a small part of the sixth are taken up by *A Survey of the ROAD from/OXFORD to BRISTOL,/* . . . (in a box at the top of column 4); *A Survey of the ROAD from/OXFORD to CAMBRIDGE,/* . . . occupies the remainder of the plate.

*The Universal Magazine . . . Vol. XLIX . . .* [Issue for December, 1771].

**Copies**   BL PP 5439. (RGS Portfolio 202).

Chubb CCXXVI.

## 10

**DAVID HENRY and R. CAVE**      **1766**

Size: the plates vary in size; see below.

Edward Cave (1691–1754) was originally a printer and, in 1731, he founded the *Gentleman's Magazine*.[1] In the following year *The London Magazine* was begun in retaliation, largely because Cave's periodical was made up from pieces culled from other contemporary journals, a practice which caused a good deal of anger. However, the success of *Gentleman's Magazine* can be measured from the sales having reached 10,000 at the end of the first eight years and over 15,000 in the 1740s. Cave, who used the pseudonym 'Sylvanus Urban', was a friend of Samuel Johnson, who wrote an account of his life and was, for some years, a regular correspondent for the journal. Unlike *The London Magazine* in which a series of county maps appeared from 1747 (**32**) Cave's work had no county maps.[2] On Cave's death David Henry (1710–1792) joined the enterprise as a partner with Cave's son. He had probably been a worker in Cave's premises from 1731 but, following his marriage to Cave's sister in 1736, seems to have risen in the world. His literary qualities extended to the editing and production of the account of the voyages of Captain Cook.[3]

From the issue of January 1765 there began a series of road strip maps, which appeared at two monthly intervals up to and including September – at the end of the first year a sixth was issued as a supplementary item. From April, 1766 bi-monthly issue was resumed until December, with another supplementary plate bringing the year's total to six. By then the notes on the title-pages, which described the contents, recorded the numbering of the plates; forthcoming issues were to be made occasionally. The next plate did not appear until February, 1769 and there was a further gap until the supplementary map for 1774; the gap had been occupied with a subsidiary series of 18 maps of canals or intended navigations, none of which features the county, of course; the navigation from Chesterfield to Stockwith appears in the issue for May, 1772, the plate showing a small part of the area east of the Trent, including Gainsborough. The issue for February, 1775 carried a general road map for England; its cover recorded that it was the 13th in the series (there had already been 14!) and that there were only four more to appear. A further five appeared in 1775, in fact, while an isolated plate appeared in January, 1782 (see below).

Kingsley has suggested that Thomas Jefferys may have engraved the plates detailing Sussex roads.[4] He had engraved other maps for *Gentleman's Magazine*. The two main Lincolnshire plates bear the signature of J. Gibson, presumably the same Gibson responsible for the maps, which appeared in a small atlas in 1759 (**36**). The engraver of the little plate of 1782, which is clearly outside the main series, remains unknown.

### EDITION

There are no pointers or compasses.

(i)    *The ROAD from LONDON to YORK.* (Ca, between the outer frame and the tops of columns 4–7). *Gent: Mag.* (Ea, OS). Size: 17.9cm x 29.7cm. *Drawn & Engrav'd by J. Gibson.* (Ee, OS). There are 10 columns, separated by single vertical lines, numbered I–X in the upper right hand corner of each strip and covering the whole route in the title. Lincolnshire's part of the Great North Road appears in columns V–VII. *Another Road from London to York is thrô Edmonton, Hodsdon, Ware, Buntingford, Royston, Caxton, Huntingdon, & falls into this Road Column IV. mark'd thus * (Ae–Ce, OS).

*The Gentleman's Magazine, And Historical Chronicle. Volume XXXVI. For the Year*

ROAD from LINCOLN to the EASTERN COAST. {Gent.Mag. 1782.p.17.

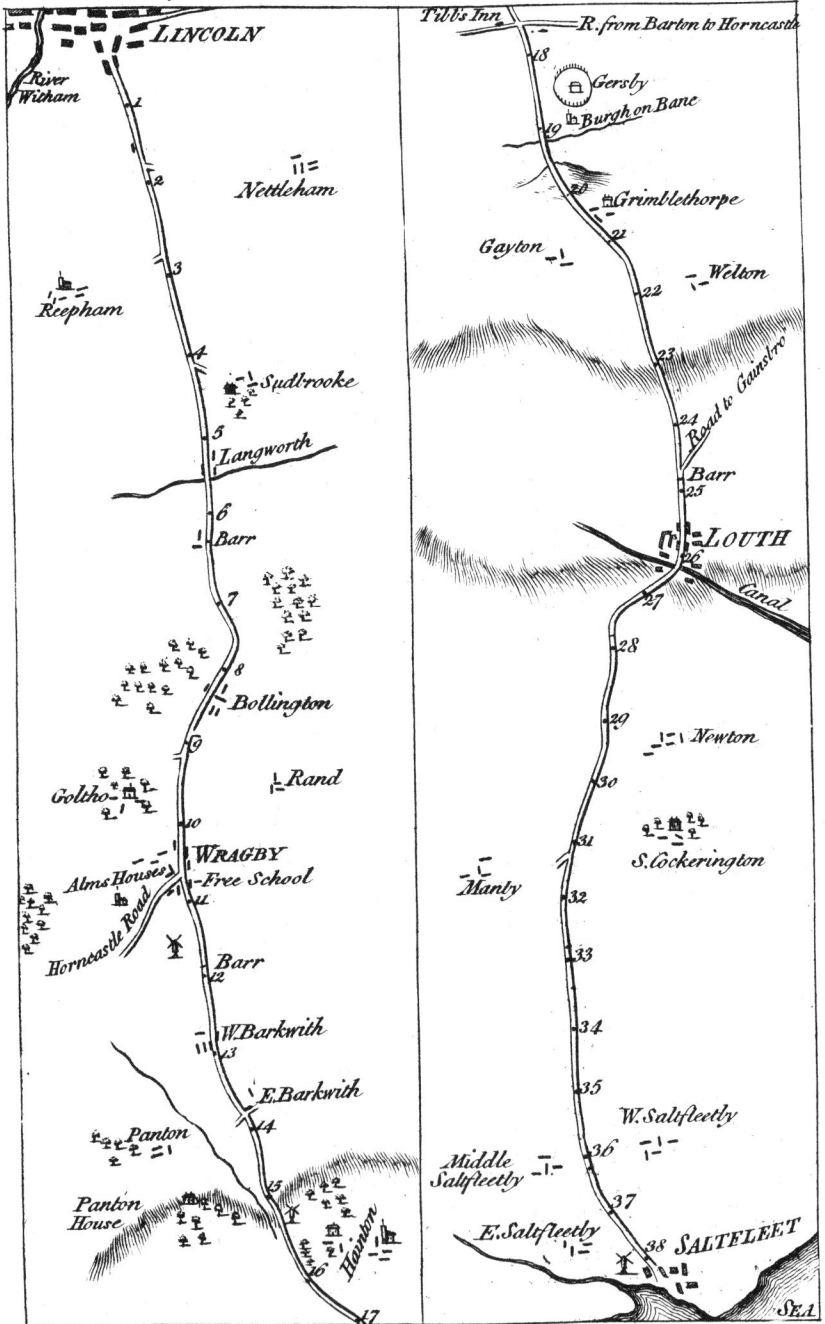

LINCOLN

River Witham

1

2

Nettleham

3

Reepham

4

Sudbrooke

5

Langworth

6

Barr

7

8

Bollington

9

Rand

Goltho

10

Alms Houses

WRAGBY
Free School

11

Horncastle Road

Barr

12

W.Barkwith

13

E.Barkwith

14

Panton

Panton House

Hainton

15

16

17

Tibb's Inn

R. from Barton to Horncastle

18

Gersby

19 Burgh on Bane

20

Grimblethorpe

Gayton

21

22

Welton

23

24 Road to Gainstro

Barr

25

LOUTH

26

canal

27

28

29

Newton

30

31

S.Cockerington

Manby

32

33

34

35

W.Saltfleetly

36

Middle Saltfleetly

37

E.Saltfleetly

38 SALTFLEET

SEA

12. 1782   APPENDIX MAP 10 – DAVID HENRY AND R. CAVE.

*M.DCC.LXVI . . . By Sylvanus Urban, Gent. London: Printed at St. John's Gate, for D. Henry; and sold by F. Newbery, the Corner of St. Paul's Church-Yard Ludgate Street.*

The cover of the individual part includes: *With a new and accurate Map of the Road from London to York; being the Eighth Map of the Series. Measured from the Royal Exchange . . . June 1, 1766.*

**Copies** CUL R 904.20.36. (LRL Map 52).[5]

*The ROAD from LONDON to FLAMBOROUGH HEAD, passing thrô PETERBOROUGH, LINCOLN &c./commencing at Stilton in the YORK ROAD.* (Aa–Da, between the outer frame and the top of columns I–VII). *From LONDON to BOSTON/commencing at Pekirk* (sic) *Column I.* (above the top of columns VIII–IX). *From LINCOLN to/BOSTON see yᵉ bottom/* (above the top of column X). *Gent: Mag:* (Ea, OS). *J. Gibson delin. et sculp:* (Ee, OS). *At Kilham Column VII. is the Road from London to Scarborough. viz. over the Woulds. to Hunanby 8 Miles – to Seamer 6¾ – to Scarborough 4.* (Ae–De, OS). Size: 18.3cm x 29.9cm. There are ten columns, numbered I–X in the upper right hand corners. The road from Stilton to Lincoln occupies the first 3 columns, with the continuation to Barton in columns 4–5. The further extension to Flamborough ends a third of the way up column VIII, at which point a horizontal rule separates the beginning of the route from Peakirk to Boston (top of column IX). The final column (X) shows the road from Sleaford to Boston (the opposite way round from that usually found) reading up the strip.

*Gentleman's Magazine . . . Volume LXVI . . . Year M.DCC.LXVI . . .*

The cover of the individual part includes: *With a new and accurate Map of the Roads from London to Flamborough Head . . . Also from London to Boston . . . and from Lincoln to Boston, commencing at Stopford* (sic) *. . . Being the 11th of the Series, which will be continued occasionally. December, 1766.*

**Copies** CUL R 904.20.36. (LRL Map 52).

*ROAD from LINCOLN to the EASTERN COAST.* (Aa–Da, OS). *Gent Mag./1782. p. 17./* (Ea, OS). Size: 17.4cm x 10.8cm. In two strips, which (very unusually) read down the map, the road is shown from Lincoln (Aa) through Wragby, Hainton and Louth to Saltfleet (Ee).

*Gentleman's Magazine . . . Volume LII . . . Year M.DCC.LXXXII . . . Printed by J. Nichols, for D. Henry, late of St. John's Gate; and sold by E. Newbery, the Corner of St. Paul's Church-Yard, Ludgate-Street. 1782.*

Part of the cover of the individual division reads: *Illustrated with an accurate Plan of the great Road of Communication between the Counties of Nottingham, Leicester, and Lincoln . . .* [Issue for January, 1782].

**Copies** CUL R 904.20.52; Bod Gough Maps 16 Fol. 3; LRL Stack 052. (LRL Map 925)

### References

1 Plomer, 1932, pp. 47–48.
2 Hodson, 1989, pp. 149–150.
3 Plomer, 1932, pp. 122–123.
4 Kingsley, p. 365.
5 LRL Stack 052 is the reference for a set of the work, but the maps from the 1766 volume are missing. Possibly the two maps mounted together on card as LRL Map 52 were originally a part of this set.

Not in Chubb.

## 11

**THOMAS KITCHIN**                                                    **1767**

Size: each plate varies – see below.

The work of Kitchin has already been described in connection with the county maps he prepared between 1749 and 1769 (**31, 32, 38** and **43**). The plates are still based on those of Ogilby as reduced by Senex; one piece of evidence being that the river at Spalding has been erroneously entitled the Witham throughout on plates showing the road from London to Boston. The plates are clearly based on the later editions of Senex, although, curiously, some details that are included by Senex are not repeated on these new plates; an example is the exclusion of the alternative calculations of distance at Grimsby on plate 86.

### EDITION

Each strip has a simple pointer.

(i)   *7* (Aa and Ea, OS). There is no title. Size: 14.7cm x 19.5cm. The six strips are on separate scrolls, which show the Great North Road from Stilton to Babworth (Notts.), i.e. mileages 71–140. *The Miles in this Itinerary are those through Huntingdon continued, but the Miles* (sic) *Stones placed upon the Road are Measured through Hatfield & Bugden & are/4 Miles more/* (Ae–Ee, OS).

*38* (Aa and Ea, OS). *The ROAD from LONDON to Boston in LINCOLN SH./Commencing at Stilton in the Berwick Road Plate 6/Stilton 71¼/Peterborough 78¼/Peakirk 85/Crowland 90/Spalding 99½/ Gosberton 106/Setherton 110¼/Boston 116/With a Branch from Boston to/ Swineshead 7/Kekington* (sic) *13½/Sleaford 19½/Lincoln 37½/* (Ca, in a box at the top of strips 3–5; the places and their mileages are arranged in two columns, headed by Stilton and Spalding). Size: 15.4cm x 20.1cm. There are seven strips on separate scrolls, the third, fourth and sixth being narrower. The first three and a half columns show the road from Stilton to Boston (mileages 72–116), and the remaining strips show the road from Boston to Lincoln through Sleaford. *Pl. 44 shews y<sup>e</sup> Road from Peterborô to Lincoln, leaving the Road at Peakirk at 85 Mile.* (sic) *Pl. 44 shews y<sup>e</sup> Road from London to Sleaford at 19 Mile, and Pl. 88 shews from Nottingham thrô Lincoln to Grimsby/* (Ae–Ee, OS).

*44* (Aa and Ea, OS). *The ROAD from LONDON to FLAMBOROUGH in YORKSHIRE/Commencing at Stilton in the Berwick Road Pl. 6./And shewing the Way to Stilton through Temsford./*[a list of 18 names in three columns headed by *Stilton, LINCOLN* and *Beseck*, with mileages from London] (Ca, in a box at the top of columns 3–5). Size: 16cm x 20.1cm. There are 7 strips on separate scrolls. The route starts at *Temsford* and continues to Lincoln through *Born* and Sleaford. *Plate 67 shows the Road from Cambridge thro' S<sup>t</sup>. Neots to Northampton and Coventry.* (Ae–Ab, OS – reading up the plate). *Another Road to Lincoln is from Panton Pl. 7 thrô Ancaster.* (Ec–Eb, OS – reading up the plate).

*45* (Aa and Ea, OS). There is no title. Size: 15.6cm x 20.2cm. Seven strips on scrolls continue the road from Lincoln through Barton to Flamborough Head (mileages 130–214).

*88* (Aa and Ea, OS). *The Road from NOTTINGHAM to GRIMSBY,/in LINCOLNSHIRE/Containing from Nottingham to/Newark 17/ LINCOLN 31¾/Walton 38½/Market Raising 48½/Stanton 54¼/ Briggesley 61¾/Grimsby 67/* (Ca, in a box at the top of columns 3–4). Size: 15.1cm x 20.7cm. There are six strips on separate scrolls.

*Kitchin's Post-Chaise Companion, Through England and Wales; Containing All the Ancient and New Additional Roads, With Every Topographical Detail relating thereto. By Thomas Kitchin, For the Use of Travellers, on One Hundred and Three Copper Plates. London: Printed for John Bowles, at N° [space], in Cornhill; Carington Bowles, at N° 69, in St. Paul's Church-Yard; and Robert Sayer, at N° 53, in Fleet-street. 1767. Price 7s. 6d.*

**Copies** BL 118 b 28; Bod Map C.17.e.2a; RGS 260 g 20; Manchester RL BR912.42 K1.

*Kitchin's Post-Chaise Companion . . . London: Printed for Robert Sayer, at N° 53, in Fleet-Street; John Bowles, at N° [space], in Cornhill; and Carington Bowles, at N° 69, in St. Paul's Church-Yard, 1767. Price 7s. 6d.*

**Copies** CUL Atlas 7.76.11; Leeds RL 912.42 K468.

*Kitchin's Post-Chaise Companion . . . London: Printed for and sold by Carington Bowles, at his Map and Print Warehouse, No. 69, in St. Paul's Church-Yard* [1770].[1]

A further issue in 1780 by Carington Bowles.[2]

### References

1 Chubb (CXLII, under Senex) gives the quoted title and imprint for a further use of the plates. No copy has been found.
2 Kingsley, p. 366, refers to a copy in BL. No further detail is given and the copy has not been identified.

Chubb CCXXXI (and CXLI under Senex).

## 12

**THOMAS KITCHIN**                                          **1771**

Size: 12.1cm x 14.9cm (the plates are not all quite the same size).

It is not clear why Kitchin prepared the present work when only four years earlier the plates described above had been issued. There are suggestions that those plates had passed to Carington Bowles by 1770 (see above) and perhaps Kitchin felt there was still a market for a new version of Ogilby/Senex. Whatever the reason Kitchin only used the new plates for one edition and, after his retirement in 1776, they seem to have passed to Carington Bowles also and later to Bowles' son and his partner Carver. Kitchin's plates had simple headings describing the routes delineated; under Bowles the areas surrounding the strips were so thoroughly altered that it is not immediately realised that the same basic plate is involved. The six strips originally had single lines between and around the whole map area. Bowles provided a wider frame to create a panel at the top and another, narrower, at the bottom, with further material added outside the frames, including pairs of plate numbers for the single plates.

The work of Carington Bowles has been noted in connection with the county maps he issued from the 1760s, usually making use of plates prepared by Bowen or Kitchin (see, for example, **31 (iii)** and **34 (vii)**). Bowles (c.1724–1793) was apprenticed to his father John and freed in 1752, when he joined his father as Bowles and Son. By 1763 he had assumed control of the business; his father died in December, 1762. Thereafter he carried on a successful trade as a print and map seller.[1] The plates are still based on the Senex reductions of Ogilby. Here the plates show further compression of the details in their predecessors. The work was issued in two volumes, the title-pages being identical. Carington Bowles's son, Henry Carington Bowles (1763–1830), issued another edition with his partner Samuel Carver probably in the 1790s.[2]

## EDITIONS

(i) *6* (Ea, OS). *LONDON to Berwick continued*. (Ca, OS). *The Miles in this Itinerary are those thrô Huntingdon continued, but the Mile Stones on the Road are measur'd thrô Hatfield & Bugden, & are 4 miles more*. (Ae–Ee, OS). There are 6 strips separated by single vertical lines, showing the road from *Walmsford* [= Wansford] (near Peterborough) to Doncaster (mileages 80–157).

*36* (Ea, OS). *LONDON to Boston, commencing at Stilton, Plate 5/with a Branch from Boston to LINCOLN*. (Ba–Da, OS). There are 6 strips separated by single vertical lines. The road from Stilton to Boston occupies the first three columns and the lower third of the fourth; the remaining strips depict the road from Boston to Lincoln, via Sleaford.

*41* (Ea, OS). *LONDON to Flamborough, commencing at Stilton, also to Stilton thro Temsford*. (Aa–Ea, OS). *London to Temsford see Plates 20 & 46 —— Plate 5 shews another Road to Stilton*. (Ae–Ee, OS). There are 6 strips separated by single vertical lines. Strips 1–2 show the road from *Temsford* to Stilton through St. Neots (mileages 52–75); the upper part of column 2 and the other 4 strips show the road from Stilton to Lincoln (mileages 71 (i.e. counting the distances from London through Huntingdon to [130]).

*42* (Ea, OS). *LONDON to Flamborough continued*. (Ba–Da, OS). There are 6 strips separated by single vertical lines. At the top of the sixth column is a note on the road from Driffield to Scarborough *over the plea/sant Woulds*. Lincoln to Barton occupies the first two strips and the lower part of the third.

*78* (Ea, OS). *King's Lynn/to Yarm^{th}. contin^d*. (Aa, OS). *Nottingham to Grimsby*. (Ca, OS). There are 6 strips separated by single vertical lines. The first section shows the road to Yarmouth, covering the route between milestone 47 and the coast (milestone 69). From half way up the second strip to the top of the sixth the road from Nottingham to Grimsby is delineated.

*Ogilby's Survey Improv'd: Or Kitchin's New and Instructive Traveller's Companion, for the Roads of England and Wales. Laid down in a Plain Intelligible manner, with all the Towns, Villages, &c. thereon and the Distances in single Miles on each Road. 1771. Printed for & Sold by T. Kitchin, at N° 59 Holborn Hill London. Price 6^s.*

**Copies**   RGS 260 C 24; NLS Newman 633; Nottingham UL Special Collections r/G 5514.

(ii)   Size: 13.4cm x 14.7–14.85cm. approx.

On each plate, the strips (already enclosed in a box) are now enclosed within a new outer frame, thus producing a panel at the top, which is filled with the names of the principal places on the road. At the bottom, a narrower panel contains (mainly) references to other plates. Simple pointers have been added to each strip. Plates 11/12, 71/72, 81/82 and 83/84 are in Vol. I; plate 155/156 is in Vol. II.

*11* (Aa, OS) and *12* (Ea, OS). *Stamford 14/Coltsworth 13/Grantham 8½/Newark 14/ Tuxford 13/Barnby Moor 10½/Bawtrey 5½/ Doncaster 8/* (in the upper panel arranged in four pairs divided by single short lines, (Ba–Da). *VOL/I*. (Ae, OS). *The Miles in this Itinerary are those thrô Huntingdon continued, but the Mile Stones on the Road are measur'd thrô Hatfield & Bugden, & are 4 miles more*. (Ae–Ee, in the lower panel). This plate corresponds to plate 6 above.

*71* (Aa, OS) and *72* (Ea, OS). *LONDON to Boston/ and from thence to/LINCOLN* (Aa, in the panel, with a bracket (}) at the right). *To Stilton\* 71/Peterborough 7½/Crowland 11½/Spalding 10/Boston 16/ Swineshead 7/Sleaford 12½/Lincoln 17½/* (in the upper panel, in four pairs, separated by single lines, (Ba–Ea). *VOL. I.* (Ae, inside the lower panel). * *London to Stilton, see p. 9 & 10.* (Ce, in the lower panel). This plate corresponds to 36 above.

*81* (Aa, OS) and *82* (Ea, OS). *LONDON to/Flamborough* (Aa, in the panel with a bracket [}] at the right). *To Tempsford 52/Bugden 10/Stilton 13/Peterborough 7½/Deeping 10½/Bourn 6½/Sleaford 18/Lincoln 17½/* (Ba–Ea, in the panel in 4 pairs, separated by single lines). *VOL. 1. London to Temsford see Plates 39 & 91* [space] *Plate 10 shews another Road to Stilton.* (Ae–Ee, in the panel). This plate corresponds to 41 above.

*83* (Aa, OS) and *84* (Ea, OS). *Spittle 12/Glanford Br. 12/ Barton 10/ Hull by water 6/Beverley 9/Driffield 14/Kilham 5½/Bridlington 7½/ Flamborough 5/Flamborō H^d. 3/* (Aa–Ea, in the upper panel, the names arranged in 5 vertical pairs). *VOL. 1.* (Ae, in the lower panel). This plate corresponds to plate 42 above.

*155* (Aa, OS) and *156* (Ea, OS). *Hadsko 17/Yarmouth 9/* (Aa, in the panel and separated from the other material in the panel by a double line, which continues down the plate and surrounds the road to Yarmouth). *NOTTINGHAM/to Grimsby/}/ Newark 17/Lincoln 15/ Walton 7/Market Raisen 10/Briggesley 13/Grimsby 15½/* (Ca–Ea, in the panel, the place-names arranged in pairs separated by single lines). *VOL. II.* (Ae, in the lower panel). This plate corresponds to 78 above.

*Bowles's Post-Chaise Companion; Or, Travellers Directory Through England and Wales: Being An Actual Survey Of All The Principal, Direct, and Cross Roads, both Ancient and Modern; With The Distances expressed in single Miles according to Measurement: Exhibiting the several Towns, Villages, Post-Stages, &c. On Or Near The Roads; Together with the Circuits of the Judges, and an exact Alphabetical List of all the Fairs, as settled since the Alteration of the Style. In Two Volumes. Vol. I* [II]. *London: Printed for the Proprietors Carington Bowles, At his Map and Print Warehouse, No. 69, St. Paul's Church Yard.* [1781].[3]

**Copy** Derby RL (Local Studies Dept. 3962).

**(iii)** All plates have *BOWLES'S POST-CHAISE COMPANION.* in a panel, reading upwards (Ad–Ab, OS) and *London: Published 2. Jan^y 1782.* reading downwards (Ec, OS).

*11* and *12*. The mileages in the upper panel have been altered at Barnby Moor (now *11*) and *Bawtrey* (now *4*). The top of strip 5 has been re-engraved; the site of mileage *142* has been lowered, *143* moved to the east of the road, *144* is now sited at *Barnby on/the Moor* and *145* (with *Tarworth*) is now 2mm. lower. All the mileages in strip 6 have been re-engraved and reduced by one from Doncaster southwards.

*71* and *72*. The mileages in the upper panel have been altered at Peterborough (now *7*), Swineshead (now *6*) and Sleaford (now *11½*). All mileages on the route from Boston to Lincoln have been re-engraved except *1* and there are now 35 miles to Lincoln (instead of 37).

*81* and *82*. Mileages in the upper panel have been altered for Peterborough (now *6½*), Deeping (now *8½*) and Bourn (now *7*). In the lower panel *Pages* and *page* have replaced Plates and Plate. In strip 1 *to St. Neots* has been deleted (Ae). Most mileages in the route to Stilton have been re-engraved; this shows especially at the top of strip one, where the final figure is *65* (previously 66). Where the new route starts at Stilton (Bb) all figures begin from the base mileage 75 and subsequent figures have been altered and spaced out correctly and now run from 75–132. The references to other roads have all been removed at Stilton, Peakirk, Folkingham, the latter's name being removed in the tidying up process.

*83* and *84*. The mileages now read: *Spittle 12/Glanford Br. 12/ Barton 10½/Hull by water 7/Beverley 9/Driffield 13/Kilham 5/Bridlington 7½/Flamborough 5/Flamborō H^d. 3/*

(Aa–Ea, in the upper panel). All mileages have been re-engraved, starting from Lincoln *(133)*, formerly 132 and reaching 215 at Flamborough (previously 214). In column 1 *Midge Inn* has been added, between milestones 138 and 139. In column 3, on the by-road north from Hessle, *Kirk Ella, Willerby, Skidby* and *Bent/ley* have been added. In the Humber, broken lines are drawn across the river from Barton to Hessle *(4)* and to Hull *(7 Miles)*, the mileages being as noted.

*155* and *156*. The mileage in the upper panel has been altered at Lincoln (now *17½*). The mileages in strips 3–6, from north of Newark, have been re-engraved and by Grimsby the final total is 70 (formerly 67).

*Bowles's Post-Chaise Companion . . . An Actual Survey Of All The Direct and Principal Cross Roads, With The Mile-Stones expressed as they stand at present: Exhibiting the several Towns . . . Fairs, as settled since the Alteration of the Style. The Second Edition, Corrected, and greatly Improved; with Additions. In Two Volumes. Vol. I* [II]. *London . . . Carington Bowles . . . M DCC LXXXII.*

**Copies**   BL Maps C.24.b.34; CUL Atlas 7.78.2–3; Bod G.A. Gen Top 8° 6–7.

(iv)   The reference to the date of issue (Ec, OS) has been deleted on all plates.

*Bowles's Post-Chaise Companion . . .* [the punctuation varies from the second edition] *London: Printed for the Proprietors, Bowles And Carver, At their Map and Print Warehouse, No. 69, St. Paul's Church Yard.* [1794].[4]

**Copy**   Brighton RL 912.42 B 68.[5]

**References**
1   Hodson, 1984, pp. 186–187.
2   Hodson, 1984, p. 188.
3   On the last plate of strip maps in each volume is engraved: *Publish'd as the Act directs, 4 June, 1781,/by Carington Bowles, London./* (Ee, in the lower panel).
4   Kingsley, p. 368 suggests 1790. As noted above Bowles and Carver's partnership began in 1793 and it seems likely that they made use of the inherited plates early in their regime.
5   Only Volume II has survived. Kingsley, p. 368, describes this copy as third edition, but the title-page is as transcribed above.

Not in Chubb.


# 13

## ROBERT SAYER and JOHN BENNETT                                                1775

Sizes vary – see below.

Sayer and Bennett became the owners of the plates prepared by Kitchin (originally in 1748–1749) and re-used by them in *The Small English Atlas* (**31**). The present roadbook is a new engraving still, however, based on Ogilby/Senex.[1] There is evidence of use of later information in the notes at the foot of plate 37 on side roads, which do not form part of the information recorded by Ogilby and his followers. The information on the Saltfleet road could have come from Kitchin in the map for *The London Magazine* (1751). This was the first time places like Holbeach, Donington and Gainsborough were mentioned in road books.

## EDITION

(i) *6* (Aa and Ea, OS). There is no title; the six strips, each separate rectangles, show the road from Stilton to Babworth (Notts.), i.e. mileages 70–[139]. *Publish'd as the Act directs by R. Sayer & I. Bennett 16 Jan<sup>ry</sup>. 1775.* (Ce–De, OS, below the note on mileages to Berwick). Size: 14.4cm x 19.9cm. *The measured Distance from London to Berwick thro' Huntingdon and Wakefield is 331 Miles; thro' Huntingdon and York is 340 Miles; and thro' Hatfield, Bugden,/and York 344 Miles.* (Ae–Ee, OS). An explanatory note at the foot of strip 1 reads: *The road to Stilton thro'/Bugden is 74 Miles/* (Ae); another at the foot of strip 2 has: *The Miles Stones* (sic) *on this/Road are reckoned thro'/Bugden which makes/them 4 to* (sic) *many./* (Be).

*37* (Aa and Ea, OS). *From LONDON to BOSTON,/commencing at Stilton. see Plate 6./Stilton 71/Peterborough 77/Peakirk 84/Crowland 89/Spalding 100/Gosberton 106/Southerton 110/Boston 116/with a Branch from Boston to Lincoln/Swineshead 7/ Kekington* (sic) *13½/Sleaford 19½/Lincoln 37/* (Ca, in a box at the top of columns 3–4; the places and mileages are arranged in two columns headed by Stilton and Spalding). *Publish'd as the Act directs R. Sayer & I. Bennett 16 Jan.<sup>ry</sup> 1775.* (Ba–Da, OS). Size: 15.3cm x 20.35cm. *The Mile Stones on this Road are number'd 4 too much, as the Road is reckoned thro Huntingdon. A Road goes from Spalding 11½ miles to Holbeach.* A Road goes from Gosberton 7 miles/to Donington From Boston a Road extends to Saltfleet as follows: to Tattershall 14 miles, Horncastle 8 more, Louth 13½ more, & Saltfleet 8 more, in all 43 miles from Boston.* (Ae–Ee, OS). There are 7 rectangular separate strips. The first 3½ strips show the road from Stilton to Boston (mileages 71–116), while the remainder of strip 4 and all of 5–7 show the road from Boston to Lincoln through Sleaford.

*43* (Aa and Ea, OS). *From LONDON to FLAMBOROUGH,/ commencing at Tempsford. 52 miles from London, see Plate 47/Stilton 70/Peterborough 82/Market Deeping 90/Bourn 98/Sleaford 116/Lincoln 132/Redbourn 150/Glamford Bridges 156/Barton 186/Hull 173/Beverley 182/Driffield 196/Burlington 208/Flamborough 214/* (Ca, in a box at the top of strips 3–5; the places are in three columns, headed by Stilton, Lincoln and Beverley). *Publish'd as the Act directs by R. Sayer & I. Bennett 16 Jan.<sup>ry</sup> 1775.* (Ba–Da, OS). Size: 15.8cm x 20.5cm. The 7 strips show the road from *Temsford* to Lincoln (mileages 52–132).

*44* (Aa and Ea, OS). There is no title; the plate shows the road from Lincoln to Flamborough Head in 7 separated strips of unequal widths (mileages 132–216). *Publish'd as the Act directs by R. Sayer & I. Bennett 16 Jan<sup>ry</sup>. 1775.* (Ba–Da, OS). Size: 15.1cm x 20.5cm. *From Glamford Bridge a Road goes to Burton Strather* (sic) *12 miles* (Ae–Be, OS). *From Kilham a road goes to Scarborough as follows; to Hunnanby 9, to Seamour 6½ more,/and from thence tis 4½ to Scarborough, in all 221 miles from London./* (Ce–Ee, OS). At the top of column 7 appears a note on the road from Flamborough to Scarborough and an alternative route from Driffield. At the foot of column 4 is a note on the route from Hull to Patrington and the intervening mileages.

*86* (Aa and Ea, OS). *From NOTTINGHAM to LINCOLN/and GRIMSBY./Newark 17/Lincoln 32/Walton 38½/Market Raising 48½/Stanton 54/Thurganby 57½/Briggesley 61½/Grimsby 67/* (Ca, in a box at the top of columns 3–4; the names are in two columns, headed by Newark and Stanton). *Published as the Act directs by R. Sayer & J. Bennett 16 Jan<sup>ry</sup>. 1775* (Ca, OS). Size: 14.9cm x 20.25cm. There are 6 separate strips showing the road from Nottingham-Grimsby (mileages [0]–[67]). *A Road goes from Nottingham to Southwell 14 miles. A Road goes from Lincoln to Gainsborough thrô Stretton 18 miles. A Road goes from Market Raising to Castor 7½* (Ae–Ee, OS).

*Jeffery's Itinerary; Or Travellers Companion, Through England, Wales, and Part of*

*Scotland, Containing All the Direct and Principal Cross Roads; With The Addition of every New Road, carefully collected from the actual Surveys hitherto published. Improved With Many Thousand Names Of Places More Than Are In Any Similar Publication. To Which Are Added Copious Indexes to all the Roads and Places mentioned in the Work, with their exact Distance from London. London, Printed for R. Sayer and J. Bennett, Map and Print Sellers, No. 53, Fleet-Street. MDCCLXXV.*

**Copies**    BL Maps C.24.a.12; CUL Atlas 6.77.6; Bod Vet A5 d 292; Leeds UL W 248.[2]

*The Roads of England Delineated; Or, Ogilby's Survey, Revised, Improved, and Reduced to a size portable for the Pocket, By John Senex, F.R.S. Being An Actual Survey of All the Principal Roads Of England and Wales; Distinctly Laid Down On One Hundred And One Copper-Plates; On Which Are Delineated All The Cities, Towns, Villages . . . Throughout The Road. With the Additions of some Roads newly drawn, which were omitted by Mr. Ogilby, and several necessary Corrections made in Others; together with a great Number of explanatory References, by which this Edition of the roads is rendered of more general Use to Travellers.* [No imprint or date].[3]

**Copy**    BL 118 b 27.

**References**

1    Kingsley, p. 364 thought these states were those of Le Rouge (q.v.) with the titles and directions re-engraved in English.
2    In all these examples plates are double-sided. Mr. C.A. Burden owns a copy in which the plates are single-sided.
3    Kingsley, p. 364 regards the title-page as spurious; it is printed on a curious brown paper, which, on watermark evidence, was probably made in the Low Countries. Additionally it certainly fails to describe accurately the contents or fresh approach of Sayer and Bennett. The lack of imprint and date adds to its curiosity. It is likely that the title-page was added to the plates by a private individual; the closeness of the title-page to that of Senex in the 1757 edition even suggests that it comes from an edition, which has not otherwise survived. The volume lacks the preface found in the 1775 version.

Chubb CXLIV–CXLV (both noted under Senex).

# 14

## MOSTYN JOHN ARMSTRONG                                              1776

Size: 15.1cm x 7.7cm (the plates vary slightly in width; measurements are taken to the inner frame).

The work and life of Armstrong has been noted under the consideration of his father, Andrew Armstrong, the first to produce a map of Lincolnshire at the one inch to a mile scale (**44**). The production of this single road-book for the Great North Road fits in well with what we know of the son's activities. Having helped in the survey of Northumberland (1766–1769, published 1769) and having access to his father's papers in connection with that of Durham (1768) and then, again with his father, the survey of the Lothians (1773) the move south to survey Rutland and Lincolnshire meant that the son already had much of the necessary material to hand for his book.

Armstrong says as much in the introductory material to the 1776 editions. He says the work was first advertised for subscription in 1773 and was intended to include the routes to Edinburgh by way of Newcastle and also through Carlisle. 'But to have included these two great roads . . . in One Volume would have increased it to a size by far too large for the pocket, and required a more general sale than could be supposed to indemnify the expences (sic) of the publication.' So Armstrong 'humbly presented' it as a specimen of what could be possible if the demand proved to be forthcoming. 'The publisher therefore informs the purchaser, that 300 miles of the road were taken from surveys made by Capt. Armstrong and himself; and that the remainder was laid down from his own observations, adopting such actual surveys as he could assuredly confide in'.

The strips show the road as far as Edinburgh, while the majority of earlier volumes, in following Ogilby, only delineate the road as far as Berwick. In that way this survey provided the would-be traveller with fresh material not found in previous road-books.

It is clear from a comparison with the finished result of his father's survey of the county that the son has included much detail that he probably did obtain from his own observation, bearing in mind that his father's survey of Lincolnshire was only beginning as the son's book was being issued. Although his later history in Norfolk might suggest that diligence and the fulfilment of promised (subscribed) work were not a large part of his character this production puts his reputation in a much better light. After its completion he produced *A Scotch Atlas* in 1777 (with two further editions in 1787 and 1794).[1]

The comparison with his father's large scale map points up several matters. Firstly, the larger scale map of Andrew Armstrong has fewer details than the son's, which is half an inch to one mile. Secondly, the mileages along the roads are not continuous but refer to the distances between the main places; Andrew Armstrong's map gives the mileages as measured from London taken, it is assumed, from the available road maps and, of course, the milestones themselves. Oddly, the mileages between places are given in the accompanying text in the son's book, twice. For instance, we are told that it is eleven miles from Stamford to Witham Common and then, on the next line, that it is eleven miles from Witham Common to Stamford! Finally, although the son includes greater detail (names and sites of houses, details of woods, etc.) the father names rivers crossing the route, information not given by Mostyn John.

The three main map plates each face a page of letter-press, which contains the data on mileages on the main road and the distances to places along the cross roads. A small part of the county also appears on Plate 12 – the area east of Misson (Notts.). Information is also provided about the major landowners in each village passed along the way and details of the inns in each town or village. Interestingly, although the Ram Jam House is marked on the map it was not then regarded as a stopping place for travellers.[2] The plates are signed by Pyle, presumably the same S. Pyle who engraved Andrew Armstrong's Lincolnshire county plates. The work went through a number of editions, with a slightly complicated bibliographical history.

## EDITIONS

(i) The plates all have inner and outer frame lines; outside them is a leafy decoration, which breaks parts of the outer frame line. There is only one 'strip' of road, reading up the plate. There are no titles to the plates.

*Plate 8*. (Ea, OS). *Pyle sculp.*[t] (Ee, between the inner and outer frames). The map shows the road starting at Burghley (mile-post 13 from Stilton) as far as Witham Common (the most northerly place is actually Gunby, west of the road). Market Overton is the most westerly point shown and Uffington the most easterly. Compass (Bb).

*Plate 9*. (Ea, OS). *Pyle sculp*[t]. (Ee, between the inner and outer frames). The plate continues the road northwards to the *Barr*, at the fourth milestone between Grantham and Newark. To the west the map stretches to Denton and eastwards to Burton [Coggles]. Compass (Ca).

*Plate 10* (Ea, OS). *Pyle sculp*[t]. (Ee, between the inner and outer frame lines). The plate continues the Great North Road and reaches to Markham (Notts.). The western-most place marked is Kilvington (Notts.) and Stubton is the furthest east. Compass (Cc).

*An Actual Survey of the Great Post-Roads Between London and Edinburgh, By Mostyn John Armstrong, Geo$^r$. London Printed for, and Sold by the Author, and the Booksellers. Price 7$^s$. 6$^d$. Harrison sculp$^t$. Published according to Act of Parliament 1$^{st}$. of June, 1776.*

**Copies**   BL 195 a 7; CUL Atlas 7.77.4; Bod Vet A5 e 1084; Guildhall L. S 388/1; York RL 912.[3]

*An Actual Survey of the Great Post-Roads Between London and Edinburgh, With the Country Three Miles, on each Side, Drawn on a Scale of Half an Inch to a Mile. By Mostyn John Armstrong . . . Published . . . 1$^{st}$. of Aug$^t$. 1776. [1777].*[4]

**Copy**   RGS Fordham 182.

**(ii)**   *Plate 8.* (Ea, OS). *Stamford.* (Ba, OS) and *Witham common* (Da, OS) have been added. *Esden* (west of Stamford) has been changed to *Easton*. Three roads leaving Stamford have been named; viz. *Oakham R$^d$. Uppingham R$^d$.* and *Lincoln R$^d$.* The facing text has been altered; semi-colons after the names of landowners have been replaced by full stops. After Woolthorp, *in ruins* has been added. Under the heading *Post-Road* mileages have been altered, that for Edinburgh by Berwick being increased by 2 (in both cases) to 300 and 289. Extra places and mileages have also been added to 3 of the 4 lines under the heading *CROSS-ROADS*. In the last line *Black-Bull* has been hyphenated.

*Plate 9. Grantham.* (Ba, OS) and *Colsterworth.* (Da, OS) have been added. The roads leaving Grantham have been labelled: *Nottingham R$^d$. Leicester R$^d$.* and *Lincoln/R$^d$.* The last has also been added to *High Dyke Roman Road.* The text page opposite has a similar addition to the mileage to Edinburgh by Berwick as on page 8 and the mileage to Hull has been added to the fifth line under *CROSS-ROADS*. The elongated dash (–) has been replaced by *Ditto*, in the list of places under Grantham.

*Plate 10. Newark* has been added (Ca, OS). Side roads leaving Newark have been labelled: *Lincoln R$^d$. Beckingham R$^d$. Nottingham R$^d$. Leeds R$^d$.* and *Mansfield R$^d$.* (the last two are branches of the same road). Under the crossed swords on *Marston/Moor*, the date *2 July 1644* has been added. On page 10 the mileage to Edinburgh has again been increased by 2 (to 265) and all but one line under *CROSS-ROADS* has been extended by the addition of more places and their mileages. In the last line both inns have been hyphenated.

*An Actual Survey . . . Aug$^t$. 1776.*

**Copies**   CUL Atlas 7.77.13; RGS F 150; Nottingham UL G 5514; Grantham RL L 388.[5]

**(iii)**   Plate 10 has had the following addition: *5 May 1649* preceded by crossed swords has been added above *Cas. in Ruins* at Newark. The plates are otherwise unchanged but the text pages have been re-set and altered. Page 8: the punctuation after the names of land-owners has been varied and is generally that of state (i); Mrs. Wyndham (line 11) has been altered to *Mrs. Wingfield*. Rev. Mr. Butter (line 13) has been altered to *Rev. Mr. Foster*. Under *Post-Road*. the name *Edin.* is now given in full and since a larger type-face has been used the third line now spreads over into an extra line. Page 9: *M.P.* has been added after *Thorold, Bart.* (line 2 of the list of places and the chief landowners); *Penyman, Esq;* has been altered to *Pennyman, Esq;* (against Little Paunton, in line 11). Under *CROSS-ROADS* the first and fifth lines have been extended; the mileage to Hull has been changed to 65 (third line). Page 10: Under *CROSS-ROADS* the mileage to Grimsby has been changed to 61 in line 1; (in line 3) *Worcester, 102* has been added; further place-names and mileages have been added (line 6; and to line 8, which leads to the creation of a new ninth line).

*(Third Edition) An Actual Survey . . . Price ˢ ᵈ . . . Published according to Act of Parliament 1ˢᵗ. of Janʸ. 1777.*

**Copy**   BL Maps C.27.a.7.⁶

*(Third Edition) An Actual Survey . . . 1777.* [1783].⁷

**Copy**   Guildhall L. S 388/1.

(**iv**)   The plates are unchanged; one alteration has been made to the text on page 10. Under *CROSS-ROADS* Ancaster has been deleted and *Sleaford 17.* has been added (line 4).

*An Actual Survey . . . Armstrong, Surʳ. London Printed for Wᵐ. Faden Geographer to the King Charing Cross. Second Edition. Harrison Sculpᵗ. Published according to Act of Parliament October 20ᵗʰ. 1783.* [1784].⁸

**Copies**   Bod Gough Maps 124; Leeds UL W 251; Glasgow Mitchell L S 914.2 Arm.

**References**
1   Examples include: CUL Atlas 6.77.1 (1777); CUL Atlas 6.78.3 (1787); CUL Atlas 6.79.15 (1794). The history of this atlas will be fully dealt with in the forthcoming third volume of Donald Hodson's *County Atlases of the British Isles.*
2   Buxton, A. In search of Ram Jam. *Rutland Record*, 9 (1989), pp. 324–328 discusses the origins of the house and its later history.
3   Page 6 carries an advertisement for the Armstrongs' maps of 'Northumberland, Berwickshire, Three Lothians, Ayrshire, and Tweedale . . .', and a note on the address for subscriptions for the forthcoming 'New and Correct Map of the County of Lincoln, On Eight Sheets at Two Guineas'. Although dated June on the title-page the Dedication is dated August, 1776.
4   A leaf has been added after the title-page. On it (in brick-red) is a list 'of Maps Surveyed & Published by Capᵗ. Armstrong & Son. Among the items listed is: *'Lincolnshire; 8 Sheets, £2.2.'* It is dated *'London. 1ˢᵗ. Janʸ. 1777 . . .'* The RGS copy is bound with *The Kentish Travellers Companion . . .* (1776). The county map was not issued until 1779 and this note represents a further bid for subscriptions.
5   The Grantham example is differently made up and may be a proof state. The text pages are arranged as double-page spreads (with blank versos); the map plates are similarly arranged. Normally, text pages face the appropriate plates, both leaves being single-sided.
6   The BL copy has no dedication or text pages. 7s 6d has been added by hand.
7   The date in the dedication is Aug 23, 1783.
8   William Faden only became geographer to the King in 1784. (Maxted, p. 78).

Not in Chubb.


# 15

**DANIEL PATERSON**                                                     **1785**

Size: 16.2cm x 9.6cm.

Paterson (1738–1825) was commissioned into the army in 1765 and served in the Quartermaster General's department until 1812, having been Assistant Quartermaster General for the last eight years before his retirement. He was then appointed Lieutenant-Governor of Quebec.¹ In 1771 appeared the first edition of the work later better known as *Paterson's Roads*, which went through 14 editions by 1808; the 15th (1811) and later editions contain some maps, but none relating to Lincolnshire. The work below had a modest success compared with that and appeared in five editions.

In the present work the pages consist of two strips each separately numbered. Some use appears to have been made of M.J. Armstrong's survey, evidenced by the provision of the names of land-owners (and, in one case, using the spelling of Armstrong's June, 1776 edition) on either side of the Great North Road. There are no such names for the road to Boston, presumably because Armstrong did not survey that particular route. Ellis is the engraver of the title-page; this may be the same Ellis of the *English Atlas* (**40**) and he may have engraved the road strips. The plates are numbered, in volume one, 1–188, in the second 1–140, with another sequence numbered 1–30. Each plate bears its volume number (Ca, OS). All the plates affecting Lincolnshire are in the first numbered sequence of the second volume.

In 1877 the plates made a fresh appearance in a lithographic transfer with fresh headings added to each plate and a great many other plate changes. However, the numbering of the plates reverts to that of the first state. Quite how the plates arrived at the publishers, Tinsley brothers, and why they felt that the thorough revision the plates underwent was worthwhile is difficult to say. Tinsley brothers had an extensive business publishing novels, usually at the lower end of the market, albeit many of their authors were some of the best-selling of the period and, at one time, included Wilkie Collins and Harrison Ainsworth (when his career began to fade). The first work of Thomas Hardy came out with their imprint, though the firm was unwilling to risk his initial offerings without a substantial financial payment by the author. In the event neither Hardy or the Tinsleys made their fortunes from their first joint ventures.[2]

## EDITIONS

(i)   All plates have the imprint: *Printed for the Proprietor CARINGTON BOWLES, London, 3 Jan. 1785.* (Ae–Ee, OS). All plates on the left of a double-page spread have: *PATERSON'S (VOL II.) ITINERARY,* (Aa–Ea); all pages on the right have: *DIRECT (VOL II.) ROADS* (Ba–Da, OS). The numbers at the start and end of each 'title' line below are those of the individual strips; the numbers are in circles enclosed by square boxes. Each strip has a simple pointer.

*5 Stilton 14 (Peterborō 6½) Stamford 14, Coltsworth 13 6.* (Aa–Ea, in a panel at the top of the plate). In a panel below the strips are the names of landowners, the letters a–e being keys to sites on the strip maps. In order to accommodate Peterborough in the first strip the vertical line dividing the strips has a large bow to the right; at that point in strip two there is a gap in the road and the pointer is placed there (Ed). The strips show the road from milestone 72 (south of Stilton) to Coltsworth (the modern Colsterworth). As far as Stamford the mileages are those from London; thereafter they are the distances between each place in the heading.

*7 Grantham 8, Newark 14, Tuxford 13. 8* (Aa–Ea, in a panel along the top). In the lower panel are the names of landowners, the letters a–f being keyed to sites on the maps. The strips continue the route from strips 5/6 as far as West Markham (Notts.). The mileages on the maps are those between each place in the heading.

*81 Stilton to Deeping 15, Bourn 7¼, Folkingham 9. 82* (Aa–Ea, in a panel along the top). In the lower panel are the names of 2 landowners, lettered a–b and keyed to sites on the maps. *LONDON to Scarborough/by Lincoln./commencing at/Stilton./p.5./* (at foot of the first strip). The strips show the route from Stilton to Folkingham (mileages 75–106).

*83 Sleaford 9, Green Man 10, Lincoln 8. 84* (Aa–Ea, in the upper panel). In the lower panel are the names of landowners lettered a–c as keys to sites on the maps. The map continues the route from 81/82 as far as [Caenby Corner].

*85 Spittal 11½, Glanford Bridge 11½, Hull 18. 86* (Aa–Ea, in upper panel). The lower panel has 2 names lettered a–b and keyed to sites on the plate. The strips continue the

*DIRECT* (VOL.II.) *ROADS.*

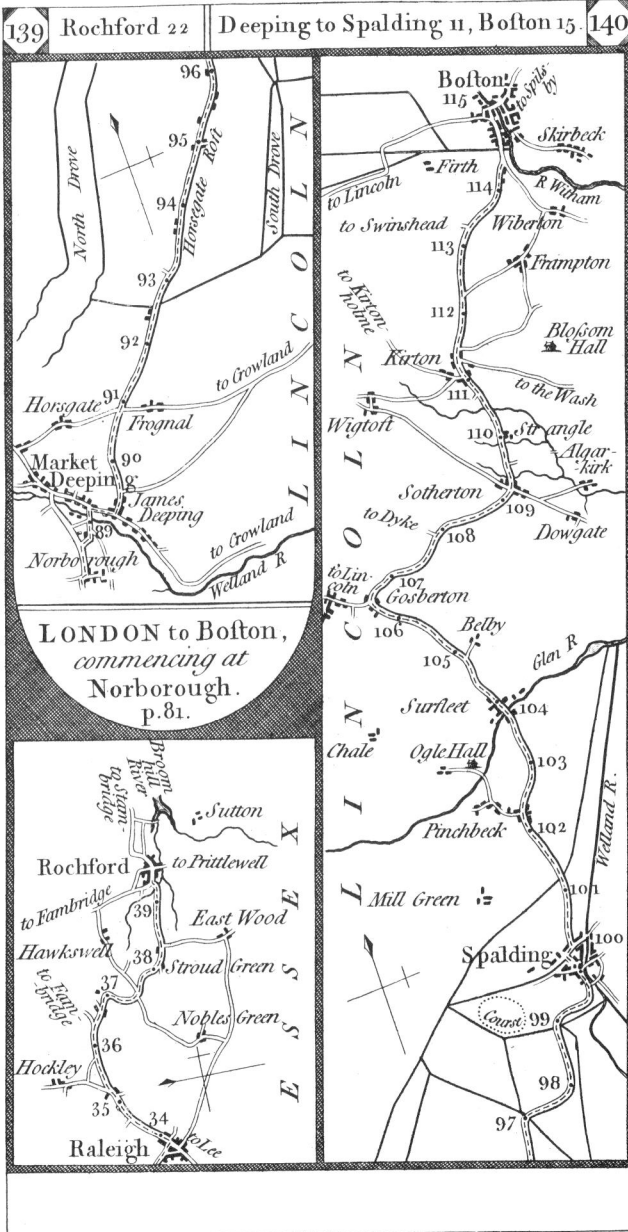

**139** Rochford 22 | Deeping to Spalding 11, Boston 15 **140**

96
95
94
93
92
91 Horsgate
Frognal
Market Deeping
James Deeping
90
89
Norborough
North Drove
Horsegate Bott
South Drove
to Crowland
to Crowland
Welland R.

LONDON to Boston, *commencing at* Norborough. p.81.

Boston 115
to Scits by
Skirbeck
Firth 114
to Lincoln
R Witham
to Swinshead
Wiberton
113
Frampton
to Kirton holme
112
Blossom Hall
Kirton
111
to the Wash
Wigtoft
110
Sutterangle
Algarkirk
Sotherton
109
to Dyke
108
Dowgate
to Lincoln
107
Gosberton
106
Belby
105
Glen R.
Surfleet
104
Chale
Ogle Hall
103
Pinchbeck
102
Welland R.
Mill Green
101
Spalding
100
Coast 99
98
97

Broom hill River
to Stambridge
Sutton
Rochford
to Prittlewell
to Fambridge
39
East Wood
Hawkswell
38
Stroud Green
to Foul bridge
37
Nobles Green
36
Hockley
35
34
Raleigh
to Lee

*Printed for the Proprietor CARINGTON BOWLES, London. 3 Jan.1785.*

13. 1785 APPENDIX MAP 15 (i) – DANIEL PATERSON.

route of plate 83/84 from *Spittal* to Barton. Broken lines cross the Humber to Hull and to Hessle, from where the road continues to Skidby in East Riding.

*139 Rochford // Deeping to Spalding 11, Boston 15. 140* (Aa–Ea, in upper panel). The lower panel is blank. The lower half of strip one has the final section of the road from Brentwood to Rochford (Essex). Above that is: *LONDON to Boston,/commencing at/Norborough./ p.81./* (Ac–Bc). The remainder of the map shows the route from Norborough to Boston, the mileages being continuous (89–115, as counted from London).

*Paterson's British Itinerary Being A new and accurate Delineation And Description of the Direct and Principal Cross Roads Of Great Britain. In Two Volumes By Capt$^N$– Daniel Paterson . . . Vol. II. London Printed for and Sold by the Proprietor Carington Bowles, At his Map and Print-Warehouse, 69 S$^t$. Pauls Church Yard. Tomkins fec. Ellis sculp$^t$. Published as the Act directs 3 Jan. 1785.*

**Copies**   CUL Atlas 7.87.7; Bod Allen 407; Leeds UL W 253; RGS D11–12; Norwich RL R 912 (S.B.3.24–25); Liverpool Athenaeum 910.3

(ii)   The numbering of the plates is now continuous. Vol. I has strips 1–340; i.e. 1–186 from the original volume one, new plates (187–198) and 199–340 (from the original volume two). Vol. II has 404–433 (the original final group in the original volume two), with new plates 434–449. The headings to the plates and the form of numbering remain the same, except that since the plates affecting Lincolnshire are all in volume 1 this is shown in the centre of each page's heading. All imprints are in the form: *Printed for the Proprietors BOWLES & CARVER, London, 6, Jan 1796.* (Ae–Ee, OS).

*203 Stilton 14 (Peterborō 6½) Stamford 14, Coltsworth 13 204* (Aa–Ea, in upper panel). A side road and *Ailesworth* have been added (5 miles north of Peterborough) in strip 203. In strip 204 at milestone 3 a park has been drawn enclosing a castle symbol; a by-road and *Pickworth* have been added, east of Exton park; *New Inn* has been added at milestone 7 and a new road, parallel to and west of the Great North Road runs to Coltsworth, where it rejoins the main road; *Thistleton* has been added.

*205 Grantham 8, Newark 14, Tuxford 13, 206* (Aa–Ea, in the upper panel). In the lower panel *e D. of Newcastle* replaces *e Late L$^d$. G. Sutton.* In strip 205 a park has been added at Easton, with a road along its northern edge *to Corby.* A cross road at *Gr. Paunton* continues westwards and then north to rejoin the main road south of milestone 6. Just south of milestone 4 (north of Grantham) a road leads to Marston, then on to Doddington where a new road from Westborough makes a junction. Another new road links Belton to *Gunnerby.* From milestone 2 a road curves to the west and rejoins the main road at Foston, with *Allington* added; a second road goes to Allington from a point between the first and second milestone, north of Grantham. In strip 206 *Shire Br.* has been added at the county boundary and, to the east, a road leads south from *Claypool. Langford* (north-east of Newark) has been added. The reference to the west of *N. Muscombe* has been altered to *to Leeds/P.269/* (Cc).

*279 Stilton to Deeping 15, Bourn 7¼, Folkingham 9, 280* (Aa–Ea, in the upper panel). The page reference at the foot of strip 279 has been altered: *203* (formerly 5). *Woodcroft* has been added (Ab). There is no change in strip 280.

*281 Sleaford 9, Green Man 10, Lincoln 8, 282* (Aa–Ea, in the upper panel). In strip 282 *Meer* (Ed) and *Grange* (Ec) have been added, the network of roads near Waddington extended and parks and castle symbols added at Harmston, Burton (by Lincoln) and Glentworth.

*283 Spittal 11½, Glanford Bridge 11½, Hull 18. 284* (Aa–Ea, in upper panel). In the lower panel the first name is now *Sir Th. Whichcote* (formerly Sir Chr. Whichcote). In strip 283 *Grange* and symbols, near Snitterby, have been added. The park near *Retburn* has been 'stippled' and *Rev./M$^r$. Carter* added. Horton has been changed to *Storton* (south of Scawby). There is no change in strip 284.

*337 Rochford//Deeping to Spalding 11, Boston 15. 338* (Aa–Ea, in the upper panel). The reference (Ac) for the London to Boston road is *279* (formerly 81). *Risegate Eau* has been added in strip 338.

*Paterson's British Itinerary . . . London Printed for . . . Bowles & Carver, At . . . No. 69 St. Paul's Church Yard. Published as the Act directs 6 Jan. 1796.*

**Copies**    CUL Atlas 7.79.17; RGS 260 D 13–14; Guildhall L. S 388/1.

**(iii)**    The date has been removed from the imprints to each plate. No other changes have been noted.

*Paterson's British Itinerary . . . By Daniel Paterson, Esq . . . The Second Edition Improved. Vol. I . . . London . . . Bowles & Carver . . . Published 6 Jan 1800.*

**Copies**    CUL Atlas 7.80.19–20; Leeds UL W 254.[3]

*Paterson's British Itinerary . . . 1803.*

**Copies**    CUL Atlas 7.80.31; RGS Fordham 3.

*Paterson's British Itinerary . . . 1807.*

**Copy**    RGS Fordham 5.

**(A)**    The numbering of the plates has reverted to that of the first edition. Each plate has *BRITISH HIGH ROADS* as a heading (Ba–Da, OS). The strips (still in pairs) are separately numbered with new descriptive headings within an upper panel. All the names in the lower panel have been revised. None of the pages that include roads passing through the county bears the name of engraver or printer.

*5 LONDON TO EDINBURGH VIA NEWCASTLE & COLDSTREAM 6* (Aa–Ea, in the upper panel). In the lower panel are: *a. G.C. Fitzwilliam Esq.* [below strip 5) and *b. Thorpe Hall/C.J. Strong, Esq.* (below strip 6). In strip 5 (which covers from Connington, south of Stilton to [Stamford]) *Walcot* has been changed from the former *Walcote/Noel Esq$^r$.* Also in strip 5, the park and the owner's name at *Ufford* have been deleted. The spelling *Burghley/House* has been adopted. In strip 6 the mileages along the road are now continuous with those on previous strips and as measured from London (previously a new series of numbers began at Stamford northwards); the Oakham road from Stamford is now simply labelled *Oakham* and the *Guash R* has been altered to *Wash R. Snow Esq$^r$.* has been changed to *J. Handley Esq* (north of Clipsham) and *Heathcote Bar$^t$.* to *Lord Aveland* (below Stocken Hall). *Suiston* has been changed to *Sewsten, Coltsworth* to *Colsterworth* and *Woolsthorp* has been deleted.

*7 LONDON TO EDINBURGH VIA NEWCASTLE & COLDSTREAM 8* (Aa–Ea, in the upper panel). In the lower panel (Ae–Ee) appears: a. *Stoke Hall, C. Turnor Esq./b. L$^D$. Brownlow* (below strip 7) and c. *Kelham Hall/J.H. Manners Sutton. Esq./* (below strip 8). In strips 7 and 8 the mileages continue from those in strip 6 (i.e. from 103 to 139). *Sir H.A.H. Cholmondley M.P* replaces *Cholmondley Esq$^r$* below Stoke. *Paunton* in the two villages south of Grantham has been changed to *Ponton* and *Gunnerby* has been altered to *Gonerby*. In strip 8 *Foss way* is spelt thus; before the spelling was *Fofs* (north of Newark). The reference to another plate number has been deleted, below *to Leeds* (north-west of Newark). The mileage from London at Tuxford has been increased from 137 to 137½ and an extra mileage added above the name of the town.

*57 LONDON TO HULL & SCARBOROUGH 58* (Aa–Ea, in the upper panel). In the lower panel (Ae–Ee) the wording is now: a. *Thorpe Hall, C.J. Strong Esq. b. Milton Ho, G.C.*

*Fitzwilliam. Esq.* At the foot of strip 57 the earlier reference to the route's description and the location of the earlier part of the route has been replaced by part of the route from *Lit Stukeley* to *Stilton*. Below Stilton *For Route from Wheat/Sheaf Inn to Stilton, see/ROUTE 1/* has been added with the mileage *74*. The lake and *Whittlesea/Meer* have been deleted. South-east of Peterborough *Farcett* has lost its final t. In strip 58 the road from Bourn westwards is now labelled *to Colsterworth* (instead of *to Coltsworth*). Changes to spelling of place-names include: *Cawthorpe* (previously without the final letter), *Haconby* (previously *Heckingby*) and *Laughton* (previously *Lowton*); *to Caseby* has been deleted (south-west of Folkingham).

*59 LONDON TO HULL & SCARBOROUGH 60* (Aa–Ea, in the upper panel). In the lower panel (Ae–Ee): *a Sir Th. Whichcote.– b. Bloxholm Hall, Right Hon. R.A.C.N./Hamilton. c. W^m. Markham, Esq./* In strip 59 several place-names have been altered: *Walcot* (previously *Wallcot*), *Osbournby* (*Osbornby*), *Threekingham* (*Threkingham*), *Scott/Willoughby* (*Scot/Willoughby*), *Ermine Street, a Roman Road* (*High Street, . . .*). In strip 60 *Meer*, *Grange* (north of Lincoln) and *Scampton* have been deleted. Place-names altered are: *N. Hykeham* (previously *N. Hykham*), *Boultham* (*Boatham*), *Aisthorpe* (*Aistrope*), *Cammeringham* (*Cameringham*) and the i in *Cainby* has been crudely altered to an e. Below Summer Castle *Sir Cecil Wray* has been changed to *John Dalton Esq*. All mileages north of Lincoln are continuous with those from the south.

*61 LONDON TO HULL & SCARBOROUGH 62* (Aa–Ea, in upper panel). In the lower panel (Ae–Ee): *a. Sir H.H.A. Cholmeley. b. Elsham Hall, Sir J.D. Astley*. The mileages throughout are continuous from those in strips 59–60. Place-names altered are: *Bishops/Norton* (previously *Bishop/Norton*), *Redbourne* (*Retburn*), *Hibaldstow* (*Hibalstow*) and *Brigg* (*Glanford/Br.* but the alteration is so crudely done that the new form is almost unreadable). *Atterby*, *Grange* (south of Waddingham), *Rev./M^r. Carter* and *Storton* have been deleted. In strip 62 the mileages are continued from those in strip 61. There is only one mileage at Barton; previously the mileage from London was given as well as that on the road from Lincoln (34). The mileage at Hull has been re-engraved and placed in the Humber. In Yorkshire *Spring H^d.* has been deleted.

*Part 1. North And North-Eastern Routes. British High Roads. Arranged For The Use Of Tourists. Illustrated by Forty-one Maps, On a Scale of 3/8 of an inch to the mile. In Four Parts. Each Part Complete In Itself. Published By Tinsley Brothers, 8, Catherine St., Strand, London, W.C. 1877 . . .*

**Copies**   CUL Atlas 7.87.10; Bod G.A. Gen Top. 8° 376.

**Reproduction**
(i) Strips 57–60 of the plates in state A are shown, in reduced form, on p. 10 of *Lincolnshire Past & Present*, No. 16 (Summer, 1994) as Fig.2. of the article: The Wheels Of Chance, by Nick Lyons. Unfortunately the caption from fig. 1 has been printed below.

**References**
1   Kingsley, p. 345; Fordham, H.G. Paterson's Roads. *The Library*. 4th series, Vol. 5 (1925), pp. [333]–356.
2   Gittings, R. Young Thomas Hardy. (London, 1975), pp. 107, 143–144, 155–159; Sutherland, J.A. Victorian Novelists and Publishers. (London, 1976), pp. 46, 159, 218–225.
3   Whitaker put this edition as 1796; the lower part of the title-page is missing.

Not in Chubb.

## 16

## ROBERT LAURIE and JAMES WHITTLE

Size: 29.8cm x 25.6cm.

Laurie and Whittle relied very largely on the plates of Robert Sayer for their stock materials when they set up in 1794. Shortly after that date volumes appeared containing the plates of Bickham (**30 (iii)**), Jefferys and Kitchin (**31 (iv)**), Emanuel Bowen (**34 (vii)**), and J. Ellis (**40 (ii)**). They were, in addition, very much in the forefront of publishing in the field of navigational charts and, after amalgamations, the firm still continues as Imray, Laurie, Norie and Wilson. Robert Laurie (1755–1836) had a reputation as an engraver, having set up on his own in 1779. In partnership with James Whittle he acquired the business of Sayer and Bennett, as noted above. Laurie retired in 1812 and two years later his son Robert Holmes Laurie took his place; Whittle died in 1818, when the firm became R.H. Laurie.[1] Between 1822 and 1825 all atlas plates were disposed of except for those of Benjamin Baker, which the firm issued as *New and Improved English Atlas* (**53 (ii)**). Meanwhile they had initiated the present work themselves, using the services of Nathaniel Coltman to compile a new road-book in 1806. Its popularity is shown by the number of editions that came out over more than 40 years of life. Coltman worked for the Post Office[2] and had helped Laurie and Whittle in the preparation of the plates for *Welsh Atlas*, first issued in 1805.[3] He is also noted for producing a map of Monmouth-shire, which has been dated to 1790[4], a very large single sheet map of Sussex (1799)[5] still being reissued in 1827, and, in 1831, Laurie and Whittle's map of London.[6] His final work seems to have been a map of York in 1833.[7]

The plates reflect the great amount of new material gathered together by John Cary, while surveying the country's roads on behalf of the Postmaster-General in the 1790s. The bulk of this information, apart from its incorporation in Cary's own maps was, in respect of roads, put out in *Cary's Survey of the High Roads* . . . (1790, with later editions to 1810), a roadbook that covers the south of England only, and *Cary's New Itinerary*, a work that went through many editions from 1798 until 1828. The latter describes in words the routes of all main and cross roads, with details of mileages, inns and posting places and a wealth of other data useful to the stage-coach traveller. A number of general maps are included, but the strip map in the Ogilby format has become largely a thing of the past. Coltman has prepared clear maps, which, with the discreet use of colour, cover larger areas of the country than hitherto. The scale is 7½ miles to an inch. The maps are strictly utilitarian, concentrating on roads and distances, but have their own distinctive elegance. Details concerning the sites of churches and the type of landscape through which the road passes – a feature of the earlier road maps – are excluded.

## EDITIONS

(i) All plates have the imprint *Published 12th. Feby. 1806 by LAURIE & WHITTLE, No. 53 Fleet Street London.* (Ce, OS); the engraver *E. Jones sculpt.* (Ae, OS) and a heading between the double inner frame and double outer frame (Aa–Ea), with the plate number in a circle (Ca); *(ROADS MEASURED FROM HICK'S HALL)* – on plates 19 and 20; *(ROADS MEASURED FROM SHOREDITCH CHURCH)* on plate 23 (Ce, between the inner and outer frames); and *Scale of Miles/5–0–15* [20 = 5cm] (Dd). The routes read upwards from the bottom, with the Great North Road and other main roads shown as double lines, cross roads as single lines and very minor roads as broken lines.

[19] *Commencement of the Roads to GLASGOW, and EDINBURGH, as 19 far as Stamford, with Roads to Uppingham & Market Deeping.* (Aa–Ea). Pointer (Dc).

[20] *Continuation of the Roads to GLASGOW, and EDINBURGH, as far as Abberford,*
*20 and YORK, with Roads to HULL, (by York) and Hull, (by Lincoln) and Barton Ferry.*
(Aa–Ea). Shows the roads north of Stamford and Market Deeping to York and Bridlington
and as far west as Chesterfield. Only the mouth of the Humber is outlined. There is no
pointer.

[23] *Commencement of Road to EDINBURGH (by Ware), with Roads to 23 CAM-*
*BRIDGE, Wisbeach, Lynn, Burnham Market, and Wells.* (Aa–Ea). Pointer (Aa). A small
portion of the coast in the [Wash] is shown. The Great North Road is shown as far as
Alconbury Hill. Holbeach and Spalding are the places furthest north in Lincolnshire.
Roads to Dereham and Bury St. Edmunds are delineated.

*Laurie And Whittle's New Traveller's Companion Exhibiting A Complete And Correct*
*Survey Of all the Direct and principal Cross Roads in England, Wales, and Scotland, as*
*far North as Edinburgh, and Glasgow: by Nath<sup>l</sup>: Coltman. 1806. Published 12<sup>th</sup>. Feb<sup>y</sup>.*
*1806, by Laurie & Whittle, N<sup>o</sup>. 53, Fleet Street, London.*

**Copies**  CUL Atlas 6.80.2; NLS Newman 637; Glasgow Mitchell L. B 109898; Cambridge RL 912.42.

(ii)  Plate 19: *Yaxley/78* has been added with a line to the Great North Road north of
*Normans Cross*. Plate 20: The coastlines of Lincolnshire and Yorkshire have been added,
that of Lincolnshire being extended to touch the scale-bar. The road from Lincoln to
*Market Raisin*, the mileage, *Nettleham/130, Welton/139¾, Snarford* and *Cold/Hanworth*
have all been deleted. The road from Hull to South Cave has been changed from a broken
line to a single continuous line. A broken line and *Froding-/ham* have been added between
Great Driffield and Hornsea. Plate 23: The coast in Norfolk has been drawn in. In
Hertfordshire *Baldock,* a broken line to Royston, *B. to R. 8 M.* and *Odsey Grange* have
all been added.

*Laurie & Whittle's New Traveller's Companion . . . 1806 . . .*

**Copies**  BL Maps 16.b.31; Bod Map C.16.d.58.

(iii)  Plate 19: many new branch roads are shown by broken lines with directions, e.g.
*to Spalding p. 20* (twice, with connecting lines to Crowland and to Market Deeping).
Several place-names have been added: *Clifton, Odsey Grange, Royston* (Herts.), and
extra plate references, e.g. *p 23* at Enfield. The river at Enfield has been extended and
the boundary between Hertfordshire and Middlesex has been extended eastwards. Plate
20: many new branch roads are shown by broken lines with directions, e.g. *to Leicester/p.*
*16* from Melton Mowbray. The Oakham-Nottingham road has been upgraded (from
single to double lines) and several new roads leading away from Nottingham have been
added. The road from Lincoln to Langworth is reduced in thickness, while the continuation to *Market Raising* has been strengthened. Plate 23: many new roads, shown by
single lines with directions have been added, e.g. *to Donnington/p. 20.* and *to Market*
*Deeping/p. 20.* (both from Spalding). A new portion of main road (double lines) has been
added south-west of Crowland, with *to Peterborough/p.19* and also from Alconbury Hill.
The road from Newmarket to Bury S<sup>t</sup>. Edmunds has been upgraded from a broken line
to double lines. Stars have been added at Holbeach and Swaffham. Places have been
added: *Watton* and *Chipping* (Herts.) and *Wimblington* (near March). At Whittlesea 77
has been added.

*Second Edition Corrected to 1807 of Laurie & Whittle's New Traveller's Companion . . .*
*and Glasgow: to which have been added the Roads to Perth & Aberdeen; by . . . Coltman.*
*1806. Published . . .*

**Copies**  NLS EMGB b.1.22; Manchester John Rylands UL R 108351.

*Second Edition Corrected to 1807 of Laurie & Whittle's New Traveller's Companion . . .
1806. Published . . .* [1808].[8]

**Copy**   CUL Atlas 6.80.18.

(**iv**)   County boundaries are more clearly marked and are coloured.

Plate 19: *HUNTINGDON S*. has been re-engraved; it is now shaped like a segment from
the top of a globe; previously it formed a curve reading up the plate. *CAMBRIDGE S*.
has been added (east of St. Neots) and south of its other appearance. Plate 20: Roads to
Theddlethorpe (from Alford and from Grimoldby) and from Louth to Wragby are now
marked – with broken lines. *For commencement of these Roads see p. 19*. has been added
(Be). A road, marked by a single line, from Lincoln to Snelland, and *Nettleham/136* have
been added. The *Bain River Navi[n]*. has been drawn in, from Sleaford to a point south of
Tattershall and from Tattershall to Horncastle. Other canals link Boston to Spalding,
Louth to the sea, and the Humber to *Market Raisin*, with a branch labelled *Caistor Canal*.
The Boston-Spalding-Market Deeping road has been raised to main road status. Among
places newly named are: *Branston, Uffington* and *Rothwell*. Similar additions appear
elsewhere on the plate, especially in Yorkshire. Plate 23: Part of the canal from Spalding
to Boston is marked. The road from Cambridge to Lynn has been raised to main road
status. Several places have been named, e.g. *Terrington/S[t]. John 95¾* and *Walpole/S[t].
Peter 94½*. Among new roads are two from Docking, to South Creak and to Brancaster,
shown by single lines. Canals have been added and rivers extended, particularly in the
Downham Market area.

*Third Edition, Corrected to 1809, of Laurie & Whittle's New Traveller's Companion . . .
to which have been added the Roads to Perth & Aberdeen, Greenock, Irvine, Ayr, Port
Patrick, Wigton, &c.– By . . . Prefixed, also, is a New General Map, which exhibits the
Whole at one View. Published . . . 1806 . . .*

**Copies**   CUL Atlas 6.80.8 and 6.80.12; Bod Johnson Maps 13 and Map C.16.d.3.[9]

(**v**) Plate 23: The broken line from the Great North Road east to Thorney has been
continued to Wisbeach and *T. to W. 14 M[s]*. has been added.

*Fourth Edition . . . Corrected to 1810 . . . Laurie And Whittle's New Traveller's Com-
panion . . . 1806 . . .*

**Copies**   BL Maps 16.b.32; NLS EMGB b.1.4.

*Laurie And Whittle's New Traveller's Companion . . . Accompanied By A New General
Map, Which Exhibits The Whole, At One View; And An Index Villaris, Distinguishing All
The Cities . . . Distances From London, &c.&c. The Fifth Edition, Corrected and Greatly
Improved. London: Printed By Harding And Wright, St. John's Square, For Robert Laurie
And James Whittle, Map, Chart, and Print, Sellers, No. 53, Fleet-Street. 1810.*

**Copies**   BL Maps 16.b.5; CUL Atlas 6.81.7; Liverpool UL L.2.2; Nottingham UL G
5514.

*Laurie And Whittle's New Traveller's Companion . . . Exhibits The Whole At One View
. . . The Sixth Edition, Corrected And Greatly Improved . . . 1811.*

**Copies**   Leeds UL W 257; NLS EMGB b.1.10.

*Laurie And Whittle's New Traveller's Companion . . . 1812.*

**Copies**   BL Maps 16.b.33; CUL Atlas 6.81.6.

(**vi**) Plate 20: The road from Newark to Worksop and the continuation to Doncaster
through Tickhill has been upgraded to main road status. Plate 23: *London & Cambridge
Junction* [canal] has been added from Bishop Stortford to Cambridge.

*Whittle and Laurie's New Traveller's Companion . . . The Seventh Edition, Corrected, And Greatly Improved. London: Printed (By Rider And Weed, Little Britain) For James Whittle And Richard Holmes Laurie, Map, Chart, And Print, Publishers, No. 53, Fleet Street. 1813.*

**Copies**   CUL Atlas 6.81.9; Bod Vet A6 d 98; NLS Newman 642.

**(vii)**   While the original 1806 imprint remains unchanged *A New Edition, 1815.* has been entered below it on all plates (Ce, OS).

*Whittle and Laurie's New Traveller's Companion . . . The Seventh Edition . . . 1814.* [1815].

**Copies**   BL Maps 16.b.35; CUL Atlas 6.81.4; Bod Allen 412; Guildhall L. S 388/1.

*Whittle and Laurie's New Traveller's Companion . . . The Eighth Edition, Corrected, And Greatly Improved . . . 1817.*

**Copy**   CUL Atlas 6.81.15.

*Whittle and Laurie's New Traveller's Companion . . . The Eighth Edition . . . 1818.*

**Copy**   NLS EMGB b.1.32.

**(viii)**   The imprint on all plates now reads: *London: Published by R.H. Laurie, N°. 53, Fleet Street.* (Ce, OS).

*Laurie's Traveller's Companion: Exhibiting A Complete And Correct Survey . . .* [the rest of the title is as developed from the 1806 edition, but all words are in capital letters] *Ninth Edition, Corrected, And Greatly Improved. London: Printed For R.H. Laurie, Map, Chart, and Print, Publisher, No. 53, Fleet Street. 1822.*

**Copy**   The private collection of Mr. D. Webb.

*Laurie's Traveller's Companion . . . Ninth Edition . . . 1823.*

**Copies**   CUL Atlas 6.82.23; Birmingham RL A 912.42.

*Laurie's Traveller's Companion . . . 1824.*

**Copy**   RGS 260 E 18.

**(ix)**   The engraver's name has been removed from plates 19 and 20. Plate 19: The names of a number of landowners have been added, e.g. *W^m. Crawshay Esq^r.* (below Wood Green in Middlesex). *(For continuation of these Roads see page 20.)* has been added (Ca). *Wiboston* has been corrected to *Wibaston* (Beds.). *Whittering* (sic) has been added south of Stamford. Several new roads have been added, especially in Bedfordshire. Plate 20: *(For continuation of these Roads, see p. 21.)* has been added (Aa). A road has been added to *Skegsness* (sic) from Croft. Other new roads in Lincolnshire are Broughton to Brigg and *Hibalstow* to *Harps/well*. Places added include *Santon, Skegsness* (sic) and *Kirton/151* [in Lindsey]. The Lincoln-Newark road is now shown by double pecked lines (instead of a single pecked line). *Baynards/Leap* has been re-engraved as *Bayards/Leap*; *Flinborough* altered to *Flixborough;* and *Cawkswell* is now *Cawkwell. Summer Road to Lincoln* (i.e. the road from Lincoln through Ancaster to the Great North Road) has been relabelled *Improved Road to Lincoln. Humber Side* has been entered in the Humber (north-west of Spurn/Head. Further changes of a similar nature have been included in other counties. Plate 23: Several place-names have been added, e.g. *Bury* (south of Ramsey), *Bourn Br.* (south-east of Cambridge), three on a new road from Godmanchester to St. Neots, etc. *Setchy* (Norfolk) is now *Setch.* A new main road has been added between Newmarket and Bourn Br. and continues *to Epping.*

*Laurie's New Traveller's Companion, And Guide through Roads of England And Wales, Including Great part of Scotland; With A General Map, and an Index Villaris, &c. An*

*Improved Edition, Corrected to the present time. London, Published by R.H. Laurie, N°. 53, Fleet Street. 1828.*

**Copy** CUL Atlas 6.82.9.

*Laurie's New Traveller's Companion . . . 1830.*

**Copy** NLS Newman 644.

*Laurie's New Traveller's Companion . . . 1832.*[10]

*Laurie's New Traveller's Companion . . . 1834.*

**Copy** CUL Atlas 6.83.28.


(**x**)  Plate 19: Below the scale * *Denotes where Post Horses may be had./* [symbol] *Mail Coach Roads.* has been added (De).

*Laurie's New Traveller's Companion . . . 1836.*

**Copy** BL Maps 16.b.34.


(**xi**)  Plate 20: A note similar to that added to Plate 19 (see state (x) has been added (Ee, below the scale). Plate 23 A note similar to that added to plates 19 and 20 has been added (De, below the scale).

*Laurie's New Traveller's Companion . . . 1838.*

**Copy** The private collection of Mr. D. Webb.


(**xii**)  Plate 19: The Hertford to Broadwater road has been upgraded to main road status. Plate 20: Railways, marked by single dark lines, have been added from Hull to Leeds and from York to Chesterfield, with a branch from Leeds joining it, east of Wakefield. Plate 23: Below the note on post-horses a third line *Railways* [thick dark line] (De) has been added. One railway has been marked, but by ++++++, from [London] to Cambridge and labelled *London &/Cambridge/Railway* at its southern end. Places added include: *Littlebury, Saffron/Walden, Newport*; a new road has been created by extending the road from Cambridge southwards through Littlebury and Newport, with *to Epping p. 24.* at its southern end; *to Epping* (near Cambridge) has been deleted. Other minor roads have been added in the Littlebury area. The road from Lynn to Wells has been upgraded to main road status.

*Laurie's New Traveller's Companion . . . 1846.*[11]

**Copy** CUL Atlas 6.84.14.


### References

1  Hodson, 1989, p. 187.
2  Tooley, p. 126.
3  Moreland and Bannister, p. 176, who also state that Coltman flourished from 1806–1886, which can not be correct.
4  Michael, D.P.M., op. cit., pp. 52 and 86–87.
5  Kingsley, pp. 113–115 (no. 59 in his catalogue).
6  Darlington and Howgego, op. cit., p. 212 (catalogue no. 334).
7  Tooley, p. 126.
8  The general map is dated 1808. Curiously, the plates are in a slightly hybrid state; all the changes of state (iii) are included except that, on plate 19, the Welwyn-Uppingham road is not of main road status; similarly, on plate 20, the Oakham-Nottingham road has not been upgraded.
9  The second Bod example is ascribed to 1810 locally, possibly because plate 25 (one of the new Scottish plates) is dated 1809.

10  Fordham, 1924, op. cit., refers to various other editions. in his personal copy of that work (in RGS) he notes an edition of 1832; no copy has yet turned up.

11  The general map is dated July 20th 1840. This may imply that there was another edition between 1838 and 1846, but no copy has yet been found of such an intermediate edition. Curiously, none of the railways proliferating north of London is shown on Plate 19; but perhaps their inclusion would obscure almost entirely many roads.

Chubb CCCXLLIII (1810 edition only).

**17**

**EDWARD MOGG**                                                                                           **1815**

Size: 20.25cm x 13.8cm.

Mogg published *Paterson's Roads* in its 1811 edition; as noted above (**Appendix 15**) this work had no road maps relating to Lincolnshire. Mogg first came to attention with the publication c. 1805 of a circular map of London and the area twenty four miles around, a work still being published in 1836; later versions extended the area covered by ten more miles. He was still flourishing in 1848.[1] In 1808 *A survey of the roads from London to Brighton . . .* appeared; in spite of its short title it covered most of the south of England.[2] In 1814–1815 Mogg began the issue in parts of a new work (see below), intended to cover the whole of England and Wales. The first parts were a revised re-issue of the 1808 work.[3] The first part was expected to have 14 numbers and cover the southern division. There would be two further parts 'in about Twelve Numbers each'. The scale was one inch to a mile. If issue did begin in 1814 it seems likely that the second part, which contained the Great North Road, came out in 1815 since issue of the individual numbers was monthly. When the whole was finished is not quite clear, since there are two title-pages in the completed work, dated 1816 and 1817.

The Great North Road appears in the fourth number of Part 2. Each number appears to have cost 5/-. Each plate is divided into two strips separately numbered, in the style of *Paterson's British Itinerary . . .* The plates themselves seem indebted to Laurie and Whittle, marked by the reference at the foot of each plate to their being measured from Hick's Hall. Nevertheless there is a wealth of new detail, particularly in the minor roads to and around the villages near to the main roads. The borders to the plates are a variation of the Ordnance Survey's keyboard pattern. The use of the word railways in the title must refer to tramways since railways in the modern sense were still some years ahead.

**EDITION**

(i)    Each plate has the legend *LONDON to EDINBURGH* (Ba–Da, in a panel between the inner and outer frame. Each plate has *ROADS MEASURED FROM HICK'S HALL* (Ce, in a panel between the inner and outer frame). Each plate has two strips separately numbered, both having a simple pointer. Parks or great houses usually have the name of the owner included.

*58* (Ba, OS) and *59* (Da, OS). Strip 58 shows the Great North Road from milestone 78 to *Wittering/Lodge* (milestone 84). Strip 59 shows the continuation and reaches Ticken-cote (milestone 92). The border is broken at the right to permit the inclusion of the side roads at Stamford to Bourne and Market Deeping and the name and owner of *Burleigh Hall/& Park/Marquis of Exeter/.*

*60* (Ba, OS) and *61* (Da, OS). Strip 60 continues from Tickencote and reaches to the

Lincolnshire boundary at *Bull & Swan* (milestone 99, north of Morkery Wood). Strip 61 carries the road to milestone 106, one mile north of Stoke Park.

*62* (Ba, OS) and *63* Da, OS). These strips continue the road north through Grantham and out of the county into Nottinghamshire.

*A Survey Of The High Roads Of England and Wales. Part The Second. Comprising The Counties Of . . . Exhibiting The Seats Of The Nobility And Gentry . . . The various Branches of Roads . . . Together With The Actual Distance . . . From The Main Road, Rivers, Navigable Canals, Railways, Turnpike Roads . . . Arranged By . . . Edward Mogg. London: Published By Edward Mogg, No. 51, Charing Cross. 1817.*

**Copy**   CUL Atlas 6.81.3.[4]

*A Survey Of The High Roads . . . 1828.*[5]

### References
1   Moreland and Bannister, p. 177.
2   Recorded by Kingsley, p. 373.
3   Kingsley, p. 373, suggests the parts were probably issued in 1814–1815. The CUL Atlas 6.81.3 has paper watermarked 1815 (strip 68/69). This example comprises the original parts with their title-pages, none of which is dated.
4   BL Maps 16.b.34 consists of the first ten parts (strips 1–162 of volume one only), with the cover of the original part one included. The Guildhall L. (S 388/1) has Part 1 only, bound as a single volume. CUL has Part 1 and numbers 1 and 4 of the second part, which ends with plates 60/61. It is not possible, therefore, to describe fully the contents of strips 62/63. The numbering of the plates in part two begins again at 1.
5   RGS Fordham 73 is an example of Part I only, dated thus; the general map (dated 1826) and the title-page both bear Mogg's new address at 14 Great Frederick Street. Part II may have come out simultaneously.

Not in Chubb.

## 18

### CHARLES SMITH                                                      1826

Size: 15.6cm x 9.2cm (the plates vary slightly in width).

The producer of this octavo volume is the same Smith responsible for *New English Atlas*, which went through many editions between 1801 and 1846 (**56**) and its smaller version from 1822 (**82**). The strips are all individually numbered and arranged three to a page. They follow the practice of Laurie and Whittle in showing the point of origin of each plate's measurement at the foot of each plate. Earlier models have been used for the information in each strip and the approach is closely related to that of Senex and his successors. The order of the roads throughout the book is idiosyncratic; as will be seen from the description below, the strip showing the Great North Road section reaching Stamford appears on the same plate as the end of a cross road from Yorkshire to Cumberland. Names of towns are given in capital letters with mileages from London; other stage coach halts are given in roman letters (lower case) with mileages; all other places are in italic, but there are no miles along the roads.

The plates were used in four editions between 1826 and 1835, but the scarcity of all issues suggests that sales had never been a serious challenge to Laurie and Whittle.

**EDITIONS**

(i)   All plates have 3 strips, each individually numbered and headed by the name of the chief town through which that strip passes. Each plate has the initial measuring point at the foot. When a new route begins within the plate its description is inserted at the starting point in a rectangular box.

*88/KESWICK* (Aa, OS); *89/BALDOCK* (Ca, OS); *90/STAMFORD* (Ea, OS). *From Hick's Hall* (Ce, OS). *Gardner sc.* (Ee, OS). The first strip shows the continuation of the road from Sheffield to Maryport (strips 86/87), starting at Gate Side, south of Ambleside. *TO EDINBURGH/ by Baldock Stamford Newark/Doncaster Durham Berwick/and Dunbar./* (Ce). Strips 89 and 90 show the road from London to *Stoke P*[ark], north of *Coltersworth.* Only strip 90 includes Lincolnshire.

*91/DONCASTER* (Aa, OS); *92/BOROUGHBRIDGE* (Ca, OS); *93/ DURHAM* (Ea, OS). *From Hick's Hall* (Ce, OS). There is no note of the engraver. Strip 91 shows the continuation of the Great North Road from *Gr$^t$ Ponton T.G.* to Doncaster; the other strips continue the road to Newcastle.

*109/ROYSTON* (Aa, OS); *110/PETERBOROUGH* (Ca, OS); *111/ LINCOLN* (Ea, OS). *From Shoreditch Church.* (Ce, OS). *TO SCARBOROUGH BY/Royston Huntingdon Lincoln/Barton Ferry and Hull./* (Ae). Strip 109 shows the road from London to Arrington by way of Buntingford; strip 110 continues through Huntingdon to Folkingham, with the road terminating at Lincoln, half way up strip 111. At Lincoln a double line separates the continuation from Lincoln to a point just north of *GLANDFORD/ BRIGGS/152.* The engraver's name is not given.

*112/SCARBOROUGH* (Aa, OS); *113/SPALDING* (Ca, OS); *114/ YORK* (Ea, OS). *From Shoreditch Church.* (Ce, OS). The engraver's name is not given. Strip 112 carries on the route from strip 111 to Scarborough. *TO G$^T$. GRIMSBY BY/Spalding Boston & Louth./* (Ce, i.e. at the foot of strip of 113, which shows the road from Peterborough to Partney). Strip 114 carries the route from Dalby to Grimsby. *TO YORK* (Ec, in a panel, half way up strip 114); the remainder of that strip depicts the road from *Ferry/Br./177* to York.

*Smith's New Pocket Companion To The Roads of England and Wales And Part Of Scotland. Engraved On Forty-Three Copper-Plates. Comprehending The Routes From London To Every Considerable Town In England & Wales, And The Principal Cross Roads. With A Copious Index, Containing The Distances From London, The Market-Days, And Principal Inns. To Which Is Annexed, The Routes Of The Mail-Coaches, And The Heights Of The Principal Mountains In England & Wales. London: Charles Smith And Son, 172, Strand. 1826.*

**Copies**   BL Maps 15.a.14; CUL Atlas 7.82.19.

(ii)   Strips 88–90. In strip 88 a new road between Keswick and Cockermouth has been added; in strip 90, below HUNTINGDON 65 has been added.

Strips 91–93. In strip 91 Gonerby has been changed to *Gunnerby.* Strips 92–93 remain unaltered.

Strips 109–111. In strip 109 the description of the route has been altered to: *TO EDINBURGH BY/Royston, Doncaster & Durham./* (Ae). At Royston the direction *to Cambridge/13 M./* has been replaced by the road itself, the names of four villages, *CAMBRIDGE/50½* and *Cam R.* In strip 110 *thence to Edinburgh pp./90–1–2–3–4–5 & 6/* has been added above Alconbury Hill; above that, in a box, is: *TO SCARBOROUGH/by Peterborough./.* The road continues as before from Stilton to Folkingham, with the following additions: the mileages have all been increased by 4 (at Stilton, Peterborough, Market Deeping, Bourn and Folkingham); in addition, the mileage attached to the direction *to CORBY/(101)/* has also been increased by 4 to *105*; other mileages have not been changed. In order to incorporate the addition in strip 109 of the Cambridge road the triple line separating the strips has been bent; this has led to the re-engraving of *Aslackby* on the east side of the road. On strip 111 all mileages have also been increased by 4, *SLEAFORD* being *115* and *LINCOLN 133*.

Strips 112–114. All mileages in strip 112 have been increased by 4, with the exception of $G^T$. *DRIFFIELD/192* (increased to 193). In strip 113 all mileages have been increased by 4 from Spalding (now *101*) to Spilsby and Bolingbroke and including those on the by-road, with the exception of that for the road to Thorney. In strip 114 the mileages are 4 miles higher on the road from Dalby to Grimsby (168). On the road from Ferry Br. to York the mileages have all been decreased by 4 on the main road, reading now from 173 to 195.

*Smith's New Pocket Companion . . . In England & Wales, And A Copious Selection Of Cross Roads. With An Index Containing The Distances From London, The Market Days, and Principal Inns. Accompanied by A General Map of England & Wales. Pickett sc. London. Charles Smith & Son, 172, Strand. 1827.*

**Copies** CUL Atlas 7.82.28; Guildhall L. S388/1; RGS 262 B 15.

*Smith's New Pocket Companion . . . 1830.*

**Copy** Bod Vet A6 e 356.

*Smith's New Pocket Companion . . . 1835.* [1836].[1]

**Copy** Bod Allen 416.

**Reference**

1 The general map is dated 1836 and the paper used has 1836 in the watermarks. An edition in 1835 is, of course, possible.

Chubb CCCCII (1826 edition only).

# 19

## THOMAS TELFORD                                                    1827

Size: 64.9cm x 350.8cm.

Thomas Telford (1757–1834) is renowned as a civil engineer, who produced a series of notable works, ranging from the masonry bridge at Montford in Shropshire (where he was surveyor of public works from 1787) to the aqueduct at Pontcysyllte and other works on the Ellesmere Canal. Commissioned by the government to survey the public works needed in Scotland he became responsible for the building of the Caledonian Canal (1802–1823) and all the associated roads, bridges and buildings, including churches and harbours. From 1819 to 1826 he was overseeing the building of the Menai suspension bridge, work that overlapped with St. Katherine's Dock in London and the survey of the Great North Road described below.[1]

He was involved in schemes for fen drainage and, in 1810, issued *Plan of the proposed Stamford Junction Navigation from Oakham . . . to Stamford and Boston . . . and from Stamford to Peterborough . . . Surveyed under the direction of Tho$^s$. Telford, by Hamilton Fulton and drawn by W.A. Provis. 1810.*[2]

The work on the Great North Road is a remarkable piece of work when one remembers that it covers the road from London to Morpeth and was finished when Telford was 70 years old. It seems to prefigure our modern system of by-passes and road straightening to ease or shorten journeys between the main towns along the route. Two instances suffice; red lines are used to indicate proposed new roads or deviations and two of these indicate a new line for the road west of Stamford, i.e. very close to the line of the present A1 road; and, further north, a similar westward by-pass at Stretton, the present line of the A1, which isolates the village from the rest of Rutland. In fact, Telford and his staff took nearly eight years to complete the survey and the basis of the work was a proposal to the government of the day that by adopting his proposed straightening and short cuts the distance from London to Edinburgh would be shortened from 391½ to 362¼ miles. With journey times of the day very slow such a saving was not to be ignored. However, the government had no money for such a plan and, with the coming of the railways, the scheme was forgotten.[3] The sheet, where it passes through the county, extends eastwards to Spalding. The map is known in two versions; a draft version was drawn in 1826[4] and 'signed' by Telford in 1827; the second fuller version is described below. The final version is also known in several forms; in one continuous strip, in four or six separate sections; in the four section case Lincolnshire appears on the second part; in its six section form Lincolnshire occurs on the third and fourth sections.

## EDITION

(i)  *LONDON AND MORPETH/MAIL ROAD./INDEX MAP/Surveyed by Tho[s]. Telford/1827/SCALE/11 Miles* [= 25.5cm]/ (Ea). *Printed at the Lithographic Establishment Quarter Master, Gen[ls]. Office Horse Guards/Drawn by L Hebert/* (Ee). The map extends from London at the left hand edge to Morpeth at the right hand edge, i.e. north is to the right. The Great North Road is yellow, with suggested new roads, by-passes, etc. in solid or broken red lines.

**Copies**    BL Maps 1210 (10); CUL Maps c.36.82.2; Bod Maps (E) C17:4 (19).[5]

**References**
1   DNB; *Chambers Biographical Dictionary*, 5th edition. (Edinburgh, 1990), pp. 1439–1440.
2   BL Maps C.10.c.24 (53); LAO 2 Anc 5/25/2 & 3 Cragg 11/35.
3   Rolt, L.T.C. Thomas Telford. (London, 1958), p. 133.
4   BL Maps 1210 (5) is an example; it is in two sheets and only covers the road from East Retford to Morpeth.
5   Skempton, A.W. *British civil engineering 1640–1840: a bibliography of Contemporary Printed Reports, Plans and Books.* (London, 1987), item 1543, p. 227 records a copy in the Science Museum L.

## 20

**LETTS and SON**                                                          **1834**

Size: 9.9cm x 6.3cm., when folded up.

The history of the firm of Letts has been referred to under the maps prepared by J. and C. Walker with reference to two transfers made by them in the 1880s (**105M** and **105N**). The little road-books of Letts occur in two formats; in one the map pulls out as a continuous strip and in the second the long strip has been cut up in shorter pieces and bound as a booklet. While the maps are numbered on their outer covers the lists that occur on the outside of the back cover usually carry no numbers. According to booklet no. 1 the seventh in the series would cover the part of the Great North Road from Peterborough to York. The list on the back of no. 16 may be a confirmation of this but the title is changed.[1] All of the items seen suggest that the strips only deal with roads from London, i.e. cross roads are not dealt with; three of them (numbers 31–33) cover different routes to Paris.

Kingsley suggested 1834 as the date of issue, since the population figure given in no. 32 (the road to Brighton) is that of the 1831 census and the 'Coach List' is 'corrected to 1834'.[2] A similar statement appears in no. 26, which also deals with another route to Brighton, and also in no. 30.[3]

## EDITION

**(i)** *The British Roadster, or Stage Coach Companion, no. 7* [YORK and LEEDS from LONDON; or, PETERBOROUGH to YORK]. [1834].

### References

1 CUL Maps d.G.01.1. The cover of no. 16 suggests that the map deals with: *YORK and LEEDS from LONDON*.
2 Kingsley, p. 375.
3 CUL Maps d.G.01.1–, Maps d.G.01.52 and Maps d.G.01.58 together represent nos. 1, 3, 5, 8–9, 16, 19, 23, 26–27, 30 and no. 8 again with a variant cover title; a single copy of no. 32 is in BL and Guildhall L. A7.2, no. 18 is an example of no. 15. No copy of no. 7 has yet been found.

Not in Chubb

## 21

### GALL and INGLIS                                                          c.1893

Size: 14.5cm x c.240cm., when opened out fully.

Gall and Inglis was an Edinburgh firm, which had its origins in 1810 when James Gall (1783–1874) set up his own independent printing business. For many years the output was concentrated on Bibles and religious commentaries. When James' son, also James, joined the firm in 1838 business had increased and also included bookselling and publishing. In 1834 the firm had been the first to publish a book of the Bible (St. John's gospel) anywhere for the blind. However, in 1847, as the son resigned to become a minister of the church Robert Inglis joined the firm. Shortly after, in 1850, the firm began to issue a series of atlases starting with the *Edinburgh Imperial Atlas* and astronomical works. Robert's son James Gall Inglis (Robert had married the daughter of the founder) joined the firm in 1880 and the involvement with maps of all sorts really took off; the firm had purchased in 1877 G.F. Cruchley's stock.[1] Reference has already been made to editions of county maps put out by Gall and Inglis after this acquisition (**57K–N; 144A, 153**). As the demand for cycling maps increased in the 1880s and 1890s the firm began its series of maps that extended beyond county boundaries. One result was the series of maps of main roads, starting naturally with those from Edinburgh but soon expanding to other areas. The Williamson map (**144**) is an example of their practice of producing maps centred on focal towns.

The strip maps of the Great North Road are based, like Williamson's cycling map, on the Cary plates for *Improved Map of England And Wales* of 1832. The first version takes a broad slice of the country on either side of the road in a long folding strip. Later versions included, along the side of the unfolding sheet, profiles of the country through which the road passes with mileages from London clearly marked. They were the work of Henry

Robert Gall Inglis, who joined the firm in 1887, then in his elder brother's hands; both brothers died in 1939 and the firm continued with James Inglis' son.

## EDITIONS

(**A**) [Pages] *12* to *14* (in upper edge, the numbers being enclosed in circles) cover the Great North Road from North Nottinghamshire to Water Newton (Northamptonshire). The main road as far south as Wansford is coloured brown; there is no other colour. The Great Northern Railway is shown by a single black line; other lines are shown by double uncoloured lines. There is a fold every 8.5cm; the whole strip forms a booklet of slightly larger size. At the three main towns on the road, Newark, Grantham and Stamford, a rectangular box (superimposed on the map face) bears the mileages of other important towns on the route.

*The Great North Road Map FROM LONDON TO EDINBURGH* (cover title). On the first leaf, when the map is folded up, the preliminary material is headed: *The 'Great North Road' And 'East Coast Route' Map Compiled by H.R.G. Inglis. In this Map the North is at right angles to the lettering . . . Gall & Inglis, 25 Paternoster Sq., London, E.C.; And 20 Bernard Terrace, Edinburgh.* [1893].[2]

**Copies**   BL Maps 1.a.68; CUL Maps D.33.89.1; Glasgow Mitchell L 912.42 GRE (B 151271).

(**B**)   The width of the map strip has been reduced to 10.9cm and the remaining space has been taken up with a profile of the road, coloured brown. The profile is covered by a grid, which marks off the miles; in the margin above the grid the distances are shown at 5 mile intervals from London thus: *105 MILES*. The grid also shows height above sea-level at 100 feet intervals, from sea-level to 800'. The numbers of the [pages] have been lost by the inclusion of the profile as have the boxes with mileages to the more important towns. In the earlier version a blank wedge had been inserted on the map from the eastern edge, the point ending at Easton. In this version the wedge's point has been replaced by a spanner-shaped device (in Stocken Wood), on which has been printed *Alteration in direction of Map*. From the western edge a shorter pointed wedge has been added, terminating at the tip of the opposite wedge. The earlier edition showed railway 51 dividing into two westward-pointing branches; the more southerly of these (passing north of the Ram Jam Inn) has been deleted. Railway 22 has been deleted also.

*Strip Maps. No 10 Price 1/- The 'Great North Road' Map. (Part 1. – London To York.) By H.R.G. Inglis. With a Large Scale 'Contour' Plan of the Road. Scale: Half an Inch to a Mile. Gall & Inglis, 25 Paternoster Square, London, E.C.: And 20 Bernard Terrace, Edinburgh.* [Cover title]. In a shield appears: *LONDON to YORK, HARROGATE, or LEEDS.* (Ea). Below the price is a drawing of a cyclist. (Aa). [1900].[3]

**Copies**   BL Maps 2.aa.65; CUL Maps D.33.90.6.

Issued as a supplement to *Cyclist* magazine. The cover title is: *CYCLING STRIP ROUTE MAP London to York, Leeds & Harrogate* (Aa) with the main part taken up by a cyclist reading a map. *1/- NET* (Ae).[4]

### References
1   Gall & Inglis, publishers 1818–1960 (cover title). (Edinburgh, 1960).
2   BL's copy was received 19 JUL 1893; the other libraries also ascribe their copies to 1893. Red and blue covers both exist.
3   The BL copy was received 29 NOV 1900.
4   No example has been seen, but the cover and part of the strip are illustrated in: Nicholson, T.R. *Wheels on the Road* (Norwich, 1983), p. [43]. The portion of the strip shown is of the Lincolnshire section, northwards from Grantham.

# ADDENDA

**2**   A great deal of reliance has been placed in the attribution of the initials W.B. on the work of David Kingsley. In a letter to *IMCoS Journal* (Autumn, 1993, p.35) he has now expressed some doubt on the identification of Bowes' forename as William. He quotes an article, which appeared in *Journal of the International Playing-Card Society* (Vol. XVI, no. 2. November, 1992), in which the design of the playing cards by Bowes is shown to be based on designs for the court cards prepared by Guillemme Acart in Rouen, possibly in 1567 or even earlier. The name GILLAM ACART occurs on the Jack of Clubs in the Bowes' map cards and this led to the assumption that Bowes was called William. There is some evidence that the engraver of the cards may have been Augustine Ryther; one of the cards has *A. Ryt. s.* which may be an abbreviation for Augustine Ryther sculpsit. See *Map Collectors Circle*, no. 87 (1972), p. 5.

**57**   A dealer in Stamford offered (10/95) an example of the map dated 1801 cut round the county's borders, mounted on wood and cut up as a jigsaw puzzle. The title portion of the map was stuck to the outside of the box containing the jigsaw. It is not known which state of the map was involved.

**72**   I am grateful to Mr. Eugene Burden who has allowed me to see his recently acquired example of the hunting map (referred to in note 7), issued by Stanford, which covers all the county except the areas north-west of Gainsborough. It was issued in January, 1877 and is number 10 in the hunting series; the map's cover bears also a small map outlining the sheet's coverage. One extra interesting addition to the projected county railways shown on the sheet – it includes the earlier two-pronged branch from Louth to Mablethorpe and Saltfleet/North Somercotes, with a continuation to Alford where the East Lincolnshire Railway is rejoined, the Louth-Bardney branch and the link from *Barkstone* to Sedgebrook, is a railway projected to Crowland. The line, which was never built, leaves the Peterborough to Wisbech line at Eye and finishes at a point due east of the Abbey.

**130 I**   Further examples of *Philips' Handy Atlas* also dated 1885 have been found by Mr. Tony Burgess and Mr. Eugene Burden. The title-page and map of one copy are identical with those of 1884 (**130 H**), with the obvious change of date on the title-page. A second, also dated 1885 and with the title-page in the 1884 form, shows that the railway changes around Gainsborough, described in **130 I**, have been added but not the additions warranted by the 1885 Redistribution of Seats Act. The names of the parliamentary divisions, described in **130 I**, were a later addition issued in a new atlas after the summer of 1885. The title-page was altered then to that quoted in the text.

**130 M**   An example of an 1894 *Philips' Handy Atlas* . . . has been found; the contents and the Lincolnshire map are as described for 1893.

**144**   Although it is correct to state that the firm of A.H. Swiss is not known to have issued a separate hunting map in its own series relating to the Belvoir and Cottesmore hunts a map covering Belvoir country was prepared by Swiss. The map, which appeared in T.F. Dale's *The History Of The Belvoir Hunt* (London, 1899), is a lithograph prepared by Gall and Inglis 'By the courtesy of Mr. A.H. Swiss, Devonport'. The map shows the area south of Navenby to Essendine Station and reaches almost to Nottingham in the west and Spalding in the east, taking in, therefore, part of the Cottesmore country also. Local examples of the book are Lincoln RL and Grantham RL (both L 799.2); Grimsby RL (X000:799.25 Dal).

Since the map covering the hunts of the northern part of the county is numbered 15 in Swiss' series and the cover of the example in BL states that maps numbered 16 and 17 were 'in active preparation' it is possible that maps for both the Belvoir and the Cottesmore were prepared and issued by Swiss; there is no geographical consistency in the numbering of the Swiss maps, however, and numbers 16 and 17 may have related to other areas of the country.

# BIBLIOGRAPHY

Adams, B. *London Illustrated, 1604–1851*. (London, 1983).

Albert, W. *The Turnpike Road System in England 1663–1840*. (Cambridge, 1972).

Almon, J. *Memoirs of an eminent bookseller*. (London, 1790).

Awdry, C. *Encyclopaedia of British Railway Companies*. (Wellingborough, 1990).

Bairstow, M. *Railways in East Yorkshire*. (Halifax, 1990).

Barker, K. and Kain, R.J.P. *Maps And History in South-West England*. (Exeter, 1991).

Bennett, S. and Bennett, N. (eds.) *An Historical Atlas Of Lincolnshire. (Hull, 1993)*.

Beresford, M.W. *Lost Villages of England*. (London, 1954).

Beresiner, Y. *British County Maps*. (Ipswich, 1983).

Bonacker, W. *Kartenmacher aller Länder und Zeiten*. (Stuttgart, 1966).

Briggs, A. (ed.) *Essays in the History of Printing . . . The House of Longman. 1724–1972*. (London, 1974).

Briscoe, A. Daly. *A Tudor worthy*. (Ipswich, 1979).

Britton, J. *[Autobiography]. Part 2. A Descriptive Account of the Literary Works . . .* (London, 1849).

Brown, P.A.H. *London Publishers and Printers, c.1800–1870*. (London, 1982).

Bryan's *Dictionary Of Painters and Engravers; new ed.* (London, 1927).

Burden, E. *Printed Maps of Berkshire 1574–1900. Part 1 – County Maps; 6th revision*. (Ascot, private circulation, 1993).

Campbell, A.K.D. *Outstanding Children's Books*. (Swansea, 1990).

Capp, B. *Astrology and the Popular Press*. (London, 1979).

Chambers, B. *Printed Maps and Town Plans of Bedfordshire 1576–1900*. (Bedford, 1983).

Chubb, T. *The Printed Maps in the Atlases of Great Britain and Ireland . . . 1579–1870*. (London, 1927).

Cox, H. *The House of Longmans, 1724–1924*. (London, 1925).

Cupit, J. and Taylor, W. *The Lincolnshire, Derbyshire & East Coast Railway*. (Oxford, 1984).

Darlington, I. and Howgego, J. *Printed Maps of London, c.1553–1850*. (London, 1964).

*Dictionary of National Biography*. (London, 1880–1900).

Dow, G. *The Alford and Sutton Tramway*. (Chislehurst, 1947).

Dunston, G. *The Rivers of Axholme*. (London, 1912).

Eden, P. (ed.). *Dictionary of Land Surveyors . . . 1550–1850*. (Folkestone, 1975–1979).

Ellis, S. and Crowther, D.R. (eds.) *Humber Perspectives*. (Hull, 1990).

Elvin, L. *Lincoln As It Was. Vol. III*. (Lincoln, 1979).

Evans, I.M. and Lawrence, H. *Christopher Saxton, Elizabethan mapmaker*. (Wakefield & London, 1979).

Fordham, F.G. *Cambridgeshire Maps . . . 1577–1900*. (Cambridge, 1908).

Fordham, F.G. *Hertfordshire Maps, with supplement*. (London, 1914).

Fordham, F.G. *John Cary, Map, Chart and Print-seller . . . 1754–1835*. (London, 1925).

Fordham, F.G. *Notes on British and Irish itineraries and road books*. (Hertford, 1912).

Fordham, F.G. *The Road-books and Itineraries of Great Britain, 1570–1850*. (Cambridge, 1924).

Fordham, F.G. *Studies in carto-bibliography*. (Oxford, 1914).

Foster, C.W. and Longley, T. *The Lincolnshire Domesday and the Lindsey Survey.* (Lincoln, 1924). Lincoln Record Society – 19.

*Gall & Inglis, publishers, 1818–1960.* (Edinburgh, 1960).

Gardiner, L. *Bartholomew 150 years.* (Edinburgh, 1976).

Goode, C.T. *The Great Northern & Great Eastern Joint Railway (March to Doncaster).* (Hull, 1989).

Goode, C.T. *The Railways of North Lincolnshire.* (Hull, 1988).

Gough, R. *Anecdotes of British Topography.* (London, 1768).

Gough, R. *British Topography.* (London, 1780).

Grinling, C.H. *History of the Great Northern Railway, 1845–1922.* (London, 1966).

Harley, J.B. *Maps for the Local Historian . . .* (London, 1972).

Harrod, W. *History of Stamford.* (London, 1785; 2nd ed. 1822).

Harvey, P.D.A. *Medieval Maps.* (London, 1991).

Harvey, P.D.A. *The History of Topographical Maps.* (London, 1980).

Harvey, P.D.A. and Thorpe, H. *The Printed Maps of Warwickshire, 1576–1900.* (Warwick, 1959).

Heal, A. *English Writing Masters.* (London, 1931).

Hearne, T. *Remarks and collections; ed. by C.E. Doble.* (Oxford, 1885–1921).

Heawood, E. *Watermarks of the seventeenth and eighteenth centuries.* (Hilversum, 1950).

Hind, A.M. *Engravings in England in the sixteenth and seventeenth centuries.* (Cambridge, 1952–1955).

Hind, A.M. *A History of Engraving & Etching from the fifteenth century to the Year 1914.* (London, 1923).

Hobden, M. and Hobden, H. *John Harrison and the problem of longitude; 3rd ed.* (Lincoln, 1990).

Hodson, D. *County Atlases of the British Isles Published after 1703. Volume 1: Atlases published 1704–1742 and their subsequent editions.* (Tewin, 1984).

Hodson, D. *County Atlases . . . Volume II: Atlases published 1743–1763 . . .* (Tewin, 1989).

Hodson, D. *The Printed Maps of Hertfordshire 1577–1900.* (London, 1974).

Holmes, C. *Seventeenth Century Lincolnshire.* (Lincoln, 1980).

Hyde, R. *Printed Maps of Victorian London, 1851–1900.* (London, 1975).

*The Itinerary of John Leland; ed. by T. Hearne. Vol. VI.* (London, 1711).

Joby, R.S. *East Anglia (Forgotten Railways, 7): 2nd rev. ed.* (Newton Abbot, 1985).

Jolly, D. *Maps in British Periodicals.* (Brookline (Mass), 1990–1991).

King, G.L. *The Printed Maps of Staffordshire, 1577–1850; 2nd ed.* (Stafford, 1988).

Kingsley, D. *The Printed Maps of Sussex 1575–1900.* (Lewes, 1982).

Kleerkooper, M.M. *De Boekhandel te Amsterdam voornamelijk in de 17e eeuw.* ('s-Gravenhage, 1914–1916).

Knight, C. *Shadows of the old booksellers.* (London, 1865).

Koeman, C. *Atlantes Neerlandici.* (Amsterdam, 1967–1970).

Leach, T.R. and Pacey, R. *Lost Lincolnshire Houses. Vol. 4.* (Burgh le Marsh, 1993).

*Lincolnshire At The Opening of the Twentieth Century.* (Brighton, 1907).

Longman, C.J. *The House of Longman.* (London, 1936).

Ludlam, A.J. *The Horncastle and Woodhall Junction Railway.* (Oxford, 1986).

Ludlam, A.J. *The Louth, Mablethorpe And Willoughby Loop.* (Oxford, 1987).

Ludlam, A.J. and Herbert, W.B. *The Louth to Bardney Branch; 2nd enlarged ed.* (Oxford, 1987).

Lynam, E. *British Maps And Mapmakers.* (London, 1944).

Mack, J.L. *The Border Line from the Solway Firth to the North Sea . . .* (Edinburgh, 1926).

Martin, R.B. *Tennyson The Unquiet Heart.* (Oxford, 1980).

McKerrow, R.B. (ed.). *A Dictionary of Printers and Booksellers in England, Scotland and Ireland . . . 1557–1640 . . .* (London, 1910).

Mann, S. and Kingsley, D. *Playing Cards Depicting Maps of the British Isles and of English and Welsh Counties.* (London, 1972).

Maxted, I. *London Book Trades 1775–1800 . . .* (Folkestone, 1977).

Michael, D.P.M. *The Mapping of Monmouthshire: A Descriptive Catalogue of pre-Victorian maps . . . from . . . 1577 . . .* (Bristol, 1985).

Millburn, J.R. *Benjamin Martin, author, instrument maker, and 'county showman'.* (Leyden, 1976).

Moir, D.G. *The early maps of Scotland to 1850.* (Edinburgh, 1973).

Moore, J.N. *The Historical Cartography of Scotland.* (Aberdeen, 1991).

Moreland, C. and Bannister, D. *Antique Maps; 2nd ed.* (London, 1986).

Munter, R. *A Dictionary Of The Print Trade In Ireland, 1550–1775.* (New York, 1988).

Myers, R. and Harris, M. (eds.). *Aspects of the English Book Trade.* (Oxford, 1984).

Myers, R. and Harris, M. (eds.). *Author/Publisher relations during the eighteenth and nineteenth centuries.* (Oxford, 1983).

Nichols, J. *Literary anecdotes of the eighteenth century.* (London, 1812–1815).

Nicholson, T.R. *Wheels on the Road: Road Maps of Britain, 1870–1940.* (Norwich, 1983).

Nock, O.S. *The Great Northern Railway.* (London, 1958).

Norton, J.E. *Guide to National and Provincial Directories.* (London, 1950).

Nowell-Smith, S. *The House of Cassell.* (London, 1958).

*Old Series of Ordnance Survey Maps, Vol. II, with an introduction by J.B. Harley and Y. O'Donoghue.* (Lympne, 1977).

*Old Series of Ordnance Survey Maps, Vol. V. Lincolnshire, Rutland and East Anglia. Introductory essay by J.B. Harley . . .* (Lympne, [1987]).

Olney, R. *Rural Society and County Government in Nineteenth Century Lincolnshire.* (Lincoln, 1979).

Padley, J.S. *Fens and Floods of Mid-Lincolnshire.* (Lincoln, 1882).

Pearson, R.E. and Ruddock, J.G. *Lord Willoughby's Railway: The Edenham Branch.* (Grimsthorpe, [1986]).

Penny, J.A. *Notes on the Monasteries and other religious institutions near the River Witham, from Lincoln to the Sea.* (Horncastle, 1918).

Pevsner, N. and Harris, J. *Lincolnshire.* (London, 1964).

Philip, G. *The Story of the last Hundred Years – A Geographical Record.* (London, 1934).

Plomer, H.R. *Dictionary of the Printers and Booksellers . . . in England, Scotland and Ireland from 1641 to 1667.* (London, 1907).

Plomer, H.R. *Dictionary of the Printers . . . from 1668 to 1725.* (London, 1922).

Plomer, H.R. *Dictionary of the Printers . . . from 1726 to 1775.* (London, 1932).

Pollard, M. *Dublin's Trade In Books, 1550–1800 . . .* (Oxford, 1989).

Popplewell, L. *A Gazetteer Of The Railway Contractors And Engineers Of Central England 1830–1914.* (Bournemouth, 1986).

Portlock, J.E. *Memoir of . . . Major General Colby.* (London, 1869).

Querand, J.M. *La France Littéraire.* (Paris, 1827).

Raistrick, A. *Yorkshire Maps and Map-makers.* (Clapham, 1969).

Rhodes, J. *The Midland & Great Northern Joint Railway.* (London, 1982).

Rivington, S. *The Publishing Family of Rivington.* (London, 1919).

Robinson, A.H.W. *Marine Cartography in Britain.* (Leicester, 1962).

Robinson, D. *The Book of Horncastle & Woodhall Spa.* (Buckingham, 1983)

Rodger, E.M. *The Large Scale County Maps of the British Isles, 1596–1850: a Union List; 2nd rev. ed.* (Oxford, 1972).

Roscoe, S. *John Newbery and his Successors, 1740–1814.* (Wormley, 1973).

Rostenberg, L. *English Publishers in the Graphic Arts, 1599–1700.* (New York, 1963).

Ruddock, J.G. and Pearson, R.E. *The Railway History of Lincoln; 2nd ed.* (Lincoln, 1985).

Shaw, G. and Tipper, A. *British Directories . . . (1850–1950) and Scotland (1773–1950).* (Leicester, 1988).

Shirley, R.W. *Printed Maps of the British Isles 1650–1750.* (Tring, 1988).

Skelton, R.A. *County Atlases Of The British Isles . . . 1579–1703 . . .* (London, 1970).

Skelton, R.A. *Explorers' Maps. Chapters in the Cartographic Record of Geographical Discovery.* (London, 1958).

Skelton, R.A. and Harvey, P.D.A. (eds.). *Local Maps and Plans from Medieval England.* (Oxford, 1986).

Skempton, A.W. *British civil engineering 1640–1840: a bibliography of Contemporary Printed Reports, Plans and Books.* (London, 1987).

Smith, D. *Victorian Maps of the British Isles.* (London, 1985).

Stanford, E. *Edward Stanford, with a note on the history of the firm from 1850.* (London, 1902).

Stevenson, E.L. *Willem Janszoon Blaeu.* (New York, 1914).

Strauss, R. *Robert Dodsley, poet, publisher and playwright.* (London, 1910).

Stukeley, W. *The Family Memoirs . . . and . . . other correspondence.* (London, 1880–1887).

Sturman, C. (ed.). *Some Historians of Lincolnshire.* (Lincoln, 1992).

Taylor, E.G. *Tudor Geography, 1485–1583.* (London, 1930).

Tennyson, A. *Letters . . .; ed. by Cecil Y. Lang and Edgar F. Shannon. Vol. II.* (Oxford, 1987).

Timperley, C.H. *Encyclopaedia of Literary and Typographical Anecdote.* (London, 1842).

Todd, W.B. *Directory of Printers and others in allied trades. London and vicinity, 1800–1840.* (London, 1972).

Tooley, R.V. *Dictionary of Mapmakers.* (Tring, 1972).

Tooley, R.V. *Maps and Mapmakers.* (London, 1949).

Tyacke, S. *London Map-Sellers 1660–1720: a collection of advertisements placed in the London Gazette 1668–1719 with biographical notes . . .* (Tring, 1978).

Tyacke, S. and Huddy, J. *Christopher Saxton and Tudor Map-making.* (London, 1980).

Van Eerde, K.S. *John Ogilby and the Taste of his Times.* (Folkestone, 1976).

Vertue, G. *Notebooks.* (London, 1930–1955).

Wakeman, G. *Aspects of Victorian Lithography.* (Wymondham, 1970).

Whitaker, H. *A Descriptive List of the Maps of Northumberland, 1576–1900.* (Newcastle-upon-Tyne, 1949).

Whitaker, H. *The Harold Whitaker Collection. (Leeds, 1947).*

Whitaker, H. *The Printed Maps Of Northampton.* (Northampton, 1948).

Wignall, C.J. *Complete British Railways Maps and Gazetteer 1825–1985.* (Oxford, 1985).

Wiles, R.M. *Serial Publications in England before 1750.* (Cambridge, 1957).

Wilson, C. *First with the News. The History of W.H. Smith, 1792–1972.* (London, 1985).

Wright, N.R. *John Grundy of Spalding, Engineer, 1719–1783.* (Lincoln, [1983]).

Wright, N.R. *Lincolnshire Town and Industry 1700–1914.* (Lincoln, 1982).

Wrottesley, A.J. *The Midland & Great Northern Joint Railway; 2nd ed.* (Newton Abbot, 1981).

Wyatt, G. *Maps of Buckinghamshire.* (Buckingham, 1978).

# INDEX

The figures in bold are the numbers of the entries in the catalogue for the works in which the person indexed has had a major involvement. Such numbers are given before any other reference to that person's work. Works that are entered as anonymous in the main catalogue are listed under that heading here. In order to help searchers to find entries for maps that lack any reference to an engraver, printer, publisher or person responsible for the drawing or design such maps are also entered under 'anonymous' with the type or individual state number in bold figures and arranged in subdivisions by map size or content, intended to reduce the degree of search required. All places outside Lincolnshire, except the larger towns, have the name of their county added, often in abbreviated form. All other places are, therefore, with the above exceptions in Lincolnshire.

Printers, engravers, artists (i.e. the person(s) responsible for the drawing of a map), surveyors and booksellers have the following abbreviations after their name: pr., eng., art., sur., bks. Engraver is used whether the map has been prepared by cutting copper or steel plates, wood blocks or by a lithographic process or later derivatives. The term publisher is a comparatively modern term; however, the abbreviation pub., has been adopted to represent the person(s) responsible for the issue of a map or book, whether in a single entrepreneurial role or as an agent, usually acting in a shared undertaking with others, in the issue of maps, books or atlases. Bks. has been adopted to cover not only booksellers but also print and map sellers. The definite and indefinite articles have been ignored in the alphabetical arrangement when they occur as the first word of a title.